Environmental Earth Sciences

Series Editor

James W. LaMoreaux, Tuscaloosa, AL, USA

More information about this series at https://link.springer.com/bookseries/8394 .

Robert Zwahlen

Assessing the Environmental Impacts of Hydropower Projects

Springer

Robert Zwahlen
Gossau, Switzerland

ISSN 2199-9155 ISSN 2199-9163 (electronic)
Environmental Earth Sciences
ISBN 978-3-030-91184-3 ISBN 978-3-030-91185-0 (eBook)
https://doi.org/10.1007/978-3-030-91185-0

This Springer imprint is published by the registered company Springer Nature Switzerland AG
The registered company address is: Gewerbestrasse 11, 6330 Cham, Switzerland

Acknowledgements

Over the years, I have worked with a large number of colleagues, and I greatly benefit from their support and input. They are far too numerous to all be listed here, and I have to apologize to all whom I forgot to mention. The following enumeration mentions those with whom I had, over shorter or longer periods, a specifically good working relationship, and I mention them in a more or less chronological order: Martin Meyer, Remo Kropf, Joaquín Nicolai, Albert Gmür, Bertil Krüsi, Irma Vila, Peter Kauz, Ian Clark, Lorenzo Zanetti, Dirk Dorn, Peter Schubarth, Anton Mauderli, Patrick Durand, Leopoldo Bartolome, Loo Ai Choo, Tunglagbayar Dorj, Pablo Durán, Guoqing Shi, Knut Sierotzki, Gianpietro Simeoni, Bernard Ruckstuhl, Marhabo Jonbekova, Reyhan Eyrice, Michael Cernea, Bernard Couson, Sven Bolomay, Sylvie Fraboulet-Jussila, Frédéric Giovanetti, Richard English, Asligul "Mavluda" Mamadatoeva, Gertrude Coulibaly, Imelda Yhr, Gary Oughton, Galina Rybinka, Pavel Kasyanov, Jennifer Cui, Dincheng (Stephen) Hu, Claudius Josephus Thomas, Kerry Connor, Pierre Biedermann, Prosper and Flavie Ngoma, Tsabang Nolé, Victor Puomogne, Wollor Topor, Vivian Castro, Joseph Okouyi, Clément Repussard, Anis Chaudry, Alexia Knezovic, Lap Thu Nguyen, Daniel Gross, Erik Helland-Hansen, Svetlana Blogoveshenskaya, Anatoly Ishuk, Alikhon Latifi, Firuz Saidov, Martin Bieri, Michiel Hartman, Anna Carlevaro, Mandy Ziranke, Nasima Sadieva, Shamafruz Mavjudalieva, Catherine Vogler, Georges Jay, Rémy Ledauphin, Victor de Genot de Nieukerken, Moses Chamfya, Bright Nyimbili, Manunart Feungpean, Herbert Weilguni, Harald Kling, Martin Lagler, François Delporte, Paul van Strijp, Stephen Syampungani, Edward Zulu, Marina Müller-Meissner, Pulak Yadav and, last but not least, Cyrill Trottmann and Britta "Lotte" Lammers.

A special thank you goes to Irina Djojoatmodjo for the carful proofreading. Of course, all remaining mistakes are entirely mine.

Most of all, I want to thank my wife Anna Rosa Zwahlen who had to put up with my numerous absences, but never complained when she had to drive me to the airport again and always welcomed me back home. All of this would not have been possible without her continuous support.

Acknowledgements

[illegible]

Contents

Acronyms and Abbreviations

ADB	Asian Development Bank
AET	Actual evapotranspiration
AfDB	African Development Bank
AKFED	Aga Khan Foundation for Economic Development
asl	Above sea level
BOD	Biological oxygen demand
BVO	Basin Water Organization
CBA	Cost–benefit analysis
CFRD	Concrete faced rockfill (dam type)
CHA	Critical Habitat Assessment
CIA	Cumulative Impact Assessment
COE	Committee of Experts
CV	Curriculum vitae
dBA	A-weighted decibels (noise unit)
DNA	Desoyxribonucleic acid
DO	Dissolved oxygen
E&S	Environmental and social
EBRD	European Bank for Reconstruction and Development
ECA	Export Credit Agency
EFR	Environmental flow release
EHS	Environment, health and safety
EIA	Environmental Impact Assessment
ELOHA	Ecological Limits of Hydrologic Alteration
EMMP	Environmental Monitoring and Management Plan
EMP	Environmental Management Plan
EOI	Expression of Interest
EPA	Environmental Protection Agency
ES	Ecosystem services
ESIA	Environmental and Social Impact Assessment
ESMMP	Environmental and Social Management and Monitoring Plan
ESS	Environmental and Social Standard

ESV	Ecosystem services value
EWE	Electrowatt Engineering
EWI	Elektrowatt Ingenieurunternehmung
EWS	Evaporation from water surface
FAO	Food and Agriculture Organization
FSL	Full supply level
GAP	South-East Anatolian Development Project (Turkey)
GBAO	Gorno-Badakhshan Autonomous Oblast (Tajikistan)
GERD	Grand Ethiopian Renaissance Dam
GFL	Great fault line
GHG	Greenhouse gas
GLOF	Glacier lake outbreak flood
GMSL	Global mean sea level
GWh	Gigawatt hours
HH	Household
HIA	Health Impact Assessment
HIA	Heritage Impact Assessment
HPP	Hydropower project
IA	Impact Assessment
ICOLD	International Commission on Large Dams
ICOMOS	International Council on Monuments and Sites
ICPR	International Commission for the Protection of the Rhine
IDB	Interamerican Development Bank
IEA	International Energy Agency
IFC	International Finance Corporation
IHA	International Hydropower Association
IPCC	International Panel on Climate Change
IRN	International Rivers Network
IRR	Internal rate of return
ITT	Itezhi-Tezhi (dam and HPP in Zambia)
IUCN	International Union for the Conservation of Nature
KGL	Kafue Gorge Lower (HPP in Zambia)
KGU	Kafue Gorge Upper (HPP in Zambia)
LMB	Lower Mekong Basin
m asl	Meters above sea level
MAF	Mean annual flow
MEI	Morpho-edaphic index
MONRE	Ministry of Natural Resources and Environment (Lao PDR)
MRC	Mekong River Commission
MW	Megawatt
MWh	Megawatt hours
NBI	Nile Basin Initiative
NELSAP	Nile Equatorial Lakes Subsidiary Action Program
NGO	Non-Governmental Organisation
NP	National Park

NT1	Nam Theun 1 (HPP in Lao PDR)
NT2	Nam Theun 2 (HPP in Lao PDR)
NTFP	Non-timber forest products
OCP	Onchocerciasis Control Program
OMVS	Senegal River Basin Development Organization (Organisation pour la Mise en Valeur du Fleuve Sénégal)
PA	Protected area
PAP	Project affected person
PDR	Peoples' Democratic Republic
PET	Potential evapotranspiration
POE	Panel of Experts
PS	Performance Standard (IFC)
RAP	Resettlement Action Plan
RCC	Roller-compacted concrete (dam type)
ROR	Run-of-river
ROW	Right of way
RFP	Request for proposal
SEA	Strategic Environmental Assessment
SIA	Social Impact Assessment
SIA	Strategic Impact Assessment
TBIA	Transboundary Impact Assessment
TEAS	Technical and Economic Assessment Study
TH	Theun-Hinboun (HPP in Lao PDR)
TL	Transmission line
TOR	Terms of reference
TWh	Terawatt hours
UMB	Upper Mekong Basin
UNDP	United Nations Development Program
UNEP	United Nations Environmental Program
UNESCO	United Nations Educational, Scientific and Cultural Organization
USD	United States Dollar
UXO	Unexploded ordnance
WB	World Bank
WCD	World Commission on Dams
WCED	World Commission on Environment and Development
WHO	World Health Organization
WTA	Willingness to accept
WTP	Willingness to pay
WWF	World Wide Fund for Nature

List of Figures

List of Tables

Part I
Approach

This book is subdivided into four main parts. Part I (Chaps. 1–6) provides an approach to impact assessment in general and to ESIA (Environmental and Social Impact Assessment). It is important to point out right from the start that the focus of here is the “E” part of the ESIA, i.e. the assessment of environmental impacts and, as indicated in Sect. 1.3, social impacts are not covered. However, the high relevance of these social impacts is clearly mentioned, and some specific aspects of such impacts are described shortly in Chap. 11. Part II (Chaps. 7–14) goes, in a more or less step-by-step way, through the process of preparing an ESIA Report. The approach used here is to list the chapters which would be required, in most cases, in such a report; however, the intention is not to provide a detailed guideline to be meticulously followed, the focus is on specific examples from actual projects for illustrating the subjects. Part III (Chaps. 15–21) picks out a number of specific issues which can—and often are—of importance when discussing environmental impacts of dam projects, but which do not necessarily neatly fit into one or the other standard chapter. Part IV (Chaps. 22–26) provides a number of case studies; here, it is not the intention to give exhaustive information on any of the projects presented, rather, each one serves the purpose of illustrating and discussing one or a number of topics, which were of special relevance in that particular case. A few specific points, which are often discussed controversially, are addressed there, like the question of sustainability of hydropower and of “dams that should not be built”. Finally, Chap. 27 tries to take a glimpse at the future and shortly addresses a number of points where environmental impacts assessment, or, more importantly, the actual mitigation of environmental impacts, could and hopefully will be improved. This, like everything else in this book, is an expression of my personal opinion, based on over 40 years of professional practice in environmental impact assessment.

I start chapters and some specific sections with a citation which, I hope, somehow illustrates the topic at hand. Here, five such citations are listed which are meant to exemplify the widely differing views people have on dams and hydropower, a difference that is a kind of leitmotiv in this book.

Dams are unique and important.
M. Rogers, ICOLD President, (Opening Address, Africa 2019, Windhoek, Namibia)

If you want to achieve climate change mitigation, you must develop hydropower.
Luciano Canale, Hydropower Specialist, the World Bank (Africa 2019)

Hydro is the largest of the sustainable energy sources.
N. Nielsen, IEA (International Energy Agency), (Africa 2019, 19-04-02)

Hydropower is a clean, renewable, and environmentally friendly source of energy.
(Berga 2016)

Unpalatable as it must undoubtedly be to the dam-building industry, there is clear evidence that building large dams is not an appropriate means of feeding the world's hungry, of providing energy, or of reducing flood damage.
(Goldsmith and Hildyard, 1984:345)

The agent causing the most immediate damage in fresh water are dams, great boosters of local economies but unfortunately chief demons of aquatic habitat destruction. Their malign effects include the barricades they create against migrating fish species, the stilling and deepening of water upstream, and pollution, all exacerbated by the usual intensive agricultural conversion that builds up around dams.
(Wilson, 2016:41–42)

Chapter 1
Introduction

Sollte also eine Heilslehre wirklich einmal die ganze Erde überziehen, so würde sie sogleich in mindestens zwei heftig befeindete Auslegungen zerfallen (die eigene wahre und die andere ketzerische).
If a doctrine of salvation should really ever cover the whole earth, it would immediately break up into at least two violently hostile interpretations (the own true one and the other heretical one).
Lorenz (1963: 364) citing Erich von Holst

It seems that we have a very strong tendency to look at things in a binary manner, by sorting all types of issues in a very either/or type of way. One example for this, which has been a heated scientific discussion for a long time, a discussion which sometimes still does not seem over, is the "nature or nurture" question concerning human capacities and abilities, i.e. the question whether something is genetically defined (= nature) or acquired by learning (= nurture); see to that e.g. the discussion of the subject by Monod (1970: 192) or the excellent book by Segerstråle (2000). For me, this was always a strange type of discussion: both are very important; nobody develops a genetically defined language, this is definitely learned, but it is just as clear that a language cannot be learned without the inherited capability for doing so (Pinker 1994): nature and nurture in combination. Another example of this either/or approach: in a recent note in Science (372(6540): 358, 2021) reporting on the analysis of the role of thumb anatomy in *Australopithecus* and early species of the genus *Homo*, lead to the conclusion that in our species "tool-using skills could have evolved not so much in conjunction with big brains as much as with stronger, more dexterous hands". I'm quite certain that both parameters had their role to play, and one would not have achieved it without interacting with the other; note that the authors of the commented study (Karakostis et al. 2021) do not come to the cited conclusion. One more such issue, which will be mentioned later on (see Sect. 21.2.3) is the discussion on traditional land management practices by indigenous peoples, which is usually considered as good and sustainable, with the implication that modern practices are not (e.g. Ellis et al. 2021). Here again: it depends on the specific case and on the circumstances. In the discussion about dam projects, the same dichotomy can often be seen: dams are invariably good for one side and invariably bad for the other

R. Zwahlen, *Assessing the Environmental Impacts of Hydropower Projects*, Environmental Earth Sciences, https://doi.org/10.1007/978-3-030-91185-0_1

side; see the citations inserted before this text. I hope that this book will be able to contribute to a more balanced view. The first chapter aims at shortly discussing its purpose and scope, but also at pointing out its limitations.

1.1 Prologue

At a very early stage in my career, it must have been towards the end of the 1970s, we once had a discussion about an irrigation project somewhere in the Sahel zone of Western Africa. I had one major concern about the approach and said something in the sense of "if this project goes ahead as planned, there is a risk that the soils will be destroyed through salinisation in about 25 years". The answer from the project economist was: "If it takes 25 years, then it is a good project, because it has a payback period of 20 years". I was quite shocked by this answer; the term "sustainability" was already in use, albeit not as prominently as it is since the publication of the Brundtland Report (WCED 1987), and I was really astonished that anybody could possibly be so short-sighted. In addition to that, since then I have come to realise that economists, when calculating the one relevant figure for any project, namely, the IRR (internal rate of return), which determines the payback period, tend to be somewhat on the optimistic side; this view is confirmed, e.g., by the recent analysis of irrigation projects in Africa carried out by Higginbottom et al. (2021).

A few years later, I was working on a feasibility study for a hydropower project in Togo. The objective there was to draft a plan for the pre-impoundment preparation of the area of the future reservoir, with the aim to make use of available resources which would otherwise be lost, namely timber, and to make sure that the reservoir after impoundment could be used for fisheries; greenhouse gas minimisation was not yet an issue at that time. During field work, we had the impression that one important aspect, which was not at all part of our terms of reference, was being utterly neglected: the fate of the human population living in this area. We voiced this concern in a discussion with the government representative who was in charge of this project, and he told us: "Oh, when the waters rise, they will leave!". Fortunately, I have not heard this in this crude form again, but the "accountant's approach", which says something in the sense of "10,000 people have lost, but 5 million have benefitted, and therefore it is a good project" still comes up from time to time.

Some years later again, in the early 1990s, we were preparing an ESIA for a dam project in Mongolia. One day, I had a discussion with a young American, who worked there for an NGO. He was against the project, because it would destroy the wonderful landscape in the project area. And while he was explaining me what a pity it would be to cause such an impact, he was munching away on a pack of chips, wrapped into an aluminium coated bag, whose manufacturing must have used a lot of energy, and which, even if it had been available in Mongolia at that time, only very few inhabitants of that country would have been able to afford. I agreed with him that the project would affect a nice landscape, but then the only thing Mongolia had plenty of was nice landscapes. For me he sounded like somebody who visits a foreign country (of course expecting all the amenities of modern life), for then to go home and tell his friends something like "oh, you know, such extraordinary, really beautiful landscapes. And the people, so poor, but so happy, friendly and hospitable!".

I do not want to pretend that these were the three key moments which triggered my decision to go into my line of work. But these and other, similar experiences kept showing me that it is important to develop projects in the best possible way, taking into account environmental and social impacts, positive as well as negative ones.

In addition to these three episodes, a few statements, along with short comments, should serve as an introduction to this book, and should also show my motivation for writing it. Here, I do not attempt to discuss any of these points in detail, I shall come back to most of them later on. As you may find out when going through the following pages, I like to phrase important points in a short and concise way—as I think it should be done in an ESIA Report—and I like to do this in the form of a list of bullet points; so why not start with that right here:

- I'm aware of the need for energy for most of human activities, and especially of the need for a reliable supply in electricity. I would certainly not limit this need to the requirements of an industrialised—or, as now is fashionable to say, a post-industrialised—society, I also, and maybe even more so, mean here societies and groups of people for which this is not, or not yet, the most normal thing in the world. If you have seen, as I have had the occasion a few times, what it means to have electricity for the first time, or to have again a reliable and sufficient electricity supply after almost twenty years without it, as I have witnessed in Tajikistan, you will understand this.
- I am just as well aware of the importance for all humankind to have a good environment. There is this saying that "Health is not everything, but without health, everything is nothing", attributed to the German philosopher Schopenhauer; this could easily be applied to the environment, since nothing can possibly happen outside of an environment (see Sect. 2.1).
- I consider hydropower as a good, reliable and renewable source of energy. Hydropower without dams of different types (see Sect. 4.1) is not possible, at least not to the extent that it is being used.
- This does not mean that every technically and economically feasible hydropower project is necessarily a good project. Unfortunately, there are too many "bad" hydropower projects (whereby I include in the term "project" not only possible future developments, but also existing structures); I'll deal with this point e.g. in Chap. 26.
- It is not possible to build and operate a dam without causing negative impacts on the environment.
- Suitable mitigation measures, if properly identified, planned, implemented and monitored, can reduce such impacts, although not completely eliminate them.
- The crucial question is: when is an impact too severe to be acceptable? Obviously, one will get different answers to that question depending on the type of impact and on whom one asks. The decision will have to be taken on a case by case basis, and unanimity will be the exception rather than the rule.

From these, I see the following main requirements in relation with dam and hydropower projects, namely:

- To prevent "bad" projects from being implemented. I'm fully aware, that when it comes to such projects, not everybody has the same notion of what is bad and what is good.
- To make good or acceptable project as good as possible, always keeping in mind that none can ever be developed without having some adverse impacts.

The role of the ESIA, as I see it, is this latter point: identifying impacts, proposing adequate measures, and showing the way to implement them. It cannot possibly do more, but this it has to do well. An ESIA can lead to the elimination of a "bad" project, but this is not the main aim of this planning instrument, and as a rule should not be expected to fulfil this role; this point is addressed in more detail in Sect. 3.1.

I hope that the work that I have done during the last 40 or so years made a contribution in this sense, and if this book helps in continuing in that way, it will have fulfilled its purpose. All the points listed above will be dealt with in more detail throughout this book.

1.2 Purpose and Scope

> *Ideally, ..., the prediction of an undesirable outcome may act as a 'self-destroying prophecy'. It follows that a risk prediction model becomes maximally useful not when it is confirmed by disastrous events but rather when, as a result of its warnings being absorbed and acted upon, the risks are prevented from becoming reality, or are minimized, and the consequences predicted by the model do not occur. This is how the predictive model acts as a self-destroying prophecy.*
>
> Cernea (1997)

Basically, an ESIA for a dam project has to do exactly this, specifically for the project under scrutiny: it has to identify all potential impacts, large and small, significant and insignificant and, hopefully, negative as well as positive ones. By the nature of the process, its main objective, it has to (or should) focus on negative impacts, and give special emphasis to significant, major, lasting impacts. It is supposed to identify them correctly and show the potential consequences. However, it does not end there: it has to identify suitable mitigation measures. This will then turn the assessment of impacts into a self-destroying prophecy as mentioned in the citation above, provided the mitigation measures are implemented properly, monitored, and adapted if required, in order to reduce the impact to an acceptable level. "Acceptable" is the aim: we have to be well aware that "no impact" is not possible, whatever we do.

When it comes to discussing dam and hydropower projects, their environmental and social impacts and the assessment of those impacts, there is quite a number of books (in addition to other publications like articles in scientific journals or increasingly, in electronical form) which can roughly be divided into the following groups; the list provided here for each of these groups just aims at giving some more or less obvious examples, but is not intending to represent a comprehensive enumeration:

1. ESIA methodology: there is any number of books entitled "Environmental Impact Assessment", "Environmental Impact Assessment Methodologies", "Environmental and Social Impact Assessment", or similar titles. These are usually textbooks dealing with "how to do" ESIAs, in part applicable to any type of project, or focusing on certain types of projects (Becker 1997; Dalal-Clayton and Sadler 2005; Eccleston 2011, Fisher 2002, 2007; Munier 2004; Therivel and Wood 2017; Therivel et al. 1992; Vanclay and Bronstein 1995).
2. Country-specific guidelines (in addition to the guidelines issued by some national environmental authorities; Duek 1982; IFC 2019; Sadar 1996).
3. Guidelines of international funding agencies (ADB 1992, 1998; Baker 2000; IFC 2002, 2012, 2018; Leitmann 1994; Lohani et al. 1997; WB 1991, 2011, 2017).
4. Anti-dam books: discussing dam impacts, either in a more or less global way or focusing on specific projects, putting emphasis on their negative direct and indirect environmental and social impacts (Goldsmith and Hildyard 1984; McCully 1996; Cans 1997; Khagram 2004; Leslie 2005; Pearce 2018; Eyler 2019).
5. Attempts for a general, objective analysis of direct and indirect, negative and positive impacts of dam and hydropower projects (WCD 2000; Tortajada et al. 2012).
6. Publications specifically dealing with resettlement issues associated with dam projects (Cernea 1999; Cernea and McDowell 2000; Scudder 2005).
7. Books dealing with one specific dam, or a specific aspect thereof (Balon 1974; Odingo 1979; Dos Santos et al. 2004).

With the exception of the books in group No. 4, which as a rule are addressing a general public, most of these books are either procedure-related (guidelines, groups 2 and 3), or are rather academic, focussing on methodologies in the different fields to be covered by an ESIA or a similar type of assessment (SIA, SEA, group 1). The two titles in group 5 do something similar to what group 4 does: attempting to assess the impacts of dam projects, however, in a balanced way, covering positive as well as negative aspects, and this in a more technical or scientific manner than group 4.

What in my view is missing completely is a book that deals, not so much with impacts of dams, but rather with the process of assessing these impacts, and this not from a theoretical, procedural or methodological perspective, but rather from the practical side, from the point of view of a practitioner who is doing this, and who has been doing it for quite some time. In the literature listed above, there is one book that does something similar, with two exceptions: its focus is on social impacts rather than on the environment, and it analyses the outcome of specific projects and how to deal with it; it is the book of Thayer Scudder, which focuses on resettlement issues as the result of large dam projects, and it mainly draws on the personal experience of the author, who was involved in a considerable number of such projects during all his professional career. In a way, I try to do something similar for the environmental aspects, obviously also dealing with impacts as such, but as mentioned focusing on their assessment. It is not intended as a general how to do text (although hopefully it will add to that as well), but the main part is practical experience, which is not in every case exactly "according to the book". By largely relying on personal experience, like Thayer Scudder, I will usually express also my personal views of certain issues, fully aware of the fact that not everybody will necessarily agree with me every time, and as he did I may, perhaps too often, use the direct personal form for indicating that.

1.3 Limitations

This book is about dams and their impacts on the environment, and, more precisely, on assessing those impacts. The focus is mostly on projects within the author's own experience, concentrating on dams and hydropower. While in some instances other fields of water use will be addressed, mainly irrigation, this undoubtedly interesting and important topic and primary or secondary purpose of many dams is here not getting the attention that it certainly deserves. Likewise, talking about "environmental effects of dams" it is often not really the dam that is of interest, but rather the reservoir; so whenever a dam is mentioned, this means this structure together with the reservoir that it created, and the objective is hydropower unless specifically mentioned otherwise. The second limitation is the fact that "environmental impacts" as used here means impacts on the physical and biological environment. Impacts on the "human environment", i.e. social impacts, cannot be left out completely, but are not the focus here. For a discussion of this important issue refer to, among others, the admirable book of Scudder (2005), and the work of Michael Cernea (e.g. Cernea 1995).

One more point needs to be mentioned: very often my involvement was in pre-feasibility or feasibility studies, and there are basically three possible outcomes of such studies, namely:

- A project idea is studied at a very early stage of project development (pre-feasibility), and then abandoned, never to be heard of again. This can happen, however, such projects never really die: once a potential project identified, the information on it still remains hidden somewhere and may stay dormant for a long period, only to pop up again suddenly at a much later date; such an example will be mentioned in Sect. 26.1.1.
- At any stage of project development, a project can be put aside for whatever reason, often because a prospective developer suddenly backs out, and is then taken up again at a later date, very often by another developer.
- A project is identified, studied, goes into detailed design and implementation, and then into operation, and all this without interruption and in a comparatively short time.

The second pattern is a very frequent one. This more often than not also means that during different development phases of such projects different entities—including different engineers and ESIA consultants—are involved. This means that there is the possibility that one is often involved in early stages of project development, but not in the continuation, re-evaluation and re-planning etc. It can therefore very well happen that information on projects (dam site, dam type and height, size of reservoir, installed capacity etc.) will be provided which does not correspond with what is now reality.

Then, a word of caution: much can be learned from mistakes, and therefore "good bad examples" are a very effective way for illustrating a topic. Throughout the book, such cases will be mentioned. Doing that, the purpose is only to show something in a way that will, in most cases, be very easily understandable, and maybe amusing on some occasions. It is by no means the intention to ridicule anybody. Therefore, in using such examples, the source of the information may not be listed, let alone the name of the author. Hopefully, this is acceptable to the reader. Some of the descriptions used will be in the form of anecdotes, and hopefully not too boring ones. Admittedly, a penchant for sarcasm may show, and some of the anecdotes may lean that way; again, no intention to ridicule anybody.

Some readers may find the text biased to a certain extent, since impacts (and therefore necessarily the discussion of corresponding basic information as well as mitigation measures) on water, in the broad sense of the term, and on biodiversity are discussed in more detail than some others. However, I do not see this as a bias, but rather as a necessity: water and biodiversity are the two components of the natural environment which are most fundamentally affected by dam projects; once again: the social impact is just as, and in many cases arguably even more, important, but this is not the subject of this book.

Another issue is the "logical" structure of the text. To some extent, this is given by the catalogue of subjects that need to be addressed in an ESIA, and which are usually listed in guidelines and TORs for such assignments. Still, it is not always very obvious and simple. To mention just one example, let's take climate change. Climate change as such is about increasing concentrations of greenhouse gases (GHG) in the atmosphere and the phenomenon of global warming resulting from it, therefore it has to be addressed under the main heading "Climate". However, when it comes to hydropower, the relevant questions are about medium- and long-term changes in river flow, influenced by trends in rainfall, shifts in precipitations falling as rain instead of as snow, and glacier melt, which would indicate that the subject belongs to "Hydrology". Finally, when it comes to the cause of climate change, namely, GHG emissions, there are again two different aspects to consider: comparison of hydropower with other ways of generating electricity, namely thermal power plants using fossil fuel, and the issue of reservoirs as potential major sources of GHGs. This means that one and the same topic may be mentioned under more than one heading. But then, ecology is a web, not a line, and so is environmental assessment.

One last point here: this book deals with environmental impacts, assessment of impacts, and to some extent in a more or less systematic way with the process of ESIA preparation and procedures. However, it is not the intention to produce a step-by-step, cookbook-style text on “how to prepare an ESIA report for a hydropower project”. Still, hopefully, a contribution will be made to improving not only hydropower and dam projects as such, but also the ESIA procedure for such projects.

1.4 My Experience

As mentioned above, the intention is, most of all, to exemplify and illustrate impacts caused by dam projects based on my personal experience. In order to give a short impression of what I am talking about Fig. 1.1 shows a map indicating the places where I have been involved in dam projects and studies related to dams and hydropower.

This map is just intended to give an idea of the distribution of these projects, which with very few exceptions, where I only commented on projects or project documents, I have all visited. Most of the dots represent ESIAs for dam projects, but there are some exceptions to this rule: there are a few multi-purpose dam projects, a few irrigation projects without hydropower, some cases where I acted as a member of a Panel of Experts and some audits or due diligence studies of existing dams and power stations. From this list it is already clear that the dots cannot in any way indicate the intensity of the studies represented or the effort involved, which can reach from a simple visit of a few days resulting in an expert opinion to a very comprehensive ESIA which involved a large team and took up to several years to complete; still, the majority of the points given in the map represent just such cases. One more difficulty here is the size of the map, or the fact that some of the countries are very small, like e.g. Switzerland, Albania and Sri Lanka, in all of which I had more than one project, but where it is impossible to place more than one dot. Another point is the

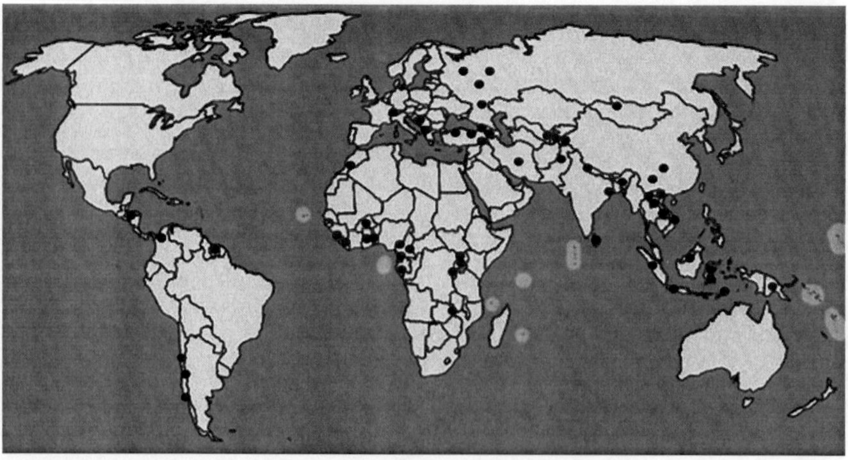

Fig. 1.1 My involvement in dam projects and related studies

nature of the projects: so the one dot in Burkina Faso stands for a project where we evaluated a total of 35 small dams for irrigation, distributed over the southern half of the country: impossible to show this here. A different case was a due diligence study of 9 major existing hydropower stations of the Volga-Kama cascade in Russia; here I put in four dots, with the aim of just to indicate approximately the geographical range of this activity.

The aim is by no means to analyse or discuss these projects in detail, but most of the examples that I will chose to illustrate specific topics will be from this list. Obviously, I have read a number of publications and reports not prepared by me, on projects in which I was not involved at all, and I'll make use of such information as well.

1.5 Sources Used

In the preparation of this book, three main types of sources were used, namely:

1. Scientific literature and other published information: articles from scientific journals, textbooks, and some articles from newspapers. This literature, to a great extent, is what was used for preparing the various reports mentioned in the text, however, this does not mean that all the references used in all the reports cited will be listed. Since the work included covers a span of by now over 40 years, some of the literature cited is rather old, and some of it may be outdated.
2. Reports prepared for specific projects; these are mostly ESIA reports or documents related to them. Mainly, reports will be used in whose preparation the author was involved, be it as specialist for a certain subject, as sole author (which is rather the exception than the rule), as team leader and main editor, or as advisor. Most if not all of the reports cited whose "author" is, in chronological order, EWI, EWE, Electrowatt-Ekono, Pöyry or AFRY, fall in one of these categories. All these reports, as a general rule, were made accessible to the public at one time, and can therefore be considered as being in the public domain; however, many of them by now may be difficult if not impossible to find.
3. Information directly taken from the internet, obviously a rather recent, but increasingly relevant source of information. While all the published literature and all the reports mentioned in the text are (hopefully, and an apology for any oversight) included in the list of references, in case a website is mentioned in the text as the source of specific information, this is not repeated there. The exception to this rule are, as mentioned, occasions when such a source is used as "good bad example", and therefore there is no mentioning of the source or the author.

References

ADB. 1992. Guidelines for the health impact assessment of development projects. ADB Environment Paper No. 11: 45 pp. + App. ADB, Manila.

ADB. 1998. *Handbook on resettlement. A guide to good practice*, 115 pp. Manila: Asian Development Bank.

Baker, J.L. 2000. Evaluating the impact of development projects on poverty. A handbook for practitioners. In *Directions in development*, 217 pp. Washington, DC: World Bank.

Balon, E.K. 1974. *Fishes of lake Kariba, Africa*, 144 pp. Neptune City: TFH Publications.

Becker, H.A. 1997. *Social impact assessment: Method and experience in Europe, North America and the developing world*, 260 pp. London: UCL Press.

Cans, R. 1997. *La bataille de l'eau*, 232 pp. Paris: Le Monde Editions.

Cernea, M. 1997. The risks and reconstruction model for resettling displaced populations. *World Development* 25 (10): 1569–1588.

Cernea, M.M. (ed.). 1995. *Putting people first. Sociological variables in rural development*, 2nd ed, revised and expanded, 575 pp. New York: The World Bank, Oxford University Press.

Cernea, M.M. (ed.). 1999. *The economics of involuntary resettlement. Questions and challenges*, 259 pp. Washington, DC: The World Bank.

Cernea, M.M., and C. McDowell (eds.). 2000. *Risks and reconstruction. Experience of resettlers and refugees*, 487 pp. Washington, DC: World Bank.

Dalal-Clayton, B., and B. Sadler. 2005. *Strategic environmental assessment*, 470 pp. London: Earthscan.

Dos Santos, G.M., B. de Mérona, A.A. Juras, and M. Jégu. 2004. *Peixes de baixo río Tocantins: 20 anos depois da Usina Hidrelética Tucuruí*, 215 pp. Brasilia: Eletronorte.

Duek, J.J. 1982. *Metodología para la evaluación de impactos ambientales*, 113 pp. Caracas, Venezuela: Contemporánea de Ediciones.

Eccleston, C. 2011. *Environmental impact assessment: A guide to best professional practice*, 286 pp. Boca Raton: CRC Press, Taylor and Francis.

Ellis, E.C., N. Gauthier, K.K. Goldewijk, et al. 2021. People have shaped most of terrestrial nature for at least 12,000 years. *PNAS* 118 (17). https://doi.org/10.1073/pnas.2023483118.

Eyler, B. 2019. *Last days of the mighty Mekong*, 365 pp. London: Zed Books.

Fischer, T.B. 2002. *Strategic environmental assessment in transport and land use planning*, 284 pp. London: Earthscan.

Fischer, T.B. 2007. *Theory and practice of strategic environmental assessment. Towards a more systematic approach*, 186 pp. London: Earthscan.

Goldsmith, E., and N. Hildyard. 1984. The social and environmental effects of large dams. In *Overview*, vol I, 346 pp. (1986): *Case studies*, vol II. (1992): *A review of the literature*, vol III, 243 pp. Report to the European Ecological Action Group (ECOROPA).

Higginbottom, T.P., R. Adhikari, R. Dimova, S. Redicekr, and T. Foster. 2021. Performance of large-scale irrigation projects in sub-Saharan Africa. *Nature Sustainability* 4: 501–508. https://doi.org/10.1038/s41893-020-00670-7.

IFC. 2002. *Handbook for preparing a Resettlement Action Plan*, 79 pp. Washington, DC: IFC, International Finance Corporation.

IFC. 2012. *Performance standard 1: Assessment and management of environmental and social risks and impacts*, 9 pp. International Finance Corporation. https://www.ifc.org/wps/wcm/connect/24e6bfc3-5de3-444d-be9b-226188c95454/PS_English_2012_Full-Document.pdf?MOD=AJPERES&CVID=jkV-X6h.

IFC. 2018. *Environmental flows for hydropower projects. Good practice handbook*, 136 pp. The World Bank Group. http://www.ifc.org/sustainabilitypublications.

IFC. 2019. *Environmental and social impact assessment guidelines for hydropower projects in Myanmar*, 114 pp. https://www.ifc.org/wps/wcm/connect/f6ee0e54-469e-435f-95ea-3e03a6c94ef2/Myanmar+HPP+ESIA+Guidelines+V8+for+consultation.pdf?MOD.

Karakostis, F.A., D. Haeufle, I. Anastoppoulou, et al. 2021. Biomechanics of the human thumb and the evolution of dexterity. *Current Biology* 31: 1317–1325. http://creativecommons.org/licenses/by/4.0/.

Khagram, S. 2004. *Dams and development. Transnational struggles for water and power*, 270 pp. Ithaca: Cornell University Press.

Leitmann, J. 1994. *Rapid urban environmental assessment. Lessons from cities in the developing world*, vol. 1. Urban Management Programme Discussion Paper No. 14, 78 pp. Washington: The World Bank.

Leslie, J. 2005. *Deep water. The epic struggle over dams, displaced people, and the environment*, 352 pp. New York: Picador.

Lohani, B.N., W.N. Evans, R.R. Everitt, H. Ludwig, R.A. Carpenter, and S.L. Tu. 1997. Environmental impact assessment for developing countries in Asia. In *Overview*, vol. I, 364 pp. *Selected case studies*, vol II, 314 pp. Manila: ADB.

Lorenz, K. 1963. *Das sogenannte Böse. Zur Naturgeschichte der Aggression*, 392 pp. Wien: Brotha-Schoeler Verlag.

McCully, P. 1996. *Silenced rivers. The ecology and politics of large dams*, 350 pp. London: Zed Books.

Monod, J. 1970. *Le hasard et la nécessité. Essai sur la philosophie de la biologie modern*, 244 pp. Paris: Editions du Seuil.

Munier, N. 2004. *Multicriteria environmental assessment. A practical guide*, 311 pp. Dordrecht, NL: Kluwer Academic Publishers.

Odingo, R.S. 1979. An African dam. Ecological survey of the Kamburu/Gtaru hydro-electric dam area, Kenya. *Ecological Bulletins* 29: 183 pp. Stockholm: Swedish Natural Science Research Council.

Pearce, F. 2018. *When the rivers run dry. The global water crisis and how to solve it*, Updated ed., 312 pp. London: Granta Publications.

Pinker, S. 1994. The language instinct. How the mind creates language, 525 pp. In *Perennial classics*. New York: HarperCollins.

Sadar, M.H. 1996. *Environmental impact assessment*, 2nd ed., 191 pp. Canada: Carleton University Press.

Scudder, T. 2005. *The future of large dams. Dealing with social, environmental, institutional and political costs*, 389 pp. London: Earthscan.

Segerstråle, U. 2000. *The defenders of the truth. The sociobiology debate*, 493 pp. Oxford: Oxford University Press.

Therivel, R., and G. Wood. 2017. Methods of environmental and social impact assessment, 4th ed. In *The natural and built environment series*, 741 pp. New York: Routledge.

Therivel, R., E. Wilson, S. Thompson, D. Heaney, and D. Pritchard. 1992. *Strategic environmental assessment*, 181 pp. Reprint 1999. London: Earthscan.

Tortajada, C., D. Altinbilek, and A.K. Biswas (eds.). 2012. *Impacts of large dams: A global assessment*, 407 pp. Berlin: Springer. https://doi.org/10.1007/978-3-642-23571-9.

Vanclay, F., and D.A. Bronstein. 1995. *Environmental and social impact assessment*. Wiley.

WB. 1991. Environmental assessment sourcebook. In *Policies, procedures, and cross-sectoral issues*, vol I, 227 pp. World Bank Technical Paper No. 139. *Sectoral guidelines*, vol II, 282 pp. World Bank Technical Paper No. 140. *Guidelines for environmental assessment of energy and industry projects*, vol III, 237 pp. World Bank Technical Paper No. 154. Washington, DC: Environment Department, the World Bank.

WB. 2011. *Strategic environmental assessment in policy and sector reform*, 222 pp. Washington: World Bank.

WB. 2017. *The World Bank environmental and social framework*, 106 pp. Washington, DC: International Bank for Reconstruction and Development/The World Bank. http://documents.worldbank.org/curated/en/383011492423734099/pdf/114278-WP-REVISED-PUBLIC-Environmental-and-Social-Framework.pdf.

WCD. 2000. *Dams and development. A new framework for decision-making*, 404 pp. The Report of the World Commission on Dams. London: Earthscan.

WCED. 1987. *Our common future*, 383 pp. Oxford: World Commission on Environment and Development, Oxford University Press ("Brundtland report").

Chapter 2
Different Ways to Look at Things

Members of these two communities drew on different stocks of knowledge or, alternatively, interpreted the same facts differently, because of their different logics-in-use.
Segerstråle (2000: 270)

The way one looks at things—from what background and with what expectations—determines, to a great extent, what one sees in a given situation. This chapter is not directly related to the ESIA or to hydropower as such, but here some points of relevance for the approach to issues at hand have to be shown. As in all sections of the book, important points are illustrated by means of examples.

2.1 Environment, Society and Economy

The graph in Fig. 2.1, representing the three main components of our everyday world, environment, society and economy, was shown to us students in the early 1970s, and can still be seen sometimes today, in this form or modified in different ways.

It can be questioned whether this is the best—or even at least a satisfactory—way to look at this important system, and this mainly for the reason shown in Fig. 2.1b: it leads easily to the conclusion that these three components are not only of the same importance, but in addition to that can be looked at independently of each other; this has been done, and in many instances is still done today, and this is especially misleading where economy is concerned (see e.g. Rees 2020, for a very recent reference to this issue). Arrows between the three sectors indicate interactions (c), but still show them as three basically separate entities. A similar model was proposed by Veen-Groot and Nijkamp (1999) for globalisation, trade and transport, three separate compartments which all have interactions, and all of them with influence on the environment. One major approach of strictly separating environment and human society was followed in classical sociology, and this still seems to be done by a good number of sociologists today (see Sanderson 2001, for a number of examples, and Irwin 2001: 4 ff., for a thorough discussion of the topic).

R. Zwahlen, *Assessing the Environmental Impacts of Hydropower Projects*, Environmental Earth Sciences, https://doi.org/10.1007/978-3-030-91185-0_2

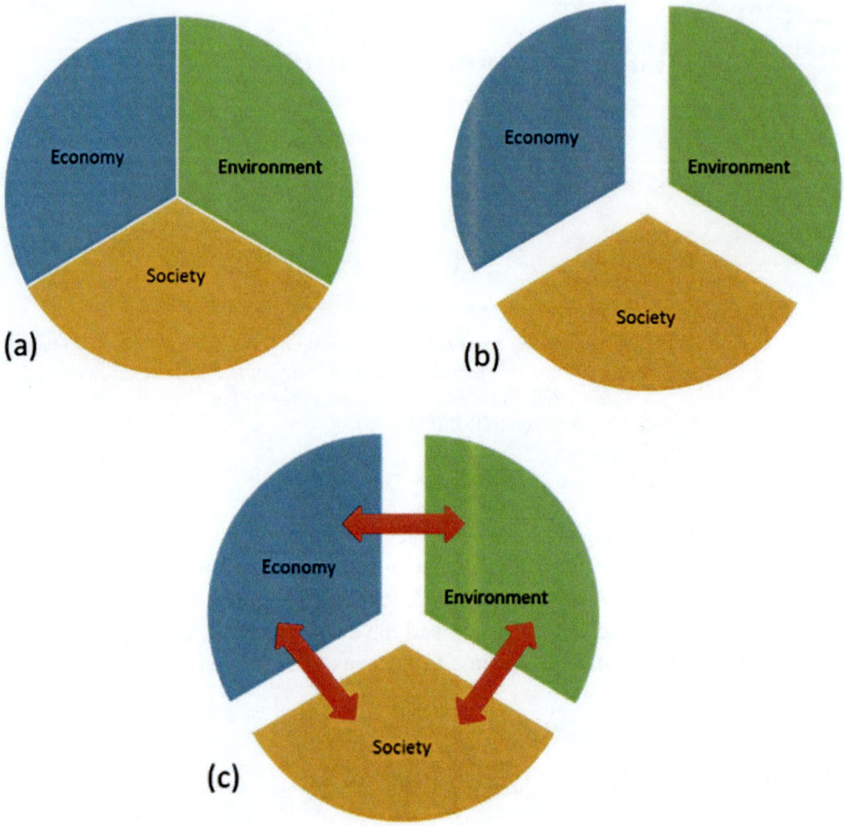

Fig. 2.1 Environment, society and economy I

A better way of looking at this is shown in Fig. 2.2.

This is a graph that can be seen quite often (see e.g. Carew-Reid et al. 1994, or Millar et al. 2019, and Roth and Valentinov 2020, for a recent discussion). Portney (2015: 7) is using the same graph, making the three components into the "three e" by replacing "society" with "equity", a definition which seems to be too restricted for our purpose. The sector where the three circles overlap is frequently labelled "sustainability" or "sustainable development"; it could just as well be labelled "conflict", this also being a question of how one looks at things. While better than the first one, it still suggests that, although there is an overlap, economy is still largely an entity in its own right, independent from and to its largest part outside of the other two.

Finally there is the way to look at it as shown in Fig. 2.3, and this is how one should try to understand the system; this is an argument also made by Bonneuil and Fressoz (2013: 40).

The so-called Brundtland Report, in a way the origin of the debate about sustainability, stated: "The environment does not exist as a sphere separate from human

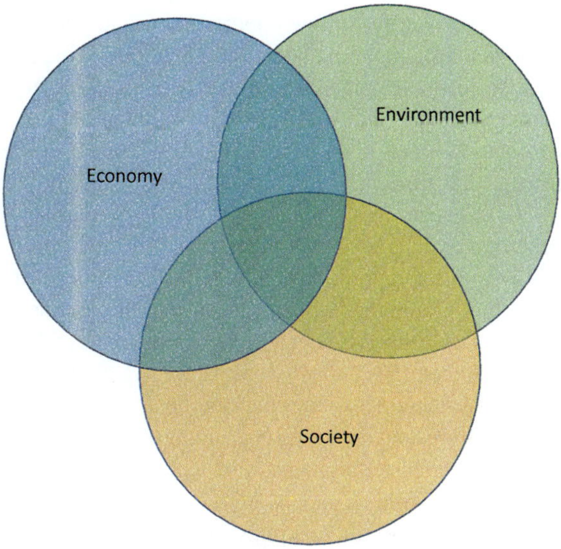

Fig. 2.2 Environment, society and economy II

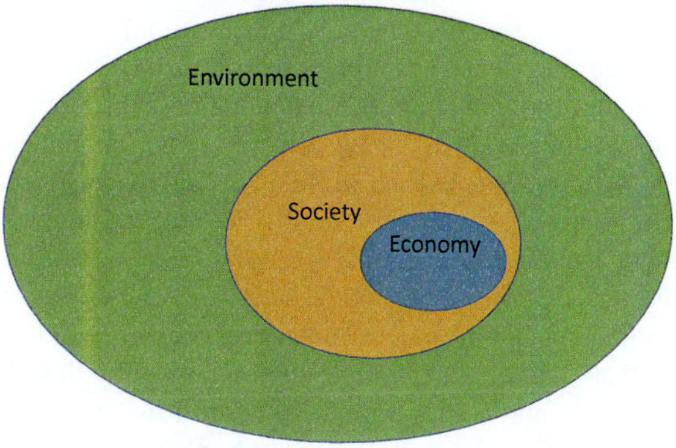

Fig. 2.3 Environment, society and economy III

actions, ambitions and needs, and attempts to defend it in isolation from human concerns have given the very word "environment" a connotation of naivety in some political circles" (WCED 1987: xi). One can agree with this in general terms, but it might be better to turn the basic statement around to see it in the following way: the human society, with all its actions, ambitions and needs, does not exist in isolation from the environment. On the contrary, it necessarily is embedded in the environment, depends heavily on it, and cannot do anything without interacting with the environment. This is best illustrated in the way Fig. 2.3 does it. It is easy to imagine an environment without human society; one only has to realise that this has been the case during almost all the time of the 13 billion years since the big bang and the approximately 4 billion years since life came into being on our earth (see e.g. Dawkins and Wong 2016: 642 ff.). Human society in the way we know it today started evolving just a few hundred thousand years ago, and until very recently (in geological times), until about 40,000–10,000 years ago, it did not have much impact on the biosphere, given its limited size and the very limited technology available then. Even so, it is generally accepted that extinction of megafauna in many places was caused by or at least accelerated by humans colonising these parts of the world for the first time, like in Australia, where this process may have started as early as 40,000 years ago (Diamond 1997: 35 ff.; Crosby 1986: 15). The same type of development has probably taken place in the Americas (Crawford 1992: 27), and this was made more likely by recent discoveries which indicate that humans may have immigrated there much earlier than thought until recently, possibly as early as 33,000 years ago (Ardelean et al. 2020; Becerra-Valdivia and Higham 2020), and even the bison (*Bison bison*) seems to have escaped this fate only just (Shaw 2021). Still, massive human made environmental changes started only with the development of agriculture, and increased in an unprecedented way by the development of industrialisation which started in the nineteenth century. So, while we can easily imagine an environment without society, it is certainly not possible to imagine a society without environment. The same applies to the next level: it might be difficult to imagine a society without economy; in fact, as soon as members of a species start exchanging any kind of goods (first of all food) in exchange for any other good (and here again probably first of all some other type of food) it can certainly be said that there is an economy, at whatever simple level (see also Bliss and Egler 2020, for a similar view). But it is absolutely impossible to imagine an economy without a society. This way to look at things also can show us that the environment has a defined size, while society and economy can expand within it, occupying an ever larger part of the environment. This concept, as well, is certainly not new; Malthus (1798) may have been the first to clearly express this, and Meadows (1972) and Ehrlich (1971) followed. Even if their predictions were not "exactly true" in the sense that technical and scientific progress enabled human population to grow beyond what was conceived as possible by these authors, the basic truth pointed out by them persists: our earth, and with it our environment, has a finite size, and it is—at least so far, maybe forever, and at least still for a long while—the only place we have. A recent look at this situation is provided by Jackson and Victor (2019), who point out the choices we have, and mainly the possibility of decoupling, i.e. the fact that e.g. carbon emission per unit of value produced is decreasing. Still, they come to the conclusion that "There may well be efficiencies to be had, but these will ultimately be constrained by thermodynamic limits, as all activity is".

That everything we do has repercussions in one way or another is shown by another development, which somehow took us "by surprise", although it has been in the making for quite some time already, but is seriously manifesting itself only now: climate change (see Sect. 8.2 for some additional comments on this).

If we now look at a project, it is easy to see that it is, first of all, a component of the economy, but that inevitably it has effects on the society and the environment (Fig. 2.4).

All this is certainly not new. However, when dealing with impact assessment of important projects like dams and hydropower, we have to always keep in mind that

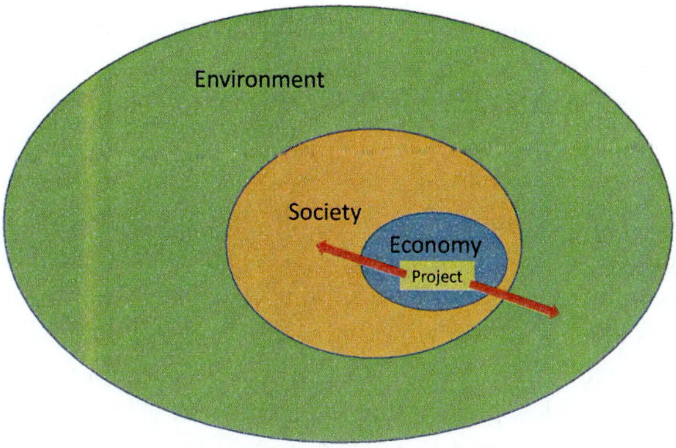

Fig. 2.4 A project as part of the system

this is what we are talking about: a project in its economic, social and environmental context and the entire range of interactions between the components of the system. Ioris (2020) asks for such an approach when he asks for "giving equal importance to the cultural, the political and the economic components of socioecological processes" in the development of dam projects.

2.2 Same Point, Different Points of View

Another issue is the fact that people can be in front of the same situation, but they see completely different things. This is illustrated here with two people who visit the Kongou Falls on the Ivindo River in Gabon (Fig. 2.5). Both are overwhelmed by the spectacular scenery. They watch the same thing, but they see something totally different.

The person on the left is a biologist. She sees the waterfall and the surrounding forest, is aware of the fact that this is located in a national park, and she imagines the

Fig. 2.5 Two people seeing the same thing—but do they?

different habitats and all the animals that live in and around the river. The person on the left is a hydropower engineer; what he sees is a river discharge Q, a difference in height Δh_1, the possibility to provide storage volume and at the same time to increase available head to Δh_2, in this way increasing the potential for generating power. The point with this example: none of these two views is wrong, they are both correct: the place is a highly valuable ecosystem (just to mention this point: the three species shown in the figure, elephant, gorilla and crocodile, actually live in this national park; for a description see Vande Weghe 2006), and the site undoubtedly has a potential for hydropower. It is an essential task of the ESIA—and therefore of the team carrying out the study and preparing the report—to collect the arguments of both sides, present them in a way that they can be understood by representatives of both sides, and to make recommendations on which view to follow, and under what conditions. Doing that, they will also have to keep in mind that the ESIA study does not make a decision, but that it has to present facts in a way that an informed decision can then be taken by those in a position—and with the mandate—to do so.

The next example is from a public consultation in Cameroon, presenting a dam project in a village which would be submerged in the reservoir. In the discussion, one of the PAPs (project affected persons) suggested to move the dam further downstream, in this way avoiding submersion of their village. The answer of the engineer was that in this case the reservoir would get much bigger, and their village would be submerged just as well. The interesting thing in this discussion was the fact that neither of the two understood the point of the other. The situation and the two "solutions" are illustrated in Fig. 2.6.

For the village inhabitant, his idea was just to move the dam, basically as planned, further downstream until it reached a point where FSL (full supply level) of the reservoir would be lower than the village, in this way sparing it. The main point of

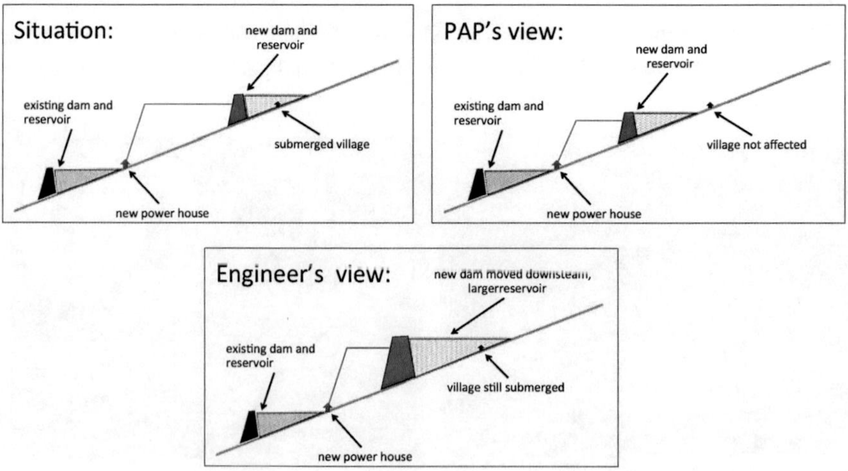

Fig. 2.6 Mutual incomprehension

the engineer was that, in order to make the project feasible he would have to reach the same head as in the original project; in order to do so, a higher dam would be required, since FSL of the reservoir would need to be maintained, and the reservoir would be much larger, still submerging the village. A short explanation then clarified the situation, and mainly made it clear for the inhabitants of the village that there would be no dam project that would not affect their village.

2.3 Islands in a Reservoir

A good practical example to illustrate different points of view is presented by some recent publications about the biodiversity on newly formed islands in reservoirs. When a large reservoir is formed submerging a rather wide, more or less flat area in generally undulating terrain, there is a high probability that after impoundment many of the original hills in this area will stick out of the water, forming islands of different sizes. Benchimol and Peres (2015; see also note in Nature 523: 7559, p. 130, 2015) investigated 37 out of the 3546 such islands of different sizes which had formed in the 3129 km^2 reservoir created by the Balbina dam in Brazil, 26 years after impoundment, by comparing larger fauna (including 35 mammal, bird and tortoise species) present on such islands with the fauna in habitats on the shore of the reservoir. They found a strong positive correlation between island surface and number of species encountered, or actually, more to what they express, a decline in the number of species still living on such islands with decreasing island size. Jones et al. (2018), investigating tree species composition and recruitment on (presumably the same) islands in the same reservoir consistently found decreasing sapling and tree densities with a decrease in island surface. Liu et al. (2019) carried out a similar study in a reservoir in China, analysing the vegetation on 29 islands in the 580 km^2 Thousand Islands Lake, which had been created by the Xin'anjiang Dam in 1959; they note that all these islands had been clear-cut before impoundment, so that the starting point for vegetation development for all of them had been similar. Their main finding was that smaller islands, especially those with a surface of <10 ha, had a larger percentage of pioneering and early succession species, while late succession species were more frequent on larger islands. Before them, Si et al. (2017) had carried out a study on birds on such islands in the same reservoir, finding that out of a total of 68 species, 57 of them were recorded on at least one of the 36 islands investigated, and 55 on two on-shore mainland plots (see Fig. 2.7).

All of these authors indicate close correlation of their findings with the principles of island biogeography as formulated by MacArthur and Wilson (1963, 1967), which basically states that biodiversity on islands, in comparison to nearby mainland, decreases with decreasing size of the islands and with increasing isolation, i.e. distance from the nearest mainland with similar conditions. This theory has since then been confirmed in numerous other studies, as e.g. Nilsson et al. (1988) or, very recently, Valente et al. (2020), who, for birds inhabiting oceanic islands worldwide show clear correlations of increasing numbers of species with increasing

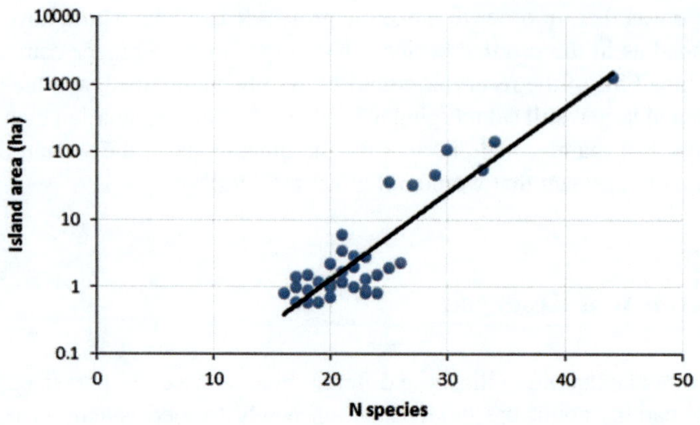

Fig. 2.7 Number of bird species related to island area. *Source* Based on data from Si et al. (2017)

island area. Isolation in the sense of distance from mainland, which is known to be a factor increasing the probability of a loss of species and reducing colonisation, i.e. immigration of new species or mainland specimens of species already present is not an important factor in the case of islands even in large reservoirs, at least not for birds, since these are usually still within easy reach of most bird species. It may be different for other groups like terrestrial mammals and reptiles. Some studies applied the island biogeography theory also on mainland habitat islands as e.g. larger and smaller forest patches in an otherwise rather sterile agricultural landscape (Diamond 1975; McCoy 1983). While these correlations are basically accepted, it has also been shown, especially for continental habitat islands or when trying to apply these principles to nature reserves, that the comparison with "real" islands is not always applicable, since in such cases vegetation in the surroundings can influence the habitat island in rather decisive ways (see e.g. Kushlan 1979; Dickman 1987; Lovejoy 1980; Margules et al. 1982; Janzen 1983). However, a recent study investigating grassland inhabiting Orthoptera (grasshoppers) in small habitat islands in a landscape dominated by agriculture shows that "despite low numbers of species per field island, due to their intrinsic small size, taken together, a system of small field island increased species richness and included a large proportion of the regional species pool"; there as well, the number of species per plot tended to increase with increasing plot size (Riggi and Berggren 2020). In any case, for islands which are actually land areas completely surrounded by water, the theory is valid, and in this sense the results of the four studies cited above do not come as a surprise, but rather as a confirmation of the theory.

However, two of these studies, the two from Brazil, which were carried out by the same group, come to additional conclusions. Benchimol and Peres (2015) state: "Based on species-area models, we predict that only 0.7% of all islands now harbour a species-rich vertebrate assemblage consisting of ≥80% of all species. We highlight the colossal erosion in vertebrate diversity driven by a man-made dam and show that the biodiversity impacts of mega dams in lowland tropical forest regions have been severely overlooked", or, as the note published in Nature mentioned above puts it "They estimate that, in the 26 years between the dam's construction and their survey, isolation has led to an overall species loss of 70% across all islands created by the dam, with smaller islands suffering the most. Just 25 of the 3546 islands are likely to host 80% or more of the animals that they looked for. Such negative impacts have not been generally considered, and biodiversity impacts should be better assessed before future hydropower projects go ahead, the authors suggest". Jones et al. (2018) point in the same direction stating that "results suggest that insular tree communities are unstable, with rare species becoming extinction-prone due to reduced tree recruitment and density on islands, potentially leading to future losses in biodiversity and ecosystem functioning across Balbina's >3500 reservoir islands". To put it simply, their position is that the creation of islands in a reservoir leads to a depletion of biodiversity, many species going to extinction on these—and especially smaller—islands. For me this is a rather strange and, in my view wrong, argument. While I do not doubt in any way the facts that they have found, namely, less species and a simplified forest vegetation on smaller islands, my conclusion would be the other way round: where the original habitat is submerged, 100% of the forest inhabiting species are lost, whereas on all islands left in the reservoir, some species will survive, on the largest 0.7% ones, meaning about 25 islands in this case, up to 80% or more of the original fauna. The facts remain the same, but the conclusions are different, depending on how you look at them. And when I read one of the recommendations put forward by Jones and colleagues, asking that "dam development strategy explicitly considers (a) dam location, aiming to minimize creation of small (<10 ha) and isolated islands ...", I am almost tempted to fall into sarcasm recommending that "next time build a higher dam, the chance is good that the number of small islands will be reduced".

2.4 The Necessity to Understand

People preparing an ESIA need not be (and usually are not, and should not be) hydropower engineers. Still, they need to have a basic understanding of the way a hydropower plant works, and, more importantly, an understanding of how the power plant which is the object of the study will work; this, for obvious reasons, includes people who revise and comment on such studies. Two examples may be sufficient for explaining this point.

In a recent case, and ESIA for a hydropower project in Laos, there was a group of experts who had to comment on the report. Important point: the project at hand was a run-of-river (ROR) project, which aimed at keeping the water level in the reservoir constant at all times (not possible during flood events, where it would be somewhat higher), and which had no "live storage", i.e. no capacity for retaining water during times of low electricity demand for then increasing the output during times of high demand; one could call this a "true ROR" scheme. This was described in much detail in the ESIA. One comment received was approximately as follows: "It is not just inflow-outflow equivalence, but also retention time. In any case, inflows will not equal outflows at any time when water levels are rising or falling, which is necessary seasonally to adjust levels. Retention time in the reservoir is also slower than the transit time along the natural river, especially in the dry season". This statement is wrong on more than one account. One point that could be made is that during reservoir impoundment (the level of the reservoir, once the dam in place, will be higher by about 15 m than the original average river level) outflow would be less than inflow, due to the volume of water retained in the reservoir; however, given the volume of the reservoir in relation to the volume flowing in the river, this would be of very short duration, reducing river discharge in the downstream area for a few days if filling should take place in the dry season, to just a few hours if done during the wet season. But once that full supply level (FSL) or normal operation level of the reservoir will be reached and maintained, there is physically no other possibility than to release exactly as much water from the dam as is flowing into the reservoir from upstream. What will change due to the volume flowing at any time is the way how this water will be discharged: through turbines at reduced capacity as long as inflowing water is below full processing capacity of the turbines, at full capacity as long as enough water is available, and partly over spillways or low-level gates in the dam during times where more water reaches the reservoir as can be used by the turbines. As a side note one can mention here that such reservoirs also do not have any potential for flood reduction, since there is no storage to be filled. The second problematic statement is that "retention time is slower"; it might be said that retention time is somewhat longer, since flow speed in the reservoir is slower, given the larger volume of water in the reservoir than what would be there in the river under natural conditions. Undoubtedly, such a reduction in flow speed has consequences for different important aspects, e.g. sedimentation of suspended solids or drift of organisms like fish larvae or benthos insects.

In another study an extremely severe impact on forests surrounding the future reservoir was predicted. Statements made are as follows: "The dam will create a huge water mass … the level of water evaporation and humidity in the atmosphere will increase, especially in summer which is originally meant to be a drier season … the evaporating water will stagnate in the upper zone of the valley … it will accumulate exponentially … it will condense on all vegetation reflecting in major changes to the bio-climatic conditions. … The level of fungus in the soil and on the vegetation will increase encouraged by the higher humidity … resulting in disappearance of classes of organisms … Trees that characterize the system … their exposure would weaken their vigour and increase attack of diseases that would be of insect or fungal sources …", and finally: "The distinctive conifer of this region *Arceuthos drupacea* will be directly influenced by the change in conditions propagated by the presence of the dam lake. As a reminder, this forest is the single of its type in Lebanon; its endangerment implies the threatening of a national natural heritage". A truly disastrous situation in the making. Unfortunately, these statements were all made without support by any figures and under a number of clearly wrong assumptions.

A few facts about the project: the 97 m high dam (height above river bed, excluding foundations) is located in a narrow gorge. It created a reservoir with an FSL of 839 m asl and a total storage volume of 38 Mm^3. The reservoir has a surface area at FSL of 1 km^2. The minimum operation level is 772 m asl, and the average water volume turbined annually is 95 Mm^3. This shows that the reservoir is very small, the water volume flowing through it being almost three times the storage volume. The climate is extremely seasonal, with as good as all precipitation falling in winter, partly as snow in the higher parts of the catchment. The effect of evaporation of such a small reservoir on the surroundings will be, if at all, hardly measurable, and if, will be restricted on the immediate surroundings of the reservoir. Even weak winds (which, according to this same report, actually exist) would carry any additional moisture away; difficult to see why or how it should "accumulate exponentially". One fact that the authors have completely misunderstood (or ignored) is the fact that there is a very considerable drawdown of up to 67 m (from 839 to 772 m asl). This will mainly be the case in summer; this means that a large part of the reservoir area, formed by steep rock cliffs, will be exposed during that time. It might therefore be possible that humidity in the immediate surroundings of the reservoir will even be less than it is now. This, and the fact that the study often mixes up reservoir area (1 km^2) and catchment area (242 km^2) clearly shows that the authors did not really understand the functioning of this specific project.

One more point: *Arceuthos (Juniperus) drupacea*, said to be under direct threat from the project, was actually not found in the area affected by the project (Pöyry 2017). However, it does occur in the Jabal Moussa Biosphere Reserve, which is located close to the project site, without however being affected by it (Thomé 2013). It grows usually at higher altitudes (from 500 up to 1900 m asl), and is, although not very frequent, rather widespread, from Greece over Turkey and Syria to Lebanon (Eckenwalder 2009: 290), it is listed as least concern by IUCN. Here, one wonders whether this was just a lack of information or actually part of the study whose aim obviously was to "kill" the project (see Sect. 3.11.5.3).

References

Ardelean, C.F., L. Becerra-Valdivia, M.W. Pedersen, et al. 2020. Evidence of human occupation in Mexico around the last glaciation maximum. *Nature* 584 (7819): 87–92. https://doi.org/10.1038/s41586-020-2509-0.

Becerra-Valdivia, L., and T. Higham. 2020. The timing and effect of the earliest human arrivals in North America. *Nature* 584 (7819): 93–97. https://doi.org/10.1038/s41586-020-2491-6.

Benchimol, M., and C.A. Peres. 2015. Widespread forest vertebrate extinctions induced by a mega hydroelectric dam in lowland Amazonia. *PLoS ONE* 10 (7): e0129818. https://doi.org/10.1371/journal.pone.0129818.

Bliss, S., and M. Egler. 2020. Ecological economics beyond markets. *Ecological Economics* 178. https://doi.org/10.1016/j.ecolecon.2020.106806.

Bonneuil, C., and J.-B. Fressoz. 2013. *L'événement anthropocène. La terre, l'histoire et nous*, 304 pp. Paris: Editions du Seuil.

Carew-Reid, J., R. Prescott-Allen, S. Bass, and B. Dalal-Clayton. 1994. *Strategies for national sustainable development. A handbook for their planning and implementation*, 203 pp. London: Earthscan, IUCN, and IEED.

Crawford, M.H. 1992. *Antropología biológica de los Indios Americanos. Colección Indios de América*, 314 pp. Madrid: Editorial MAPFRE S.A.
Crosby, A.W. 1986. *Ecological imperialism. The biological expansion of Europe, 900–1900*, 368 pp. In Studies in environment and history. Cambridge: Cambridge University Press.
Dawkins, R., and Y. Wong. 2016. *The ancestor's tale. A pilgrimage to the dawn of life*, updated ed., 771 pp. London: Weidenfeld and Nicolson.
Diamond, J. 1997. *Guns, germs, and steel. The fates of human societies*, 480 pp. New York: Norton.
Diamond, J.M. 1975. The island dilemma: Lessons of modern biogeographic studies for the design of natural reserves. *Biological Conservation* 7: 129–146.
Dickman, C.R. 1987. Habitat fragmentation and vertebrate species richness in an urban environment. *Journal of Applied Ecology* 24 (2): 337–351.
Eckenwalder, J.E. 2009. *Conifers of the world. The complete reference*, 720 pp. Portland: Timber Press.
Ehrlich, P. 1971. *Die Bevölkerungsbombe*, 191 S. München: Hanser.
Ioris, A.A.R. 2020. Socioecological economics of water development in the Brazilian Amazon: Elements for a critical reflection. *Ecological Economics* 173. https://doi.org/10.1016/j.ecolecon.2020.106654.
Irwin, A. 2001. *Sociology and the environment*, 210 pp. Cambridge: Polity Press.
Jackson, T., and P.A. Victor. 2019. Unraveling the claim for (and against) green growth. *Science* 366 (6468): 950–951.
Janzen, D.H. 1983. No park is an island: Increase in interference from outside as park size decreases. *Oikos* 41 (3): 402–410.
Jones, I.L., C.A. Peres, M. Benchimol, L. Bunnefeld, and D.H. Dent. 2018. Instability of insular tree communities in an Amazonian mega-dam is driven by impaired recruitment and altered species composition. *Journal of Applied Ecology* 56: 779–791. https://doi.org/10.1111/1365-2664.13313.
Kushlan, J.A. 1979. Design and management of continental wildlife reserves: Lessons from the everglades. *Biological Conservation* 15: 281–290.
Liu, J., D.A. Coomes, G. Hu, J. Liu, J. Yu, Y. Luo, and M. Yu. 2019. Larger fragments have more late-successional species of woody plants than smaller fragments after 50 years of secondary succession. *Journal of Ecology* 107 (2): 582–594. https://doi.org/10.1111/1365-2745.13071.
Lovejoy, T.F. 1980. Los espacios abiertos intermitentes: Areas minimas para la conservación. *Parques* 5 (2): 13–15.
MacArthur, R., and E.O. Wilson. 1963. An equilibrium theory of insular zoogeography. *Evolution* 17: 373–387.
MacArthur, R.H., and E.O. Wilson. 1967. *The theory of island biogeography*, 203 pp. Princeton, NJ: Princeton University Press.
Malthus, T. 1798. *An essay on the principle of population*, 91 pp. London: Digital: J. Johnson, St. Paul (word version of the text).
Margules, C., A.J. Higgs, and R.W. Rafe. 1982. Modern biogeographic theory: Are they lessons for nature reserve design? *Biological Conservation* 24 (2): 115–128.
McCoy, E.D. 1983. The application of island-biogeographic theory to patches of habitat: How much land is enough? *Biological Conservation* 25: 53–61.
Meadows, D. 1972. *Die Grenzen des Wachstums. Bericht des Club of Rome zur Lage der Menschheit*, 180 S. Stuttgart: DVA.
Millar, N., E. McLaughlin, and T. Börger. 2019. The circular economy: Swings and roundabouts? *Ecological Economics* 158: 11–19.
Nilsson, S.G., J. Bengtsson, and S. As. 1988. Habitat diversity or area per se? Species richness of woody plants, carabid beetles and land snails on islands. *The Journal of Animal Ecology* 57 (2): 685–704.
Portney, K.E. 2015. Sustainability. In *MIT Press essential knowledge series*, 234 pp. Massachusetts Institute of Technology.

Pöyry. 2017. *Jannah dam and lake project. Update of environmental impact assessment reports.* Eaux de Beyrouth et du Mont Liban. Unpublisehd Report.

Rees, W.E. 2020. Ecological economics for humanity's plague phase. *Ecological Economics* 169. Article 106519. https://doi.org/10.1016/j.ecolecon.2019.106519.

Riggi, L.G.A., and A. Berggren. 2020. Small field islands systems include a large proportion of the regional orthopteran species pool in arable landscapes. *Journal of Insect Conservation* 24: 695–703. https://doi.org/10.1007/s10841-020-00245-z.

Roth, S., and V. Valentinov. 2020. East of nature. Accounting for the environments of social sciences. *Ecological Economics* 176. https://doi.org/10.1016/j.ecolecon.2020.106734.

Sanderson, S.K. 2001. *The evolution of human sociality. A Darwinian conflict perspective*, 392 pp. Lanham: Rowman and Littlefield.

Segerstråle, U. 2000. *The defenders of the truth. The sociobiology debate*, 493 pp. Oxford: Oxford University Press.

Shaw, J.H. 2021. Neither stable nor pristine: American bison populations were long influenced by humans. *Therya* 12 (2): 171–175. https://doi.org/10.12933/therya-21-1112.

Si, X., M.W. Cadotte, D. Zheng, et al. 2017. Functional and phylogenetic structure of island bird communities. *Journal of Animal Ecology* 86 (3): 532–542.

Thomé, G.H. 2013. *Guide des arbres de la réserve de la biosphère du Jabal Moussa*, 104 pp. Liban: Association pour la Protéction du Jabal Moussa.

Valente, L., A.B. Phillimore, M. Melo, et al. 2020. A simple dynamic model explains the diversity of island birds worldwide. *Nature* 579 (7797): 92–96. https://doi.org/10.1038/s41586-020-2022-5.

Vande Weghe, J.P. 2006. *Ivindo et Mwanga. Eaux noires, forêts vierges et bais. Les parcs nationaux du Gabon*, 272 pp. Libreville, Gabon: Wildlife Conservation Society.

Veen-Groot, D.B., and P. Nijkamp. 1999. Globalisation, transport and the environment: new perspectives for ecological economics. *Ecological Economics* 31 (3): 331–346.

WCED. 1987. *Our common future. World commission on environment and development*, 383 pp. Oxford: Oxford University Press ("Brundtland report").

Chapter 3
Approach to ESIA

It is better to be approximately right than exactly wrong.
BUWAL (1990)

This chapter describes the ESIA process from the point of view of somebody involved in preparing such reports, and it describes the process from the early stages of project identification and acquisition. Different issues like stakeholders involved in the process, ESIA report structure, the bidding process and the composition of the team are discussed. Then types of criticism of the ESIA, often uttered and from different sides, are addressed. Here again, a number of good or, in this case maybe mainly not so terribly good, examples serve the purpose of illustrating the points raised. The main aim of the chapter is to show that the ESIA has an important role to fulfil, but that an awareness of potential and very real risks in preparing such reports is required.

3.1 Origins of Water Management

Water management is a human activity of vital importance since the development of agriculture, which started independently in at least three different regions, the fertile crescent in the Near East, in East and South East Asia, and in Central America. The first efforts for managing water resources were directed at irrigation. There is a theory that irrigation systems are at the origin of the formation of feudal states, since their development from small isolated systems to ever larger and more complex ones required a centralised administration for planning, implementing and operating; this is rather convincingly discussed by Gao (1999: 373 ff.) for the Dai culture in Yunnan province, Southern China, but contested by Diamond (1997) as a general model of state origin. Interestingly, although maybe on a side-line to the subject of this book, there is a similar question concerning the development of religion, namely the question whether moralising high gods (i.e. gods which actually cared about social behaviour of mankind and punished misbehaviour, as opposite to deities who might have punished disrespect towards them, but seemed not to care about

R. Zwahlen, *Assessing the Environmental Impacts of Hydropower Projects*,
Environmental Earth Sciences, https://doi.org/10.1007/978-3-030-91185-0_3

moral behaviour) or "broad supernatural punishment" (i.e. supernatural entities who punished misbehaviour, without however being "high gods") had developed before the raise of large and complex societies, and were almost a prerequisite for this type of social development; a recent investigation of this subject indicates that complex societies preceded the development of such moralising gods (Whitehouse et al. 2019), which could indicate that belief in such deities may have facilitated the cohesion (and potentially control) of such societies, without however being a condition for their emergence.

Dams supporting these irrigation systems were mostly low structures which served as intakes, but usually not for water storage. Very often, these simple dams were swept away during floods, and often each year during the high-flow season, and had to be rebuilt every year. Their potential for changing the river ecosystem in any way were very limited. Next in line was river training for drainage and flood protection. Being Swiss, I have to mention here the "Juragewässerkorrektion", a massive project undertaken in the late nineteenth century for redirecting the Aare River, regulating the three lakes of Murten, Neuchâtel and Biel-Bienne, and draining the vast swampy area that surrounded these lakes, transforming it into Switzerland's most fertile agricultural area (Vischer 2003: 105 ff.; Nast 2006). By doing so, it undoubtedly caused a major environmental impact, literally destroying Switzerland's largest wetland, but protecting the local population from frequent and often severe flooding and turning agriculture into a productive activity, taking the local population—my mother's ancestors among them—out of poverty. Not surprisingly, this as well as other, very numerous river training projects were all done under the name of "Korrektion" (correction): man took measures for correcting, in a way, mistakes of nature, in these cases with the main aim of reducing the risk of floods, and often with the additional aim of making more land available for agriculture.

The construction of large dams for energy generation, and often with the additional or main aims to store large quantities of water and to gain head, is a very new development, which started not much earlier than the late nineteenth century, not much more than a hundred years ago. Leslie (2005: 3 ff.) mentions the Hoover Dam in the USA, whose construction started in 1931 and which was commissioned in 1939, as the "first dam of the modern area" and the start of modern dam construction. However, other, albeit lower dams had been built before (e.g. the Rheinfelden hydropower plant on the River Rhine on the border between Germany and Switzerland, see Sect. 13.2).

The positive effects of such projects were clearly evident. Some of their potential negative effects were anticipated, like e.g. risks for fisheries or drinking water supply, while others, as e.g. increasing erosion further downstream, were completely ignored and had then to be corrected in their turn by additional measures, as e.g. described by Vischer (2003: 61 ff.) for the case of redirecting the Kander, a notoriously dangerous mountain river, into the lake of Thun (Switzerland) in the early eighteenth century.

3.2 The Early Days of ESIA

Caminante, no hay camino, se hace camino al andar.

Joan Manuel Serrat, Spanish singer-songwriter, in his album Dedicado a Machado, 1969

When I started to be involved in preparing ESIAs (or, more precisely: EIAs; the emphasis at that time was entirely on environmental impacts, the "S" was added only considerably later, although it is a very important—and undoubtedly in many cases *the* most important—topic to be addressed), with my first such study being done for the Ayago-Nile HPP in Uganda in 1978/79, there was little available in terms of experience with such studies, the respective legislation was largely still amiss, only rudimentary if any guidelines or standards were there to be consulted and followed (not to mention the fact that, even if they would have been there, they would have been harder to find in this "pre-www-era"!), and there were no older and experienced colleagues around whom one could have asked. Still, with a solid background in ecology I think it was not too hard to identify the relevant problems and impacts. For more about this project, see Sect. 26.1.1. This gave all the colleagues who started to be active in this line of work the possibility to be "making the way while going", as Joan Manuel Serrat put it in his song mentioned above.

One problem I had to face at that time, and I am sure this was the case for other early environmental scientists in the hydropower business, was the attitude and the self-assured position of the engineers, and mainly of those with much experience. As I see it, there are three major groups of professionals who are convinced that they know, while all others don't: the medical doctors, the lawyers, and the engineers. Therefore, when discussing about a dam project, the engineers had it all covered, and why should they listen to a young person who was not even an engineer and who tried to tell them what to do (or worse: what not to do). Luckily, this situation has changed in a very considerable way. Young engineers today are very much aware of the fact that a dam is not only, to simplify it, about concrete, but has to consider a wide array of components, from financial, legal and engineering to environmental and social. As far as I know, all students in civil engineering today get a background on all these other topics, and I hope that I have made my contribution to this with my lectures. Even so, the original engineering attitude is still around. So e.g. in a recent congress (Africa 2019: Water Storage and Hydropower Development for Africa), the overall aims of ICOLD were listed as "Engineering, Construction and Operation" of dams. This is certainly too restrictive. At the very least, "Planning" should be added to this list of topics, ahead of engineering (and I sternly refuse the idea that "engineering" completely covers this decisive phase of project development), and today one might even want to add "decommissioning" as one additional important aspect (although, admittedly, I have my own point of view on this topic, see Chap. 13). The struggle certainly has to go on.

3.3 The Main Actors, Stakeholders

The stakeholders involved in the development of an HPP—and probably that of any major infrastructure project—can be roughly subdivided into three groups as illustrated in Fig. 3.1.

The three groups and the main actors within them are briefly described here.

The **Interested Parties** are the ones who have a direct interest in the project and/or the power to influence it in one way or another:

- The government, depending on the situation the national and/or a provincial government, usually have the ultimate power to accept or to refuse a project; in the case of HPPs, the government, in that case usually presented by its Ministry of Energy (or equivalent), can also be the project proponent.
- The project proponent, either a governmental or a private organisation, is certainly the most relevant stakeholder; no project without a project proponent.
- Financing institutions are regularly involved in such projects. The most well-known such organisations in the field of hydropower are the multilateral development banks, mainly the World Bank (including IFC), but also the regional

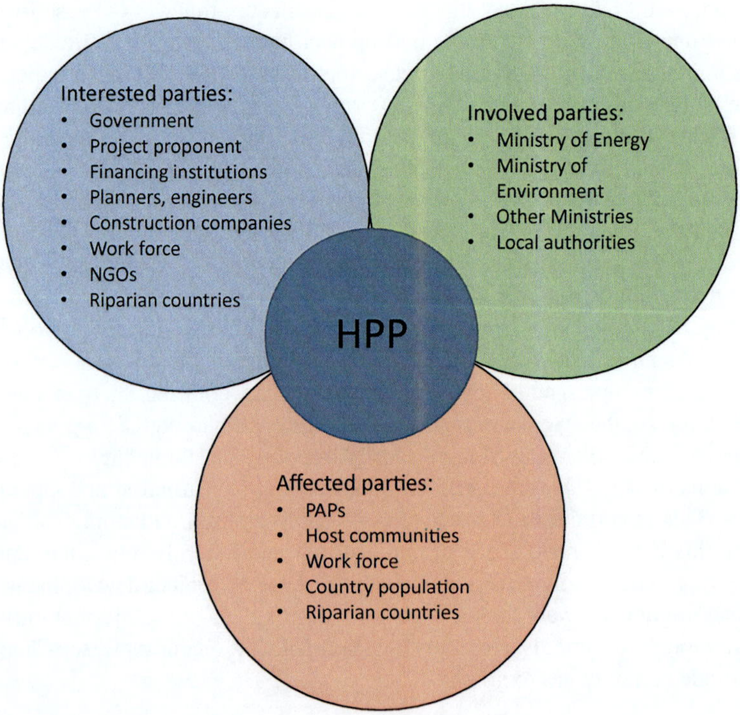

Fig. 3.1 Stakeholders in an HPP

development banks, i.e. Asian Development Bank (ADB), Interamerican Development Bank (IDB), African Development Bank (AfDB), and European Bank for Reconstruction and Development (EBRD). However, internationally active private or state banks are also involved in hydropower financing. In terms of environmental and social issues the development banks have their E&S guidelines, standards and safeguards, while many of the private banks adhere to the Equator Principles, which largely follow the same lines.

- Planners and engineers are obviously interested in such projects, since this is their field of professional activity, and their work influences the projects to a very large extent.
- Contractors, construction companies: these are mainly large construction companies as well as the manufacturers of electro-mechanic equipment, but also smaller local companies which can get involved in large construction sites, mainly as subcontractors of larger companies.
- The work force, here understood as people who are willing to work for such projects, either in large numbers during the construction phase or in smaller numbers during operation of the plant, may be a special case in this group. While they have certainly an interest in such a project, their potential for influencing it, especially during the relevant stages of project development and licensing, is very limited at best. For this reason the work force is included in two groups in Fig. 3.1.
- NGOs: this here refers to NGOs that are or can be involved in hydropower projects in one way or another. These are mainly three groups, all of which can be organised at the local, national or international level, namely, organisations dedicated to nature and environmental protection, organisations interested in human rights and related aspects, and those with a very specific focus on hydropower and dams; any NGO active in this wider area may not always be attributable clearly to one of these three categories, there can be a wide margin of overlap. At least some of these groups, the most well-known certainly being the International Rivers Network (IRN; https://www.irn.org/), certainly have a vital interest in dam projects, since this is their *raison d'être*.
- Riparian countries are countries sharing a river, and they have an interest in what is being done there by other countries, mainly by those located upstream from them. They can influence projects via governmental bodies and political activities. In a number of cases, such countries have come to agreements on how to use a common river, as is e.g. the case for countries in the Aral Sea basin in Central Asia (with the rivers Amu Darya and Syr Darya: Kyrgyzstan, Tajikistan, Kazakhstan, Turkmenistan and Uzbekistan, however without Afghanistan, at least so far; see Chap. 23). There are also some river basin organisations that deal with aspects of river and river basin management and make sure that projects are at least brought to the attention of and discussed among riparian countries. The best-known of these organisations are probably the Mekong River Commission (MRC; http://www.mrcmekong.org/), which involves Thailand, Laos, Cambodia and Viet Nam, and the Nile Equatorial Lakes Subsidiary Action Program (NELSAP; http://nelsap.nilebasin.org/index.php/en/) that includes Burundi, DR Congo, Egypt, Ethiopia, Kenya, Rwanda, South Sudan, Sudan, Tanzania and Uganda.

Involved Parties are organisations which according to their function have to participate in a specific way in the development process of such projects, without necessarily having a direct interest in them.

- Ministry of Energy: usually involved in one way or another in the development process of an HPP; as mentioned above, this ministry can also act as project proponent.
- Ministry of Environment: this is the organisation to which, as rule, the EPA (Environmental Protection Agency; the name used here is a generic one, as for the ministries listed) belongs, i.e. the competent authority for the ESIA process. While here only the national level organisation is explicitly mentioned, depending on the type or size of the project and/or applicable legislation, provincial and local authorities will often be involved in the process.
- Other ministries (e.g. in charge of public works, labour conditions and social security, transport, finance etc.) can be involved.

Affected Parties are those who, one way or another, are affected by the project, but have little if any influence on it. The most important members of this group are the following:

- PAPs: the Project Affected Persons. Under this heading we understand mostly the local population in the project area, and here again foremost the people who are affected in a very direct way, i.e. those who have to be relocated due to the project and/or who lose all or part of the resources their livelihood depends on; these are usually the people living in the future construction and dam site, and in the area of the future reservoir.
- Host communities: these are the settlements and their population where the PAPs will be reinstalled.
- Work force: as mentioned above, this is a group of people with a vested interest in such projects, but with little if any power in it. One example to show this: in case a contractor is being changed, for whatever reason, the workers employed by this company may or may not be re-employed by the new contractor, but they have no influence on that.
- Local, regional and national population: usually interested directly or indirectly in the project for getting (at least temporary) employment, either as workers on the construction site, in the form of indirect opportunities, e.g. by providing services to the construction work force, or benefiting indirectly from increased economic activities due to the project. Here, we can also mention the entire population that will benefit from an improvement in electricity supply. Involvement of this large group, if any, is usually very marginal.
- Riparian countries: while above they are included in the group of interested parties, they can also be affected, meaning that they are exposed to consequences of a project without having any influence on it. Laos is a member of MRC mentioned above, but this organisation covers only the lower Mekong basin. China has developed a number of dams on the Lancang (as the Mekong is called in China) and is

continuing to do so. Laos, which experiences the consequences very directly (see Chap. 22) is being informed at best, but has no influence at all on this development.

3.4 Structure

3.4.1 Basic Structure of an ESIA

Here, the most important advice to give is: closely follow the structure as specified in the ESIA guidelines of the respective country or agency which will receive and evaluate the report. They are usually very similar to one another, more or less closely following the outline as proposed by the World Bank (see following section), but they may contain some donor agency or country-specific items. The main reason for this recommendation: the report will have to go to the "competent authority", which will check and either approve or reject it. I had some experiences with such authorities which I would not necessarily qualify as "competent", but they will still decide, and the impression gained in some cases was that this process consists mainly in the use of their decreed structure as a checklist, for seeing if all the items listed there are covered. This can go into real details, as illustrated by an ESIA prepared for a dam project in a rather remote area, for a dam site located in a narrow, uninhabited, gorge-like valley more than 20 km away from the nearest road or human settlement. The only noise one could hear, maybe apart from the occasional bird, was the noise of the river in this steep and rocky valley (a noise which, by the way, would no longer be there once the dam would be built). The ESIA guideline asked for a baseline noise investigation, and the competent authority insisted on this point, so that is what was done. In another occasion, which was not about a dam, but about water supply, waste water treatment, waste management, drainage and flood protection of three small towns I was in charge of reporting on compliance with environmental and social safeguards. The only comment I ever received on my report was that they needed two separate reports, one environmental and one social. Since in that case environmental and social issues were so completely intertwined and overlapping, both aspects had been dealt with in one report. However, following the request from the funding agency, it had to be separated in two, which resulted in an approximately 80% overlap between the two reports, with no gain whatsoever in terms of additional information.

3.4.2 Comparison of Different Approaches

Table 3.1 provides the basic content of an ESIA as originally proposed by Goodland et al. (1993: 305) and then provided in the World Bank's Operational Policy of 1999 and the Bank's recent (2017) Environmental and Social Framework. It can easily be seen that the basic content is still the same, although complexity, or in some

Table 3.1 Basic content of an ESIA

Goodland et al. (1993: 305)	WB 1999: OP 4.1, environmental assessment, Annex B	WB 2017: environmental and social framework. The World Bank, Washington DC., p. 25 ff.[a]
EA reports should be concise and focused on the significant environmental issues. The detail and sophistication of the report should be commensurate with the potential impacts. The target audience should be project designers, implementing agencies, proponents and financiers. The EA report should include	The EA report should include the following items (not necessarily in the order shown)	Indicative outline of ESIA Where an environmental and social impact assessment is prepared as part of the environmental and social assessment, it will include the following
(a) Executive summary: concise discussion of significant findings and recommended actions	(a) Executive summary. Concisely discusses significant findings and recommended actions	(a) Executive summary Concisely discusses significant findings and recommended actions
(b) Policy, legal and administrative framework within which the EA is prepared. The environmental requirements of any cofinanciers should be explained	(b) Policy, legal, and administrative framework. Discusses the policy, legal, and administrative framework within which the EA is carried out. Explains the environmental requirements of any cofinanciers. Identifies relevant international environmental agreements to which the country is a party	(b) Legal and institutional framework Analyses the legal and institutional framework for the project, within which the environmental and social assessment is carried out, including the issues set out in ESS1, paragraph 26 Compares the Borrower's existing environmental and social framework and the ESSs and identifies the gaps between them Identifies and assesses the environmental and social requirements of any co-financiers

(continued)

Table 3.1 (continued)

Goodland et al. (1993: 305)	WB 1999: OP 4.1, environmental assessment, Annex B	WB 2017: environmental and social framework. The World Bank, Washington DC., p. 25 ff.[a]
(c) Project description in a geographic, ecological, social, and temporal context, including any off-site investments that may be required by the project (e.g. dedicated pipelines, access roads, power-plants, water-supply, housing, and raw material and product storage facilities)	(c) Project description. Concisely describes the proposed project and its geographic, ecological, social, and temporal context, including any offsite investments that may be required (e.g., dedicated pipelines, access roads, power plants, water supply, housing, and raw material and product storage facilities). Indicates the need for any resettlement plan or indigenous peoples development plan (see also subpara. (v) below). Normally includes a map showing the project site and the project's area of influence	(c) Project description Concisely describes the proposed project and its geographic, environmental, social, and temporal context, including any offsite investments that may be required (e.g., dedicated pipelines, access roads, power supply, water supply, housing, and raw material and product storage facilities), as well as the project's primary suppliers Through consideration of the details of the project, indicates the need for any plan to meet the requirements of ESS1 through 10 Includes a map of sufficient detail, showing the project site and the area that may be affected by the project's direct, indirect, and cumulative impacts

(continued)

Table 3.1 (continued)

Goodland et al. (1993: 305)	WB 1999: OP 4.1, environmental assessment, Annex B	WB 2017: environmental and social framework. The World Bank, Washington DC., p. 25 ff.[a]
(d) Baseline data: dimensions of the study area and description of relevant physical, biological, and socio-economic, conditions, including any changes anticipated before the project commences. Current and proposed development activities within the project area (but not directly connected to the project) should also be taken into account	(d) Baseline data. Assesses the dimensions of the study area and describes relevant physical, biological, and socioeconomic conditions, including any changes anticipated before the project commences. Also takes into account current and proposed development activities within the project area but not directly connected to the project. Data should be relevant to decisions about project location, design, operation, or mitigatory measures. The section indicates the accuracy, reliability, and sources of the data	(d) Baseline data Sets out in detail the baseline data that is relevant to decisions about project location, design, operation, or mitigation measures. This should include a discussion of the accuracy, reliability, and sources of the data as well as information about dates surrounding project identification, planning and implementation Identifies and estimates the extent and quality of available data, key data gaps, and uncertainties associated with predictions Based on current information, assesses the scope of the area to be studied and describes relevant physical, biological, and socioeconomic conditions, including any changes anticipated before the project commences Takes into account current and proposed development activities within the project area but not directly connected to the project

(continued)

Table 3.1 (continued)

Goodland et al. (1993: 305)	WB 1999: OP 4.1, environmental assessment, Annex B	WB 2017: environmental and social framework. The World Bank, Washington DC., p. 25 ff.[a]
(e) Environmental impacts: the positive and negative impacts that are likely to result from the proposed project should be identified and assessed. Mitigation measures, and any residual impacts that cannot be mitigated, should be identified. Opportunities for environmental enhancement should be explored. The extent and quality of available data, key data-gaps, and uncertainties associated with predictions, should be identified/estimated. Topics that do not require further attention should be specified	(e) Environmental impacts. Predicts and assesses the project's likely positive and negative impacts, in quantitative terms to the extent possible. Identifies mitigation measures and any residual negative impacts that cannot be mitigated. Explores opportunities for environmental enhancement. Identifies and estimates the extent and quality of available data, key data gaps, and uncertainties associated with predictions, and specifies topics that do not require further attention	(e) Environmental and social risks and impacts Takes into account all relevant environmental and social risks and impacts of the project. This will include the environmental and social risks and impacts specifically identified in ESS2–8, and any other environmental and social risks and impacts arising as a consequence of the specific nature and context of the project, including the risks and impacts identified in ESS1, paragraph 28
(f) Analysis of alternatives: proposed investment design, site, technology, and operational alternatives, should be compared systematically in terms of their potential environmental impacts; capital recurrent costs; suitability under local conditions; and institutional, training, and monitoring, requirements. To the maximum extent possible, for each of the alternatives, the environmental costs and benefits should be quantified, and economic values attached where feasible. The basis for the selection of the alternative proposed for the project design must be stated	(f) Analysis of alternatives. Systematically compares feasible alternatives to the proposed project site, technology, design, and operation—including the "without project" situation—in terms of their potential environmental impacts; the feasibility of mitigating these impacts; their capital and recurrent costs; their suitability under local conditions; and their institutional, training, and monitoring requirements. For each of the alternatives, quantifies the environmental impacts to the extent possible, and attaches economic values where feasible. States the basis for selecting the particular project design proposed and justifies recommended emission levels and approaches to pollution prevention and abatement	(g) Analysis of alternatives Systematically compares feasible alternatives to the proposed project site, technology, design, and operation—including the "without project" situation—in terms of their potential environmental and social impacts Assesses the alternatives' feasibility of mitigating the environmental and social impacts; the capital and recurrent costs of alternative mitigation measures, and their suitability under local conditions; and the institutional, training, and monitoring requirements for the alternative mitigation measures For each of the alternatives, quantifies the environmental and social impacts to the extent possible, and attaches economic values where feasible

(continued)

Table 3.1 (continued)

Goodland et al. (1993: 305)	WB 1999: OP 4.1, environmental assessment, Annex B	WB 2017: environmental and social framework. The World Bank, Washington DC., p. 25 ff.[a]
(g) Mitigation plan: feasible and cost-effective measures which may reduce potentially significant adverse environmental impacts to acceptable levels should be proposed, and the potential environmental impacts, capital and recurrent costs, and institutional training requirements of those measures, duly estimated. The plan (sometimes known as an "action plan") should provide details on proposed work-programmes and schedules, to ensure that the proposed environmental actions are in phase with engineering and other project activities throughout implementation. The plan should consider compensatory measures if mitigation measures are not feasible and cost-effective		(f) Mitigation measures Identifies mitigation measures and significant residual negative impacts that cannot be mitigated and, to the extent possible, assesses the acceptability of those residual negative impacts Identifies differentiated measures so that adverse impacts do not fall disproportionately on the disadvantaged or vulnerable Assesses the feasibility of mitigating the environmental and social impacts; the capital and recurrent costs of proposed mitigation measures, and their suitability under local conditions; and the institutional, training, and monitoring requirements for the proposed mitigation measures Specifies issues that do not require further attention, providing the basis for this determination
		(h) Design measures Sets out the basis for selecting the particular project design proposed and specifies the applicable EHSGs or if the ESHGs are determined to be inapplicable, justifies recommended emission levels and approaches to pollution prevention and abatement that are consistent with GIIP

(continued)

Table 3.1 (continued)

Goodland et al. (1993: 305)	WB 1999: OP 4.1, environmental assessment, Annex B	WB 2017: environmental and social framework. The World Bank, Washington DC., p. 25 ff.[a]
(h) Environmental management and training: the existence, role, and capability, of environmental units at the on-site, agency, and ministry, levels should be assessed, and recommendations made concerning the establishment and/or expansion of such units, and the training of staff, to the point that EA recommendations are implemented (i) Monitoring plan regarding environmental impacts and performance. The plan should specify the type of monitoring, who would do it, how much it would cost, and what other inputs (e.g. training) are or may be necessary	(g) Environmental management plan (EMP). Covers mitigation measures, monitoring, and institutional strengthening; see outline in OP 4.01, Annex C	(i) Key measures and actions for the Environmental and Social Commitment Plan (ESCP) Summarizes key measures and actions and the timeframe required for the project to meet the requirements of the ESSs. This will be used in developing the Environmental and Social Commitment Plan (ESCP)

(continued)

Table 3.1 (continued)

Goodland et al. (1993: 305)	WB 1999: OP 4.1, environmental assessment, Annex B	WB 2017: environmental and social framework. The World Bank, Washington DC., p. 25 ff.[a]
Appendices References: written or published materials used in study preparation. This is especially important in view of the large amount of unpublished data that is often used Record of inter-agency/forum/consultation meetings: including lists of both invitees and attendees. The record of consultations to obtain the informed views of the affected people and local NGOs should be included	Appendixes (i) List of EA report preparers—individuals and organizations (ii) References—written materials both published and unpublished, used in study preparation (iii) Record of interagency and consultation meetings, including consultations for obtaining the informed views of the affected people and local nongovernmental organizations (NGOs). The record specifies any means other than consultations (e.g., surveys) that were used to obtain the views of affected groups and local NGOs (iv) Tables presenting the relevant data referred to or summarized in the main text (v) List of associated reports (e.g., resettlement plan or indigenous peoples development plan)	Appendices List of the individuals or organizations that prepared or contributed to the environmental and social assessment References—setting out the written materials both published and unpublished, that have been used Record of meetings, consultations and surveys with stakeholders, including those with affected people and other interested parties. The record specifies the means of such stakeholder engagement that were used to obtain the views of affected people and other interested parties Tables presenting the relevant data referred to or summarized in the main text List of associated reports or plans

[a]In this column, positions (f) and (g) were switched in order to follow the sequence of topics given in the older sources
ESS = Environmental and Social Standards

cases maybe just precision of the description of specific items, like e.g. mitigation measures, has increased, although some points, most noticeably on mitigation, were shortened and simplified in the 1999 WB version.

Most national guidelines, when specific structures of ESIA Reports are provided, closely follow the structure of the WB 1999 guideline. However, in these cases the guidelines were not taken as such, but as a well-defined structure to be followed; the sentence "not necessarily in the order shown", which was there in the 1999 version, but is no longer included in the more recent one, was largely ignored. This has led to the fact that today, worldwide, most if not all "standard" ESIAs follow this structure, taking the required points (a) to (h) of the 1999 outline as titles of the main chapters of such a report. The basic structure was originated by Robert Goodland, who, to the best of my knowledge, was the first environmental specialist of the World Bank, and who wrote the Bank's first ESIA guidelines. However, he had intended this, as clearly stated, as points which should be included in such a study, and not as a strict structure for EA reports; I had the occasion to verify this point once when I discussed EIA report structures with him, on the basis of a report prepared by us which obviously covered all the points mentioned, but did not follow this structure.

Annex 1 includes an example of a structure as provided in a national ESIA guideline: the case of Lao PDR (MONRE 2016). As can be seen from this Annex, this structure rather closely follows the one outlined in the WB OP, and in addition to that it is extremely detailed. This is certainly an attempt to provide a "one and for all" ESIA structure, i.e. a structure that can be followed by every ESIA for any type of project. This is basically a good approach, however, it has a number of weaknesses and risks. Without intending to give a detailed analysis or discussion of this specific guideline, a few of these risks and weaknesses are addressed here; similar examples can be found elsewhere, i.e. in the guidelines developed by environmental protection authorities of other countries, the case of Laos was chosen just because of familiarity with this one.

These are some of the issues:

- If it was meant as an indication of what could or might have to be included in an ESIA Report, then it is certainly a good approach, because it gives the preparer of such a report a rather clear and comprehensive list of topics that might have to be covered in the report. However, there is the risk that the staff of MONRE (the competent authority for such studies in Laos) will not see this as a guideline which needs to be adapted in every case to the conditions and requirements of the project at hand, but rather as a checklist which enables them to easily control whether the study is "complete and according to relevant standards" or not. If this is the case, this will inevitably lead to the requirement of describing, albeit shortly, items which in a specific case might not be relevant at all. The result will be ever more bulky reports with ever more irrelevant content, aimed only at fulfilling bureaucratic requirements, which cannot be in the interest of any of the stakeholders (with the possible exception of paper manufacturers). Note that Goodland's sentence "should be concise and focus on the significant environmental issues", a very relevant point, was omitted in most of the recent guidelines.

- Even with the best effort it might be difficult to come up with a structure which covers really all possible cases and topics to be included in a study. In this example, e.g., there is no mentioning of climate, although this is an important issue especially for an ESIA for a hydropower project, in two ways, namely (i) as a description of the situation in the project area (with precipitation and temperature as the two most relevant factors, and evaporation as a possible third), and (ii) climate change for a discussion of its potential middle and long term impact on water availability, and the role of hydropower for climate change as such.
- The structure also provides some built-in problems: In Chap. 4 (which could be named "Baseline") there is a section on biological components with 7 sections. Chapter 5 on impact assessment has a section labelled "Biological impact", with 9 sections, which however do not match the 7 sections in Chap. 4.
- Chapter 4 also has a section on physical components, which among others has a sub-section hydrology. There is no corresponding section and sub-section in Chap. 5, which means that there is no place foreseen for describing and assessing impacts on hydrology, which is a core issue in any dam project.

I prefer a somewhat different structure, similar to the one that was already proposed by De Santo (1978: 125), which is as shown in Box 3.1.

Box 3.1 Structure of an ESIA Report for a hydropower project (modified from Zwahlen 2003: 292)

Executive Summary

A Baselines

1 Introduction
2 Administrative and Legal Framework
3 The Project
4 The Study Area

B Sectoral Studies

I The Physical Environment
5 Geology and Soils
6 Climate
7 Water
II The Biological Environment
8 Vegetation*
8.1 Approach
8.2 Prevailing Situation
8.3 Impacts
8.4 Mitigation Measures

8.5 Conclusions
9 Terrestrial Fauna
10 Aquatic Fauna
11 Protected Areas
III The Human Environment
12 Settlements and Population
13 Land Use
14 Economy
15 Public Health
16 Cultural Sites, Archaeology

C Synopsis

17 Main Impacts
18 Mitigation Measures
19 Comparison of Alternatives
20 Resettlement and Compensation Planning
21 Watershed Management
22 Environmental and Social Monitoring and Management

D Annexes

Annex 1: References
Annex 2: (any other material not to be placed in the text)

*The structure inserted under this section would apply to all the sectoral studies.

The first observation that has to be made concerning this outline is that it should by no means be considered as a rigid structure; on the contrary, it should be adapted in any case to the project at hand and the specific requirements presented in that case. Secondly, the outline does not provide any further subdivision beyond the main chapters, with the exception of what is shown for Chap. 8, a structure that will basically have to be repeated for all the main chapters in this section. Other subdivisions will be required.

The main difference between the two approaches is that in the first case we have a kind of "horizontal" structure:

- Baseline: physical environment, biological environment: vegetation, fauna, socio-economy, etc.
- Impacts: physical environment, biological environment: vegetation, fauna, socio-economy, etc.
- Mitigation: physical environment, biological environment: vegetation, fauna, socio-economy, etc.

On the other hand, in the second approach, we have a "vertical" structure:

- Water: approach, prevailing situation, impacts, mitigation, conclusions.
- Vegetation: approach, prevailing situation, impacts, mitigation, conclusions.
- Protected areas: approach, prevailing situation, impacts, mitigation, conclusions.
- Human population: approach, prevailing situation, impacts, mitigation, conclusions.

The overall result of a report, i.e. identification of impacts, mitigation measures etc., will not (or at least should not) be influenced by the choice of approach. However, I see the following main advantages of the second ("vertical") structure:

- In any complex ESIA, there will be a team of specialists involved (see following section). If this approach is chosen, the various specialists will be able to write their respective chapter in one go, from describing the approach (i.e. the degree to which existing information can be used, type and extent of field work, etc.) to the description of the prevailing situation (= baseline), the identification of impacts and the subject-specific mitigation measures, without having to split it into separate sections which will fit into separate chapters of the report. An additional advantage of this approach is the fact that each specialist, already when doing the field work, is constantly confronted with the question of impacts that will be caused by the project. This helps respecting one, too often neglected, ESIA rule as formulated by Goodland (see Table 3.1), namely that such a report should "be concise and focused on the significant environmental issues. The detail and sophistication of the report should be commensurate with the potential impacts."
- Likewise, when the competent authority has to evaluate a report for coming to its conclusions, there will usually be a number of specialists involved as well, who in most cases will not want, and do not have the time to, read the entire report. The fisheries expert will concentrate on this subject, which in one case will be found in one or two ("fish" under biology, and "fisheries" under economy) compact sections of the report, while in the other he will have to look for it in at least three chapters, and there again in more than one sub-chapter of each. The same may apply for many of the affected persons, who will be interested in a few specific topics.
- One additional issue that can be seen in too many reports: the "horizontal" approach very often leads to endless repetitions, as some points will be mentioned under baselines, then again under impacts, and a third time under mitigation, a fact that also increases the volume of the reports without any gain in additional information. In addition to that, very often one can find some items described in detail in the baseline chapter, not to be mentioned again anywhere under the headings of impact and mitigation, leaving one wondering why this was mentioned in the first place. This is even exacerbated by a relatively new development, namely the fact that in many places nowadays a separate Baseline Report has to be prepared, which describes the prevailing situation more or less independently from any considerations about the project and its potential impacts.

Having sad this, and in spite of the advantages of the second approach: as mentioned earlier, when preparing an ESIA for a project do follow the structure as foreseen in the guidelines applicable in that place and time. I do not think that what I have just written will lead to a change in all these well established procedures!

3.5 Team Composition

The question of how to form the team for preparing an ESIA seems quite straightforward: one needs the specialists which are required for the specific issues to be dealt with, and this again depends on the project at hand. Still, however simple this may sound, there are a number of problems. Before going shortly into that, I have to specify one additional point. There are roughly speaking two types of ESIAs: on the one hand the ones for projects which are developed locally, and where the ESIA (along with the other studies required) are done locally or on a national level. On the other hand, there are the projects which are usually larger, more complex, and for this reason more costly, requiring bi- or multilateral financing; for such projects, planning including EISA is normally announced on an international level, and studies are usually done by companies or groups of companies which are active on an international level. Here, I am only dealing with this second type of study.

Not too infrequently, e.g. in WB sponsored projects, the TOR include a list of specialists (often labelled as "key experts") which are required for this specific project, and often this list also indicates the number of person-months to be attributed to each of them. This is a valuable help for the preparation of the proposal: you know that you have to include suitable CVs for each of these key experts, in part because most probably you will need them anyway, but also for a much simpler reason: there is a number of points attributed to each of these key experts, and not including any of them (or, just as bad, a CV which does not convincingly identify the person as an appropriate candidate for the position attributed to him or her) will make you lose so many points that the probability of getting the assignment is very small indeed. This means that you need a good candidate for each position announced in that way, and if you do not have her in your team you have to find her elsewhere. It is not a good idea to put in any specialists just because they are available, if they do not fit the description provided in the TOR. On the other hand, quite often you will come to the conclusion that not every type of specialist that will be required is listed in the TOR; the way out here is obviously to include such persons as non-key specialists.

The TOR often encourage, or even ask for, the inclusion of local/national experts into the team, wherever possible also in the role of key experts. There are many good reasons for actually doing that, like knowledge transfer—both ways!—or simple economic reasons. Maybe most evidently there is a great benefit for including local biologists, of whatever specialisation which may be required, in the team, since they are more familiar with the local biodiversity than a non-specialised expat would be. This also applies to the social specialists, especially in situations, which is often the case, where the affected local population has to be addressed in their own local language. In addition to that, local experts are usually familiar with the national administrative and legal setting and the required procedures. One potential difficulty needs to be mentioned as well. A local expert is usually much more exposed than an expat one. Let's assume that we are dealing with a kind of prestige project which is promoted by the (usually very powerful) Ministry of Energy, or any other powerful political and/or administrative body, a project which however would have very serious environmental or social impacts, and where for that reason the ESIA should come to the recommendation that the project should not go ahead. It might be very difficult for a local expert to come to that conclusion, and to voice it, since he would be very well aware that he will have to go on living and working there, and that to oppose such a project might come close to professional suicide. Especially in such cases it is important that the local specialists provide their input, mainly for the description of the prevailing situation, but that the conclusions very clearly are the responsibility of the (foreign) consultant, and especially of the team leader for this specific study.

With this I come to the role of the team leader for an ESIA. The importance of this position is often already reflected by the fact that the TOR as mentioned above attribute a large number of points to this position, very often so high that not having a suitable person for this would mean the loss of so many points that it would actually not seem worthwhile even to prepare and submit a proposal. Formally, the most important qualification is often to have "at least 15 years of experience in similar assignments". One thing that is really important is that somebody, to make a good job as team leader, needs to have a good overview of all the relevant topics, in order to be able to decide what, in the specific case at hand, will be relevant and what might not be, and to integrate all the different contributions into one comprehensive report, a task that also includes being able to carry out a quality control of the contributions of the different experts participating in the work. This by no means intends to say that the team leader should be an expert in all relevant topics; nobody can possibly claim to be that, but in my experience it helps if he is an expert in one field. There is one additional requirement, especially in the case of an ESIA for a dam and hydropower project: the team members, and definitely at least the team leader, need to understand how such a project works, since without this understanding it will not be possible to really identify and assess the impacts that would be caused by such a project. This may sound very basic and trivial, but experience shows that this is not necessarily the case.

Having said this, the question would be: should the team leader be an engineer, or should she be an ecologist or "environmentalist"? The answer to this question is possibly the fact that, as a relatively new development (well, "relatively new" at least for somebody my age), there is a specialisation called "environmental engineer". The question now is: are they "environmentals" or "engineers"? One criticism could be that they are neither, since they get "the surface of everything", without going into the depths of anything (according to the definition that a generalist is one who knows nothing about everything, while a specialist is one who knows everything about nothing). However, the specialist will usually very easily admit that he knows next to nothing in fields outside of his own, also because he is aware of how much effort is required for becoming a specialist, while the generalist may easily have the impression that he knows everything, and this also due to the fact that he does not know how much effort would be required to know "everything" even in a very narrow field.

As I see it, there is a basic difference in approach between an ecologist and an engineer (or whatever denomination), as was illustrated in Fig. 2.5: the ecologist tries to understand the functioning of the ecosystem. Doing impact assessment, she may come to the conclusion that the impact is too important, and there is no solution for sufficient mitigation, and the recommendation therefore must be not to implement the project. The engineer (and this includes, to a great deal, environmental engineers just as well as civil engineers), by their training, are oriented towards finding technical solutions for problems, and they tend to come to the conclusion that there is such a solution for every problem. While this attitude is the good approach, and even required, for important fields like e.g. waste water treatment, where one has a massive problem with potentially very serious environmental consequences in the absence of a good solution, and where a good solution is certainly of a very technical nature, it may not always be the best for addressing environmental impacts caused by "greenfield" projects, as HPPs usually are. On the other hand, the ecologist, with all her knowledge about functioning of natural ecosystems and the potential consequences of interfering with these systems, may not see that there actually is a solution. The conclusion, in the end, is certainly that any ESIA is ultimately an endeavour of a very interdisciplinary nature, and that this should be reflected in the team composition.

3.6 The Bidding Process

The responsibility for preparing ESIAs which are the subject of this book, i.e. ESIAs for large dam and hydropower projects, is always the project proponent's. Some of these, mainly in the case of state organisations, i.e. national entities responsible for preparing, planning, implementing and operating hydropower projects, have their own in-house capability for the preparation of such studies, or are in the process of developing it. Still, a large number of ESIAs, especially in the case of projects financed or co-financed by multilateral organisations like the World Bank or the Asian Development Bank, are announced publicly and attributed to a private company or a consortium of such companies based on a competitive process which is usually done in two steps. In the first step, based on the public announcement of the project, interested bidders have to submit an Expression of Interest (EOI). These will then be checked, and a shortlist of usually 4–6 bidders will be established, who will be invited to submit a detailed technical and financial proposal. It is not intended here to present a handbook on how to make a bid for such a project; however, I have prepared or helped preparing a large number of EOIs and proposals, and in a few occasions I was "sitting on the other side", i.e. I helped in evaluating EOIs and proposals, and some hints on what to do and what not to do may be useful.

The first two points to make are valid for both stages, the EOI as well as the full proposal:

- Take it seriously: even for an EOI, which for many reasons must be done rather quickly, it is not sufficient just to tell somebody who happens to have a little bit of time to put together a number of reference sheets and CVs, the more of each the better, and send that with a cover letter to the prospective client. If you do not take it seriously, then better leave it, it's a waste of time.
- Think: try to put yourself in the place of the ones who are going to read your contribution and who will have to evaluate it. What do they want to find there? Important point: try not to make them angry, because if you do, you are out!

3.6.1 Preparation of an EOI

For the preparation of an EOI one may or may not have detailed TORs, often however one will get just a rather short description of the project and the tasks at hand, without much detail. Here, the most important points to deal with are basically three:

1. Experience of the company: it is important to show that the company has the required experience; in case one intends to do the study as a consortium or joint venture, this has to be demonstrated for all partners. References have to be selected carefully. Example: in case the Request for Proposal (RFP) is a specific hydrological study to be carried out in the framework of a hydropower project, it is of no use to include 20 project sheets that demonstrate that the company has great experience in building dams. Concentrate on the subject at hand.
2. Competent staff: ultimately, this is even more important than the experience of the company, although the two usually go hand in hand. In most cases, the RFP will contain a list of key staff members who will have to participate in the study. In such a case, it is important to present a team, each team member with a complete CV that corresponds to the request. It is not of much use to submit a table containing a complete staff list, most of which will have no relevance whatsoever for the specific study to be done; if one still wants to submit such a list, it should be put in Annex; nobody ever reads Annexes. The best, as I see it, is to propose one valid candidate for each position proposed. If one thinks that this list, for whatever reason, has to be modified, then an explanation has to be provided; however, in my experience it is better to do that, if required, in the proposal and not in the EOI. But the main point is: if, e.g., 6 key experts are specifically asked for, do not include 20 CVs in your EOI, expecting that the ones who check it will look for suitable candidates in this crowd. They will not, you make them angry, and if you make them angry …
3. Understanding of the problem: The EOI is not the place and the time for submitting a complete methodology presenting the detailed approach one proposes to follow; very often one may not even have all the information necessary to do that. However, it is still important to show that, in addition to general experience, you understand the particular question or situation to be addressed in this specific case. Usually, a short text of one or two pages will do, but without that an EOI may not be chosen.

3.6.2 Preparation of a Proposal

The points mentioned above largely also apply for the preparation of a proposal. Again, one will have to present the experience of the company (in addition to legal requirements asked for, like official documents as proof of the registration and the financial performance of the company, etc.), one will have to present experienced and capable staff, and a methodology, this time very detailed. As said above, no complete handbook or guideline intended, only a few comments and hints.

1. Comments to the TOR: this is a standard chapter to be included in a proposal. Clarifications that can be asked for during the time given for preparing the proposal go in the same direction, and both are very important, and this also since TOR are not always prepared as one would wish for. I have often thought that if we, as consultants, would write our proposals to the same standard as sometimes TOR are being prepared, we would be out of business in no time. Just a few examples. In one case, the TOR foresaw an approach in three stages. The first stage would be to prepare a resettlement policy for the country. In a second stage, this policy would have to be approved by the national parliament, and in a third stage this policy would then have to be applied on a specific project; the time schedule proposed for the entire assignment included exactly 0 days for the parliamentary approval process. The second case was a technical as well as EHS (Environment, Health and Safety) audit of nine large hydropower stations spread over basically the entire European part of Russia; for the field work required to do this, the TOR proposed a time span of 11 days. Not even considering the size of the plants to be audited, it would not have been physically possible to do this given the enormous distances between the plants. These are obviously extreme examples, but it is in any case required to also check TOR on their consistency, and to propose changes or ask for clarifications if there are problems.
2. Staff: here, very often the TOR provide a specific list of key experts. This is helpful, but it can lead to difficulties, if e.g. you can cover all the capabilities required and asked for, but not exactly with the proposed staffing list. One way out of this is to ask for a change in advance (see above), or you can add non-key experts to your team, e.g. for covering a specific capacity that may not be covered, or not sufficiently, in the key team. Just changing the list of key experts without at least a very good explanation however is not a recommended approach, and be it just for one reason: in most cases, in the evaluation of a proposal points are being attributed to a number of key points in the proposal, and usually a considerable number of these points, and for very good reasons, are foreseen for the key staff. If one such key staff member is missing in your proposal, you risk losing so many points that your proposal has no chance to succeed.
3. Approach: I was often asked, in cases of package proposals where engineering and E&S topics had to be addressed, to prepare the environmental and social part of the proposal. Not too rarely in such cases the person who had the overall

responsibility for the proposal, and who was always a civil engineer, told me that "you certainly have something similar that you can simply copy-paste", meaning that I should not waste too much time on this aspect of secondary importance. Well, this is never the case. While, as certainly in every discipline, there is some baseline which is always applicable, as here e.g. a basic description on how impacts are identified and classified, the important part is always that such a proposal has to be project specific, and the approach submitted has to show that the specific conditions of the project, which are never exactly as in any other, were understood and will be taken into consideration.

4. Schedule: this is another important point which has to be given consideration. Not too rarely, a schedule is proposed which is clearly impossible to follow right from the start. The impression here is that at times TOR are prepared, then the finalisation of the documents for request of proposals is delayed for whatever reason, but the schedule is never updated. Just one example to illustrate this: a call for an EOI was sent out dated 2020-01-21 with the following specifications: deadline for clarification 2020-01-30; deadline for submission of EOI 2020-02-11; assignment expected to start in February 2020; "the consultants are now invited to submit a complete Technical and Financial Proposal". Besides the fact that in this case the timeline was impossible, it was also not clear whether an EOI or a full-fledged proposal had to be submitted. But also the schedule proposed for proposal preparation, contract negotiations and implementation of the assignment is rarely if ever followed. There usually is a strict deadline for submitting a proposal, which e.g. reads like this: "the proposal has to be submitted on date xx, before 10.00 a.m. (local time) at location yy". If you are 10 min late, you are out. But this is about the only "hard" deadline in the process, since from then on it rarely goes according to schedule. Often, a deadline is included saying when you should be notified about acceptance or non-acceptance of your proposal. Sometimes this actually happens, but it is the exception rather than the rule. It could well be that you submit your proposal and then you never hear about it again (an approach which I find inacceptable, but what can you do …). Another possibility is that after almost 3 months you receive a letter asking you to extend the validity of your proposal by another 3 months, and this can be repeated several times. In at least one case, I did not hear anything about a proposal until more than a year later when I received a phone call asking me if I would be prepared to modify and resubmit a proposal (which of course I was). Besides making planning very difficult for a consultant, this can also have very practical consequences. For one project, we had submitted our proposal according to the schedule in the TOR. For whatever reason, however, the start of the work was delayed for several months, we could only start in August instead of in May as originally planned. This may not seem to be a point of great relevance at first, but it did have consequences in this case. When the Bank's social expert joined us for a part of the field work I informed her that, since in the meantime the rainy season had started, there was at least one village that we were not able to reach. Here answer was that I had signed a contract and promised to do the work in the time foreseen! I told her that I was not willing to

risk the lives of members of my team, but more importantly I could tell her that in the proposal, which was an integral part of the contract, I had written that if for any reason the work would have to be carried out during the rainy season instead of as planned in the dry season, it might be possible that some parts of the study area could not be accessed.

5. Budget: sometimes, a budget available for the study is indicated. Going beyond that will obviously eliminate you, even if your technical proposal was accepted. Here, I just want to point out one issue that I have observed repeatedly: if you know the situation, and if you know what you have to do for responding to the requirements specified in the TOR, you will frequently realise that the budget—often related to the time schedule proposed—is far from being realistic. I observed this especially in RFPs where there was an important social component. Reading the TOR, one can find the whole vocabulary: participatory approach, gender sensitive, comprehensive socio-economic survey, prior and informed consent, etc. However, when looking at the budget, in terms of time as well as money, one wonders how, when and with what kind of resources one is supposed to do that. This can give the impression that the TOR had been made by a social specialist (who knew what needed to be done), while the budget was prepared by somebody else (who did not understand how much work this requires). What to do in such a situation? I consider this, in a way, as being the devil's alternative for a consultant. There are different ways to deal with such a situation, none of them the only one to be followed in every case, but some definitely worse than others, which are outlined here shortly:

    a. Proactive: before preparation of the proposal, you make observations/requests to the client, hoping for a more realistic budget.
    b. Naïve: you promise to do everything asked for in the TOR, accepting the proposed budget; you will bleed, since you will be expected to act accordingly.
    c. Honest: you ignore the budget and make your own, stating the costs you actually think it takes; you are out, at least in most cases. Even if your technical proposal reaches a top score, when the financial proposal is opened it will be found that you did not comply with the rules.
    d. Furtive: you include in your proposal all kinds of caveats, hidden as best you can, and use them afterwards for claiming for all "additional" work, of which you knew right from the start that it would have to be done; this, unfortunately, has a good chance of success, but might not go down well with many clients, meaning that it might be bad for future assignments from the same client. I say unfortunately, because there is the very real possibility that one of your competitors will be successful by applying exactly this strategy.
    e. Realistic: you list what can be done for the budget, and clearly exclude things that cannot be done, pointing out their importance; this may lead to success, depending on the client's reaction to this.

In my view, approaches (a) and (e) are the best ones, even if neither does necessarily lead to success; I have experienced both outcomes.

Following these points is, obviously, not a guarantee that your EOI or proposal will be successful. However, ignoring them will in all likelihood make it fail.

3.7 Who Should Do ESIAs?

Complex ESIAs, and every such study for a large dam project qualifies as that, are done by a team of specialists as detailed above. In most cases, the contractor mandated for carrying out this type of work is a company, either a small to medium size company specialising in such studies, or a large engineering and consulting company which has a corresponding department. Very often it is not one company, but a grouping of such entities. This can have two main reasons, namely (i) companies join, because they do not have all the required specialists, or (ii) international companies join forces with a local company, for the reasons I'll be mentioning below.

The strongest players in this market are certainly the large engineering and consulting companies. Historically, they started with doing the engineering part of the planning work, and when then environmental aspects got more important, and especially after the now almost universal requirement for having an ESIA before any technical project can get the go-ahead and be implemented, many of them went into hiring the required specialists and finally created the respective specialised departments.

During approximately 20 years, from the early 1970s, when EIAs started to become mandatory, until at some point in the 1990s, the World Bank, which was certainly the first and the most persistent promoter and developer of EIAs, used to make comprehensive packages in the style of "technical and environmental feasibility study for HPP xyz". Then this changed, and a new approach was chosen, which is still followed today. Whenever the World Bank finances or co-finances a hydropower project, or a study for developing such a project, it announces two different packages, one for the technical aspects and one for the environmental and social part, and the corresponding mandates are awarded to two different companies. The main reason for this was presumably the concern that, if an engineering company involved in planning and implementing hydropower projects was entrusted with the entire package, priority would be given to the usually much larger and, especially when it goes into further planning stages like detailed design and implementation, much more lucrative aspects of engineering; maybe there was also the concern, or possibly even confirmed cases, that in the interest of obtaining a large contract environmental studies would be influenced in a way as not to "put a hindrance to development".

I have done it both ways, and I think that it is not easy to say which of the two approaches is better. There is certainly less risk of some kind of influence being taken on the outcome of an ESIA as just described when the two studies are being done by separate, legally independent entities. On the other hand, I often had the experience that when a conflict between engineering and environmental aspects is at hand, it is much easier and more effective when this can be discussed with a colleague who sits just next door, rather than when the other party is located far away, possibly even on a different continent and in a different time zone. Modern communication technology has made exchange in such cases much easier, but even so the direct contact, and also the direct access to all kinds of planning documents, from the technical as well as from the environmental side, is often very helpful. The "all in one hand" approach may also have direct and practical advantages for the formulation of at least some of the mitigation measures that may be required. Quite a lot of these require some engineering, and it is much less embarrassing if, before the ESIA report is finalised an out, a colleague from the engineering department will point out that a certain good idea proposed for mitigating an environmental impact may just not be technically feasible, rather than when one has to be told that e.g. by the authorities in charge of checking the report.

On some occasions a university institute is mandated with the preparation of an ESIA. I have come across a number of studies prepared in this way, and I was usually not impressed. Therefore it is certainly no big surprise to hear that I do not think that this is a good idea. I have three main reasons for this.

First, universities have the duty to form specialists in a large number of different professions, among them environmental and social specialists, some of whom may later on go into preparation of environmental and social impact assessment studies. The role of universities, however, should not be to carry out this type of work and in this way competing with the very specialists that they trained.

Second, in some cases there was a problem with the team that was set up for a specific study, and the weight that some aspects were given, and some not, sometimes giving the impression that the chosen approach was not to ask "what are the main issues to be addressed, and what kind of specialists do we need to do that", but rather "let's see what kind of specialists we have in the institute, and what part of the budget we can allocate to them".

The third observation is linked with the way scientific research institutions work. They make scientific research, obviously, and this more often than not produces, in addition to a few answers, a long list of additional questions that deserve to be or have to be investigated. I do not want to be misunderstood here: this is the correct way to do scientific research (in addition to being in the direct interest of the scientists). However, in the case of ESIAs the main requirement is to come to conclusions and answers, and this, admittedly, very often on the basis of incomplete data.

Finally, what I have noticed a few times was the lack of a competent, experienced team leader; for the importance of this position refer to Sect. 3.5. In such a case, the reports sometimes give the impression of a number of specialised studies which were just put together at random, many of them with an extensive baseline, but without the subsequent and very important sections: no impact assessment results from the baseline analysis, and no mitigation measures, leaving the reader to wonder why the analysis was done in the first place.

Such observations can also be made sometimes in studies which have been prepared by specialised companies, but I have the impression that this is less often the case. However, let me repeat: this is based on impressions, not on a statistically relevant sampling and analysis. And one important point needs to be added: for specific questions to be addressed in such a study, the participation of university experts can be very valuable, specifically if these are specialists from a university of the country in which the project is located.

3.8 The Use of Local Knowledge

The first EIA-type report I ever prepared, for Ayago-Nile HPP in Uganda, back in 1979/80, was entirely a one-man show: I collected published information, went to Uganda where I talked to a number of relevant persons (who nowadays would be called stakeholders), visited the site (the fact that I could make two visits was not due to the original plan of the assignment, but entirely to political circumstances), and wrote my report; for some details on this see the Ayago case study (Sect. 26.1.1). Such an approach would be completely impossible today, since the complexity of these investigations has greatly increased; what I had done at that time today might be accepted as a scoping and screening study, but not more. Presently, every EIA or ESIA is being prepared by a team of specialists.

Here, the aim is just to discuss shortly the importance of using local knowledge for preparing such studies. This is being done, and should be done, on two levels, namely (i) having local specialists in the team and (ii) interviewing inhabitants of the affected area.

3.8.1 Local Experts

Having local experts in an ESIA team is clearly required for different reasons. The necessity for this is especially high where it comes to biological aspects, i.e. botany (vegetation and flora) and zoology (here usually with a focus mainly on fish, mammals and birds, occasionally reptiles and amphibia, and rather exceptionally other groups of animals). The main and obvious reason is that local biologists know flora and fauna of their country, while an expat consultant, unless he or she happens to be a specialist of the area in question, is not. Especially in tropical areas, with their often extremely rich biodiversity, it would be strictly impossible to have field work done by somebody who is not already familiar with this situation. Of course, there is an economic side to this as well: an all expat team of specialists, given their usually higher rates and additional travel costs, would simply not be competitive. One additional point is the fact that in this way, local specialists, in addition to being able to apply their knowledge in their field of work, get an on-the-job training in doing such studies.

This raises the question whether any foreign experts should be involved at all, and if yes, what their role should be. Why not have everything made locally, as Linaweaver (2003) postulated? The answer to this question is probably a little bit too complex as to be given here in an exhaustive way (and World Bank, which is mostly criticised in Linaweaver's paper, which basically demands that decisions on local projects are being made locally, would have a hard time vis-à-vis its financing bodies to push such a practice through). The following story may serve as an illustration, while at the same time highlighting a number of other issues. Before going into it I have to point out that I have not met all the persons involved, but that the events have happened pretty much in the way as told, even if maybe in some aspects they are a bit simplified here.

The project in question was Houay Ho HPP in southern Laos. Dam and reservoir are located on the Boloven Plateau, at an elevation of about 800 m asl, the power house is situated in the valley of the Xe San River, a left bank tributary of the Mekong, at an elevation of 100 m asl. The horizontal distance between water intake structure in the reservoir and power house is only about 2.8 km, while the difference in elevation is 700 m, which is used as head by the HPP. The two sites are separated by an extremely steep escarpment. The first in the series of events to be described is that Mr. A, in an office presumably in Zurich, at some point in the early stage of project planning, took a 1:50,000 map and a felt pen, drew a thick zig-zag line on the escarpment between dam and power house and labelled this "access road". Quite some time later, on the site where construction of the dam had started (via a gently sloping road reaching the plateau from the opposite direction) Mr. B, a foreman on site, took exactly this plan, gave it to Mr. C, a bulldozer operator, and told him: "go and build this road". So Mr. C started his work by removing the lush tropical vegetation on the short stretch leading from the dam to the edge of the plateau. When he reached the edge and had a look at the escarpment the may have started to realise that this would not be an easy job; however, the entire area was covered with forest and thick undergrowth, the terrain was certainly not clearly visible. So he started by pushing first the remaining trees which were in his way, and then an increasing amount of loose earth material over the edge, material which accumulated on the uppermost part of the steep mountain slope. After having pushed probably some hundreds of cubic meters over the edge he came to the conclusion that it could not be done; he went back to his boss, Mr. B, and told him that. Mr. B was very busy and answered that they had this plan, and that he should try harder. So Mr. C went back and tried harder, pushing another several hundred, or maybe several thousand, cubic meters over the edge, without however ever reaching a point where it would have been possible to enter the escarpment as such and start building the road there. Finally, he went back to Mr. B again and told him that it was really not possible, and that he better had a look for himself, which Mr. B finally did. I do not know what exactly happened then, but I can imagine three possible reactions: (i) Mr. B looks and says "yes, you are right; stop immediately, I'll go to my boss and tell him that we need a different solution"; (ii) Mr. B looks at the situation and starts shouting at Mr. C "look at that! how could you possibly make such a mess? now I have to go back to my boss and tell him not only that we need a different solution, but I'll also have to explain what happened here"; (iii) like (ii), but in addition he immediately sacked Mr. C for being able to demonstrate his boss that he was capable of handling a difficult situation, and that he did not tolerate reckless behaviour. This is speculation, but the rest is not. A different alignment was chosen, much longer, but feasible. But the mess had been made and was still there, no way of undoing it. At the onset of the rainy season, the entire mass of loose material clinging to the top of the escarpment was flushed away, a mass of mud flowing down the steep slope, in the end just reaching the main road in the valley and blocking it; luckily, no assets like fields or buildings were damaged, and no persons came to harm. A few months later, I had to go to Seoul, in order to finish a report on environmental risks and impacts of the project, in the offices of our client, a large South Korean company. The report, among a lot of other issues, also described this event. In a meeting with the project team Mr. D, one of the directors of this company and responsible for the project, shouted at me, since in his view I was unjustly accusing the project of something which was not related to it, pointing out that if one carefully looked at the topography on the map one would see that it was not possible that material from up there could possibly have reached the valley road, since between the two points, on the slope, there was a small raise in terrain; my answer was that I had never said that it had come down in a straight line; it had actually been diverted, still a stretch above the valley floor, by a small crevice in the escarpment, which it then followed, but this did not change the basic facts. In this meeting, there were about 10 other persons, all of them involved in the project and all of them employees of the company. At least some of them, and very certainly the company geologist who was also present, knew the situation very well, but none of them said a word. I could not blame them: it was not directly their problem, and they knew better than to contradict their boss, certainly not on something that was not of a major concern to them. One point if favour of Mr. D: after this discussion, the matter was settled, the report, including the description as it was, was approved. Mr. D invited me for dinner, where we had good discussions, and when I still joined him with the second bottle of whatever type of Korean brandy we were drinking, I think I had gained his respect.

After this short excursion, coming back to the subject addressed above, there are two points of relevance.

The important role of the team leader for the preparation of an ESIA was mentioned above. When including local specialists (well, any specialists) in such a study, there should be somebody who can judge the relevance of the contribution of any specialist, the value of the resulting conclusions and recommendations, and the quality of the information provided, the latter obviously without being able to go into all details. Just two examples concerning quality of the information: in one study in Zambia, the list of reptiles reportedly having been found in the project area included *Lacerta agilis*; this is a European lizard, whose range of distribution definitely does not include southern Africa. In another study, in Cameroon, the local wildlife specialist on the team reported the occurrence of the African manatee, *Trichechus senegalensis*, in the reservoir area of the project under examination. While it might be possible that this species lives or occurs occasionally in the estuary of the river in question, some 100 km downstream of the dam site, it was absolutely impossible that it could ever have reached the area of the future reservoir, the dam site being located just upstream of a series of waterfalls of considerable height. While the lizard would most probably have gone unnoticed, or might have raised a few eyebrows, in the worst case throwing some doubts on the quality of the report (see the "good bad example" on fish in the Mono River, Table 3.3), the second case, if it had been taken seriously e.g. by the competent authority, or if it had been taken up by an NGO opposing dam projects, could either have caused problems for the development of the project (threatened species, IUCN Red List category: vulnerable, and a project that would certainly have a negative impact on it!), and/or could have caste serious doubt on the quality of the report.

Now, if a competent team leader (or any other type of quality control) is required, this of course does not mean that this has to be done by a foreign expert. On the contrary, it is certainly important to develop local knowledge. There is, however, one other aspect where, in my opinion, the participation of a foreign expert is definitely an advantage, or maybe even a necessity, at least for the time being and in a number of countries, as already mentioned above.

National legislations concerning ESIAs stipulate, and I do not know of any exception to that rule, that the project proponent is responsible for carrying out the required studies and preparing the reports, usually by hiring a consultant specialised in this matter. The project proponent, very often, is the Ministry of Energy (or however the denomination in the respective country is), and if not, this Ministry at least has a vested interest in the project, e.g. having identified a potential site for a developer, planning to include the energy production expected from this project in the future supply to the national grid, and/or envisaging the sale of the energy to a neighbouring country. In case such a project would now cause negative impacts of a magnitude that would mean a no-go, a local consultant would find himself in a very difficult situation: he would have to come to a conclusion that, as he is well aware, his client would not appreciate at all. In addition to that, he would be aware of the fact that his conclusions might finally not be backed up by the Ministry of Environment, authority in charge of examining ESIA reports and issuing—or not—the required permit; it is just a fact that the Ministry of Environment, in many cases, has a much weaker position—politically as well as economically—than the usually very powerful Ministry of Energy. In such a case, it might well be that a local consultant would almost commit professional suicide if opposing such a project. For a foreign expert, however, the biggest risk that he may face is that he will not get another assignment from this particular client or in this particular country, as was my position in the story told above. Too bad, but not really a catastrophe. The local consultant however, if he was merely a specialist providing specific information, while the conclusions were made by someone else, would not be in the line of fire. For this reason, and especially for complex and possibly contested projects, it might be a good idea in many cases to have an expat consultant being responsible for such studies. The statement made by Pearce (2018: 99) "An assessment that recommends the dam not to be built is virtually unheard of. The author would never be hired to write another" is wrong on both accounts, but especially for national consultants in many places this is still a clear risk. That this is not a new type of pressure, and certainly not an invention of the ESIA era, is demonstrated e.g. by the way people who uttered critical observations on or pointed out potential negative consequences of the High Aswan Dam Project risked their careers or had to fear worse consequences, as described by Hughes (2009: 176).

3.8.2 Language

One final reason for using local expertise is language. This is obviously most important for the social part of impact assessment, but for other aspects as well it can be of relevance. Being able to communicate in several languages is certainly an advantage, but there are limits to this for everybody, the number of languages is just too high; 3000–4000 according to Katzner (1995), but depending on the source up to 10,000 (Crystal 1998: 284) different languages are still spoken by groups of people as their main language. One additional problem here is that many of these are often used by only small groups of people, in geographically well-defined areas. An example from Tajikistan, encountered when working on the Pamir 1 HPP (see Sect. 25.2): in the capital, Dushanbe, people speak Tajik (an Iranian language of the Farsi Group), and at that time as good as everybody also spoke Russian, with very few able to communicate in English. In the project region, GBAO (Gorno-Badakhshan Autonomous Oblast), the language is Shughni or, as most inhabitants call it, Pamiri, a language of the Iranian language group as well, related to Tajik, but still quite different (Grimes 1996: 782). This was spoken in the capital of the province, Khorog, as well as in Bulunkul, the small village located near the project site at an elevation of almost 4000 m asl, a 4 h drive from Khorog. However, in the next village, Alichur, only about 20 km further into the high plateau, the language is Kirghiz, which is a Turkic language. In such a situation, an interpreter is indispensable, and a good one is invaluable. Working with translators, while often unavoidable, can be quite frustrating. One problem is often that people who work as interpreters, just because they happen to speak e.g. English in addition to the local language(s), but do not have any formal training as translators, tend to take over the discussion instead of actually translating; in this case, the translator decides what information you get: the one he thinks is relevant for you; in my experience, this is more often the case with male translators than with female ones. I remember one specifically difficult experience in China, where the translator that had been made available by the client first turned out to have a rather limited knowledge of English, and where in addition to that it then became evident that he did not understand the local Chinese dialect (which may actually have been a different language). So, when I wanted to know something from a local inhabitant, he translated my question first into Mandarin, which was then translated by one of our group into the local dialect, and for the answer it was the other was round, leaving me to wonder what type of information I was getting and, maybe worse, what kind of potentially relevant information I would never get. Therefore, where direct and extensive communication with local inhabitants is required, and where understanding of nuances can be essential, the work is best done by a capable native speaker.

3.8.3 Knowledge of the Local Population

Traditional knowledge is widely recognised as very relevant in different areas, the most important potentially being the field of ethnobotany, i.e. knowledge of local peoples on plants, and here mainly plants with medicinal properties (see e.g. Green et al. 1999). Such knowledge is also of importance in the framework of an ESIA, especially when done in places with high biodiversity, where people at least to some extent depend on using traditional resources. Information can—and has to—be obtained mainly with two objectives.

The first is assessment of biodiversity. Trying to get this information only by own field work would very often just be too time consuming, and in most cases far from complete anyway (being aware of the fact that "complete" is restricted to the few groups of organisms considered in any case, and even there may be far from that, see Chap. 10). One frequently used approach, in addition to observe catches, is to ask local fishermen about fish living in the waters around their villages, often by showing them tables illustrating species known to occur in the wider area, as a means of establishing a species list of fish in a specific and well defined area, e.g. the parts of rivers and streams which would be turned into a reservoir by the project. A similar approach can be used for terrestrial animals, including birds: hunters usually know very well what species live in the surroundings of their villages, and often have a rather good knowledge about their frequency. A recent study carried out in the Amazon region came to the conclusion that using indigenous knowledge for assessing abundance of a number of species revealed results that were very close to information gained by transect searches over 10 years, and the value of this approach was confirmed in other places as well (Fiennes 2020; Learn 2020). As mentioned above for specialists, information gained in this way also has to be checked critically. There are several potential pitfalls. For one, local fishermen and hunters will best know the species which are of importance for them, i.e. the ones they usually catch, and this may mean limited or no information on species of conservation concern. Then, it might be difficult to match the local name of a species with certainty to a scientific name, and finally, even a very good picture may not always be a very accurate way for identifying a species. Bessesen and González-Suárez (2021) point out the importance of using such knowledge, but at the same time also mention potential limitations. This should by no means be taken as a reason for not using this type of information, but like always it will be the responsibility of the respective specialist to check it critically.

The second aspect is the one of traditional use of plants and animals, to a large extent plant material used in this way, be it for food, medicine or any other purpose, usually referred to as NTFP (non-timber forest products). When a population affected by a project will have to be relocated and may therefore lose access to the forest, it loses a part, and in some cases a very relevant part, of its livelihood. In such a case, compensation for the loss has to be provided, and this can be done accurately and in a fair way only if the information on types of products and their importance for the household economy are known, information that can only be obtained from the local people themselves.

There is one additional aspect that has to be mention here, albeit very shortly. There was—and still is to some extent—a type of "romantic" idea on how indigenous peoples deal with their environment in a very sustainable way, while the "modern, industrial" societies do not. I have addressed this issue, pointing out that this "sustainability" is often just due to the small number of people in question, and by the lack of technical equipment at their disposal (Zwahlen 1996). Charest (1995: 376), dealing with the resistance of indigenous populations against mega-projects in their traditional lands, seemed almost a little bit ashamed when observing that "[w]ith more and more direct involvement in the management and exploitation of their forests, Indian peoples of Québec are apparently in contradiction with some principles of their traditional ecological philosophy". O'Hara et al. (1993) describes a case of massive soil erosion due to pre-Colombian agriculture in Mexico, Dore and Nogueira (1994) point out that traditional methods, under present conditions, are not necessarily a guarantee for sustainable land use in Brazil, and Diamond (2005) describes numerous cases where even very traditional societies were not acting in a sustainable way at all. In a study on shifting cultivation practice in the Peruvian Amazon, Coomes et al. (2021) convincingly demonstrate that farmers prefer to clear secondary (i.e. fallow regrowth) rather than primary forest, since the latter requires much more effort; they then observer that households whose head has received more education tend to cut primary forest more often, a "question that requires further research". But they also mention that such households "have higher rates of chain saw ownership"; maybe this is the explanation: they do it because they can, because they have the means to do it. Undoubtedly, in many places and in very different environments, traditional methods have enabled a sustainable, long term use of the available resources. The main point to be made here is that a traditional approach is not necessarily a guarantee for sustainability, and, maybe more importantly, that under changing conditions activities which have been sustainable may not be so any longer; these aspects are discussed again in Sect. 21.2 and Chap. 22.

3.9 Purpose and Limits of ESIAs

Sometimes, in public statements made e.g. by an NOG which opposes a certain project one can read sentences in the style of "although an ESIA had been prepared, the project still goes ahead"; also see the statement of Pearce (2018: 99) cited in Sect. 3.8.1. In my view, this reflects a fundamentally wrong picture of the ESIA and the role it has to play in the project cycle.

The ESIA has a number of generally well-defined objectives, which are:

- To understand and describe the project; "understanding" here means two things. First and mainly, the ESIA or, more to the point, the persons involved in its preparation, must understand what aspects of a project are relevant for finally assessing its impacts. This must be reflected in the description of the project, which should focus on these relevant aspects and can leave away a lot of information which may be important from an engineering point of view, but irrelevant from an E&S point of view. Just one simple example: in a recent ESIA report on a hydropower project which involves a considerable amount of tunnelling, one finds three or four times the information that the tunnels have a horseshoe-shaped cross section, but there was no information on the expected amount of excavation material, which will have to be disposed of somewhere; the first of these two pieces of information, in our context, is totally irrelevant, the second is definitely not. Second, while the authors of such studies usually are not, and should not be, hydropower or dam engineers, they must understand the basic principles of how such a project works. One sentence often encountered, that "the project will produce xy MW of electricity", already casts a doubt in this respect: installed capacity (MW) and energy output (MWh) are certainly related, but it is not the same; see also 2.4.

- To understand the legal framework by which the project will be governed.
- To analyse, understand and clearly describe the prevailing situation of the area which will or is likely to be affected by the project; this covers the natural as well as the man-made aspects of the environment. In order to be able to do so, and especially for being able to clearly identify those components of the environment which are at risk of being affected, it is definitely necessary to have a clear picture of the project and its potential consequences.
- To identify all relevant impacts (and possibly, especially in a case where an impact could usually be expected, clearly stating why in a specific case there is none).
- To identify potential and feasible mitigation measures for all cases where mitigation is required and possible.
- To make statements on overall unavoidable impacts and on compliance of the project with the applicable legal framework, including any relevant standards.

These are the relevant points, admittedly somewhat simplified; more detailed descriptions of the single steps will be given in other sections of this book, especially in Chap. 7.

While in most guidelines it is clearly stated that negative as well as positive impacts have to be identified and described, there are some limits to that. First of all, by its very nature, an impact assessment will have to put the main focus on negative (adverse, undesirable, legally not acceptable) effects, and it will have to propose mitigation for such cases. However, there is a number of things that an ESIA may mention shortly, but where it will not go into any details. Such an aspect is the ultimately first and main purpose of an HPP: generating electricity. It is not the role of the ESIA for a hydropower project to assess in detail the benefits the population which will receive this energy will gain from this, or the effect on the economy of the country; note that on the other hand it is certainly an objective of a social impact assessment to point out cases where a directly affected population will not be supplied with electricity, and possibly to outline the ways for doing this and the benefits that could be gained from such a measure. One additional point that is not the objective of an ESIA for an

HPP is the expected economic performance of the project, the lack of which is often listed as one of the main problems of hydropower in general, or at least of many such projects which have not lived up to their expectations in this respect (e.g. WCD 2000).

Remains the question whether an ESIA should lead to the rejection of a project. In order to discuss this question, one thing has to be stated very clearly first: the ESIA is not a decision on whether or not to proceed with a project, it is the basis which has to provide the arguments for the environmental authority to come to such a decision. The ESIA has to present the facts, and it will come to conclusions and possibly recommendations, but it does not decide. Still, in case there is anything that can or has to be considered as a no-go for a specific project, this needs to be analysed and stated clearly in the ESIA.

This definitely means that, yes, an ESIA could actually lead to a decision not to grant the permit for implementation for a project (of any type, not limited to hydro). Still this, in the usual procedure, should not be the rule, but a rare exception, and this for the following reasons.

The first point is the original identification of suitable sites for HPPs. This, as expressed in Sect. 4.2.1, in my view is the weakest point in the entire project appraisal procedure. If done properly, such a first selection should be done in a way as not only to exclude obviously unsuitable sites, but rather as not to include them in any list of potentially suitable sites worth evaluating in the first place. Unfortunately, this may happen rather often for sites which are not suitable mainly from hydrological and geological points of view, but rarely if ever for environmental and social reasons.

But even if this is not the case, there is still one more possibility. In an ESIA procedure that is done as it should be done, there still is the so-called screening and scoping process. During this early phase, normally a preliminary examination, including a site visit, is carried out. If there is a condition in the prospective project area which is likely to lead to a rejection of the project, such a condition should be identified at this stage, and it is very much possible that then a further, more detailed analysis would no longer be required, time and money could be used for a different option. One small example: once, still quite in the early days of ESIA, I had a phone call from somebody who intended to build and operate a small hydropower plant in the Swiss Alps. I got a short description of the project, which would have consisted in building a small weir with intake structure, a headrace structure of a few 100 m, and a small power plant which would only be used in winter for the production of artificial snow (snow cannons) in a ski resort. The caller ended with the statement that they would need an EIA for this project, and the question was whether I could do that. I pointed out that this project was facing one major problem: it would use water, and potentially leave a part of the river with very little or no water at all, in winter, when the natural discharge is at its lowest, and that there was very little chance of getting a licence in case this small stream had any importance as fish habitat, and I suggested that they discuss this first with the authority in charge before taking any further steps. A few days later I had another phone call from the same person who thanked me for this advice: the suggested discussion had made it very clear that there would have been no chance for obtaining such a permit, EIA or no EIA.

It is quite certain that this approach could lead—and presumably has led—to similar conclusions in other, more complex projects, and could possibly save a lot of time, preventing unnecessary costs and frustration. There are also a number of issues that act against such a solution, of which just the most obvious ones are listed here.

- In the case mentioned above, my answer potentially cost me a contract for carrying out an EIA. A consultant will rather be interested in carrying out a detailed investigation instead of just "killing" a project in such an early stage.
- The ESIA usually is not the first stage in the evaluation process of a project. Once it comes to this stage, money has already been spent, and a project developer might not be ready to abandon a project only based on an expert opinion.
- The potential project often is limited to one site, with limited options for changes, and an alternative site might not be easily available.
- There is still the chance that a detailed investigation would come up with a solution that would make a project, which seemed to be hopeless at first sight, feasible.
- Finally, it cannot be excluded that the interest in the implementation of the project, e.g. from an economical or political point of view, would prevail over environmental and social concerns.

Even with all of this, I maintain that elimination of a project on environmental and social grounds should be done at a very early stage of project development (Zwahlen 1989), and that doing this based on a full-fledged ESIA should be an exception. The main reasons that such assessments are "virtually unheard of" may be the fact that, as mentioned above, a full ESIA may—and usually should—not be required for coming to the conclusion that a certain project idea is not worth being developed any further, or that in the case of such a "negative" outcome the study will not be widely publicised. There is, however, another option: as mentioned several times throughout this book, once a project has been identified it cannot be "dis-identified" again, and even if a project has been rejected once, it can be studied again, potentially with a different outcome.

3.10 Comments to Reports

Every report will have to be submitted for inspection, and comments will be received. I clearly remember one occasion when we had a meeting with the official in charge of an environmental protection agency to whom we had recently submitted an EIA report. He said: "This reports contains everything that I need". My guess is that I remember this because it is a rare event; in most cases one will get questions and comments, which will require an updating of the report. In the normal process of report preparation there will usually be several stages, where drafts of reports will be submitted, here with the aim of getting comments for actually improving the report and in this way reducing the risk of having to face serious criticism at a later stage.

There are five main types of stakeholders usually involved in the process of commenting, namely (i) the developer of the project, (ii) the financing institutions,

(iii) the environmental authority, (iv) NGOs, and (v) the public, and here mainly the directly affected people and communities. The type of questions and observations one will get depend on their author, not all of the five groups mentioned are likely to point out the same items. It is here also not the intention, and would certainly not be possible, to provide a complete list of potential questions, and maybe even answers to them, but it should be possible to shell out a few types of questions one has to expect. This list is discussed shortly, with some comments on each of the types. One of the aims of this list is also to reflect on possible motivations for such questions or comments, which can be very different between these five groups.

Omissions: there can be different reasons for omissions, and different reasons for pointing this out. An EPA is likely to point out an omission if the person checking the report takes the official guideline of the agency and uses this to check whether all the issues or subjects listed there have been dealt with in the report. There can be one good reason for such omissions, namely, if a specific subject is just not relevant in a specific case, but in order to avoid such observations it is in any case better to list all these issues and to explain (in not too few words) why one or the other point is not important. Really serious major omissions should not occur, because this would really disqualify the report (and its authors). On the other hand, as explained in different parts of this book (see e.g. Sect. 5.6), it is never possible to cover everything, so omissions are inevitable. This, of course can be used as an argument. Assuming that you are opposed to a project and you have an occasion to pronounce yourself on the ESIA, then one easy way to find an argument is to look for omissions. Just listing, as an example, some groups of animals which have not been included in the study may perhaps not result in a very relevant objection, but it will certainly offer a number of points for discussion. One type of comment that goes in the same direction is the argument of a specialist who thinks that "his" subject has, perhaps not been completely omitted, but did not get the attention it should have.

Mistakes 1: under this category, serious mistakes are understood. This is of course something that every author of such a report, and every specialist involved in its preparation, tries (or should try) to avoid. Depending on one's position, the definition of what is a serious mistake will of course vary. To make an example: not everybody may think that the example shown in Table 3.3 is a serious mistake; I do. And if one notices in a report (as I have in one case) that the author does obviously not know the difference between reservoir area and catchment area, then serious doubts about the qualification of the author are certainly justified. Here maybe one example of a really serious mistake once encountered, not in a report, but in an official document needed for the preparation of a report. The issue there were rubber tree plantations in the area to be submerged, and more precisely the compensation to be paid for them. Using the guideline issued by the Ministry of Agriculture resulted in absolutely fantastic amounts of money the project would have to pay. Going into the details of these guidelines led to recognising that its author had made not one, but two mistakes: first, he had taken the number of trees per hectare, but then stated this as the number per acre (wrong by a factor of 2.5); second, the value of an average yearly harvest per tree was assumed to be the monthly revenue, wrong by a factor of 12. This resulted in an amount 30 times the "real" value. Obviously, if a mistake of that type should be detected in a report it needs to be corrected.

Mistakes 2: this means irrelevant mistakes. Concerning the definition: see above. However, if there is a mistake which has no influence whatsoever on the result of the study, as for instance a typo which does not change the sense of a statement, then it is good to point such a thing out during proof reading (or when checking a draft as mentioned above), but not in an official comment on a report where it is about major results and outcomes. Here again an example. In a document commenting on an ESIA report on behalf of the government there was the following statements: first in the introductory part, it said "it is not better documents but better decisions that count". In the parts with detailed comments, then: "The plural of fish in English is generally 'fish', not 'fishes', e.g., 'he caught 10 fish'. 'Fishes' applies when referring to species of fish, e.g. (...). Most of the usage of 'fishes' in the text should be 'fish'". Yes, agreed, the statement is correct, but the meaning in the text was clear every time, no misunderstanding possible, and one has to accept the fact that probably the majority of such reports are being prepared by non-native English speakers. This, with all due respect, is a type of "elementary school teacher's comment".

Different opinions: this, for obvious reasons, can be a serious issue. Depending on where you stand, you will have a different point of view, e.g. on the importance of an omission as mentioned above (see also Fig. 2.5). So, e.g., one could write: "half of this important habitat will be destroyed", or "half of this important habitat will be maintained". Mathematically, these two ways of expressing a fact are the same, but psychologically they are not. The first type of sentence might be used when the focus is on the severity of the impact, the second one when it is on the effectiveness of minimisation or mitigation measures. Usually, an NGO opposing a project will prefer the first type, while the project proponent, in all likelihood, will go for the second one.

Demonstration: with this a type of comment is meant that can really be a nuisance: the instances where a reviewer needs to demonstrate that he knows better. In this respect I have experience mainly with WB sponsored projects. In such a project, unless it is a small study of short duration, there will usually be several rounds of meetings with and/or revisions by Bank staff. And here the WB has a habit, which may be dictated by necessity, but which can really result in problems for the consultant preparing the ESIA report (and, resulting from this, also for the project proponent): the constant change of specialists involved in the process. This, in my experience, usually causes two types of issues. First, of course, the person who is new on the project does not know what has happened so far (and apparently usually also does not have the time to read any documents that had been produced in the past), and this will lead to a tendency of always coming back to certain points that one had considered as closed a long time ago. The second issue, however, can be even more of a problem: more often than not, the new person will feel a need to justify his or her presence, and this will usually be done by pointing out something that apparently everybody else had not been aware of so far. It is not meant to say here that such a thing—oversight of an issue—is absolutely impossible, but it is certainly not the rule.

3.11 Biased ESIAs

Unfortunately, governments and dam builders have invariably turned the EIA process into a bureaucratic formality, merely another regulatory hurdle which developers must jump before they can get their project approved.

McCully (1996: 54)

3.11.1 Criticism of ESIAs: Cuts Both Ways

In the public discussion, there are two main types of criticisms of the ESIA, namely, (i) on the anti-dam side, e.g. NGOs which oppose as good as any project as soon as the word "dam" appears in it, or persons or organisations opposing a specific project, who will very often accuse the ESIA in a very general way as being a paper tiger or plainly an instrument for greenwashing such—in their understanding invariably bad—projects, and on the other hand (ii) persons or organisations favouring dam projects, be it as project proponents, contractors for such projects, national energy agencies etc., who will consider the ESIA as a process aiming at or being misused for preventing such—in their view invariably good—projects from being implemented.

Quite clearly, the ESIA as such is an instrument aiming at carefully scrutinising projects by taking into account environmental and social aspects which, before its introduction, were largely neglected, a fact which had led to the implementation of a large number of projects which then turned out to have unexpected negative environmental and social effects, effects which in many cases could have been reduced to an acceptable level, or could have been mitigated or compensated, if only the knowledge and the will had been there to first identify these impacts, and to take then the required mitigation measures. Doing so might have, or would probably have, also led to the decision not to implement one or the other project, since the negative effects would have weighed too heavily in comparison with the advantages gained by it. Opposition arising from outcomes of such projects built up, and this in the end led to, or was certainly one reason for, finally deciding on standards and regulations for environmental impact assessments in national legislations as well as in guidelines of multilateral organisations, the first of which was the World Bank.

And one thing has to be seen very clearly: the legislation regulating the ESIA procedure is an expression of the will of the legislative body of the respective country. Still, as certainly in other fields as well, there are problems with it, for some which there is no easy solution. The most important of these issues are the following:

- Agreeing on acceptable or unacceptable impact: as discussed in Chap. 5, there are environmental parameters where it is possible to physically exactly determine what is acceptable and what is not, namely those which can be measured and for which legally binding threshold values can be defined. However, for many others, and for most of those which are of really high importance in the case of dam projects, especially biodiversity and hydrology (and even more so for social impacts), it is not possible to define threshold values. Just imagine, for example, that the "no net loss" concept sometimes used in approaches for mitigating impacts on biodiversity by means of a biodiversity offset (see Sect. 6.3), would actually be a standard to be reached. This, in the end, might mean that no human activity at all would be allowed any longer to take place, since it would be next to impossible to actually prove that such an activity would not cause any loss. To illustrate this: in a recent study, Chen et al. (2019) found that roads have a negative impact on propagation of main forest trees by the fact that they change the populations of small rodents along their track, an effect which has an influence on predation, hoarding and dispersal of tree seeds, which again negatively affects tree recruitment and in this way the structure of the forest. How should one ever prove that any project will not have such an effect, especially also considering the fact that with this we have not yet even come near to identifying all the secondary and tertiary effects such a change in forest structure may have on any other forest dwelling organisms? As expressed in Sect. 10.5 on biodiversity, ecosystems are just too complex to ever be analysed in all their complexity in an ESIA. This invariably leads to the fact that people tend to consider an ESIA as bad—one way or the other—if the conclusions they find there are not the ones they wanted to hear.
- Biased ESIAs: such reports, of course, do exist, and it is possible to find examples for both ways, i.e. some unduly favouring a project, and some which obviously aim at killing one.
- Incompetence: there are reports where one has to come to the conclusion that they, or some parts thereof, testify to the incompetence of those who prepared them, and maybe also of those whose duty it was to check and comment on them and who may have failed to refuse to accept them. Incompetence may be a harsh criticism, but there are very clearly many examples of weak ESIA reports, as e.g. mentioned by Fearnside (2019a), who, for the case of Brazil, comes to the conclusion that "dams have major social and environmental impacts, which are systematically underestimated in the environmental impact assessments (EIAs) submitted for licensing".

Needless to say, the two latter categories, biased and incompetence, again have to be seen under the angle of view of the beholder: what one will view as perfectly balanced and acceptable will be seen by the other as heavily biased. It may be a little bit more straightforward to judge on the question of incompetence, but even there the opinions will be divided. The following section tries to illustrate some of these issues.

3.11.2 Different Concepts

As mentioned above, for criteria where there is no clearly defined threshold value it has to be expected that different people will have different points of view. One such example was the case of Rogun HPP, when it came to making a choice on the dam height. During the project studies, two alternatives with lower dam heights had been developed and analysed in addition to the original project. Table 3.2 summarises the most important parameters for a comparison of the three alternatives.

As can be seen from the table, choosing between the alternatives meant to choose between effects of a very different type: the original (with a 335 m high dam) would have the largest energy output and by far the longest life span, while it would also require the resettlement of more than twice the number of people of the next lower alternative. This is clearly a case where no firm threshold value, e.g. in the form of "acceptable number of people to be relocated per MW installed capacity or per GWh energy produced" can be formulated, although this is certainly an important issue. In the final discussions, I had the feeling that at least one member of the Panel of Experts, and probably some members of the World Bank team accompanying the study, were in favour of alternative 2, although none of them ever clearly said so. My main argument was that in my opinion a life span or 75 years for a 300 m high dam (or even only 45 years for a 265 m high dam, which, although it was sometimes called "the low dam alternative" is still a very high dam!) is too short for justifying the overall impacts (positive as well as negative) of the project, and that therefore the 335 m alternative should be favoured. This is the alternative which is presently under construction. I do not say that my recommendation was right, and even less so that a recommendation for alternative 2 would have been wrong. But this example illustrates the decisions which sometimes have to be made, and it emphasises the fact that in certain cases there are good arguments for both possible solutions. One point needs to be added here: as said, the ESIA report gave a recommendation. The decision on which alternative to finally choose, however, is not for the author of the ESIA report to make. In such a case it would be good if one more question could be answered. As the table shows, there is also a marked difference in energy output. Assuming that the need for the energy is given, if not produced by Rogun another

Table 3.2 Main parameters of the three alternatives for Rogun HPP

Parameter	Unit	Alternative 1	Alternative 2	Alternative 3
FSL	m asl	1290	1255	1220
Dam height	m	335	300	265
Reservoir area	km^2	170	114	68
Yearly energy output	TWh/y	14.4	12.4	10.1
Persons to be resettled	N	42,000	18,000	13,000
Minimal life span	years	115	75	45

Source Pöyry (2014)

project, and presumably another HPP, would be required to generate the difference, and the question is what the impact of this project would be, or more to the point, how many people would have to be resettled for it. Finally, one additional point: this description here is simplified, since other aspects which were of importance in that case are not mentioned; so the consideration was not only the total amount of energy which can be produced, but also the flexibility (producing it when it is needed or producing it when water would otherwise have to be spilled), and the capacity for seasonal flow regulation, which was a very sensitive and important point in this case (see Chap. 23).

3.11.3 Incompetence

One of my favourite "good bad examples" is shown in Table 3.3 (left column) and appeared, quite a long time ago, in an EIA for a hydropower project in Togo, in the

Table 3.3 List of fish species in Mono River, Togo

No.	Species list: original	Species list: corrected
1	Mormyrus sp.	Mormyrus sp.
2	Gnathonemus sp.	Gnathonemus sp.
3*	Bagus sp.	**Bagrus** sp.
4	Heterobranchus sp.	Heterobranchus sp.
5	Gymnanchus sp.	**Gymnarchus** sp.
6	Distichodus sp.	Distichodus sp.
7	Citharinus sp.	Citharinus sp.
8*	Chrysitchus nigrodigitatus	**Chrysichthys** nigrodigitatus
9	Latis niloticus	**Lates** niloticus
10	Labeo sp.	Labeo sp.
11*	Clarius senegalensis	**Clarias** senegalensis
12	Tilapia zilli	Tilapia **zillii**
13	Tilapia galilea	Tilapia **galilaea**
14*	Synodontis sp.	Synodontis sp.
15	Barilus sp.	**Barilius** sp.
16	Polopterus senegalus	**Polypterus** senegalus
17	Alestes sp.	Alestes sp.
18	Hydrocyon forskali	Hydrocyon forskali
19	Hepsetus odol	Hepsetus **odoe**
20	Epyplatis sp.	**Epiplatys** sp.
21	Heterotis sp.	Heterotis sp.
22*	Schilbe mistus	Schilbe **mystus**

chapter on aquatic fauna. It is a list of fish species living in the river on which the dam was going to be built.

As can be seen, this list contains 22 species of fish. There are mainly two problems with this list. The first one is order; biologists are people who usually tend to present their faunistic lists in one of three orders. To start with, there would be the option of alphabetical listing; this, while not a terribly biological approach, may be justified by the fact that if one looks for a specific species, this will usually be found quite easily. Quite obviously, the list is not organised in that way. Then, frequency of occurrence could be used, the most frequent being at the top, the rarest at the bottom (or vice versa). This makes sense especially if dealing with fisheries, i.e. with economic importance of the species. There is no indication on frequency in the list, therefore it is not possible to know if this might originally have been a reason, and frankly, looking at the list I doubt that very much. Finally, there would be the possibility usually preferred by biologists, namely, to group the species according to their systematic relationship. In the table, species belonging to one order, the Siluriformes (generally called catfish) are marked with an asterisk: 5 species spread over the entire table, which means that this is no systematic list at all. The second problem is that out of this list of 22 species, 12 are misspelled, with a total of 14 spelling mistakes.

Why is this important? For one thing, it gives the impression that in this case somebody, who might not know the first thing about fish, somewhere found a list of species and copied it, copying any errors that might have been there already, and/or adding a number of new ones. But it can go well beyond that. Just imagine that among the staff of the competent authority who officially check the report in order to see whether it is of acceptable quality there is a fisheries specialist. This specialist, in this case, may well come to the conclusion that since the chapter covering his subject lacks the required quality, this would probably be the case for the entire report.

A different type of incompetence is the obvious lack of knowledge and skills in the preparation of the ultimately most important parts of an ESIA: the impact assessment as such, the identification of mitigation measures, and the overall conclusions on the relevance of remaining impacts and therefore possibly even on whether or not a project should go ahead—which are, as mentioned above, conclusions and recommendations, but not decisions, as these latter are to be made by the authorities in charge. Still, a report without such conclusions, or with obviously shaky ones, is not what one should have to expect from this process. As mentioned in Sect. 3.7, I have seen this type of output mainly in (a few) reports that were prepared by a university. While all the different baseline studies may be detailed and scientifically sound—albeit not always of high relevance for the subject—there is no suitable synthesis, measures are not necessarily very well chosen and explained, and there are no coherent and solid conclusions and recommendations.

3.11.4 Under Pressure

When discussing the ESIA process, one also has to ask whether—and if yes to what extent—there is pressure used on any of the stakeholders involved in the process. Only one answer, I think, is possible to that question: of course there is! The question would then be to find out how much pressure there is, to what extent it changes (or can change) the outcome, whether anything should be done about it, and what the benefit derived from such a measure would be. The different potential ways for pressure are illustrated in Fig. 3.2, which shows the project and the most important stakeholders in a project development process as well as the most important interactions between them.

The roles of the main stakeholders (mostly already mentioned in Sect. 3.3) are as follows:

- Project proponent: the entity which intends to plan, implement and operate the project in question. The project proponent can be a private entity (national or foreign) or a national agency (so e.g. the Ministry of Energy, or the corresponding entity in the respective country, can take the role of project proponent), or any combination of these.
- Engineering consultant: entity (usually a private engineering company, national or foreign, or a joint venture of several companies) mandated by the project proponent for project planning and technical design.
- ESIA consultant: entity mandated by the project proponent for preparing the ESIA and related studies. As for the technical consultant, this can be a national or foreign consultant, however, in the case of a foreign company, national experts are

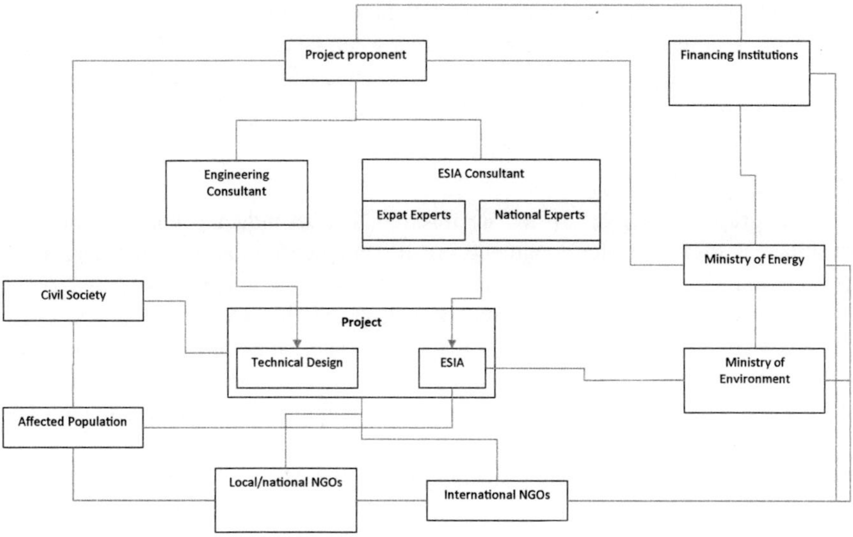

Fig. 3.2 Project and project development stakeholders

usually included in the team. It is possible to mandate one company, provided it has the required capabilities, to carry out the technical as well as the ESIA-related work, or these two parts can be, as is usually the case e.g. in World Bank financed projects, assigned to two different, independent entities.

- Financial institutions: multilateral financial institutions like World Bank, IFC or one of the different regional Development Banks, or bilateral financial institutions (commercial banks). They will usually have their guidelines or standards, whose application will be mandatory if they agree to participate in the financing of the project.
- Ministry of Energy: national entity responsible for energy supply and related projects; denomination and exact competencies can vary from one country to another. It is also possible that this Ministry acts as the project proponent in case of a project to be implemented and owned by the country.
- Ministry of Environment: this agency is the responsible body for evaluating ESIA studies, accepting or rejecting them, defining mitigation measures to be taken, and ultimately for the decision on whether the project in question can be accepted or has to be rejected on the ground of environmental and social impacts. Very often evaluation of environmental and social effects of a project are entrusted to different entities, like e.g. an agency for social welfare for social impacts.
- Civil society: this here is defined as the population of the country which is not directly affected by the project.
- Affected population: the local population of the project area which is directly affected by the project; there are different groups, the most important (but very often if not even usually the group with the least influence on the project) being the ones who will have to be relocated due to the project.
- Local or national NGOs: can have a very different attitude towards the project depending on the project as such, the impact on the local population, impact on certain natural features of the project area; such NGOs, especially in case they are opposed to the project, often enter an alliance with international NGOs.
- International NGOs: in the case of hydropower and dam projects, they usually get involved in a project if they, for whatever reason, are in opposition to it. In such a case, they often associate with local NGOs.

In the graph, only two arrows are inserted: one to indicate that the technical consultant is preparing the documents for the engineering aspects of the project, the other to indicate that the ESIA consultant prepares the documents related to the assessment of the environmental and social impact. The other direct connections indicated in the graph mostly show mutual interactions, some of which may be more in the type of one-way connections than others. Other interactions not shown there may be possible. Pressures of very different types and intensities are possible in such a situation, and it all depends very much on the type and potential impacts of the project. It is difficult if not impossible to describe all possibilities for such pressures that may arise, but it may suffice to describe just a few which can be seen, or which can be expected, very often.

The first obvious pressure may arise from the project proponent on the consultants who, after all, have been hired to do a job for their client. The project proponent's main interest is to develop the project under the best possible conditions, whereby "best" here mainly refers to economic criteria: minimising the costs while maximising the economic benefits. Environmental and social concerns, from this point of view, can be seen as cost factors, and more importantly, as avoidable cost factors, and this may lead a project proponent towards pushing the E&S consultant to paint a good picture of the project, by certainly not overstressing the negative aspects. This temptation to exert pressure on the outcome of the assessments is counterbalanced, however, by the need to have studies which will be accepted by the competent national authorities as well as, if applicable, by the financing institutions; in other words, they need a so-called "bankable study". It will also be in the interest of the project proponent to present reports which, when submitted to the public, which includes NGOs, are accepted as balanced and objective.

One other possible pressure can come from "the engineer" to "the E&S specialist", as e.g. the requirement to accept certain constraints, often also in the interest of reducing project costs, or out of fear that "exaggerated" mitigation measures will increase the costs, thereby potentially affecting the economic feasibility of the project. Concerns of this type of pressure probably also led to the practice applied by World Bank since about 20 years to award technical and E&S studies for a project to two different entities. It is not easy to say which approach in better. It is certainly the case that if two different companies are engaged, the work of each of them will be more independent from the other, since there is much less possibility to exert pressure on one or the other, e.g. in case the management of a large firm, being aware of the fact that the engineering part of a project is financially much more interesting than the E&S part, instructs its environmental and social unit not to put obstacles in the way of the engineering work. On the other hand, it is much simpler to discuss about project features, and to possibly identify better solutions—better in the sense of being sound from an engineering point of view while at the same time reducing or avoiding an environmental impact—when the respective specialists are under the same roof and have developed a way of working together.

One very serious type of pressure can be exerted by the different ministries involved in the planning process. Independently of the question whether it is a privately or publicly owned project, the Ministry of Energy very often has a vested interest in implementing it. On the other hand, the Ministry of Environment has no such direct interest, but has the obligation to, ultimately, either accept the project, by giving it the environmental licence (or whatever document may be required under the respective national legislation), or to prevent the project from being implemented by refusing to issue this document. Such a refusal, however, is often not seen as a necessary measure for protecting the country's environment—and applying its legislation for protecting the environment—but as a hindrance to development. In such a situation it has to be seen that very often the Ministry of Energy has much more political weight, including financial potential, than the Ministry of Environment, and will therefore often prevail in case of a direct confrontation between the two entities, as e.g. observed by Corlett (2019: 243). I have come across at least one case in which a dam project was refused by the Ministry of Environment, and then accepted a few weeks later, without any new studies or changes in the project as such having been made, and without any explanation of the reasons for this turn-around. Fearnside (2019a, b) notes that in Brazil, statements of licencing authorities are often ignored or overruled, and that political pressure aiming at weakening environmental legislation threatens to further weaken the environmental assessment process; this threat is certainly not limited to Brazil.

A number of hydropower projects have met considerable opposition; in such cases, there is often a link between civil society in the wider sense (i.e. the population outside of the project area and therefore not directly affected), the affected local population, and local as well as international NGOs. Prominent examples for this are e.g. Narmada in India (see e.g. Leslie 2005), Ilisu in Turkey (see Sect. 26.3) and the Three Gorges Dam in China (21.1.2); as less prominent examples Ralco in Chile and Lengarica in Albania could be mentioned. Such opposition against dam projects is always caused by environmental and social concerns; at least I have never heard about any NGO or any other group who protested because they wanted to have, e.g., a concrete arch dam instead of a proposed earthfill dam. But I also have never come across such an opposition against a project which gave the impression that it was completely unfounded; if nothing else, the opposition of the directly affected population, of the people who will lose their home, and often their ancestral lands, is certainly justified. The main thrust of such oppositions will be towards political entities (sometimes including financing institutions, especially the multilateral ones like World Bank), i.e. the authorities who will ultimately have to decide whether or not to go ahead with a project.

3.11.5 Three Cases

This section briefly discusses three examples with the aim of illustrating a number of points made above. As said in the introduction, there is no intention of blaming or making fun of a specific person or organisation, therefore in this case neither the author(s) of the reports nor the project will mentioned; it cannot be excluded, however, that a few persons will recognise one or the other of these three projects, which actually are real and not just examples invented for the sake of emphasis.

3.11.5.1 Lack of Coordination

The first example is an ESIA report prepared for a hydropower plant in Colombia. The consultant selected for the preparation of the report was a local university. Reading this report, there were mainly two concerns. The first one was about the selection of the team. The impression that was given by the report was that the very first question the team leader was asking himself was not "what type of experts or specialists are needed for preparing a report that will answer to the requirements of the assignment", but rather was "what kind of specialists do we have in the institute or university who would like to contribute to this study". This resulted in a number of specialised studies in all types of fields in some way related to the situation in the project region, and the collection of a considerable amount of information on this region, without however showing the relevance of the individual specialised investigations for the main question at hand, i.e. the identification of project impacts and the formulation of mitigation measures where required. The second problem was that quite obviously there was nobody in the team who really had an understanding of precisely these basic main questions, and no experience in using the information provided by the specialists for answering these question. In the end, one was left wondering what now the conclusions would be, and what measures could, should or might be implemented in order to mitigate the impacts this project would certainly have. This was quite a while ago; today, it seems rather unlikely that an environmental protection authority in any country would accept a study of that type.

3.11.5.2 Turning a Blind Eye

This is about an ESIA for a project in Tanzania. The project is located within a protected area, which should lead to the conclusion that potential impacts on habitats and biodiversity would have to be assessed carefully. Unfortunately, this was not the case at all. Some of the most salient weak points of this ESIA were:

- The report stated that the access road will open access to the "most extensive and pristine area where accessibility is limited …", however, it only mentions that about 50 km of access road will be required, without providing a map showing it, and without ever addressing the potential impacts of this road; to the general issues related to access roads see Chap. 15.
- The reservoir size is given as 914 km^2. However, description of the reservoir, the reservoir area and volume curve in the design report and one other source suggest that it is probably around 1250 km^2. No suitable map is provided.
- The report mentions the fact that the project is situated in an important protected area and identifies the downstream area as potentially affected. Neither of these two areas is described in the report.
- Two Ramsar sites are reported in the project area, one of which, being located downstream of the dam site, will certainly be affected. No discussion of impacts on either of these sites is provided.
- Affected habitats are not described, and impacts (e.g. on type, significance and affected area) are not described. Negative impacts are turned into positive ones, e.g. when under the heading "significant negative impacts", instead of actually describing these, it is stated that "Creation of the approximately 914 km^2 reservoir would encourage development of lake regime resulting in increased number of fish in the reservoir". Increasing the number of fish can only be considered as a positive impact in case it was planned to develop fisheries; nothing in this direction is mentioned, and it cannot be an aim to encourage fishing in a reservoir located

within a protected area, especially in this case where, according to another part of the report, the government has already decided against fisheries development.

- There are contradictory statements, e.g. on vegetation, like "The magnitude of forest loss and extent of de-vegetation is expected to be highly significant and long term", "significant loss of plant biodiversity is likely as many plant species may be lost", "… the diversity of vegetation in the project impact area would not significantly change." No statement on actual significance of the impact is provided, and no mitigation measures are proposed.
- "Loss of rare, endemic or endangered species" is listed as a very high negative impact. However, no list of these endemic and endangered species is provided, it is not indicated what "loss" means (loss of a small local population, loss of a significant proportion of the entire population or even loss of entire species), and again no mitigation measures are provided.
- Some parts of the report give the impression that the planned operating regime of the power plant was not understood. Descriptions of the operation lead to the conclusion that the plant would be operated as a peaking plant at least during part of the year. This is however not stated clearly, and positive impacts derived from this operation are most probably wrong.

This systematic ignoring of potentially very serious negative impacts, the constant stressing of possible positive effects, the complete lack of detailed baseline information and the absence of mitigation measures of any relevance can lead to a totally wrong image of the project and its consequences. On the other hand, such a report is very vulnerable to criticism, and so obviously is the entire project. Reports of that type can certainly lead to the overall negative opinion on ESIAs and the entire impact assessment process.

3.11.5.3 Out for a Kill

The Executive Summary of an ESIA for an HPP usually starts with a statement on the basic conditions for the study and a short description of the project. It then mentions the most relevant characteristics of the project site and the different parts of the study area, goes on with listing and shortly evaluating main impacts and the identified mitigation measures. Finally, it enumerates the main conclusions, and it then gives recommendations on the way forward.

Not so in this case, the ESIA for a hydropower project in Lebanon. Here, the Executive Summary started with a description of the biological situation, and it described this situation with phrases like "… important about this wealth is that it was established over millions of years of evolution under the particular conditions of the site," "… of uniqueness at the national and Mediterranean levels …", and "Its intactness, richness in microhabitats and the dense overlaying of niches has promote a hot zone of endemism, not only providing sanctuary and refuge to the endemics of Lebanon, but also to create its own local endemics." It then goes directly to the main impacts of the project stating that the "dam-lake is to be installed in the heart of the ecological system and its intact green cover.", and would lead to a number of very serious impacts like "Valuable parts of the ecological system will be severely affected; segments of community segregations by height and altitudes and the community associations therein will dissipate without further evidence" (whatever that means), "The continuous existence of a large water mass as the lake in an ecological system that was never adapted in its evolutionary history to accommodate this incident will engender a chain of long-term forces of degradation on the extant ecological and agricultural systems.", "Prevalence, proliferation and expansion of fungal species as well as insect pest communities will threaten the health of the ecosystem and agricultural system with outbreaks of diseases and epidemics.", and "The endemics will suffer the loss of habitats, and local endemics would face pressures of extinction. The value of the ecosystem as a cauldron of endemism and its evolutionary potentials will dissipate."

There are, at the very least, the following most basic serious problems with this approach:

- The description of the situation covers the entire valley "from the coastline to the peaks", i.e. from 0 to around 2000 m asl.
- Since there is no description of the project, it is not possible for anyone not familiar with the situation to get a clear picture of the potential effects. The description sounds as if the entire valley would be impacted. As a matter of fact, the dam site is located at the lower end of a rather short, steep and very narrow gorge, where a dam with a height of about 100 m would create an elongated and narrow reservoir with a surface area of about 1 km^2 (roughly 5 km long with an average width of 200 m), FSL at 839 m asl.
- The wording of the Executive Summary makes it clear that the authors of the study do not intend to provide a basis for an objective evaluation of the project, they want to destroy the project. The report confirms this impression.

To go into some more detail, there are a number of additional problems as e.g.:

- The "evolution over millions of years under the particular conditions of the site": it is well known that the situation in the Mediterranean was far from stable over such a time span. Conditions changed dramatically at the end of the last glaciation (Ellenberg 1982: 24), and the mountains of Lebanon have been stripped of their forests during historical times, by Phoenicians, Egyptians and later on Romans, who largely destroyed the cedar forests (Eckenwalder, 192).
- Endemisms: the Executive Summary insists on endemic species that will be threatened by the project. Five endemic species were actually recorded, all of which are either "endemic to Lebanon" or "endemic to the Levant region". Jabal Moussa, a protected area close to the project site but not touched by it, is home to 28 species endemic to Lebanon, three of which apparently restricted to this protected area (Thomé 2012: 140). There is no indication of a species endemic to the project area, or of a risk of a significant loss from the population of one of these species due to the project.

- The study mentions one "distinctive conifer" species, *Arceuthos drupacea*, and comes to the conclusion that "The distinctive conifer of this region *Arceuthos drupacea* will be directly influenced by the change in conditions propagated by the presence of the dam lake. As a reminder, this forest is the single of its type in Lebanon; its endangerment implies the threatening of a national natural heritage." According to Eckenwalder (2009: 290) this species, *Juniperus (Arceuthos) drupacea*, is nowhere frequent, but grows at altitudes between 500 and 1900 m in an area extending from Greece to Lebanon. It was not found in the project area as such, in this region it only grows at higher altitudes.
- Development of parasitic fungi: this, according to the report, would be the consequence of the increased humidity in the area due to evaporation from the reservoir. However, in this dry climate, with a small surface, frequent winds and a considerable drawdown in summer, there is definitely no risk of this.
- More than once, it becomes obvious in the report that the authors could not distinguish between catchment area and reservoir area, and that they obviously do not understand the way of functioning of the project.

All this is not meant to say that the project does not have any impacts. The gorge will be (or rather was, by now) largely submerged, and the impressive cliffs are not only interesting from a geological and visual point of view, but are also important habitats for a number of species, albeit not a type of habitat that is extremely rare in this area. The example just serves to point out that wrong arguments should not be used for discrediting a project. Whether this was done on purpose or out of ignorance, in either case it discredits the authors of the report.

3.12 Baseline: How Much Is Enough?

How much time and effort is required for preparing an ESIA for a hydropower project? This is a very important question, and not easy to answer. A once and for all answer is not possible, it depends on too many variables, like i.e. how much information is already available, how well known the situation in the project area is, what type of seasonality we have, mainly: do we deal with a strongly seasonal—winter-summer or dry season-rainy season—type of system, or do we have a more or less uniform situation year round; the latter is certainly the exception, and in strongly seasonal situations it is certainly required to have a clear picture of all season, where biological aspects are concerned, but usually also for socio-economic aspects, in the cases where rural populations, subsistence farmers or nomadic herders for instance, are affected. One thing can certainly be said: the Pareto principle certainly also applies to ESIAs, as illustrated in Fig. 3.3, if not in its strict 80/20 form, then certainly not far from it.

To make this very clear: this is a hypothetical example and not anything statistically verified and corroborated. The graph serves to make mainly two points:

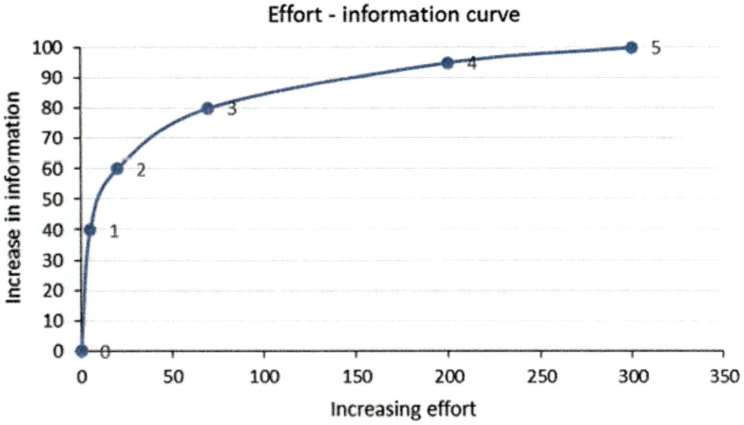

Fig. 3.3 Effort and return in ESIAs for HPPs. Stages: 1 = initial site visit, 2 = scoping, 3 = first field studies, 4 = detailed field studies, 5 = additional specialised studies

- The most important step in ESIA preparation, for gaining a very substantial amount of information, is a site visit by a small team of experienced experts who cover all relevant aspects, i.e. all fields that will potentially be affected by the project in question.
- Detailed field work is the most time consuming and therefore most costly part of the work, and gain in relevant information is often rather limited.

The point here is this: a carefully selected small team of experienced specialists needs to carry out a site visit of usually just a few days. Team composition obviously depends to a very large extent on the situation of the project area. If we deal with a project located in the middle of a National Park, without a human population that could be affected, the social specialist might not be required. On the other hand, if the affected area is densely populated and intensively used, terrestrial ecology, wildlife habitats and generally biodiversity issues may not be overly important. Time of the visit has to be selected carefully (e.g. do not go there during the rainy season if this means that relevant places will be inaccessible, and do not go in winter in the case of higher latitude projects, when everything is covered in snow; believe me: neither of these points is in any case obvious for "project managers" in air-conditioned offices e.g. in Washington DC or in the capital city of the country where the project is located!), and what exactly "a few days" means will depend very much on accessibility, type of project, and size of potentially affected areas, the size of the future reservoir and the potential of downstream impacts among the most important issues. Such a site visit, preceded by a short analysis of available documents, will provide most of the crucial information, such as impact on hydrology, i.e. changes in river flow pattern downstream of the dam which will have significant impacts in this area, impacts on biodiversity by submerging valuable habitats or fundamentally changing habitats in the downstream area, and impacts on local human populations, and this again by requiring resettlement in the reservoir area and/or affecting inhabitants in the downstream area. It will also be possible to say, at this point in time, whether one of these impacts has the potential of turning into a no-go argument for the project.

The situation is different for detailed field work (which, in most cases, is biological field work, i.e. investigation of vegetation and fauna, and/or social survey work, depending on the situation in the project area). Here, the effort, in terms of time spent in the field, and therefore in terms of costs, is usually very high for obtaining rather little, but more detailed additional information. Still, the effort is required. In the case of a dam and reservoir located in a protected or an otherwise biologically valuable and highly diverse area, after the site visit it is quite clear that the potential impact of the project is likely to be very serious. Likewise, in the case of a densely populated area, it will be clear that resettlement will be a major issue. In both cases, however, we need to know in more quantitative terms what it is going to be. Obtaining this detailed information does not change the basic impression gained about the nature and importance of the impact, but it will allow to describe it with more precision; detailed data may be required specifically for then determining the nature, extent and cost of required mitigation measures. And the more there is a need to go into details, the greater the effort for an ever smaller amount of additional knowledge. Just one example: in the case of a project in a protected area, there usually is information available on key species living in this area, which often are the main reason for the protection in the first place. This means that with next to no effort one already has data about the presence of such species in the area in question. Now, to know whether such a species occurs not only in this protected area, but specifically also in the part of it affected by the project requires a considerable effort, especially in the case of rare and not easily observed species. If in addition to that one wants to know how important this affected area is for this species—e.g. whether it needs to rely on this area or a specific part of it as a site of high importance in its life cycle, this will require even more effort. And finally, while it may be possible to demonstrate that a rare and endangered species actually does occur in a specific place, it is as good as impossible to prove that it does not occur there.

The question of how much effort is required therefore is important. Doing too little can mean that important aspects remain undetected, doing too much on the other hand can result in a major waste of time and money. As De Santo (1978: 134) put it: “Anyone responsible for the collection and/or the interpretation of data, regardless of its nature, must understand why the data is needed and what level of comprehensiveness and detail is required. In the absence of careful control over this function, great quantities of time, let alone budget, can be needlessly spent with no useful conclusion resulting from the work.” The other side of the coin, obviously, also applies: by doing too little one risks to miss important information. There is no easy way to determine how much is enough, this needs to be decided on a case to case basis.

One thing that can often be observed in ESIA studies is the fact that for the baseline description a number of items are being investigated, and information accumulated, which then turns out to be completely useless for the purpose of the study. Just to shortly mention two examples: noise assessment for the baseline situation in the case of a dam project in a remote area, where the only noise before the start of the construction work stems from the river or the occasional bird singing in the nearby vegetation, (see Sect. 8.3), and information on plankton in the river, for which I still have to see a case where this information actually leads to a relevant conclusion, or even a mitigation measure (Sect. 16.5).

This point can be illustrated with an example which, for once, has nothing to do with dams or hydropower, since it is taken from experience in developing an EIA report for the project of a quarry in the Swiss Alps. It was a special type of quarry, aiming to use an ancient rock fall site. Big blocks of rock, some of them the size of small houses, were strewn across a part of the valley floor, and the project envisaged to use these rocks. The main question for the EIA was whether in this somewhat special situation a special vegetation would have developed, which might need protection or would require some mitigation measures. There was a young botanist in the team who was charged with investigating the site, drawing up a species list and a description of the vegetation. She then reported back that basically there was nothing special about the local vegetation, that it was just what could be found everywhere in the surroundings, in meadows, forests or rocky areas. But then her eyes started to brighten and she said that she had found two species of mosses that so far had only been found two or three times in Switzerland. She was, obviously, a botanist specialising in mosses, and this was where her main interest lay. She then was surprised when she was asked what now should be done with this information. The specific question was whether these mosses had been found so rarely before because they were actually very rare, or whether so far just nobody had bothered to look for them. There was no answer to this question. The point here is: such an observation can be very interesting from a scientific point of view, but for the practical questions to be answered by an EIA they are basically useless. Within the framework of an EIA it will not be possible to carry out a specialised study covering the distribution and frequency of a rare (or not well-known) species for actually being able to answer these basic questions, this will have to be done by a scientific institution.

3.13 A Multitude of IAs

In the beginning there was the EIA, as mentioned in Sect. 3.2: the first concern, apparently, was the environmental impact of various types of projects, hydropower prominently among them. In a logical and urgently required next step "S" was added, for including social impact assessment in the process. This led to the abbreviation "ESIA", which is now generally used, although in some cases the two aspects are dealt with separately, in an EIA covering the environmental and an SIA the social aspects. I do not think that this is necessarily a useful approach, given the fact that very often the two sides are closely linked to each other, e.g. when it comes to water and its ecosystem functions and as an essential resource for the human population, or fish and fisheries, where aspects of biodiversity and economic use are intertwined;

still, the main requirement is that both topics are adequately dealt with, irrespective of the formal approach.

Since then, a number of other Impact Assessments (IAs) have been added; Fig. 3.4 provides a list of them and also indicates their position in the planning and project development process. From this graph it can be seen that these additional assessments can be put in two groups. The first of these groups contains only one such IA: the SEA or SIA (Strategic Environmental Assessment or Strategic Impact Assessment). This is a procedure which does something that the ESIA cannot do: this latter focuses on one specific project, while the objective of the SIA/SEA is the early assessment of potential impacts at the earlier and higher-level planning stages, i.e. policies, strategies and plans (Dalal-Clayton and Sadler 2005; Fisher 2002, 2007; Therivel et al. 1992). An assessment at this level will necessarily be much less specific, with less details, and usually without formulating specific mitigation measures for well-defined impacts. It mainly serves the purpose of identifying, at a very early stage, potential environmental risks of such an envisaged development and, ideally, of alternatives. The second group is different: these are all covering specific aspects of topics usually included in an ESIA (which, as mentioned above, is sometimes split up into an EIA and an SIA). Note that the ESMMP, the Environmental and Social Management Plan, is shown in the graph although strictly speaking it is not an IA, but an important plan aiming at operationalising mitigation measures and monitoring the process during specific project stages (see Chap. 12); the last stage, decommissioning, will usually require a separate ESIA (see Chap. 13).

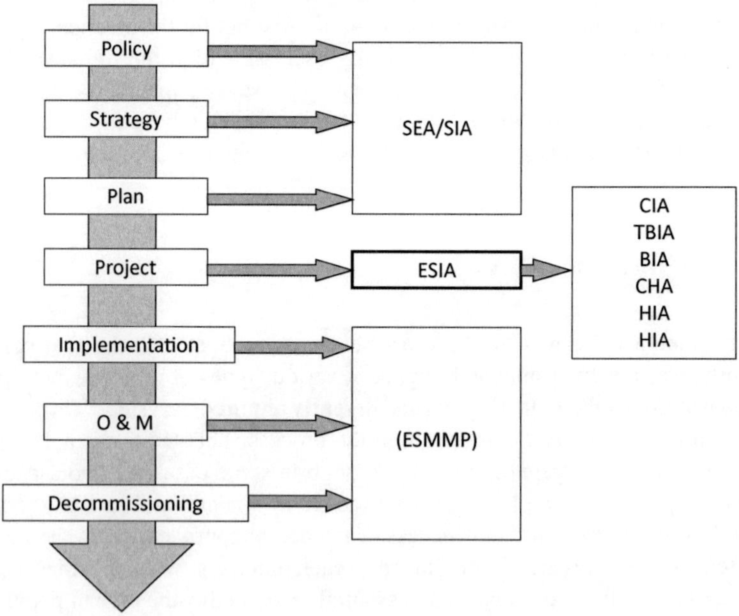

Fig. 3.4 IAs in the planning hierarchy

The list of IAs put in group 2, as mentioned, can be seen as integrating parts of an ESIA; if a project is likely to cause impacts on any of the specific topics listed there and the ESIA does not address this, then it is certainly not up to expectations. However, whether the individual point should be dealt with in the ESIA chapter dealing with the issue, or as a separate study and report, is more a question of approach (or formal requirements!) than of practical importance (see to this the respective observations in Sect. 27.2.2). The items in the list are shortly described here; note that the list is not necessarily comprehensive:

- **CIA, Cumulative Impact Assessment**: aims at identifying, in addition to the impacts of the study-specific project, impacts of other projects, existing or planned ones, and to assess their cumulative effects. These other project can be of the same type (e.g. several hydropower projects in one river basin) or of very different nature (e.g. a mining project in the catchment area of an HPP). However, it is still a project-specific investigation, since the ultimate aim is not (and cannot be) an identification of all existing and imaginable impacts in the wider project area, but to address the incremental effect of the project at hand on such developments. A good example in our context is the impact of HPPs on the Tonle Sap in Cambodia (Sect. 22.5), where each HPP in the river basin has only a small effect, while the cumulative impact will ultimately be very significant.
- **TBIA, Transboundary Impact Assessment**: the focus here is, as the name clearly says, the identification of impacts the project can have on other countries (who, in most cases, otherwise do not have a say in the process). For obvious reasons, rivers flowing from one country into another, hydropower projects have to consider this topic. An example is Rogun HPP in Tajikistan (see Sect. 23.3). CIA and TBIA are described with somewhat more detail in Chap. 19.
- **BIA, Biodiversity Impact Assessment**: in cases where a project affects an area of high biodiversity value, such a separate evaluation of this effect is sometimes asked for. Usually, however, this will be dealt with in the respective chapter(s) of a standard ESIA.
- **CHA, Critical Habitat Assessment**: the concept of Critical Habitat was introduced by the new Performance Standards of IFC (IFC 2012), namely by its PS6 dealing with impacts on biodiversity (see Sect. 6.1.3). While this PS and its accompanying GN (Guidance Note) mention a critical habitat assessment, they do not specifically ask for such an assessment as a separate study and/or report; still, sometimes there is a request for such a CHA.
- **HIA, Heritage Impact Assessment**: this IA specifically aims at identifying project impacts on items included in UNESCO's World Heritage List (https://whc.unesco.org/en/list/); this list includes objects, sites or areas of global cultural, historical or natural significance. If a project is likely to have relevant impacts on such a site, an HIA can be requested. There are specific guidelines prepared by ICOMOS (International Council on Monuments and Sites; ICOMOS 2011) which have to be followed by such an IA. The process as such can also be applied

for assessing impacts on any objects considered as important cultural or archaeological sites, it is not necessarily limited to those included in the UNESCO list.

- **HIA, Health Impact Assessment**: no, including this abbreviation twice in Fig. 3.4 was not an oversight! Health impacts can be an important topic in water resources development projects (see Sect. 10.4), and especially also in social impact assessment and resettlement planning. A specific HIA may occasionally be asked for.

This chapter was, to a great extent, about difficulties, shortcomings and pitfalls encountered in the course of ESIA preparation. A number of conclusions and suggestions on how to hopefully improve the process will be provided in Chap. 27, the final chapter of this book.

References

Bessesen, B.L., and M. Gonzélez-Suaréz. 2021. The value and limitations of local ecological knowledge: Longitudinal and retrospective assessment of flagship species in Golfo Dulce, Costa Rica. *People and Nature* 3 (2): 627–638. https://doi.org/10.1002/pan3.10219.

BUWAL. 1990. *UVP-Handbuch*. Bern, Switzerland: Bundeasmt für Umwelt, Wald und Landachaft.

Charest, P. 1995. Aboriginal alternatives to megaprojects and their environmental and social impacts. *Impact Assessment* 13 (4): 371–386.

Chen, W., J. Zhong, W.P. Carson, et al. 2019. Proximity to roads disrupts rodent's contributions to seed dispersal services and subsequent recruitment dynamics. *Journal of Ecology* 107 (6): 2623–2634. https://doi.org/10.1111/1365-2745.13221.

Coomes, O.T., Y. Cheng, Y. Takasaki, and C. Abizaid. 2021. What drives clearing of old-growth forest over secondary forests in tropical shifting cultivation systems? Evidence from the Peruvian Amazon. *Ecological Economics* 189. https://doi.org/10.1016/j.ecolecon.2021.107170.

Corlett, R.T. 2019. *The ecology of tropical East Asia*, 3rd ed., 320 pp. Oxford, UK: Oxford University Press.

Crystal, D. 1998. *Die Cambridge Exzyklopädie der Sprache*, 478 pp. Frankfurt: Campus.

Dalal-Clayton, B., and B. Sadler. 2005. *Strategic environmental assessment*, 470 pp. London: Earthscan.

De Santo, R. 1978. *Concepts of applied ecology*, 310 pp. New York: Springer.

Diamond, J. 1997. *Guns, germs, and steel. The fates of human societies*, 480 pp. New York: Norton.

Diamond, J. 2005. *Collapse. How societies chose to fail or survive*, 575 pp. London: Penguin.

Dore, M.H.I., and J.M. Nogueira. 1994. The Amazon rain forest, sustainable development and the biodiversity convention: A political economy perspective. *Ambio* 23 (8): 491–496.

Eckenwalder, J.E. 2009. *Conifers of the world. The complete reference*, 720 pp. Portland: Timber Press.

Ellenberg, H. 1982. *Vegetation Mitteleuropas mit den Alpen*, 3. Aufl, 989 S. Stuttgart: Ulmer.

Fearnside, P.M. 2019a. Impactos das hidrelétricas na Amazônia e a tomada de decisão [Impacts of hydroelectric dams in the Amazon and decision making]. *Novos Cadernos NAEA* 22 (3): 69–96. https://www.researchgate.net/publication/338057009.

Fearnside, P.M. 2019b. Represas hidroeléctricas en la Amazonia brasileña: impactos ambientales y sociales. *REB Revista de Estudios Brasileños* 6 (11): 123–138. https://www.researchgate.net/publication/330639535.

Fiennes, S. 2020. Why researchers should be using local traditional knowledge to estimate biodiversity. *The Niche (BES Magazine)*, March 2020, 20–21

Fischer, T.B. 2002. *Strategic environmental assessment in transport and land use planning*, 284 pp. London: Earthscan.

Fischer, T.B. 2007. *Theory and practice of strategic environmental assessment. Towards a more systematic approach*, 186 pp. Earthscan, London.

Gao, L. 1999. *On the Dai's traditional irrigation system and environmental protection in Xishuangbanna*, 630 pp. Kunming: Funded by the Ford Foundation, Yunnan Nationality Press.

Goodland, R.J.A., H.E. Daly, S. El Serafy. 1993. The urgent need for rapid transition to global environmental sustainability. *Environmental Conservation* 20 (4): 297–309.

Green, E.C., K.J. Goodman, and M. Hare. 1999. Ethnobotany, IPR and benefit-sharing: The Forest People's Fund in Suriname. *Indigenous Knowledge and Development Monitor* 7 (1): 8–11.

Grimes, B.F. (ed.). 1996. *Ethnologue. Languages of the world*, 13th ed., 966 pp. Dallas: Summer Institute of Linguistics, Inc.

Hughes, J.D. 2009. *An environmental history of the world. Humankind's changing role in the community of life*, 2nd ed., 306 pp. New York: Routledge.

ICOMOS. 2011. *Guidance on heritage impact assessments for cultural world heritage properties*, 20 pp. Paris: International Council on Monuments and Sites. https://www.iccrom.org/sites/default/files/2018-07/icomos_guidance_on_heritage_impact_assessments_for_cultural_world_heritage_properties.pdf.

IFC. 2012. Performance standards on environmental and social sustainability. Performance Standards PS1-8, Guidance Notes GN1-8. http://www.ifc.org/performancestandards.

Katzner, K. 1995. *The languages of the world*. New, 378 pp. London and New York: Routledge.

Learn, J.R. 2020. "Two-eyed seeing": Interweaving indigenous knowledge and western science. *Wildlife Professional* 14 (4): 18–26. http://wildlife.org/dbadmin/twp_archive/Jul_Aug_2020.pdf.

Leslie, J. 2005. *Deep water. The epic struggle over dams, displaced people, and the environment*, 352 pp. New York: Picador.

Linaweaver, S. 2003. Catching the boomerang: EIA, the World Bank, and excess accountability: A case study of the Bujagali Falls Hydropower Project, Uganda. *International Journal of Sustainable Development and World Ecology* 10 (4): 283–301.

McCully, P. 1996. *Silenced rivers. The ecology and politics of large dams*, 350 pp. London: Zed Books.

MONRE. 2016. *ESIA guidelines*. Lao PDR: Ministry of Natural Resources and Environment.

Nast, M. 2006. *überflutet—überlebt—überlistet. Die Geschichte der Juragewässerkorrektionen*, 192 S. Verein Schlossmuseum Nidau.

O'Hara, S.L., F.A. Street-Perrott, and T.P. Burt. 1993. Accelerated soil erosion around a Mexican highland lake coused by prehispanic agriculture. *Nature* 362: 48–51.

Pearce, F. 2018. *When the rivers run dry. The global water crisis and how to solve it*, updated ed., 312 pp. London: Granta Publications.

Pöyry. 2014. Environmental and social impact assessment for Rogun hydropower plant. In *ESIA*, vol. I: Text, vol. II: Annexes. Report prepared for OSHPC "Barki Tojik". Republic of Tajikistan.

Therivel, R., E. Wilson, S. Thompson, D. Heaney, and D. Pritchard. 1992. *Strategic environmental assessment*, 181 pp. Reprint 1999. London: Earthscan.

Thomé, G.H. 2012. *Guide des belles fleurs de la réserve de la biosphère du Jabal Moussa*, 140 p. Liban: Association pour la Protéction du Jabal Moussa.

Vischer, D.L. 2003. *Die Geschichte des Hochwasserschutzes in der Schweiz*. Von den Anfängen bis ins 19. Jahrhundert. Berichte des BWG, Serie Wasser, Nr. 5:208 p. Bern: Bundesamt für Wasser und Geologie.

WB. 2017. *The World Bank environmental and social framework*, 106 pp. Washington, DC: International Bank for Reconstruction and Development/The World Bank. http://documents.worldbank.org/curated/en/383011492423734099/pdf/114278-WP-REVISED-PUBLIC-Environmental-and-Social-Framework.pdf.

WCD. 2000. *Dams and development. A new framework for decision-making*, 404 pp. The Report of the World Commission on Dams. London: Earthscan.

Whitehouse, H., P. Francois, P.E. Savage, et al. 2019. Complex societies precede moralising gods throughout world history. *Nature* 568 (7751): 226–229. https://doi.org/10.1038/s41586-019-1043-4.

Zwahlen, R. 1989. Die "negative" UVP: Soll ein UVP-Bericht zur Ablehnung eines Projekts führen? *Schweizer Ingenieur und Architekt* 107 (24): 636–638.

Zwahlen, R. 1996. Traditional methods: A guarantee of sustainability? *Indigenous Knowledge and Development Monitor* 4 (3): 18–20.

Zwahlen, R. 2003. Identification, assessment and mitigation of environmental impacts of dam projects. In *Modern trends in applied aquatic ecology*, ed. R.S. Ambasht, 281–370. New York: Kluwer Academic Publishers.

Chapter 4
Dams, Reservoirs and Power Plants

Planning should avoid unnecessary expenditure and effort on projects that in the end will not be carried out.
Oud and Muir (1997: 20)

This chapter deals with the basic features of dam and hydropower projects, not from an engineering point of view, but aiming at identifying the main aspects which are of relevance for an impact assessment exercise. In the first section, a short description of the main types of dam and water intake structures is provided, mainly emphasising the difference between ROR (run-of-river) and storage plants. Two specific cases, which can consist of any combination of these basic types are discussed, namely cascades and inter- and transbasin projects. In the next section the main project stages, from site identification to decommissioning, are described, again with a focus on potential and relevant impacts during the various stages. One important point made is the fact that first identification of potential sites, where, as usually supposed, "nothing is happening yet", is in reality a very important activity with potentially far reaching consequences. Section 4.3 tries to identify and discuss shortly the impacts of dam and HPPs which can be, albeit with highly different intensity and magnitude, observed in any such project, and which therefore will have to be evaluated in detail in every ESIA. One basic point in the discussion of dam projects is the question on whether a specific project is a "good project" and therefore worthwhile implementing, or a "bad project", which therefore should not be implemented. Section 4.4 addresses these issues, providing a number of criteria which can be helpful in this respect. Here again, as in various places throughout the book, the problem of applying simplistic rules and the importance of always looking at the specific situation of a given project is emphasised. Dams were one of the first categories of large infrastructure projects which led to the conclusion that, along with all their positive effects, they can and do have major negative consequences, and they were decisive for the development of the ESIA procedure. This is the subject of Sect. 4.5.

R. Zwahlen, *Assessing the Environmental Impacts of Hydropower Projects*, Environmental Earth Sciences, https://doi.org/10.1007/978-3-030-91185-0_4

4.1 Main Characteristics of Dams and Hydropower Plants

4.1.1 Types

Table 4.1 provides a first step for characterising HPPs. The idea of this is to provide a characterisation and description of individual projects in the order of presumably rising impact whereby, as always, "impact" has to be seen as negative as well as positive effects of a project. While a higher dam has a potentially more important negative impact, e.g. in terms of land use, resettlement and influence on downstream water discharge patterns, it also has a presumably higher positive impact by producing more electricity and providing more flood protection.

These are rather broadly characterised groups. While most HPP will fit more or less easily on one of these categories, this is not necessarily the case for all of them; so e.g. it is possible that even a high dam creates only a small reservoir with limited storage capacity. On the other hand, a ROR scheme on a very large river, like e.g. the Volga or the Mekong, may create a "rather small" reservoir in terms of stored water in relation to normal river discharge volume, which however in absolute figures can still be very large. The site- and project-specific conditions will always have to be taken into account.

All the schemes described above have to be one of the following types, again in the order of increasing impact:

- No river diversion: power house and tailrace immediately below intake structure, weir or dam, no part of the river with reduced flow (however: flow always depending on the combination of turbine output and spilling).
- River diversion: power house and tailrace at some (potentially considerable) distance below intake structure/weir/dam, resulting in a stretch of river with reduced flow (potentially zero; spilling, residual flow).
- Interbasin scheme: power house and tailrace in a different river than intake/weir/dam. Permanent flow reduction in "original" river, change of flow pattern in "restitution" river (additional volumes, change in seasonal and/or short time (peaking) discharge).

Pump storage schemes, where excess electricity is used for pumping water up into a reservoir during hours of low consumption, or in general when generation is higher than consumption, which then can be used for producing electricity during times of peak consumption, can be considered as an additional type of structures and projects. Their importance will most probably increase with an increase of electricity generation from renewable natural sources like wind and solar, which produce electricity in proportion with availability of the energy source (wind and solar irradiation) irrespective of actual demand (Sect. 14.3). In such cases pump storage can be used, since water pumped up can then be turbined at peak hours, in this way making it possible to store excess electricity.

Table 4.1 Main types of HPPs

No.	Type	Comment
1	Run-of-river (ROR)	No or very small storage, no change in downstream river discharge pattern. No potential for flood protection
1.1	Lateral intake	No interruption of the river (no barrier to drift and migrations), only part of the water used
1.2	Weir	Low structure across the river; barrier to migration (usually easy to overcome with fish ladder). May also hinder, to some extent, drift, especially of river bottom dwelling organisms
1.3	Low dam	Dam of up to 10 m high, small reservoir, no peaking; dam height may be more in the case of ROR schemes on very large rivers. Barrier to migrations with possibility for fish ladder; barrier to drift
2	Intermediate	Scheme basically operated as ROR, some limited peaking (daily/weekly), but no change in seasonal pattern of downstream flow; limited if any potential for flood protection. No or very limited drawdown of the reservoir. Fluctuation in reservoir level usually smaller than natural seasonal fluctuations in the river were before the construction of the structure
2.2	Low to intermediate dam	Dam of up to 20 m high; small reservoir; some daily peaking might be possible (higher production in peak hours of the day); usually no possibility for complete shutdown. Barrier for drift and migration, but there might still be a possibility for a fish ladder
2.3	Intermediate to high dam	Dam higher than 20 m. Small reservoir with a maximum of weekly storage, capacity for peaking accordingly. Barrier to drift and migration, limited possibility for a fish ladder. Some sediment retention. Limited if any potential for flood protection
3	Storage plants	Large (in terms of height and/or width) dam with intermediate to large reservoir. At least monthly, possibly seasonal or overannual storage. Potential for peaking and at least some potential for change in seasonal flow pattern in the downstream area. Barrier to migration and drift, no possibility for fish ladder. Some to considerable potential for flood protection. Considerable sediment retention. Marked drawdown of the reservoir in the dry season possible
3.1	Seasonal storage	As above. Seasonal storage (live storage volume > average monthly inflow, but < annual inflow); allows for peaking and for some shift in flows from wet to dry season with corresponding influence on downstream flows
3.2	Annual storage	As above. Large reservoir with annual storage, i.e. live storage volume similar to average annual inflow. Allows for peaking, and for a shift from wet to dry season in downstream flows, up to inverting this pattern. Very noticeable seasonal reservoir drawdown. High potential for flood protection for the downstream area. High capacity for sediment retention

(continued)

Table 4.1 (continued)

No.	Type	Comment
3.3	Overannual storage	As above. Large reservoir with overannual storage, i.e. live storage volume ≫ than average annual inflow. High potential for changing seasonal flow pattern in the downstream area. Potential for providing water downstream in exceptionally dry years, high potential for flood protection. Considerable seasonal drawdown of reservoir. High capacity for sediment retention

4.1.2 Cascades

On the 141 km of Rhine River largely forming the border between Germany and Switzerland between the Lake of Constance and the city of Basel, there are 11 hydropower plants, each one making use of a fraction of the drop of the river between these two points, basically using the same water over and over again. The first one was built in 1894, ten more followed in the period between 1912 and 1963 (Lagendjik 2016). Such cascades of hydropower plants nowadays are probably the rule rather than the exception.

As a general observation, it is probably safe to assume that overall the environmental impact is smaller if, e.g., four dams are located in a cascade on one river than if these four dams were each located on a separate river; this, however, is a highly theoretical case: building and operating a cascade on one river does not automatically mean that by doing this the next river will be left intact. This is shown by an experience of this type in Laos, in the case of Nam Ngum (NG) II HPP. This dam is located a short stretch upstream of the Nam Ngum I reservoir, which by the time NG II was studied had been operating since over 20 years. NG II uses the upper reach of the Nam Ngum River, the main tributary of the existing reservoir. One impact was that the dam would block any migration of fish from the reservoir upstream into the river, and vice versa; since it is a high dam, a fish ladder or similar structure was not an option. For this reason the ESIA report recommended to leave intact another, considerably smaller tributary which joined the Nam Ngum about half way between NG II dam site and the NG I reservoir, with the aim of keeping at least this migration route open (EWI, 1998). The study was accepted, and, seemingly, so were the measures that had been proposed. However, shortly after NG II had received the required permits, a new project was being studied on exactly this river that had been, in a way, earmarked as an environmental offset.

Still, cascades can definitely have an advantage. The incremental impact of additional dams is usually much smaller than the impact of the first one was. If one looks at a river which is a habitat for migrating fish (as the Rhine River used to be before any dams were built there), one single dam can completely block this migration route, and additional dams will not really change this situation. On the other hand, in case fish migration should be restored, this can be much more difficult when a large number of existing structures block the way, and the measure will only be effective when all will have been rehabilitated. Such an effort is presently under way for the Rhine River.

Now, whenever a dam needs a new concession, or requires any major rehabilitation work, this can be done only under the condition that suitable fish passes are installed. One of the first to have done that was Rheinfelden HPP (see Sect. 13.2), but the aim will be reached only once all the eleven structures, plus the ones further downstream (Germany, France and Netherlands) will allow fish migration again. Still, the measures taken so far seem to slowly show results. Recently (December 2020) it was announced that for the first time in more than 50 years 4 salmons have been observed in the river near Basel. This is not yet a healthy population, but hopefully a start.

But the incremental impact is not restricted to fish migration alone. Even non-migrating organisms can be negatively affected, for instance by the fact that each dam creates another barrier, in this way cutting the population of a species into smaller and smaller sub-populations which will become genetically isolated from each other. This was one concern, e.g., for one of the not very numerous autochthonous fish species of the Chilean Andes, the endemic and endangered siluriform *Diplomystes nahuelbutaensis* in the case of the cascade on the Biobío River (Arratia 1981; EWE 1997). The extent of fragmentation in Chilean rivers and the prospective development of this situation is described in Díaz et al. (2019): in eight major river basins of central Chile, there were 4 obstacles to fish movements (waterfalls) in the natural situation, 80 were added till 2018, and an additional 377 are expected to be added till 2050, which basically means an almost complete separation of a number of species into numerous but ever smaller and ever more isolated sub-populations. River fragmentation can lead to the extinction of fish populations (Bogan et al. 2017: 366), although fragmentation can also be a natural phenomenon, in the case of the so-called intermittent rivers and ephemeral streams (Datry et al. 2017a, b). There, (p. 377 ff.) a discussion of processes at different levels, populations to ecosystems, is provided, pointing out the importance of at least temporary connectivity between parts of a fragmented river. For many organisms living in and adapted to conditions in such intermittent streams it is very important that changes between flow and no-flow conditions occur with some degree of (mainly seasonal) regularity or predictability. Artificial fragmentation due to dams and weirs, however, create conditions which are very different, especially in the case of peaking power plants which may cause very short-term and marked variations which do not follow any biologically relevant patterns (see also Sect. 9.1.3.3, peaking). An interesting case was reported recently from a 20 dam cascade on the Inn River in Austria (Seitz 2018). There, a dam with a 23 km long tailrace tunnel was added to the cascade, leaving the river between intake and power house with a discharge reduced to the residual flow. Even if this, in a way, may seem like driving out the devil with Beelzebub, this is probably still an improvement, since earlier this part of the river had been exposed to a peaking regime, strong short time water fluctuations probably eliminating all organisms from there.

Older cascades of dams usually do not have any facilities for fish migration, since at that time that was not yet a point of concern. However, for future developments of hydropower in this way, it is very important that this is being taken into account. One example of relevance is the cascade on the lower Mekong (see Sect. 22.2). The first HPP built on this stretch of the river, Xayaburi, made a very considerable effort for providing structures that are expected to allow fish migrations up- as well as downstream, and the first monitoring results seem promising. It is now important that all the other plants that will be built on this river apply the same standards, since otherwise the effort will have been in vain.

For the discussion of some aspects of specific cascades see Sects. 22.2 (Mekong) and 9.1.1.2 (Kafue River, Zambia).

4.1.3 Interbasin or Transbasin Projects

These two terms characterise projects which, by means of canals or other structures, transfer water from one river or one river basin into another. Such projects do not need to be related with hydropower, and indeed the main purpose of the probably largest such projects is usually water supply and/or irrigation, as e.g. the diversion of a considerable part of the water of the Amu Darya out of the Aral Seabasin (see Chap. 23), or major Interbasin projects in China (Shao et al. 2003; Zhuang 2016).

In the context of hydropower, two types of such water transfer schemes are possible. First, water can be taken by an intake structure in a river to then be diverted to a reservoir on a different river, in this way increasing the amount of water available for generating electricity. The second and usually more conspicuous possibility is to divert water from a reservoir to a different river; this is usually done with the aim of gaining additional head.

In both cases, the first immediate effect is a permanent reduction in flow in the river from where the water originates, i.e. downstream of the intake in the first case, and downstream of the dam in the second case. Where the water is diverted directly via power house, this has the additional effect of increasing the water flow in the receiving river, often very considerably. Such a case was e.g. Kukule Ganga (see 14.1.2), where the average flow of the receiving stream was 2 m^3/s, the input from the power house however can go up to 28 m^3/s. Two other HPPs of that type, Nam Theun 2 and Theun-Hinboun, in Laos, are mentioned shortly in Sect. 22.3. In the case of water diversion to the reservoir, such an effect will also arise, however, it will be masked by the storage capacity of the reservoir and the operation scheme of the power plant. Both effects, reduction in one case and flow increase, modified by plant operation, in the other, will have the impacts described elsewhere in this book.

One additional effect, especially in cases where the transfer is not just between tributaries of one major river, but between river basins which so far were really separated, could be the transfer of species from one river (the donor) to the recipient one, which in turn could lead to an impact on biodiversity e.g. one species, when given the possibility to extend its range to another river, would outcompete an ecologically similar, but distinct species in the receiving river.

4.2 Project Stages

Each HPP goes through a number of different stages, all of which have their characteristics in terms of effects on the environment. Here, these stages are described just very shortly; some of the issues indicated will be addressed later on, when individual components of the environment and project impacts on them are discussed.

4.2.1 Identification

In a way, at least from an environmental and social impact point of view, this can be considered as the most decisive step in the development process of any dam and hydropower project: the identification of a place as a potential site for the construction of a dam with all the appurtenant structures. Even though usually not much if any direct impact is caused during this stage, it is still the founding stone for all potential later impacts which may follow in the subsequent stages. And it does one thing, in a final, decisive and permanent way: it earmarks a site as a potential place for an HPP. Whatever the result of later studies: this information will still be there; once identified, such a site cannot be "dis-identified" (see Sect. 5.5 for additional comments in this point).

One basic problem: none of the approaches for impact assessment even addresses this stage, and none suggests a suitable approach, let alone a solution for problems potentially created at this stage. Salumu and Fuamab (2019) published a paper that allegedly aims at using a "multi-criteria decision-making approach to select potential hydropower sites", which however is not about site selection, but about comparative evaluation of already selected (and apparently rather well-investigated) sites, and in my view not a very relevant approach for the following reasons: the paper compares 20 sites; the comparison includes two criteria which contain environmental information, namely the "Safety factor" (= ecological safety, plant safety, etc.) and the "Environmental factor" (= Ecology, GHG emission reduction" etc.), without however providing further information on their content. For these two criteria, all of the projects included in the evaluation reach the same value, which is also the same for the "positive and negative ideal solution" to which the alternatives are compared. Given this, it is difficult to see in which way environment, however defined (and resettlement, which is not even mentioned) could possibly have an influence on the outcome.

4.2.2 Pre-feasibility and Feasibility

The name clearly indicates what the objective of this stage is: to see whether an identified site actually is suitable for the installation of an HPP. Gradually, more detailed investigations and analyses are being done, and with each step that produces a favourable result the project is one step closer to implementation. This is also the phase where usually the ESIA for the project is being prepared, or where this process starts.

Overall, direct environmental impacts are still small, if any, in this stage. The main risk, from an environmental point of view, however, is that depending on the type of work considered necessary in this phase, an access road may be built, and if yes this has potentially far ranging direct and—mainly—indirect consequences (see Chap. 15).

One risk here, as already mentioned, is the difficulty to accept "no" as an answer, which might give impression that an engineer who has been sent to such a place with the aim of producing a feasibility study, if then he comes back with the verdict "not feasible", may consider himself, and may be seen in this way by others, as not having done his job. If however he comes back with an answer like: "not easy, but possible", that is fine. In any case, this is the stage where "bad" projects should be clearly identified as such, and should be abandoned. There are obviously two issues with this: first, one would have to agree on what is good or bad; see to that Sect. 4.4 and Chap. 26. The second potential problem the fact that, as mentioned above, "dis-identification" of a project site is not possible: the project as such will still be around somewhere. This will be shown with some examples, as e.g. in Sect. 26.1.1.

4.2.3 Detailed Design

Once feasibility has been demonstrated, and a developer has decided to actually go ahead, detailed design will be done. This may, in addition to planning work, imply certain work on site, as e.g. more detailed geological investigations aiming at clarifying the situation for dam foundations, and the identification of sources for material needed for the construction, like e.g. suitable locations of quarries. In this stage, the final design of the dam and the power plant can still vary to a considerable extent. An ESIA which has been done in the feasibility phase will—or at least should be—updated for reflecting such changes, and for adapting mitigation measures where required.

4.2.4 Implementation: The Construction Period

This is, for obvious reasons, the period when impacts—environmental as well as social ones—happen. Some main points are:

- Access: if not yet done, access has to be prepared now, and an existing access may have to be upgraded.

- Space requirements for construction site installations including storage areas, workshops, main structures as crushing and batching plants, etc., but also offices and living quarters for the work force.
- Land that will be used temporarily, but will be changed in a significant way during this period: quarries, borrow areas, spoil dumping sites.
- People living on site; in the case of a large HPP, this can possibly mean several thousand persons living there for a number of years.
- Transport of vast amounts of material to the construction site; this can cause considerable changes in the amount of traffic along access routes.
- Construction of the HPP as such, with all the changes this brings about in the river and its surroundings.
- Preparatory work in the future reservoir area. There are mainly two such activities, extent and importance depending on the site-specific conditions, namely:
  - Relocation of the population living in the reservoir area; without going into any details here: the result should always be to have the resettlement implemented one year ahead of reservoir impoundment.
  - Pre-impoundment reservoir area clearing: this means mainly eliminating as much biomass as possible from the area to be submerged for reducing negative impacts on water quality and GHG emissions.
- Impoundment: filling of the reservoir, which means submersion of the land in the reservoir area as such, a process which can have considerable impacts on the downstream area due to the fact that water is being retained, mainly in the case of large reservoirs.

All these aspects will have been addressed, to the extent possible, in the ESIA, and measures will have been prepared which now need to be implemented. The main framework for addressing these issues is the ESMMP. It has to be pointed out, however, that this cannot be considered as a static document: if conditions change, or if some provisions turn out not to have been as accurate as necessary, the measures to be taken will have to be adapted accordingly.

4.2.5 Operation

Operation will then be, hopefully, for a long time; large dams are supposed to have a live span of 100 years or more. During this period, a new equilibrium will usually be established. Still, important changes can take place, especially in the first years after commissioning. This concerns mainly developments in the reservoir as such (water quality, fish stock), but consequences in the downstream area (e.g. floodplains, issues related to sedimentation and erosion) can be important as well. Here again, while the ESIA will have addressed these issues, monitoring and, wherever required, adaptation of measures to be taken, are of crucial importance.

4.2.6 Decommissioning

A rather recent requirement, but in the meantime integrated in many standard TOR for ESIAs for dam projects, it the question of decommissioning. Personally, I am not of the opinion that this topic is a relevant addition to general ESIA TOR, but that it is rather one of the points one has to do since the TOR and the guidelines say so, and this in spite of the fact that it does not really improve the basis for decision. This issue will be discussed in Chap. 13.

4.3 Standard Impacts of Dam Projects

For the assessment, the main project features potentially causing impacts will be taken into consideration. The main characteristic impacts are listed and described very shortly here, a more detailed discussion of each relevant topic will follow.

- Dam (or weir for smaller schemes): structure blocking the river, forming a barrier to migration (i.e. active movements of organisms, mainly fish; a rather small structure can be a complete barrier for upstream movements and at least a hindrance and a risk for downstream movements) and drift (passive downstream movements of organisms, such as fish eggs and larvae, benthic or planktonic organisms). The relevant parameters here are height and crest length of the dam or weir, while the type (e.g. arch, RCC or earth dam) is not of relevance for the magnitude of the impact. The organisms most obviously affected by such structures are long-range migrating fish (salmonids in the Northern hemisphere being the best know example). However, basically all aquatic organisms are affected by river fragmentation, since this can also hinder the less spectacular, but nonetheless essential short-range movements of many species, and it can cut populations of aquatic organisms into more or less genetically isolated sub-populations. To what extent rivers today are fragmented was shown by Belletti et al. (2020) for the case of Europe, where in 1.6 million river km they found 630,000 interrupting structures on record, and from sampling several 100 river km in the field they concluded that in total there must be about 1.2 million such structures; according to these authors, 68% are less than 2 m, and 91% less than 5 m high, but they can nevertheless block movement of organisms.
- Reservoir: the "artificial lake" forming behind a dam typically has a number of impacts (or risks), mainly the following ones:
  - Submersion of land, which can be any type of terrestrial habitat, from natural primary forest to inhabited agricultural landscapes. Note: this is often termed "inundation" or "flooding"; however, a clear distinction has to be made. Submersion is caused by a dam (which can also be of natural origin, when a river is blocked by a landslide) and is a permanent effect, reversible only if the dam is removed. Inundation or flooding, on the other hand, is caused by a

flood (natural or man-made), can be devastating especially if densely populated areas are affected, but is temporary, being reversed when the flood recedes.

- Change in aquatic habitat from lotic (river) to lentic (lake-like) conditions; this can cause an important change e.g. in the fish fauna inhabiting the affected area. It can also be a barrier to drift (see above), since e.g. fish eggs might not be transported any further in the almost stagnant water of the reservoir; they can sink to the bottom and may perish there.
- Effects on water quality, e.g. through the decomposition of submerged biomass, or triggered by the inflow of waste water. Such effects can be exacerbated by thermal stratification in the reservoir (see Chap. 17).
- Climatic effects: reservoirs as potential source of greenhouse gases (GHG), thus having an effect on climate change; influence on local or microclimate.
- Effects on downstream river discharge: while this effect is caused directly by the operation pattern of the power house, only the presence of a reservoir, e.g. of a capacity for water storage, can make noticeable changes in downstream flows possible (short-term by peaking, seasonal shift).

These effects of the reservoir generally increase with increasing reservoir size, the two main parameters being surface area and storage volume (and for the latter mainly live storage).

- Waterways: relevant for river diversion schemes and for interbasin projects. The length of the waterway as such may not be very relevant, however, the length of the affected stretch(es) of river(s) are important to determine (see e.g. the case of Shnogh HPP in Sect. 9.1.1.1, where there was a considerable difference between these two).
- Access road: this is not, strictly speaking, a part of the HPP as such, but is an important structure for providing access mainly during the construction phase. It has to be taken into consideration because it has the potential, e.g. especially for small projects in a remote and almost inaccessible area, to cause more direct and mainly indirect (e.g. deforestation) impacts than the HPP itself (see Chap. 15).
- Transmission line: here again, access, and the preparation and maintenance of the TL corridor, can have major impacts on natural ecosystems.
- Power production: this is the overall aim of an HPP and the socio-economically positive effect of such a project. Negative social impacts have to be seen not only in their absolute dimensions (such as e.g. number of displaced people), but also in relation to power production (i.e. in terms of displaced people per MW installed capacity or per GWh of annually produced electricity).

Data collection on HPPs in the pipeline will focus on these features of the individual projects, while other project details, even where available, will not be required at this stage. If for any project any of this essential information is not available (e.g. no indication of the intended operational pattern of the power house), plausible assumptions will have to be made for such cases, where "plausible" usually will mean "environmentally worst case".

I have always found it useful, as a first step in the screening and scoping process of an ESIA, to start with such a more or less standardised list of impacts in general to be expected from HPPs and to ask which of these impacts, for the specific project at hand, would certainly or probably become manifest, which of them are likely to be serious and which are likely to be of marginal importance or would most probably not occur, and where the main uncertainties in all this would be. While this is certainly only a first step, it is helpful for planning the preparation of the study; needless to say, all assumptions, including the "will certainly not occur", will have to be verified and corroborated. But it is definitely not helpful to spend a lot of effort and resources on a topic which could have been eliminated right from the start if only one had thought it through at the very beginning.

4.4 Good Dams and Bad Dams

4.4.1 What Is Good or Bad?

Dams have been and still are widely criticised, and there is an ample literature on this topic (e.g. Goldsmith and Hildyard 1984; McCully 1996; Martinez-Alier 2002: 125 ff.), Khagram (2004), Leslie (2005), Hughes (2009), Pearce (2018), Ioris (2020), this latter criticising some Brazilian HPPs, but mainly the Brazilian water resources development politics). Trainer (1990: 281) puts it that way: "The conventional approach favours big infrastructure development, the prime example being the construction of large dams. The environmental and social undesirability of dams has been well documented, notably the destructions of forest, displacement of native people, loss of silt, and destruction of food chains, such as fisheries. These high technology, expert-controlled, capital-intensive projects take up development resources that should have been devoted to enabling people to meet their own needs, and they thereby make these people more environmentally destructive as well as adding to foreign debt". Another example illustrating very well the image of dam projects, especially in the context with rural development and poverty reduction is the following paragraph from a publication entitled "Forests, trees and the eradication of poverty" (Miller et al. 2020: 104): "The development of other types of infrastructure, including the construction of hydroelectric dams and expansion of extractive industries such as mining, is often associated with pollution and rural livelihood losses due to conflicts over land and access to natural resources, although beneficial effects may include electrification and increases in incomes through the creation of low-skilled labour markets. Dams and mining activities are also often associated with increases in deforestation (Bebbington et al. 2018), and the down-grading, down-sizing and de-gazettement of protected areas (Golden Kroner et al. 2019)". Obviously, the negative impact of dam projects tend to stick, while positive effects are taken for granted without being mentioned specifically. However, it has to be stated clearly that it is rather easy, and certainly justified, to criticise dam projects: there is no dam project, however big or small, which has no negative environmental impact, and even if the benefits are high this cannot be accepted as an excuse for not mitigating negative impacts.

Unfortunately, it is easy to find bad examples of dam projects, although there are still two possibilities to consider: one is to look at the "really bad" examples, meaning dams which caused a very considerable impact without ever producing the results they were intended for, or whose negative impacts prevailed over the positive effects; such examples may not be very easy to find, and views on the relative importance of impacts—positive as well as negative ones—may differ depending on who is looking at them; concerning this point see the Aswan Dam case study in Sect. 21.1.1. The other, much easier way is to only focus on the negative impacts of dams without taking into consideration the positive ones, and this is definitely not really a challenging task, examples are plenty.

This is relevant in two ways. First, it is obviously unfortunate that dams indeed caused—and in some instances continue to cause—an important overall amount of not or not adequately mitigated negative impacts, either socio-economically, causing a real burden for the people affected by these projects and suffering from the consequences, and/or on the natural environment, here mainly meaning impacts on terrestrial and aquatic habitats of different types, and on biodiversity. Looking at such examples, one often cannot do anything but wonder "why some predicted difficulties have not in practice been avoided.", as Barrow (1988) put it. Secondly, it is unfortunate for the hydropower and dam industry, to name it this way, since it creates a high degree of resistance which then often also makes "good" projects more difficult, as described in the paper entitled "How to maintain the bad reputation of hydropower projects" (Zwahlen 2011).

In this context one thing is quite clear: opposition against dam projects, whether locally or by national or international occasions, is always founded on negative environmental and social impacts, like e.g. the destruction of natural habitats, the interruption of rivers or the loss of living space for a local human population, and it is never about technical reasons; the exception here may be concerns about dam safety, but then, this again is a more social than merely technical concern. In any case, I have never come across a case where a public movement was asking to have a double arch concrete dam rather than an RCC dam, for instance. Still, it is not an easy thing to define what a "good" and what a "bad" dam is. Ledec and Quintero (2003: 13) came to the conclusion that "a useful rule of thumb is that usually the most environmentally benign hydroelectric dam sites are on upper tributaries, while the most problematic ones are on the large main stems of rivers", based on the list of criteria shown in Table 4.2. It is easy to see that, although the points mentioned certainly can give some indications, this cannot be considered as a reliable list of criteria for making a good distinction between these two categories. As repeated several times already, it depends firstly on the specific project at hand, and secondly on the point of view of the person looking at it.

These are useful criteria, although they are too general. Taking the first one, for instance: a large reservoir is not necessarily always bad, and a small one in a narrow gorge is not necessarily good in any case. One needs still to look at the specific situation.

Table 4.2 Criteria for environmentally good and bad dams

	Bad	Good
a	Large reservoir surface area	Relatively small reservoir surface area (often in a narrow gorge with a high head and even a tunnel)
b	Much flooding of natural habitats and consequent loss of wildlife	Little loss of natural habitats and wildlife
c	A large river with much aquatic biodiversity damaged	A relatively small (often highland) river with little aquatic biodiversity at risk
d	A relatively shallow reservoir (sometimes with a fairly short useful life)	A deep reservoir which silts up very slowly
e	Few or no downriver tributaries	Many downriver tributaries
f	Water quality problems due to the decay of submerged forests	Little or no flooding of forests
g	Location in the lowland tropics or subtropics, conducive to the spread of vector-borne diseases	No tropical diseases (often due to high elevations or temperate latitudes)
h	Serious problems with floating aquatic weeds	No aquatic weed problems

Source Ledec and Quintero (2003:13)

4.4.2 Efficiency Ratios

Another approach for getting an idea on whether a project is "good" or "bad" is the use of "efficiency ratios", as e.g. the comparison of projects based on the ratios of displaced persons and area submerged per MW installed capacity; Goodland (1997: 87) shows a comparison of 50 projects based on these two ratios. In Fig. 4.1, the

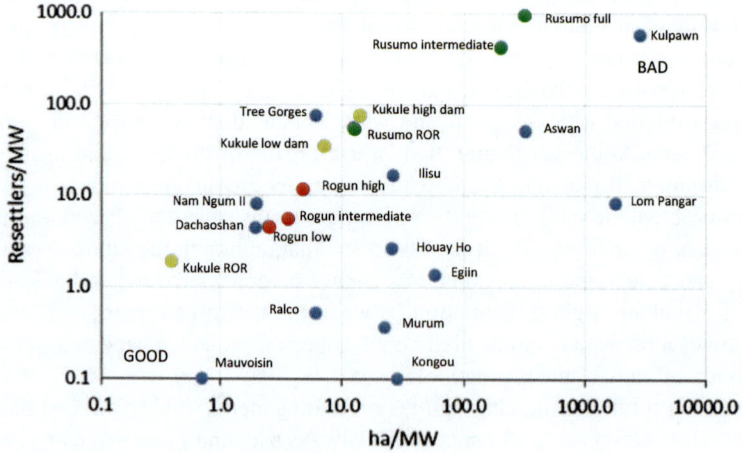

Fig. 4.1 Efficiency ratios for a number of HPPs

same is done for a number of the projects mentioned in this book. As can be seen from this graph, it provides an comparative ranking of projects ranging from"good" (both ratio values as close as possible to zero) to "bad" (both ratios showing high values).

The graph covers a good range from a "really good" (Mauvoisin) to a "really bad" (Kulpawn) project. However, once again, to make a judgement only based on this would definitely not be sufficient, one has to look at the specific situation of each project in order to come to a conclusion. This point can be explained by mentioning just a few of the projects included here. The case is quite clear for the two already mentioned: Mauvoisin, in the Swiss Alps, can certainly be considered as a good project: reservoir with small surface but a rather large volume thanks to the 250 m high dam, located in a high mountain area, where no resettlement was required, and which has a comparatively high installed capacity thanks to the high head of over 1000 m. On the other hand Kulpawn, located in a flat and densely populated area, which not only leads to a large reservoir, but in this case also a very large number of people that would have to be relocated (note that this is a project, not an existing scheme), and modest in terms of installed capacity; its value would be worse still if not the actually planned installed capacity would be considered, but only the net addition brought by the project to the grid, since its installation would come at the additional cost of reducing the productivity of an existing power plant.

This last observation already points out a potential problem with this way of illustration: in order to make any project included in the graph "look better", it would be sufficient to also include an even worse example. The graph can only compare projects, but it does definitely not give an absolute judgement, at least not unless one would try to add quality labels to the scales, e.g. by saying that for the case of ha/MW, < 10 = good, 10–100 = acceptable, 100–1000 = bad, and > 1000 = very bad, and by creating a similar scale for displaced persons per MW. I would certainly not recommend this, mainly since just by adding an "even worse" project—or by leaving the worst ones away—one could easily influence this scale. But there are some other problems. So e.g., Aswan seems to be quite bad in this evaluation. But here one has to consider that Aswan dam was built not only (or, one could argue, not mainly) for hydropower generation, but also or mainly for irrigation and flood control, two aspects that are not covered here. Likewise, Lom Pangar seems to fare rather badly in terms of area per MW, but here again, the main purpose of this dam is not power generation at the dam site, but rather providing storage for seasonal regulation of the main river in the downstream area, in this way increasing output of the entire cascade of power stations along this river (see Sect. 24.1). On the other hand, Kongou in Gabon would have to be considered as a good project, with quite a good area per MW ratio and no induced resettlement; still, the ESIA for this project came to the conclusion that it should not be implemented due to the fact that it is located in the central part of an important protected area (see 26.1.3).

As mentioned above, this type of evaluation it suitable, not for an absolute qualification of a project, but for a comparison of projects, and here very specifically for a comparison of alternatives for a project. Three cases are included in the graph, described to some extent elsewhere, for each of which three alternatives were in the discussion during the feasibility phase, namely Rogun (Tajikistan, red dots, see Chap. 23),Kukule Ganga (Sri Lanka, yellow dots, see 14.1.2) and Rusumo (Rwanda/Tanzania, green dots, see Sect. 14.1.2). In the case of Rogun, the comparative values lie rather closely together, and in this case, as explained in Sect. 3.11.2, it was still the "worst" solution that was chosen for implementation. However, in the two other examples the three alternatives show very considerable differences, and in both cases the ROR version, which rates better in terms of area as well as in terms of displaced persons, were chosen.

A different type of efficiency ratio is the power density, expressed as installed capacity per surface unit of the reservoir (MW/km^2 or W/m^2; the figure is the same, but since reservoir size is more likely to be expressed in km^2 rather than m^2, and capacity in MW rather than W, especially for large projects, the first unit seems more appropriate). In Table 4.3, the projects included in Fig. 4.1 are listed, with a number of other projects added, to show this relation.

According to EU TEG (2019), a power density of > 0.5 is acceptable to classify a project as "environmentally sustainable action". Table 4.3 shows that quite a number of the projects listed do not reach this threshold value. However, some observations are required as follows:

1. Three projects where three alternatives for each had been evaluated (see Fig. 4.1). Reducing reservoir size increased energy density, most notably for the case of Rusumo.
2. Kulpawn HPP: very large and very shallow reservoir with a very low-head power house; this project (not implemented) also ranks very badly in terms of resettlement.
3. Lom Pangar does not reach the energy density ration of 0.5; however, it has to be considered that the main role of this project is not direct energy generation, but river regulation in order to increase generation in the entire downstream cascade during the low-flow season. If the net gains achieved in this way would be included, the ratio would certainly improve.
4. Project not implemented. Main environmental issue: located in a relevant part of a protected area.
5. Project not implemented. No major environmental and social concerns.
6. Aswan High Dam: very large storage reservoir with rather low head; the reservoir has other functions in addition to power generation: storage of water for irrigation, flood protection (see Sect. 21.1.1).
7. Mauvoisin: small reservoir, very high dam (Sect. 25.1), very high head; such a layout will inevitably lead to a very large power density ratio. However, it could still be possible that due to the long deviation impacts on the downstream area (reduction of discharge in river, with potential negative impacts on biodiversity as well as on socio-economic aspects) such a project would have to be

Table 4.3 Power density

Name		Country	River	Type	Capacity MW	Reservoir km^2	MW/km^2	Obs.
Rogun	h	Tajikistan	Vakhsh	S	3600	170.00	21.18	1
	m			S	3200	114.00	28.07	
	l			S	2800	68.00	41.18	
Kukule	h	Sri Lanka	Kukule Ganga	S	150	21.00	7.14	
	m			S	100	7.14	14.01	
	l			R	70	0.27	259.26	
Rusumo	h	RWA/TZA	Akagera	S	90	280.00	0.32	
	m			S	85	171.20	0.50	
	l			R	80	9.77	8.19	
Kulpawn		Ghana	White Volga	R/I/S	36	1925.00	0.02	2
Lom Pangar		Cameroon	Lom/Pangar	S	30	540.00	0.06	3
Kongou		Gabon	Ivindo	S	150	41.97	3.57	4
3 Gorges		China	Jangtze	S	18,200	1100.00	16.55	
Ralco		Chile	Biobío	S	570	34.70	16.43	
Dachaoshan		China	Lanchang	R(S)	1350	20.00	67.50	
Houay Ho		Laos	Houay Ho	S	150	38.00	3.95	
Murum		Malaysia	Murum	S	1100	245.00	4.49	
Egiin		Mongolia	Egiin Gol	S	220	125.00	1.76	5
Aswan		Egypt	Nile	S	2100	6750.00	0.31	6
Mauvoisin		Switzerland	Drance	S	330	2.30	143.48	7
Ilisu		Turkey	Tigris	S	1200	313.00	3.83	
Uglichskaya		Russia	Volga	R/I/S	110	249.00	0.44	8
Rybinskaya		Russia	Volga	R/I/S	346	4550.00	0.08	
Nizhegorodskaya		Russia	Volga	R/I/S	520	1591.00	0.33	
Cheboksarskaya		Russia	Volga	R/I/S	1400	1080.00	1.30	
Kamskaya		Russia	Kama	R/I/S	501	1915.00	0.26	
Votkinskaya		Russia	Kama	R/I/S	1020	1120.00	0.91	
Zhigulievskaya		Russia	Volga	R/I/S	2300	6150.00	0.37	
Saratovskaya		Russia	Volga	R/I/S	1464	1831.00	0.80	
Volzhskaya		Russia	Volga	R/I/S	2420	3117.00	0.78	
Memve'ele		Cameroun	Memve'ele	S	201	19.00	10.58	
Bujagali		Uganda	Nile	R(S)	250	3.80	65.79	9
Amaila		Guyana	Amaila	S	165	23.30	7.08	10
NT1		Laos	Nam Kading	S	600	93.60	6.41	
Jannah		Lebanon	Nar Ibrahim	S	140	0.90	155.56	

(continued)

Table 4.3 (continued)

Name	Country	River	Type	Capacity MW	Reservoir km^2	MW/km^2	Obs.
KGL	Zambia	Kafue	S	750	1.85	405.41	11

Types R = run-of-river; S = storage; R(S) = ROR with small storage; R/l/S = ROR, low head, some storage; note that these are very broad categories, with considerable differences within each of them
Observations 1–11: see text

considered as negatively affecting the environment to a too great extent. This is not the case for Mauvoisin, but is a point which certainly has to be checked for projects of this type.

8. Dams and power plants of the Volga-Kama cascade. Only four of the nine plants reach the 0.5 threshold, none of them with a very wide margin. All of them are large to very large power plants, operated as ROR schemes with small margins for peaking. However, the cascade is very important for the power supply of Russia. The reservoirs are large to very large due to the flat area in the entire Volga-Kama basin, which in spite of its size does not offer any possibility for high head power plants. It can be assumed that the loss of land was the least concern at the time when they were built. However, no reliable data on project-induced resettlement are available.
9. Bujagali is a ROR scheme with a very small reservoir, most of which was river before construction of the dam; however, it gets its water directly from Owen Falls HPP, which is the "hydropower plant with the worldwide largest reservoir", Lake Victoria, with an enormous storage and regulation capacity.
10. Amayla: project not (yet) implemented. Main concern: 100 km of access road through primary tropical forest (see Sect. 15.2).
11. Kafue Gorge Lower HPP: very high ratio, which however is possible only due to the regulating capacity of the Itezhi-Tezhi reservoir upstream of it (see Sect. 9.1.1.2).

4.4.3 Conclusions

The evaluation by means of these ratios shows two things, namely, (i) the ratios are a good way for quickly comparing a number of projects in order to classify them in a range from better to worse (or good to bad) with respect to just this one criterion, and (ii) none of them is suitable for reaching a simple conclusion on what is a good or a bad project. There is no easy and formalised way to categorise projects as good or bad in absolute terms, since a large number of criteria have to be included in such an evaluation.

However there are three conditions which are quite easily verifiable and which, even if they do not necessarily in any case exclude a project, at least should raise a red flag at a very early stage, namely:

1. Large number of people affected.
2. Valuable natural habitats with high biodiversity affected, whether or not they have a status as legally protected areas.
3. Large, highly diverse aquatic systems affected; this includes the river which would be blocked by the dam, but also e.g. floodplains, deltas and related habitats in the downstream area which would be influenced by a change in flow regime.

Hopefully, in the following chapters it will be possible to show, at least to some extent, how the situation can be improved. One thing is very clear: if a dam or hydropower project at an advanced stage of planning meets with a serious—and well-founded—opposition, to the astonishment of planners and proponents, then in the very early planning stages somebody has either done a bad job, or turned a blind eye at certain conditions which were definitely there to see for all. As will be explained in more detail below, the fact that a dam project will cause a massive resettlement, threatens an especially valuable river or will affect a site of high importance for biodiversity conservation cannot possibly come as a surprise, if only one is willing to look. A synthesis of these concerns is attempted in the discussion of sustainability of hydropower (Sect. 21.2).

4.5 Why the Focus on Dams?

Dams are the main reason why fully one-fifth of the world's freshwater fish are now either endangered or extinct.

McCully (1996: 7)

Every dam, big or small, will cause an impact on the environment, and this is meant to say *every*! The following citations may serve to emphasise this point: "Beavers are known for their ability to build dams that change the environment" (Swinnen et al. 2019: 356). According to Likens (2010: 201), before beavers were reduced in numbers very substantially by hunting, "their ponds may have once contributed significantly to lake numbers on a global scale". Collen et al. (2000) cite the following impacts of beaver (*Castor canadensis*, the American, and *C. fiber*, the European beaver) dams: change from lotic to lentic conditions upstream of the dam, change in habitats by submerging land, change in aquatic fauna composition including fish, blocking fish migrations, retaining sediment, changes (positive as well as negative ones) in fishing, creating conflicts with the local human population. All these are typical impacts of man-made lakes created by dams as well. For those who are interested in knowing more on the evolutionary aspect of this specific activity, "the beaver's tale" in Dawkins and Wong (2016: 223 ff.) is a good read.

Man-made dams have all these effects, although at a much larger scale. Beavers were and are limited to damming small streams, doing this on rivers like the Nile, the Volga or the Mekong was left to our own species, and so was becoming aware of often negative consequences and trying to do something about it. Among the first distinctly negative consequences of dams that were noticed were probably the cutting off of fish migrations, mainly of salmons in North America and Europe. Dam projects undoubtedly were among the first major infrastructure projects to be scrutinised by means of EIA reports. One aspect that then led to increased opposition against dam projects, on the local as well as on the international level, was the increasing awareness that they caused sometimes massive displacement of local human populations, and that compensation, if any, was often far from satisfactory. This finally turned EIA (with the focus originally only or mainly on environmental concerns) into ESIA, where social impacts are an important, and very often the most important, topic to be dealt with. It was to a great extent the protests against such projects, and especially against dam projects with an involvement of the World Bank, which in the late 1990s, as a joint effort of WB and IUCN, led to the creation of the World Commission on Dams, which published its report in 2000 (WCD, 2000); see Schulz (2020) for a recent evaluation of the role of the WCD in the development of the dam business since then. As pointed out there, in spite of these efforts opposition against dam projects continue. In an attempt to find explanations on why and under what conditions a dam project is likely to meet resistance, Kirchherr et al. (2016) identified inadequate compensation (termed there "lack of social safeguards", which might be somewhat misleading, since this term is usually understood as the guidelines and standards to be applied for reaching a satisfactory outcome in such situations) as one main reason for opposition against a dam project. Boyé and de Vivo (2016) give a short overview of the main reasons for resistance against dam projects, and they summarise the situation very well by saying that dams "globally, … have many advantages, but locally—especially for local populations—the advantages are outweighed by drawbacks". Their main conclusion is just that for the development of dam projects today, environmental and even more so social aspects have to be taken into consideration, and that any directly affected population has to be adequately involved in the planning and implementation process. Still, this article, which does not fail to point out all positive aspects of dam and hydropower projects, e.g. for facing climate change, gives the impression that with careful planning and good public participation all obstacles can be overcome and projects can be developed successfully. What is missing is the fact that there are some projects that, even if technically and economically feasible, cannot be developed in an environmentally and socially acceptable way and should therefore, if even discussed, definitely be abandoned in a very early planning stage; about this issue see also Sect. 4.4 and Chap. 26.

References

Arratia, F.G. 1981. Géneros de peces de aguas continentales de Chile. Publicación Ocasional No. 34. Museo de Historia Natural, Santiago de Chile, 108 pp.

Barrow, C. 1988. The impact of hydroelectric development on the Amazonian environment: With particular reference on the Tucuruí project. *Journal of Biogeography* 15: 67–78.

Bebbington, A., J. Humphreys Bebbington, D. Sauls, et al. 2018. Resource extraction and infrastructure threaten forest cover and community rights. *PNAS* 115 (52): 13164–13173. https://doi.org/10.1073/pnas.1812505115.

Belletti, B., C. Garcia de Leaniz, J. Jones, et al. 2020. More than one million barriers fragment Eurpe's rivers. *Nature* 588 (7838): 436–441. https://doi.org/10.1038/s41686-020-3005-2.

Bogan, M.T., E.T. Chester, T. Datry, et al. 2017. Resistance, resilience and community recovery in intermittent rivers and ephemeral streams. Ch. 4.6. In *Intermittent rivers and ephemeral streams: Ecology and management*, eds. T. Datry, N. Bonada, and A.J. Boulton, 349–376. London: Academic Press. https://doi.org/10.1016/B978-0-12-803835-2.00013-9.

Boyé, H., and M. de Vivo. 2016. The environmental and social acceptability of dams. Field Actions Science Reports, Special Issue 14. http://journals.openedition.org/factsreports/4055

Collen, P., and R.J. Gibson. 2000. The general ecology of beavers (Castor spp.), as related to their influence on stream ecosystems and riparian habitats, and the subsequent effects on fish—A review. *Reviews in Fish Biology and Fisheries* 10(4): 439–461.

Datry, T., N. Bonada, and A. Boulton. 2017a. *Intermittent rivers and ephemeral streams. Ecology and management*, 597 pp. London: Academic Press, Elsevier.

Datry, T., R. Corti, J. Heino, B. Hugueny, R.J. Rolls, and A. Ruhí. 2017b. Habitat fragmentation and metapopulation, metacommunity and metaecosystem dynamics in intermittent rivers and ephemeral streams. Ch. 4.10. In *Intermittent Rivers and ephemeral streams: Ecology and management*, eds. T. Datry, N. Bonada, and A.J. Boulton, 377–404. London: Academic Press. https://doi.org/10.1016/B978-0-12-803835-2.00014-0.

Dawkins, R., and Y. Wong. 2016. *The ancestor's tale. A pilgrimage to the dawn of life*, 771 pp. Weidenfeld and Nicolson, London, updated edition

Díaz, G., P. Arriagada, K. Górski, O. Link, B. Karelovic, J. Gonzalez, and E. Habit. 2019. Fragmentation of Chilean Andean Rivers: expected effects of hydropower development. *Revista Chilena de Historia Natural* 92(1). https://doi.org/10.1186/s40693-019-0081-5.

EU TEG. 2019. Taxonomy Technical Report. Financing a sustainable European economy. https://ec.europa.eu/info/sites/default/files/business_economy_euro/banking_and_finance/documents/190618-sustainable-finance-teg-report-taxonomy_en.pdf.

EWE. 1997. Central Pangue: Estudio Condiciones Aguas Abajo. Cumulative Impact Assessment for Ralco and Pangue. Panque SA, Chile

EWI. 1998. Nam Ngum 2 Hydroelectric Project, Lao P.D.R.: Environmental impact assessment (EIA). The Shlapak Group.

Golden Kroner, R.E., S. Qin, C.Y. Cook, et al. 2019. The uncertain future of protected lands and waters. *Science* 364 (6443): 881–886. https://doi.org/10.1126/science.aau5525.

Goldsmith, E., and N. Hildyard. 1984. The social and environmental effects of large dams. Vol. I: Overview. 346 pp. Vol. II (1986): Case studies. Vol. III (1992): A review of the literature, 243 pp. Report to The European Ecological Action Group (ECOROPA).

Goodland, R. 1997. Environmental sustainability in the hydro industry: disaggregating the debate. In *Large dams: Learning from the past, looking at the future*, eds. Dorcey, et al., 69–101. IUCN/The World Bank, Gland and Washington.

Hughes, J.D. 2009. *An environmental history of the world. Humankind's changing role in the community of life*, 2nd ed., 306 pp. New York: Routledge.

Ioris, A.A.R. 2020. Socioecological economics of water development in the Brazilian Amazon: Elements for a critical reflection. *Ecological Economics* 173. https://doi.org/10.1002/jwmg.21601

Khagram, S. 2004. *Dams and development. Transnational struggles for water and power*, 270 pp. Ithaca: Cornell University Press.

Kirchherr, J., K.J. Charles, and M.J. Walton. 2016. Multi-causal pathways of public opposition to dam project in Asia: A fuzzy set qualitative comparative analysis (fsQCA). *Global Environmental Change* 41: 33–45. http://www.sciencedirect.com/science/article/pii/S0959378016301273.

Lagendijk, V. 2016. Europe's Rhine power: Connections, borders, and flows. *Water History* 8: 23–39. https://doi.org/10.1007/s12685-015-0140-z.

Ledec, G., and J.D. Quintero. 2003. Good dams and bad dams: environmental criteria for site selection of hydroelectric projects. Latin America and Caribbean Region Sustainable Development Working Paper 16, 20 pp.

Leslie, J. 2005. *Deep Water. The epic struggle over dams, displaced people, and the environment*, 352 pp. New York: Picador.

Likens, G.E. (ed.). 2010. *Lake ecosystem ecology. Encyclopedia of inland waters*, 463 pp. Elsevier, Oxford.

Martinez-Alier, J. 2002. *The environmentalism of the poor. A study of ecological conflicts and valuation*, 312 pp. Edward Elgar, Cheltenham, UK.

McCully, P. 1996. *Silenced rivers. The ecology and politics of large dams*, 350 pp. London: Zed Books.

Miller, D.C., S. Mansourian, and C. Wildburger (eds.). 2020. *Forests, trees and the eradication of poverty: Potential and limitations. A global assessment report. IUFRO world series*, vol. 39, 240. Vienna. https://www.iufro.org/fileadmin/material/publications/iufro-series/ws39/ws39.pdf.

Oud, E., and T.C. Muir. 1997. Engineering and economic aspects of planning, design, construction and operation of large dam projects. In *Large dams: Learning from the past, looking at the future*, eds. Dorcey, et al., 17–39. IUCN/The World Bank, Gland and Washington.

Pearce, F. 2018. *When the rivers run dry. The global water crisis and how to solve it*, 312 pp. London: Granta Publications. Updated edition

Salumu, Z.A., and M. Fuamba. 2019. A multi-criteria decision-making approach to select potential hydropower sites in the Democratic Republic of Congo. In *Paper 10.02 presented at the Africa 2019 hydropower conference*, Windhoek, Namib-ia, April 2019.

Schulz, C. 2020. Revisiting the debate on large dams. https://www.waterpowermagazine.com/features/featurerevisiting-the-debate-on-large-dams-7856450/.

Seitz, P. 2018. Gemeinschaftskraftwerk Inn. *TEC21* 42(2018): 26–35.

Shao, X., H. Wang, and Z. Wang. 2003. Interbasin transfer projects and their implications: A China case study. *International Journal of River Basin Management* 1 (1): 5–14.

Swinnen, K.R.R., A. Rutten, and H. Leirs. 2019. Environmental factors influencing beaver dam locations. *Journal of Widllife Management* 83 (2): 356–363. https://doi.org/10.1002/jwmg.21601

Trainer, F.E. 1990. Environmental significance of development theory. *Ecological Economics* 2 (4): 277–286.

WCD. 2000. Dams and development. A new framework for decision-making. The Report of the World Commission on Dams. Earthscan, London, 404 pp.

Zhuang, W. 2016. Eco-environmental impact of inter-basin water transfer projects: A review. *Environmental Science and Pollution Research* 23: 12867–12879. https://doi.org/10.1007/s11356-016-6854-3.

Zwahlen, R. 2011. How to maintain the bad reputation of dam projects. In *Dams and reservoirs under changing challenges*, eds. A. Schleiss, and R. Boes, 761–768. Proceedings of the 79th annual meeting of ICOLD, Lucerne, Switzerland.

Chapter 5
Impacts

> *... if there is no one in the forest observing a falling tree, then the impact on the ground will make no sound, if we define "sound" as the vibration of air stimulating the hearing apparatus of sentient beings.*
> Shermer (2018: 73)

Everybody knows what "impact" means; the meaning is so obvious that very few if any of the numerous handbooks, methodologies or guidelines on how to prepare an environmental and/or social impact assessment gives an explanation of what an impact is (Leopold et al. 1971; De Santo 1978; Duek 1982; WB 1991; Therivel et al. 1992; Leitmann 1994; Sadar 1996; UNEP 1996; Becker 1997; Lohani et al. 1997; Baker 2000; Fischer 2002; Dalal-Clayton and Sadler 2005; Fischer 2007). The exceptions to this rule are the IFC Performance Standards as well as the new World Bank environmental and social guidelines which provide the following definition: "Environmental and social impacts refer to any change, potential or actual, to: (i) the physical, natural, or cultural environment, and (ii) impacts on surrounding community and workers, resulting from the business (IFC)/project (WB) activity to be supported" (IFC 2012: 1; WB 2017: 3); the wording is almost the same in the two guidelines, WB just replaced the word "business" used by IFC with "project". Still, it may be worthwhile to give this term, which is used so often, a few thoughts. The chapter starts with a discussion of the concept "impact", in general as well as specifically in the context of an ESIA. It then continues by discussing a number of issues associated with identification of impacts and with determining their relevance. In the last part of the chapter, some attempts to evaluate impacts in monetary terms are presented and discussed shortly, pointing out risks and problems for doing so.

5.1 What Is an Impact?

The word "impact" basically has two different meanings: the first is the impact of an object when it hits another object at great speed. One such impact was definitely of major importance for us: the impact of a large meteorite which hit earth about

R. Zwahlen, *Assessing the Environmental Impacts of Hydropower Projects*,
Environmental Earth Sciences, https://doi.org/10.1007/978-3-030-91185-0_5

65 million years ago, and which led to the extinction of the dinosaurs (at least of the large ones, the birds being the surviving part of this group that dominated earth for a few 100 of millions of years), which made way for the development of the mammals, including our own species. The other meaning is, according to the Cambridge Dictionary: "a powerful effect that something, especially something new, has on a situation or person". Now, "something" can basically be anything, and as a matter of fact nothing can happen or be done without having an impact on something. A term with similar content might be "progress" in an evolutionary context (see the discussion of this in Dawkins and Wong 2016: 682 ff.); however, "progress" is usually associated with a development or change perceived as positive and desirable, while "impact", especially in the context of an environmental and social impact assessment, is most often associated with a negative, undesirable effect.

Homo sapiens is certainly the species with the most marked impacts on its environment, but definitely not the only one. Just a few examples may suffice for showing that other species have impacts—be they considered as negative or positive—on their environment as well.

One of the first, and certainly one of the most far-reaching, and arguably most decisive impact life had on earth was the production of oxygen and its release into the atmosphere; this, as Dawkins and Wong (2016: 648) put it "polluting waste product of green bacteria", which must have been, in the beginning, very toxic to life, and to which life only adapted later on and in a way that made many life forms (all of what we usually consider as "higher organisms") dependent on it. Another major impact of life on earth was the massive accumulation of mostly tiny skeletons of different types of animals which formed thick layers of limestones, some of which are now forming entire mountain ranges.

But also on a smaller scale animals have impacts on their habitats, and in this way change living conditions for a large number of other organisms. Lacher et al. (2019) actually speak of "environmental engineers", animals which actively change their environment, like e.g. elephants who destroy trees and in this way create savannas (or prevent savannas from being reverted to forests), to the benefit of other savanna dwelling species, or burrowing animals, some of which create large underground systems which are also inhabited by numerous other species. The dam-building activities of beavers have been mentioned already (Sect. 4.5), but other animals, although at a smaller scale, do this as well, as e.g. the West African Goliath frog [*Conraua goliath*), the largest living frog species, which builds small dams to create ponds to protect its offspring from rapidly flowing water and predatory fish and for this purpose moves stones with a weight of up to 2 kg (Schaefer et al. 2019).] One group of animals which are very decisive as environmental engineers are the termites. In the southern African Miombo forest, termites create enormous mounds, several meters high, covering a considerable percentage of the entire surface, and providing space for a specialised vegetation and presumably associated fauna (Cole 1986: 139). But more important than this is the ability of termites (or rather of symbiotic microorganism living in the guts of termites) to digest wood. While this is a very important function for maintaining tropical ecosystems, it is certainly not (or not totally) to the advantage of us, since wooden structures are constantly being destroyed by this otherwise beneficial activity. Sánchez-Montoya et al. (2017: 299 ff.) point out the importance of what they call "landscape engineers", specifically for the special case of intermittent and ephemeral streams.

Like for animals, every human activity necessarily has an impact on the environment. When humankind left the Garden of Eden—and my guess is that this was at the moment we started to invent and use tools which go beyond those chimpanzees are using—we stopped being just another species in an African ecosystem and started to put an ever increasing mark on our environment. One early and decisive moment was certainly the invention of bow and arrow, which allowed us to kill from a distance and which therefore turned us into much more efficient hunters than we would have been based only on our physical strength and our teeth and claws. Fire and clothing were other important steps, enabling us to live in colder climates. When *Homo sapiens* started to migrate out of Africa, our species was already well equipped for leaving a lasting mark wherever it went, e.g. by exterminating much of the megafauna in newly colonised continents. This was most marked, or is best known, for continents which were colonised rather late, i.e. the Americas and Australia, where much of the megafauna went extinct after the arrival of man (Crosby 1986: 217 ff., 273 ff.; Diamond 1997: 35 ff.; Ehrlich 2002: 170 ff.; Lyons et al. 2019); a recent study of the development of ruminants (Chen 2019) showed a marked demographic bottleneck for many of those species, "massive population decline events that occurred in most ruminant species, starting from ~ 100,000 to 50,000 years ago, which was temporarily and spatially concurrent with the increased human activities on different continents during this period". Further development steps like domestication of the dog, domestication of other animals, and the domestication of plants, i.e. the development of agriculture, then increased our influence on the environment, not least because each of these steps allowed for a further increase of the human population. It is probably not just a coincidence that the Neanderthal disappeared about 39,000 years ago, shortly after "modern humans", i.e. our direct ancestors, had appeared in the Middle East and Europe (Tattersall 1999: 198 ff.); the role which domesticated dogs might have played in this process is described by Shipman (2015). A recent study analysing archaeological evidence from all over the world shows that probably as early as 3000 years ago earth had already been changed to a great extent due to land use, i.e. activities of hunter-gatherers, farmers and pastoralists (ArchaeoGLOBE Project 2019). With the start of industrialisation, the impact of mankind on the environment of the entire earth has taken on a dimension that led to declaring it as the start of a new area, the end of the Holocene and the beginning of the Anthropocene (Bonneuil and Fressoz 2013: 17).

Intensity is also relevant, not only the type of impact as such. A good example is slash-and burn cultivation; this was, during millennia, the most sustainable way of using forests of different types, but mainly tropical forests with their soils which quickly wash out and lose fertility under cultivation. However, it was sustainable only as long as the human population was small, punching just tiny holes into the forest canopy, and giving these places then sufficient time to recover and to restore fertility. Once population pressure got too high, very large parts of forest were used and fallow periods were too short for complete recovery, this practice turned into a threat

(Tucker 1988; Harris 1995; Zwahlen 1996; see also Sect. 21.2.1). Presently, even the collection of non-timber forest products (NTFP), usually seen by conservationists as a sustainable, non-destructive way of utilising tropical forests, has increasingly negative impacts on biodiversity, due to population growth and shrinking of forest areas, which leads to an increasing pressure on these products (Corlett 2019: 190). In this context, see also the comments on the effects of fisheries (Sect. 22.4).

5.2 Environmental Impacts

It gets much clearer, and necessarily more restricted, when we ask for "environmental impacts" which can be defined as "the effects that the activities of people have on the environment". A useful way of illustrating what "project impact" could mean is shown in Fig. 5.1.

This theoretical example could be, e.g. NOx concentrations related to a highway project. The long flat curve would show concentrations without the project. The important points here are that there was already a contamination before the project started (baseline), which would increase slowly, and the curve may or may not flatten over time. At the start of the construction, local concentrations would increase due to construction activities, the amount 1 representing the impact due to construction. A further increase would then be at the start of operation, due to traffic, amount 2 being the impact of the project during operation. This would then be a permanent impact, which however does not yet say anything about its importance (concerning this point, see Sect. 5.4).

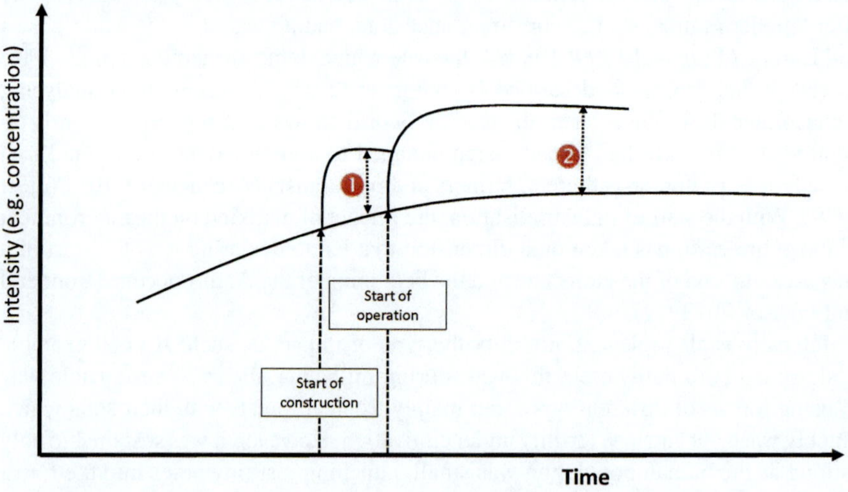

Fig. 5.1 Graphical representation of a project impact. Redrawn after BUWAL (1990(31)); further explications in the text

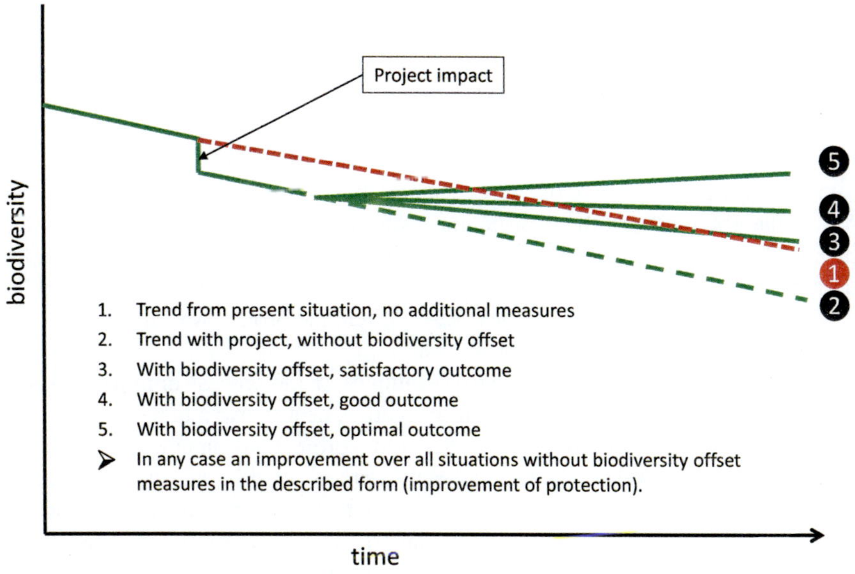

Fig. 5.2 Impact on biodiversity and effect of mitigation. Additional explanation in the text

One more way to illustrate not only an impact, but also the expected effects of mitigation measures is shown in Fig. 5.2, which was developed for illustrating project impact and mitigation outcome for a biodiversity offset program related to a hydropower project in Laos.

In this project, the reservoir is located in a protected area which is considered as important for biodiversity conservation on a national level. Reservoir impoundment will submerge a part of this area, permanently destroying the terrestrial habitats presently in place there, mainly secondary forest, some of which however may still qualify as primary forest. Under the prevailing situation without the project, the area was legally protected but received little if any actual protection, and was under high pressure from encroaching agriculture, NTFP collection, and poaching, some for food, some potentially for the trade with protected species. This led to an increasing loss of biodiversity, indicated by the solid line in the left part of the graph; without any change in the situation, this development was anticipated to go on, as indicated with the dashed line (1); it is important to acknowledge, as already indicated in Fig. 5.1, that the prevailing situation or baseline is by no means a steady state situation, but one which is changing independently of the project. The project would lead to an additional loss, caused by the submersion of habitats mentioned (project impact). Without any mitigation measures the trend would then continue as before, although on a lower level, indicated by dashed line (2). The plan was therefore to provide actual protection of the remaining part of the protected area, which was still by far the largest part, and which, as previously verified, contained all the core areas, i.e. the most valuable zones in terms of biodiversity. Depending mainly on the effort put in, different outcomes are possible, illustrated in the graph. A first possibility would be that the loss in biodiversity would be slowed down (3); a second possible outcome would be that further loss of biodiversity could be stopped, and the situation maintained at a stable level (4). In an ideal case, biodiversity might even increase over time, by recovery and internal spread of species that had become very rare, or even by the re-establishment of some species which had disappeared from this area (5). As the graph shows, the expectation is that even the first version would, over time, lead to a situation which is better than what would have been the result without measures or, in that case, without the project; this is why even this result could be considered as satisfactory. As this text was written, the project was under way, and it remains to be seen what the outcome will be.

Usually, as demonstrated in these two examples, in the process of impact assessment, "impact" is mostly understood as a negative, destructive and therefore undesirable effect a project or any activity related to a project has on any component of the environment. However, "impact" as such is not necessarily negative, it actually does not have a positive or negative connotation, this will only become clear when a specific impact is discussed. This is an important point to consider, since ESIAs deal mainly with "negative", i.e. undesirable or damaging effects of an activity or a project, and in many reports this word is only used in this way. This way of looking at things is understandable if one considers only environmental impacts. Here, we tend to look at the natural, undisturbed situation as good (and I basically tend to agree with that), and any change brought to it by human activities as disturbance, i.e. negative and undesirable. Looking at it this way, it is difficult to conceive how any impact could be positive. However, this way of looking at things would only be applicable in in the case of real "greenfield projects", i.e. projects located in a landscape completely natural, untouched by human activities. Such projects, while they still exist, get increasingly rare. As soon as the social component is included in an impact assessment, the aspect of positive effects gets more important.

Therefore, just saying that something has an impact is, as such, not yet a qualification. On the contrary, any project, overall, should have positive impacts which outweigh the negative ones. In hydropower, the one main positive impact is the production of electricity, from which people, and the local and national economy, will benefit. It can have additional positive impacts, like providing or improving flood protection, making water available for other purposes, like e.g. irrigation or drinking water supply, or, hopefully increasingly, on mitigation of climate change effects. Each dam project has negative impacts, like e.g. the interruption of a river, or the submersion of terrestrial ecosystems and inhabited areas. In some cases, an impact can have positive (desirable) as well as negative (undesirable) consequences, and depending on the point of view the definition of what is desirable or not may be different. An access road can lead to indirect negative effects on surrounding natural habitats (undesirable), but can improve e.g. market access for a local population (desirable), and the draining of a swamp can destroy a rich natural habitat, which would be considered as negative, but can turn swamp land into fertile agricultural land and contribute to the elimination of water related diseases, both of which would be looked at as positive effects. A comprehensive ESIA has to pay specific attention to the negative impacts and the possibilities to avoid, minimise, mitigate or compensate them, but positive effects have to be addressed just as well.

5.3 Definition of "Impact" in ESIAs

Impact assessment has been defined as "…the process of identifying the future consequences of a current or proposed action" (Becker 1997: 2). When we set out to do an ESIA for a dam project—or for any other project—it is obviously not enough to say that, as was described above, "every activity has an impact". There needs to be a definition, or at least a more or less common agreement, on what has to be considered as an impact, and what has not.

Such an agreement is not easy to reach, because finally in every case it is a matter of definition. The most simple cases are probably the ones for which there is a clear legal basis for a parameter which can be measured with precision. Contaminating water in a way that the pH goes above 9 or below 5 is a legally determined impact, i.e. is usually defined in water quality standards (which can vary to some degree from one country to another, whence the need to check the applicable standards on a case to case basis), and if this happens measures need to be taken. Between these two threshold values, variation is considered as "normal", no measures required. This, quite obviously, is a mere definition, although in most cases, like the mentioned pH, but also for other water quality parameters like oxygen content or heavy metals, as well as for noise, based on observations which had the aim of defining a level below which no damage on target organisms can be observed. Under such circumstances it is clear: if the expected outcome or value is within the legally defined threshold values, it is not an impact, no measures are required, and if they go beyond this threshold measures must be taken which are suitable to make the outcome compliant with legal standards. It should be noted, however, that it is permitted and desirable to minimize an impact even if it is within the legal boundaries. So e.g., if a machine is emitting a noise level of 85 dBA in an environment where 90 is permitted, it is still allowed, and certainly a positive point, if the noise level can be reduced.

To reach a clear definition is usually much more difficult in other, non-physical parameters, especially in biological ones. How much of what type of habitat can be lost for the impact still to be considered as minor, no mitigation required? There is usually no clear legal definition, no threshold for such questions, which could be simply applied. It always depends on the situation, and it is only logical that under such circumstances different people, with a different background and often a different agenda, come to completely different conclusions. It is not possible to give such a definition here, but many examples under various headings will deal with this problem, or will at least indicate how the question was addressed in specific cases.

5.4 Assessment of Impacts

Once impacts as such have been determined, it will usually be required to determine their importance, relevance or significance. As simple as this may sound, there are some pitfalls in the process. These, and a way of how to deal with this issue, is briefly discussed here.

One approach often used to assess environmental impacts of a project consists of the determination of their magnitude according to three characteristics, grouping the impacts in three categories for each of them, as follows:

1. Intensity
   a. Low: effect on environmental parameter in question is weak or barely perceptible, not changing the parameter in a fundamental way.
   b. Medium: the parameter will undergo strong, clearly perceptible changes.
   c. High: the parameter will suffer very strong, potentially deleterious changes due to the project.
2. Extent, geographical range
   a. Local: spatial effect will be very restricted, to the project site as such or to any other, limited space within the study area.
   b. Regional: the effect will be felt over a larger area
   c. Global: the effect will be felt over a very large area, far beyond the project site, and in an extreme case on a global level.
3. Duration
   a. Short: the effect will be manifest only for a very short time.
   b. Medium: the effect will be manifest for a considerable time span.
   c. Long: the effect will be manifest for a very long time, the induced change will potentially be permanent.

This list already shows one basic problem: the qualifications used (local, regional, short, long) are not clearly defined and may have to be defined on a case to case basis. Given the wider range of different parameters, impacts and expected effects, this is often not an easy thing to do.

One additional important aspect of impact identification is its reversibility. This, however, does not have to be expressed as a separate impact characteristic, since it is implicitly included in the terms of intensity and duration. Reversibility is related mainly to two independent variables, namely, duration of the impact and resilience of the affected environmental parameter. A few examples may highlight this:

- Vegetation destroyed by submersion at reservoir impoundment: this will be of long duration, at least for the total life span of the project, which is usually supposed to be around 100 years in the case of a large dam (or potentially longer, if later on replaced by a new, similar structure). For all practical purposes, this is equivalent to non-reversible.

- In the case of an endemic species which could be found only in one location affected by a project, and which would become extinct due to this impact, this would be an example of a permanent, irreversible impact. While this might be a rare event, it is by no means theoretical; an example could be the spray zone toad (*Nectophrynoides asperginis*; see Sect. 18.3).
- Vegetation on land used temporarily during construction: this is a reversible impact, given the usually high resilience of the vegetation. The duration required for recovery would depend on a number of variables, most of all on the local climatic conditions (vegetation regrowth will usually be more difficult and therefore slower under arid than under humid conditions), the way of site rehabilitation (e.g. availability and use of topsoil) and the type of vegetation to be replaced (usually quite fast in the case of secondary vegetation, but potentially very long in the case of e.g. primary forest). The system in question, the vegetation, is resilient, but specific components of it, namely mature trees present in a primary forest, take a long time for becoming re-established.
- Noise due to a project installation, e.g. a breaker: this can be a very high impact (noise levels far beyond acceptable thresholds), but it is highly reversible, only dependent on duration of the impact: it will stop every time the machine is stopped, and it will cease to exist as soon as the structure has been removed.

This brings us to the question of determining the significance of an impact, which is a combination of the three impact characteristics mentioned above. One way to do that would be a "quasi-mathematical" approach, by defining points to each level of intensity of the three characteristics listed above, then determining the intensity of the impact for each parameter, adding the points obtained and defining impact significance accordingly. This could result in an evaluation matrix as illustrated in Table 5.1.

As can be seen from this table, this approach would result in 27 possible cases, yielding a range of ranks ("sum" in the table) between 3 (for low impact in all three categories) to 9 (for high impact in all three categories). This could then be translated into three degrees of relevance: low (1, green), medium (2, yellow) and high (3, red). One first difficulty here is the decision on how to define these categories: the sum values show a range of 7 steps (3–9), which makes it impossible to distribute it evenly on 3 categories. The way shown here, which is a mere definition without any "scientific" justification, assumes that the two lowest values (3 and 4) qualify as low significance, the two highest ones (8 and 9) as high, and the remaining four (5–7) as medium significance. Obviously, it would also be possible to define the limit between the categories in a different way.

The second, and much more serious problem is given by the fact that impacts, while it is more or less well possible to characterise them with the three categories with three intensities each, are still very different in nature. Attributing a significance in the way indicated in the table above might be seen as an attempt of making them easily comparable with each other; this, however, is not really the case. This is especially important since the significance of an impact is an indication on whether a mitigation measure would be required or not. In fact, one could argue that for

Table 5.1 Determination of impact relevance

Cases	Intensity			Extent			Duration			Sum	Relevance		
	Low	Med.	High	Loc.	Reg.	Glob.	Short	Med.	Long		Low	Med.	High
	1	2	3	1	2	3	1	2	3		1	2	3
1	1			1			1			3	1		
2	1			1				2		4	1		
3	1				2		1			4	1		
4		2		1			1			4	1		
5	1			1					3	5		2	
6	1				2			2		5		2	
7	1					3	1			5		2	
8		2		1				2		5		2	
9		2			2		1			5		2	
10			3	1			1			5		2	
11	1				2				3	6		2	
12	1					3		2		6		2	
13		2		1					3	6		2	
14		2			2			2		6		2	
15		2		1					3	6		2	
16		2				3	1			6		2	
17			3	1				2		6		2	
18			3		2		1			6		2	
19	1					3			3	7		2	
20		2				3		2		7		2	
21			3	1					3	7		2	
22			3		2			2		7		2	
23			3			3	1			7		2	
24		2				3			3	8			3
25			3		2				3	8			3
26			3			3		2		8			3
27			3			3			3	9			3

impacts of significance 1 no mitigation would be required, for significance 2 some measures, while for significance 3 very serious measures would have to be taken, or this should even be considered as a no-go for the project. This, however, would be a too simplistic approach, as the following examples may show:

- Case 1: the local population of a common and very widespread species would be exterminated by the project. This would correspond to the case 5 in the table: low intensity (very little of the total population of the species affected), low extent (very localised effect), but long duration (permanent impact), resulting in a sum of 5, which would translate into a medium significance, i.e. requiring at least some mitigation. Easy to see, however, that if the species affected would be the house sparrow (or even the common rat), such a mitigation would definitely not be required.

- Case 2: noise of the breaker. High intensity (noise levels far above legally defined threshold), but low extent (very localised impact) of short duration (limited to the working time of the machine); this corresponds to case 10, and also results in a significance of 2. However, in this case mitigation is absolutely required: workers who are exposed to the noise must be equipped with hearing protection, and wearing it must be assured.
- Case 3: an endemic, very localised species affected. This would correspond to case 21: high intensity (entire species), low extent (very localised), long duration (permanent effect), resulting in a sum of 7, significance 2. However, in such a case very far reaching mitigation measures might be required.

These examples show the difficulty of an entirely formalised approach to impact evaluation. Being aware of this situation, it will in most cases be preferable to define significance of the impact with an argumentative approach (shortly discussing, for every specific case, the reasons why to consider a specific impact as relevant enough for requiring mitigation measures) rather that to try and follow an entirely formalised procedure.

One last point to mention, before actually going into the impact assessment as such, would be the method chosen, not for scaling, but actually for identifying impacts. The important item to point out here is the fact that, given the wide range of (potentially) impacted media to be considered, there is no unified, overall, simple methodological approach available; it has to be very parameter-specific. Again, a few examples may show the extent of the possibilities to be considered.

- The simplest case is certainly the one concerning specific physical characteristics of the environment, which can be measured, and for which well-defined, legally binding thresholds are available. This is the case e.g. for noise. Depending on the characteristics of the receiving environment, different values are applicable. At the same time, there are many values from experience (measured noise emissions of different types of equipment), and there are good models available for calculating noise diffusion. In such a case if, based on this type of experience and models, values higher than the legally defined thresholds are to be expected, mitigation will be required. A similar case is water quality, e.g. pH: if there is a risk of contaminated water with higher pH values than allowed, measures will have to be taken to lower this risk for reaching values below the acceptable threshold.
- For assessing the amount of land lost to impoundment, it is required to have good, reliable topographical maps and a good hydraulic model which indicates the future level of the reservoir as well as its fluctuations (including, if of relevance, flood events that might be influenced by the project). In addition to that, mapping of the present vegetation and land use will be required. This shows that a combination of methods are required for assessing this type of impact.
- In the case of social impact, and here mainly loss of assets, in addition to good maps as just mentioned, detailed field work, here in the form of a socio-economic survey, will be required.
- The most difficult issues to adequately address are very often the impacts on the biological components of the environment. Just a few indications must be

sufficient for pointing out the problems at hand. In the case of often highly mobile, and often nocturnal, animals, it may be very difficult to actually exactly know the prevailing situation; in fact, it may be rather easy to prove that a certain species is present in a certain location, but it is close to impossible to prove that a certain, very rare and elusive species is not present. Here, more often than not, information obtained from the scientific literature, from similar studies, from the local population as well as expert opinions may be required for addressing the questions at hand in the best possible way and within the time frame available.

5.5 Impacts in Stages of the Project Cycle

As described in Sect. 4.2, dam projects go through different stages, namely, identification, preparation and planning, construction, operation and decommissioning. Each of this stages has its specific impacts. The decisive phase here is certainly the construction period. This is the time when impacts actually happen: opening of access roads, preparation of the construction site (which in larger projects can cover a quite considerable area including workers' camps, offices, workshops, storage areas, borrow areas which can comprise, depending on the type of the project, borrow areas for filling material as well as quarries, spoil dumping sites, etc.), construction activities as such including transport of material to the site, etc. Here reservoir impoundment is also counted as part of the constriction period; this is the time when the reservoir is filled, and depending on the situation, mainly the size of the river in relation to the size of the reservoir, this can take anything from a few days to several years, and in the latter case often starts long before the end of construction activities, the reservoir being filled gradually while the dam is growing. Such a case is Rogun HPP in Tajikistan, presently under construction (Chap. 23). Compared to this, impacts during operation are often much less dramatic, being restricted to issues related to reservoir drawdown and filling, and to effects on discharge in the downstream area (which, however, can be quite decisive and of long duration, especially concerning shifts in seasonal discharge patterns or when a power plant is operated as peaking plant, see Sect. 9.1.3.3). Decommissioning can be another high impact period, but is also of limited duration (see Chap. 13).

So it is clear: the most decisive phase for environmental (and social) impacts is the construction period. Nevertheless, one could ask whether this is really so. The construction period is certainly the most obvious one, when things which until then were only plans become reality, and where the physical changes in the environment are apparent on a day to day base. Still, as mentioned before, it could be argued that the really most decisive moment in the development cycle of a dam project is not this, but rather one which is never addressed in an ESIA, and which, unfortunately, is not covered in any of the international guidelines: it is the first identification of a place as a potential site for a dam (of whatever purpose).

This, at first sight, may seem far-fetched, but in reality it isn't: once a place has been identified as a potential dam site, it does not go away ever again, it cannot be "dis-identified", even if subsequent studies show that it is not really suitable. It is still around, somewhere hidden on a shelf in an archive or in the bottom drawer of a desk, or now of course on the hard disk of a computer somewhere in an office, ready to reappear at any moment. And there are numerous such cases (see a few examples below). The big problem here is the fact that preliminary identifications of potential dam and hydropower sites have always–at least in the cases that I know of–been done by a team of usually two specialists, namely, one dam engineer and one geologist. There can be no doubt that these two experts are required for this purpose, but it is not sufficient: in order to avoid the choice of sites which have to be qualified as bad or unsuitable, it would be required that environmental as well as social aspects are integrated in such evaluations from the very start. As is the case for engineering and geology, a detailed assessment is certainly not required at this stage; however, a short visit by suitably experienced E&S specialists would provide enough information for identifying at least those potential sites which, if selected, would invariably lead to major conflicts for environmental and social reasons. A few examples may suffice for illustrating this point:

- In the 1960s, the Murchison Falls HPP in Uganda came under strong criticism from an environmental point of view, namely, the presumably extremely negative impact it wold have had on the Murchison Falls National Park, the largest national park in Uganda, and it was finally abandoned. Shortly afterwards, a new site, Ayago-Nile, was identified, and in 1979/80 I participated in the feasibility study for this project. The conclusion was that the impact on the national park would be very similar to the impact the original project would have had. For different reasons, Ayago-Nile was not built. However, it resurfaced and has been re-studied, and it seems to be high on the list of power projects to be implemented in Uganda (see Sect. 26.1).
- In 1978, I participated in a study of a cascade of 9 dams in the upper Rhine valley in Switzerland, upstream of Lake Constance. This was an old project which had literally been taken out of the bottom drawer to be re-evaluated. The river had been trained long ago on the entire stretch that would have been used; still, one of the impacts would have been that the last bit of this river where at least on one side it was still free-flowing, conserving a small but ecologically important rest of the original floodplain, would have been lost. Ironically (to me, at least), the "killer argument" was then the risk that the increase in water level in a part of the river might have risen the groundwater level in its surroundings, and this in turn might have negatively affected the golf course of a tourism resort. The project was re-shelved, was then taken out again for yet another evaluation about 20 years later, only to be re-shelved again; but it is still there.
- Much more recently, in 2014, I acted as an observer in a case study aiming at applying the IHA Sustainability Protocol on a number of hydropower projects. One of the projects evaluated was Kulpawn HPP in Ghana, which had the following main features: a 25 m high and 9000 m long dam would create a reservoir with a surface of 1025 km^2, serving a power plant with an installed capacity of 36 MW which would produce a net amount of 25 GWh/y (gross production

of 166 GWh/y, but creating a loss of 141 GWh/y in and already existing HPP); it would require the relocation of 2760 HH with a total population of 21,000, whereby it had to be kept in mind that these were 1993 figures, the population presumably having increased in the meantime. It became immediately clear that this project had been studied twice before, and both times the studies had clearly demonstrated that it was not feasible, mainly for economic reasons, it having a negative rate of return (which in non-economic language means that it would never have been possible to repay the construction costs from the revenue generated by selling the electricity produced by it). The conclusion of the new assessment was just as negative, and this time mainly for social reasons, given the very massive resettlement it would require. Makes you wonder when it will be investigated next time.

As a general rule, two criteria should, if not eliminate a site right from the start, then at least lead to a careful evaluation (which, in this stage, does not mean a complete environmental and social impact assessment) before classifying them as suitable potential sites for a hydropower project, namely (i) if it would lead to a massive resettlement (whereby "massive" has to be seen in relation to the potential of the site) and (ii) if it is located in a protected area. Considering this latter point, a few cases are described in Sect. 26.1.

5.6 The Problem of System Boundaries

The question of system boundaries is important, and there three dimensions have to be discussed: one is the geographical dimension, i.e. the question of how one defines the area of influence of a project. If that is being done too narrowly, one might exclude right from the start the possibility to even identify some relevant impacts; on the other hand, if by choosing a too wide area, one might not only do a lot of unnecessary work, one might also in a way dilute an impact by identifying no effect on the scale of the area chosen and overlooking a relevant impact on a small fraction of the area. The second one is the question of topics or aspects of the natural environment to cover in an ESIA. Finally, the third question is the one of secondary or indirect impacts: how far do we want to go–or can we go, given constraints in budget, time and available information–in identifying and taking into consideration such impacts. For all three questions there is no straightforward and easily applicable answer (or, one could in this context also cite one of my favourite corollaries to Murphy's Law: "Complex problems have simple, easy to understand, wrong answers" [Grossman's misquote of H. L. Mencken, Bloch 1980: 17)]. It is important to realise that the problem exists, but then feasible and pragmatic solutions have to be found. These issues are illustrated with an example each.

It is probably safe to say that the first comprehensive EIA (and here specifically the "S" is left away, since it was really about environmental impact) which was ever prepared in Switzerland was the one for the Curciusa dam project. A considerable number of specialists were involved in the assessment (EWI 1987; Zwahlen 1988). The project (which finally was not implemented, but mainly for economic and not environmental reasons) foresaw the construction of a 122 m high dam and the creation of a reservoir of rather modest dimensions (total volume 60 M m^3, out of which 60 m m^3 live storage, reservoir surface 1.1 km^2, FSL at 2205 m asl), which would have provided additional water to an existing cascade of hydropower plants. The main purpose, as for most if not for all HPPs in mountainous areas with cold winters, would have been to make additional water available in winter, when energy consumption is highest while natural river discharge is at its lowest. One specific point in this project was the fact that it was a transbasin project, and this in the probably most extreme way imaginable: the dam would have been located in the mountains in the South-Eastern part of Switzerland, on a small tributary called Areua, in the headwaters of the Rhine River. The water from this creek flows into one of the main branches of this major European river, called Hinterrhein, and it finally ends in the North See. However, the water stored in the reservoir would then have been diverted to an existing cascade of power plants, which drains southwards, to the Ticino River and finally to the Po River, ending in the Adriatic Sea, part of the Mediterranean Sea.

As Fig. 5.3 shows, a deviation over a short distance would cause water from a very small catchment area to cross the continental divide, diverting it from the North Sea basin to the Mediterranean basin. Theoretically one could argue that this has an impact from the dam to the North Sea and from the outlet of the deviation tunnel to the Adriatic Sea, but for practical reasons this had to be limited. It was decided to assume a potential downstream impact to the point where it would no longer be possible to detect the difference caused by the project, i.e. where the diverted amount

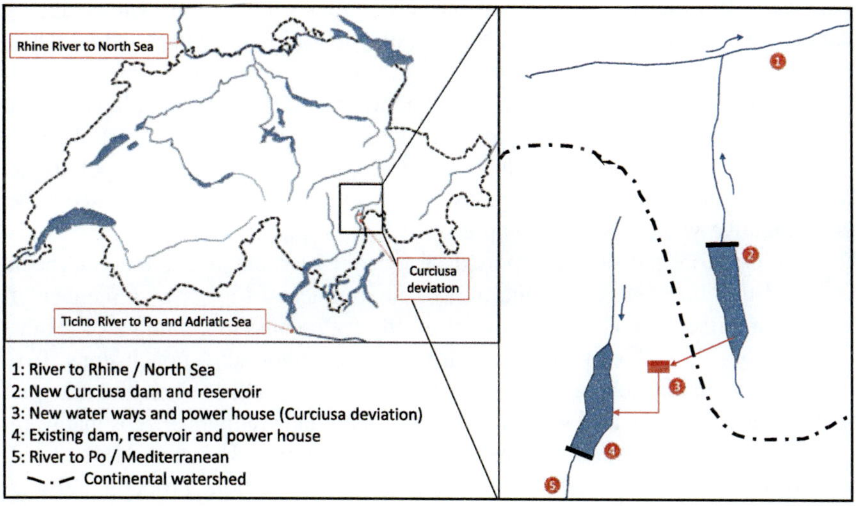

Fig. 5.3 Location of the Curciusa deviation

would be smaller than the natural year to year difference in flow, and this effect, given the small stream diverted, did not reach very far.

The second point can probably be illustrated best with aspects of the biological environment: when natural or near natural habitats are affected, both terrestrial and aquatic, how far do we have to go in our investigation, i.e. what groups of organisms need to be covered? Wilson (2016: 90 ff.) cites a case of a long term study in the 2114 km^2 Great Smoky Mountains National Park, which has the objective to record all species of living organisms in this well-defined area. Up to 2014, spending a total of 50,000 working hours a group of specialists recorded a total number of 18,500 species, among which 471 species of vertebrates (mammals, birds, reptiles, amphibians and fish), 2188 higher plants (flowering plants, mosses and ferns), and, as the largest group, 8442 species of insects, and nobody would want to assume that all species had been found and identified yet. This shows quite clearly that it is absolutely impossible, in any even very detailed ESIA, to identify all species of plants and animals living in the area affected by any project. We need to limit ourselves to certain relevant groups, which must fulfil the following requirements: (i) they must be likely to be affected by the project in question, (ii) they should be rather easy to identify, (iii) their ecological requirements have to be known, and (iv) they should have a key position in the respective habitat or ecosystem in the sense that if they can exist there all–or at least most–other species living there can continue to live there as well. Examples for these conditions are rather easy to find. For hydropower projects, it is in most cases obvious to concentrate on fish as a group to be investigated, since their habitat is going to change; they also usually fulfil the other conditions. Still, even if the relevant key groups are duly identified and taken into consideration, this does not say that all impacts on all groups of organisms will be adequately covered in any case. But the sheer number of species, and the often lacking knowledge of their ecological requirements, make it impossible to aspire to a complete biological inventory.

For addressing the third question, that on indirect effects, an example is chosen which has nothing, or very little, to do with dams. The question of the overall environmental impact of different products was addressed in a recent newspaper article (Weber 2018), and the main message was that it is very often not easy to find out the best alternative without a very comprehensive and sophisticated lifecycle analysis of the products in question. It reported on a series of such studies that had been done, and that had determined the overall environmental impact of a number of products in terms of total CO_2 equivalent, and which came to a number of results that contradicted–or at least did not confirm–usual common sense conclusions. Two findings were that plastic bags are at least not worse than paper bags, and that one way cups, even when coated with plastic, fared better than individual cups that were used over a longer period, if the impact caused by washing the latter was included in the balance. I have no doubt that the calculations were done properly and that therefore the results are mathematically correct, but I do not think that this makes them true, and this because of a question of, precisely, system boundaries. A paper bag, when thrown away, will decompose rather quickly, and the multiple use cup will, at the end of its life cycle, at worst cause very little waste (and will usually not

Fig. 5.4 Problems with plastic waste. Left: plastic bags, silt and sand make a combination that blocks culverts of drainage channels with a mass almost as solid as concrete. Drainage water, often contaminated with sewage, cannot flow away, and in the rainy season everything is flooded, which also presents a serious health issue (slum area in Eldoret, Kenya, dry season: no flooding at that time, but sewage remains blocked in channel instead of being drained away). Centre: Fish traps in the Mekong catch plastic bottles just as well as fish (Lao PDR). Right: Plastic waste in the trash rack of a hydropower plant (Republika Srpska, Bosnia and Herzegovina)

be discarded just anywhere, as is unfortunately often the case with one-way cups). Much of the plastic waste, however ends up somewhere on the ground, from where it sooner or later reaches a river and is finally washed into the seas, where it persists for a very long time. A recent study has shown that by now even animals living in the deepest sea trenches ingest plastic debris with their food (Jamieson et al. 2019). A few problems with plastic waste, which are not covered by a lifecycle CO_2 analysis, are shown in Fig. 5.4.

While, as said, this example did not have much to do with dam projects, the issue of indirect impacts will be further emphasised, e.g. in the Section on access roads (Chap. 15).

5.7 Monetising Impacts

5.7.1 Cost–Benefit Analysis

The Cost–Benefit Analysis (CBA) is an economic instrument for evaluating projects of any kind by comparing expected costs to expected benefits, in monetary terms. This last condition is also the shortcoming of the approach for its application to a more overarching and comprehensive evaluation of a project. The issue is that all qualities and dimensions which cannot directly be expressed in monetary terms, often included under "externalities", cannot be included in such an evaluation in a really comparative manner. Attempts were and are made for expanding the method by including these aspects, but defining monetary values for assets like biodiversity, aesthetic or cultural values remains, at the very least, contested (Johansson and Kriström 2011a, b; Dasgupta 2021: Box 12.1. p. 305; and the discussion—or rather harsh criticism—following the latter, e.g. Spash and Hache 2021, and https://www.nature.com/articles/d41586-021-00616-9).

The ESIA, in a way, is "the opposite" of a CBA: it takes into account and evaluates all the aspects of project related effects which are important from an environmental and social point of view, most of which however cannot be expressed in terms of money, while economic and financial aspects, i.e. hard cost facts, are not included. Here, no attempt is made to discuss advantages and disadvantages of CBA in an exhaustive way. Only some aspects of such attempts to evaluate environmental parameters by adding a price tag to them are discussed shortly and illustrated with one practical application.

5.7.2 Willingness to Pay (WTP) Analysis

Willingness to accept (WTA) is the minimum amount of money that a person is willing to accept in exchange for putting up with something negative, like e.g. pollution or loss of integrity of a specific site or landscape. Conversely, willingness to pay (WTP) is defined as the maximum amount a person accepts to pay for avoiding something undesirable, e.g. to keep water from being polluted or to maintain a specific site in its natural state, or alternatively for receiving a certain good or service.

A WTP application often used in feasibility or similar studies consists in asking potential consumers on how much they would be prepared to pay for receiving a commodity as e.g. electricity or, as is done often in such studies in the context of rural electrification, indicating the price of the commodity and asking whether they would accept to pay this price; this approach was applied e.g. in studies in Tajikistan and Laos (EWE 1999; EWE and SMEC 2000; EWE and PA 2002). The result is then used as a means for estimating economic feasibility of the project.

In environmental economics WTP is often used, as indicated above, for evaluating the intrinsic value of a specific site or situation, and especially for also taking into account values which, as such, cannot directly be expressed in monetary terms. Such values can be, e.g., attachment of an individual to a place, aesthetic values, value attributed to biodiversity etc. Situations where such investigations are often undertaken include projects of nature conservation (question: how much would you be willing to pay for protecting such and such an area, by e.g. preventing a road from being built through it?). The value obtained in this way can then be used for comparing different options, like e.g. the "with project" and the "without project" cases.

The WTP approach, as said, is a way for estimating the intrinsic value of a case, a scenario or a site, but it is not more than that. It has several shortcomings and should therefore in no case be considered as a firm basis for economic decision-making. The most important of these shortcomings are the following:

- It is often assumed that values reached by applying a WTA or a WTP approach are basically the same. However, studies have shown that especially for valuing environmental goods WTA consistently reaches substantially higher values than WTP (Brown and Gregory 1999).
- The result of a WTP study can be influenced by the approach chosen. Voltaire et al. (2017) demonstrated in such a study that the WTP value obtained was consistently higher if the amounts to be contributed were presented in a descending order (in this case from EUR 100 to 0) than if they were presented in an ascending order.
- If WTP is used e.g. for evaluating the economic feasibility of a rural electrification project, the persons interviewed may see themselves in a difficult situation: they realise that they will not be able to pay the bill, since very simply the income is not sufficient. On the other hand they want to have electricity and they fear that if they admit that they will not be able to pay the project might not go ahead. The obvious way out of this situation is to agree with the proposed tariff, i.e. to express their willingness to pay and to worry about paying once the situation materialises. This dilemma, the fact that the real question is the ability to pay rather than the willingness to pay, became very clear in the studies in Tajikistan and Laos mentioned above. The difference between ability and willingness to pay is also emphasised in Guyatt et al. (2002) or in Shah et al. (2012). It can be assumed that the situation in Mongolia would be similar; there, for the city of Ulan Baatar, WTP for air quality improvement, specifically for a reduction in PM_{10} and $PM_{2.5}$ (fine dust, particulate matter) concentrations in winter, was found to be 4.8 M MNT (Mongolian Tugrig) per person and year, in a situation where average annual income is 5.7 M MNT (Sanduijav et al. 2021), which really seems to be on the high side as well, certainly beyond the ability to pay of most inhabitants, even though air pollution in winter actually is very bad.
- In the second case, WTP for a nature reserve or for keeping a site without change, the problem is different. Here the people become very aware of the fact that they will never actually have to pay anything, and certainly not the amount they may state, since it is a hypothetical question which will not be applied in that way. This obviously means that the interviewees, while still stating a concern, may come up with an amount which might not truly reflect an actual "monetary value" (see e.g. Garikipati 2005; Garfeld et al. 2016); as Johansson and Kriström 2011a, b: 61) put it: "It is not necessarily true that a real world implementation of one or other scenario would be financed by local residents in the way proposed in the contingent valuation study". This fact is probably the strongest argument against the validity of WTP values. On the other hand, Onwujekwe et al. (2001) had found that 76% of persons asked about WTP for buying insecticide-impregnated bednets to a proposed price actually bought the nets at that price when they were offered later on, which led to the conclusion that WTP approach is suitable for market research in health care.

One application of a WTP study is described here very shortly, the one for the Jannah Dam and Lake Project in Lebanon. In the original ESIA for this project, residents were asked about their WTP to keep the small valley, in which the dam was going to be built, intact. A total of 283 responses from households in the villages around the project site were obtained, all located within close reach of the selected site (14.9 min. in average). 159 of the respondents stated a willingness to pay, with monthly rates ranging from 5′000 LBP (91 cases) to > 50,000 LBP (1 case), resulting in an average monthly WTP of USD 1.83 per HH, or USD 21.96 (or approximately 22) per year. So far, this can be considered as a correct way of approaching the issue. However, the WTP values obtained were then extrapolated first to the entire population of the villages close to the project site, in a second step to the population of the Jbeil Caza area, and finally to the Beirut and Mt. Lebanon governorates. Since in this entire area the number of households amounts to about 435,000, assuming this value led to a WTP amount, or "worth of keeping the site intact", of about USD 9.5 million per year.

However, this approach is definitely not acceptable. If in such a study an attempt is made to extrapolate WTP beyond the direct area affected by the project, it has to be taken into consideration that people, as a general rule, will value a site or situation (and therefore express WTP) if they have or expect to have a direct benefit from it. This means that WTP decreases with the distance of persons from the site in question, i.e. from the site to be (hypothetically) protected.

This fact, decreasing WTP with increasing distance from the site, has been demonstrated in other such investigations. An example is provided in Fig. 5.5, which shows that WTP decreased over a distance of 20 km from about £ 9 to less than £ 2, which corresponds to a reduction to about 20% over a distance of 20 km. This is logical: people value situation, goods or sites from which they have (or are likely to have) a benefit; why should anybody be willing to pay for the conservation of a site he/she might not even know and might never visit?

Without a detailed investigation it was impossible to know exactly what in the case of the Jannah Dam and Lake Project the distance-decay function would be. Using the curve in Fig. 5.5 it could be assumed that WTP in the Jbeil Caza area would be

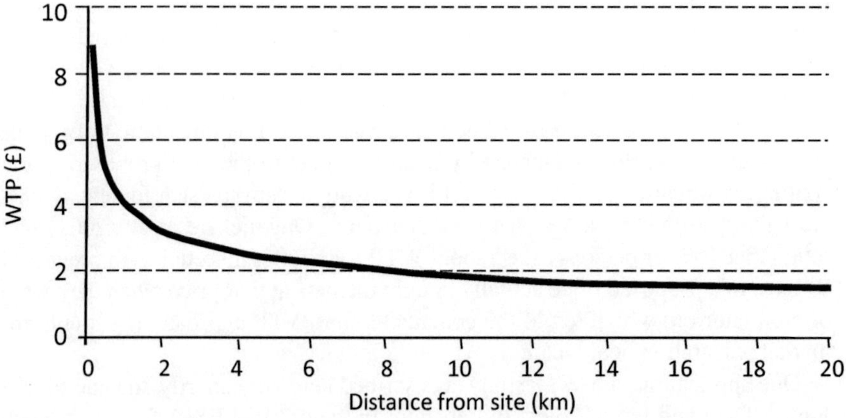

Fig. 5.5 Distance-decay function of WTP for a case of river protection. Based on Hanley 2011

Table 5.2 WTP applying a distance-decay function

Site	HH	WTP/HH*y	WTP (%)		WTP (USD/y)	
			High	Low	High	Low
Survey sample	283	21.96	1	1	6215	6215
Project area (6 villages)	3680	21.96	1	1	80,823	80,823
Jbeil Caza	15,070	21.96	0.4	0.2	132,371	66,185
Beirut and Mt. Lebanon	415,548	21.96	0.15	0.05	1,368,816	456,272
Total	434,298	21.96	0.17	0.05	1,582,009	603,280

Notes HH: for avoiding double counts, the number of HH in the lower administrative unit were deducted from the number in the higher one. WTP "high" and "low" fractions applied as explained in the text. The WTP % values in the "Total" row reflect the fact that the fraction of the directly affected (and therefore interested) population in the entire population is small
Source Pöyry (2017)

about 0.4–0.2 (40–20%) of that recorded in the project area (distance Jannah site to Jbeil roughly 20 km), while for the Beirut—Mt. Lebanon area it might be 0.15–0.05 (15–5%), Beirut being at a distance of roughly 40 km from the site. An estimate of the WTP value based on these assumptions is shown in Table 5.2.

Using this approach, WTP for the Jannah site, or the value attributed to it, can be estimated as being in the order of magnitude of M USD 0.6–1.6 per year. Since a very considerable part of this value stems from the extrapolation of the WTP value to the population of Beirut, without a survey corroborating this, it is rather likely that this estimate is still substantially too high. But even the direct extrapolation mentioned above would still have been too low to outweigh the expected benefit from the project, which was roughly calculated as being between 36 and 69 million USD per year.

5.7.3 Ecosystem Services

The concept of ecosystem services (ES) is based on the insight that intact ecosystems provide a number of different services which are important for human welfare (which includes, but is not limited to, economical aspects). Classical cost–benefit analyses (CBAs) had taken into consideration only immediate economic benefits (e.g. the value of timber produced by a forest), without taking into consideration values like water regulation, effects on climate, biodiversity, aesthetic values, etc. This approach is criticised e.g. by Temper and Martinez-Alier (2013), who point out that evaluating forests in terms of their Net Present Value (NPV) does not include cultural and other values and therefore does not contribute to forest conservation and tends to exclude forest dwelling people from the decision making process.

In an attempt to make ecosystem services as a whole measurable, i.e. to be able to express them in monetary terms, efforts were made to determine all these values provided by different ecosystems, and to put a price on them. As was said for the WTP approach above, there is no unanimity on the question of usefulness of the ES approach; see e.g. Fisher and Brown (2014) or Presnall et al. (2015) for a discussion of pros and cons for this approach. One especially controversial point is the difficulty of including cultural values, like e.g. aesthetics or the concept of holy places in such a system (Temper and Martinez-Alier 2013). One other obvious point related to this is the difficulty of objectively determining the overall value of specific ecosystems; see e.g. Yi et al. (2017) or Boerema et al. (2017), who demonstrate to what extent the result of an evaluation can depend on these ES values. Still, some studies show that the approach can be useful for getting an estimate of the value of certain systems, sites or situations in comparison with a potential alternative (e.g. Yi et al. 2017; Raynaud and Lanzanova 2017).

One set of values which is often used is the one proposed by Costanza et al. (1997, 1998 and an update published 2014). These studies had identified the following fields, to which ecosystems contribute to very various degrees:

1. Gas regulation
2. Climate regulation
3. Disturbance regulation
4. Water regulation
5. Water supply
6. Erosion control
7. Soil formation
8. Nutrient cycling
9. Waste treatment
10. Pollination
11. Biological control
12. Habitat/refugia
13. Food production
14. Raw materials
15. Genetic resources
16. Recreation
17. Cultural.

They then attempted to evaluate the contribution of major ecosystem types to all of these fields, coming in this way to an overall contribution or environmental service of the respective ecosystem types expressed in USD/ha*y. A comparison of the ESV figures provided in the 1998 and the 2014 studies shows a partially marked increase in these values. This is mainly due to a more detailed assessment of services in the more recent study. It can therefore be assumed that the "high" values, reflecting a more recent analysis, provide a more realistic estimate.

An application of these ESVs is again illustrated with the example of Jannah, Lebanon. The two sets of figures were applied in Table 5.3 for calculating the ESVs for the project area. Overall values are provided for the entire area studied in the ESIA (Area A, covering a total of 1249.9 ha), and the "lost" area, i.e. the area that will be directly occupied by the project, and where therefore the ecosystems presently existing will be completely replaced by a totally different type of land use (reservoir and area directly used for construction, Area B, with a size of 107 ha.)

The conclusions from this analysis are:

- Assuming the values for 2014, the total ESV for the study area is estimated as M USD 5.65 per year. This corresponds to the total of the services provided by this area in the 17 aspects listed above.
- Out of this, ESV of a total of M USD 0.74 will be lost due to the Jannah Dam and Lake Project. This corresponds to 13% of the value of the study area, provided by 8.5% of its surface. The fact that the affected area exhibits a somewhat higher ESV per ha than the entire study area is due to the higher percentage of wetlands in the directly affected area, a type of ecosystem with an exceptionally high ESV. This was seen as an indication of the importance of the proposed mitigation measures for protecting or upgrading wetland areas (riparian vegetation) in the downstream area of the dam.
- With the available information it was not possible to make an estimate of the ESV of the entire project catchment area, but the loss caused by the project would have been only about 0.5% in terms of area, with the loss in ESV presumably being in the same proportion.

5.7.4 Comparison of Expected Gains and Losses

Although the analysis carried out for the Jannah project was not a comprehensive cost–benefit analysis, it is still interesting to directly compare the values obtained by the different approaches chosen. This is done in Table 5.4.

Table 5.4 makes it immediately clear that the expected revenues from the project by far out-weigh the calculated losses, whether these latter are expressed in terms of willingness to pay or ecosystem services.

5.7.5 Conclusions

The analysis of the Jannah project briefly discussed in the preceding sections is definitely not—and was not intended to be—a comprehensive CBA. The concerns about the attempts to express value of a site in monetary terms, be it by applying a WTP approach or by determining ESVs, have been clearly stated. For these reasons, it is said here once more that interpretation of the results has to be made carefully, and that the figures provided should not just be accepted as firm and un-shakeable truth. Still, a number of project-specific conclusions can be reached from these evaluations.

- The annual revenues to be generated by providing electricity and water are far greater than the losses incurred by the project. As for the "values" of ecosystems,

Table 5.3 Ecosystem Values of the project area

Ecosystem	Area		ESV (USD/ha*y)		Value of ecosystem services for the area			
	Area A [ha]	Area B [ha]	low[1]	high[2]	Study area (A)		Lost area (B)	
					A low	A high	B low	B high
Forest								
Pinus brutia conifer forest	475.0	21.4	302	3137	143,450	1,490,075	6463	67,132
Pinus brutia mixed forest	331.6	34.3	302	3137	100,143	1,040,229	10,359	107,599
Forestry (charcoal production)	7.8	2.9	302	3137	2356	24,469	876	9097
Wetlands								
Riparian habitats	50.5	14.5	19,581	25,681	988,841	1,296,891	283,925	372,375
Hygrophilus thickets / tall herbs	2.3	2.3	19,581	25,681	45,036	59,066	45,036	59,066
Rocks								
Dolomite and lime cliffs	15.4	5.4	150	1600	2310	24,640	810	8640
Dolomite and lime blocky scree	77.0	8.3	150	1600	11,550	123,200	1245	13,280
Urban and cultivated land								
Agriculture and villages	286.7	17.9	92	5567	26,376	1,596,059	1647	99,649
Roads	3.6				0	0	0	0
Total	1249.9	107.0			1,320,062	5,654,629	350,360	736,838

Explanations

- Ecosystems: Vegetation types identified in the ESIA;
- A: Area mapped in the ESIA, covering the project area as such and its immediate surroundings;
- B: Area which will be occupied by the project (dam, reservoir, area permanently changed by construction activities);
- ESV: Ecosystem Service Values. For rocks, the sources used indicate an ESV of 0; while this might be true for bare rock, it is certainly not adequate for the cliffs and blocks in the project area, where values related to production are certainly low, but those e.g. related to biodiversity are considered as high. For this reason, a value corresponding to half the value for forests was assumed.

– low: values from Costanza et al. (1998)

– high: values from Costanza et al. (2014)

- Value of ESV: calculated overall ESVs for the Areas A and B, for lower (1998) and higher (2014) ESV unit values, respectively.

Table 5.4 Expected gains and losses from the Jannah Dam and Lake Project

Item	Low*	High*
Yearly revenue expected to be generated by the Project (from electricity, drinking water and irrigation)	36.18	68.93
Compensation costs (compensation to be paid to PAPs)	32.2	32.2
Willingness to pay	0.60	1.58
Ecosystem services	0.35	0.74

* All figures in M USD/year *with the exception of compensation costs, which are once to be paid costs in M USD*

it could also be said that these commodities, electricity and water, can easily be expressed in monetary terms, just by applying the tariffs used, but that this does not reflect their overall "true value". One additional value would be, e.g., the undoubtedly very positive impact of sufficient and secure drinking water on the public health situation of the population serviced.

- Even if the losses are comparatively small, this does not mean to say that they are without value. The fact that people state a certain WTP for the protection of their environment indicates that this is a value to be taken into consideration. It has also been shown that some of the affected habitat types, and here foremost the wetlands (riparian and hygrophilous vegetation) have a considerable value, even if the areas affected are small.
- From this it can be derived that the proposed mitigation measures, e.g. for protecting and/or actively creating wetland habitats as an offset for the losses in the area downstream of Jannah dam, are meaningful and should be implemented.

The issues illustrated here for one project, Jannah, in Lebanon, can certainly be applied to other dam, reservoir and hydropower projects as well. The main point to be made here is that an evaluation of project impacts should not be reduced to a simple CBA. Even without having a large set of projects analysed in this way it can be concluded that such an approach would, most likely, prove ESVs to outweigh expected revenues from power production only very exceptionally.

References

ArchaeoGLOBE Project. 2019. Archaeological assessment reveals earth's early transformation through land use. *Science* 365 (6456): 897–902. https://doi.org/10.1126/science.aax1192.

Baker, J.L. 2000. *Evaluating the impact of development projects on poverty. A handbook for practitioners. Directions in Development*, 217. World Bank, Washington DC.

Becker, H.A. 1997. *Social impact assessment: Method ad experience in Europe, North America and the developing world*, 260 pp. London: UCL Press.

Bloch, A. 1980. *Murphy's law book two. More reasons why things go wrong*, 96 pp. London: Magnum Books.

Boerema, A., A.J. Rebelo, M.B. Bodi, K.J. Esler, and P. Meire. 2017. Are ecosystem services adequately quantified? *Journal of Applied Ecology* 54 (2): 368–370.

Bonneuil, C., and J.-B. Fressoz. 2013. *L'événement anthropocène. La terre, l'histoire et nous*, 304 pp. Paris: Editions du Seuil.

Brown, T.C., and R. Gregory. 1999. Why the WTA–WTP disparity matters. *Ecological Economics* 28 (3): 323–335.

BUWAL. 1990. Handbuch Umweltverträglichkeitsprüfung UVP. Bundeasmt für Umwelt, Wald und Landschaft (BUWAL), 121 pp. Bern, Switzerland.

Chen, L., Q. Qiu, Y. Jiang, et al. (2019). Large-scale ruminant genome sequencing provides insights into their evolution and distinct traits. *Science* 364 (6446): 1152 eaav6202. https://doi.org/10.1126/science.aav6202.

Cole, M.M. 1986. *The Savannas: Biogeography and geobotany*, 438 pp. London: Academic Press.

Corlett, R.T. 2019. *The ecology of tropical East Asia*, 3rd ed., 320 pp. Oxford, UK.: Oxford University Press.

Costanza, R., R. d'Arge, R. de Groot, S. Farber, M. Grasso, et al. 1997. The value of the world's ecosystem services and natural capital. *Nature* 387: 253–260.

Costanza, R., R. d'Arge, R. de Groot, S. Farber, M. Grasso, et al. 1998. The value of the world's ecosystem services and natural capital. *Ecological Economics* 25 (1): 3–15.

Costanza, R., R. de Groot, P. Sutton, et al. 2014. Changes in the global value of eco-system services. *Digital: Global Environmental Change* 26: 152–158.

Crosby, A.W. 1986. *Ecological imperialism. The biological expansion of Europe, 900–1900. Studies in environment and history*, 368 pp. Cambridge: Cambridge University Press.

Dalal-Clayton, B., and B. Sadler. 2005. *Strategic environmental assessment*, 470 pp. London: Earthscan.

Dasgupta, R. 2021. *The economics of biodiversity: The Dasgupta review*, 604 pp. London, HM Treasury. http://www.gov.uk/official-documents.

Dawkins, R., and Y. Wong. 2016. *The ancestor's tale. A pilgrimage to the dawn of life*, 771 pp. Weidenfeld and Nicolson, London, updated edition.

De Santo, R. 1978. *Concepts of applied ecology*, 310 pp. New York: Springer.

Diamond, J. 1997. *Guns, germs, and steel. The fates of human societies*, 480. Norton, New York.

Duek, J.J. 1982. Metodología para la evaluación de impactos ambientales. Contemporánea de Ediciones, Caracas, Venezuela, 113 pp.

Ehrlich, P.R. 2002. *Human natures. Genes, cultures and the human prospect*, 531 pp. New York: Penguin.

EWE and PA. 2002. Power sector strategy study. Electrowatt-Ekono and CA Consulting for Lao PDR, ADB TA 3374 LAO.

EWE and SMEC. 2000. Power sector development project. ADB 3207 TAJ. Vol. 1: Energy sector review and policy formulation. Vol. 2: Diagnostic analysis of Barki Tajik. Vol. 3: Rehabilitation and investment program. Vol. 4: Environmental and socio-economic analysis. Vol. 5: Feasibility report. Reports prepared by Electrowatt Engineering and SMEC International for Asian Development Bank and Government of Tajikistan.

EWE. 1999. Gorno-Badakhshan energy development project. Review of Pamir 1 project (Tajikistan). IFC.

EWI. 1987. Saisonspeicherwerk Curciusa: Bericht zur Umweltverträglichkeitsprüfung (UVP). Elektrowatt Ingeinieurunternehmung AG., Zürich, für Misoxer Kraftwerke.

Fischer, T.B. 2002. *Strategic environmental assessment in transport and land use planning*, 284 pp. London: Earthscan.

Fischer, T.B. 2007. *Theory and practice of strategic environmental assessment. Towards a more systematic approach*, 186 pp. Earthscan, London.

Fisher, J.A., and K. Brown. 2014. Ecosystem services concepts and approaches in conservation: Just a rhetorical tool? *Ecological Economics* 108: 257–265.

Garfeld, S., K. Oleson, M. Barnes, M. Peng, C. Chan, and M. Weijerman. 2016. Divers' willingness to pay for improved coral reef conditions in Guam: An untapped source of funding for management and conservation? *Ecological Economics* 128: 202–213.

Garikipati, S. 2005. Consulting the development-displaced regarding resettlement: Is there a way? *Journal of Refugee Studies* 18 (3): 340–361.

Guyatt, H.L., S.A. Ochola, and R.W. Snow. 2002. Too poor to pay: Charging for insecticide-treated bednets in highland Kenya. *Tropical Medicine and International Health* 7 (10): 846–850.

Hanley, N. 2011. Environmental cost-benefit analysis and water quality man-agement. In *Modern cost-benefit analysis of hydropower conflicts*, ed. P.-O. Johansson and B. Kriström, 6–21. Cheltenham UK: Edward Elgar.

Harris, M. 1995. Kannibalen und Könige. Die Wachstumsgrenzen der Hochkulturen. DTV, München, 276 S.

IFC. 2012. Performance standards on environmental and social sustainability. Performance Stadards PS1-8, Guidance Notes GN1-8. http://www.ifc.org/performancestandards.

Jamieson, A.J., L.S.R. Brooks, W.D.K. Reid, S.B. Piertney, B.E. Narayanaswamy, and T.D. Linley. 2019. Microplastics and synthetic particles ingested by deep-sea amphipods in six of the deepest marine ecosystems on Earth. *Royal Society Open Science* 6: 180667.https://doi.org/10.1098/rsos.180667.

Johansson, P.-O., and B. Kriström (eds.). 2011a. *Modern cost-benefit analysis of hydropower conflicts*, 199 pp. Cheltenham, UK: Edward Elgar.

Johansson, P.-O., and B. Kriström. 2011b. A blueprint for a cost-benefit analysis of a water use conflict - hydroelectricity versus other uses. In *Modern cost-benefit analysis of hydropower conflicts*, eds. P.-O. Johansson, and B. Kriström, 35–64. Cheltenham, UK: Edward Elgar.

Lacher, T.E., Jr., A.D. Davidson, T.H. Fleming, et al. 2019. The functional roles of mammals in ecosystems. *Journal of Mammalogy* 100 (3): 942–964. https://doi.org/10.1093/jmammal/gyy183.

Leitmann, J. 1994. Rapid urban environmental assessment. Lessons from cities in the developing world, vol. 1. Urban Management Programme Discussion Paper No. 14: 78 pp. The World Bank, Washington.

Leopold, L.B., F.E. Clarke, B.B. Hanshaw, and J.R. Balsley. 1971. A procedure for evaluating environmental impact. *Geological Survey Circular* 645: 13. https://doi.org/10.3133/cir645.

Lohani, B.N., W.N. Evans, R.R. Everitt, H. Ludwig, R.A. Carpenter, and S.L. Tu. 1997. Environmental impact assessment for developing countries in Asia. Vol. I: Overview, 364 pp. Vol II: Selected case studies, 314 pp. ADB, Manila.

Lyons, S.K., F.A. Smith, and S.K.M. Ernest. 2019. Macroecological patterns of mammals across taxonomic, spatial and temporal scales. *Journal of Mammalogy* 100 (3): 1087–1104.

Onwujekwe, O., R. Chima, E. Shu, D. Nwagbo, and P. Okonkwo. 2001. Hypothetical and actual willingness to pay for insecticide-treated nets in five Nigerian communities. *Tropical Medicine and International Health* 6 (7): 545–553.

Pöyry. 2017. Jannah dam and lake project. Update of Environmental Imapct Assessment Reports. Eaux de Beyrouth et du Mont Liban.

Presnall, C., L. López-Hoffman, and M.L. Miller. 2015. Adding ecosystem services to environmental impact analyses: More sequins on a "bloated Elvis" or rockin' idea? *Ecological Economics* 115: 29–38.

Reynaud, A., and D. Lanzanova. 2017. A global meta-analysis of the value of ecosystem services provided by lakes. *Ecological Economics* 137 (184–194).

Sadar, M.H. 1996. *Environmental impact assessment*, 2nd ed., 191 pp. Canada: Carleton University Press.

Sánchez-Montoya, M.M., M. Moleón, J.A. Sánchez-Zapata, and D. Escoriza. 2017. The biota of intermittent and ephemaral rivers: amphibians, reptiles, birds and mammals. Ch. 4.6. In *Intermittent Rivers and ephemeral streams: Ecology and management*, eds. T. Datry, N. Bonada and A.J. Boulton, 299–322. London: Academic Press. https://doi.org/10.1016/B978-0-12-803835-2.00011-5.

Sanduijav, C., S. Ferreira, M. Filipski, Y. Hashida. 2021. Air pollution and happines: Evidence from the coldest capital in the world *Ecological Economics* 187.https://doi.org/10.1016/j.ecolecon.2021.107085.

Schaefer, M., S. Tsekané Jr., F.A.M. Tchassem, S. Drakulić, M. Kameni, N.L. Gonwouo, and M.O. Roedel. 2019. Goliath frogs build nests for spawning—The reason for their gigantism? *Journal of Natural History* 53 (21–22).

Shah, A.K., S. Mullainathan, and E. Shafir. 2012. Some consequences of having too little. *Science* 338 (6107): 682–685.

Shermer, M. 2018. *Heavens on earth. The scientific search for the afterlife, immortality and utopia*, 305 pp. London: Robinson.

Shipman, P. 2015. *The invaders. How humans and their dogs drove Neanderthals to extinction*, 266 pp. Cambridge, Massachusetts, London, England: The Belknap Press of Harvard University Press.

Spash, C.L., and F. Hache. 2021. The Dasgupta review deconstructed: An exposé of biodiversity economics. *Globalisations*. https://doi.org/10.1080/14747731.2021.1929007.

Tattersall, I. 1999. Neandertaler. Der Streit um unsere Ahnen. Birkhäuser, Basel, 216 pp.

Temper, L., and J. Martinez-Alier. 2013. The god of the mountain and Godavarman: Net Present Value, indigenous territorial rights and sacredness in a bauxite mining conflict in India. *Ecological Economics* 96: 79–87.

Therivel, R., E. Wilson, S. Thompson, D. Heaney, and D. Pritchard. 1992. *Strategic Environmental assessment. Reprint 1999*, 181 pp. London: Earthscan.

Tucker, R.P. 1988. The depletion of India's forests under British imperialism: planters, foresters and peasants in Assam and Kerala. In *The ends of the earth. Perspective on modern environmental history*, ed. D. Worster, 118–140. Cambridge: CUP.

UNEP. 1996. *Environmental impact assessment: Issues, trends and practice*, 96 pp. Geneva, Switzerland: United Nations Environment Program (UNEP).

Voltaire, L., H.P.P. Donfouet, C. Pirrone, and A. Larzillière. 2017. Respondent uncertainty and ordering effect on willingness to pay for salt marsh conservation in the Brest Roadstead (France). *Ecological Economics* 137: 47–55.

WB. 1991. Environmental assessment sourcebook. In *Policies, procedures, and cross-sectoral issues*, vol I, 227 pp. World Bank Technical Paper No. 139. *Sectoral guidelines*, vol II, 282 pp. World Bank Technical Paper No. 140. *Guidelines for environmental assessment of energy and industry projects*, vol III, 237 pp. World Bank Technical Paper No. 154. Washington, DC: Environment Department, the World Bank.

WB. 2017. The World Bank environmental and social framework. International Bank for reconstruction and development/The World Bank, Washington, DC: 106 pp. http://documents.worldbank.org/curated/en/383011492423734099/pdf/114278-WP-REVISED-PUBLIC-Environmental-and-Social-Framework.pdf.

Weber, C., 2018: Nachhaltige Irrtümer. Sonntagszeitung 2018-09-03:18-19.

Wilson, E.O. 2016. *Half earth. Our planet's fight for life*, 259 pp. New York: Liveright.

Yi, H., B. Güneralp, A.M. Filippi, U.P. Kreuter, and I. Güneralp. 2017. Impacts of land change on ecosystem services in the San Antonio River Basin, Texas, from 1984 to 2010. *Ecological Economics* 135: 125–135.

Zwahlen, R. 1988. Environmental impact assessment (EIA) for an alpine power plant. In *Management of environmental impacts of hydropower in Switzerland*, eds. A. Streichenberg, G. Collet, B. Gilg, C. Zimmermann, R. Zwahlen. ICOLD Q.60, R. 13:183–203.16ème Congrès des Grands Barrages, San Francisco

Zwahlen, R. 1996. Traditional methods: A guarantee of sustainability? *Indigenous Knowledge and Development Monitor* 4 (3): 18–20.

Chapter 6
Mitigation

We are all conscious today that we are drowning in a sea of data and starving for knowledge.
Sydney Brenner, Nobel Prize laureate (2002).

Mitigation is "the action of reducing the severity, seriousness, or painfulness of something. (Oxford English Dictionary, https://www.lexico.com/definition/mitigation). Other sites present similar definitions, with somewhat different wording. In the context of ESIAs, mitigation is used as the umbrella term for exactly that: reducing adverse impacts. The widely accepted and generally applied mitigation hierarchy is to avoid, or, where avoidance is not possible, reduce or minimise, and where residual impacts remain, compensate/offset the negative consequences of project-induced impacts. Different types of mitigation are addressed, first the generally accepted mitigation hierarchy avoid—reduce—compensate, where the risk of conflicting outcomes is emphasised, then specifically some more sophisticated mitigation approaches summarised under the term "offset".

6.1 The Mitigation Hierarchy

6.1.1 Avoiding

Avoiding an impact is, quite obviously, the best way of mitigating it: if a dam is not built, all the negative impacts associated with it, or caused by it, will not arise, which would make mitigation complete. This may actually be true in certain situations, and even then, whether the conclusion is that this is true or not may, once again, depend on how one looks at things, or expressed in a different manner, what the boundaries of the system are. It is a fact of life that "doing nothing" in most cases is not really an option, since the decision not to take a specific step or implement a certain action does not really mean to do nothing, it can mean "business as usual", which very often, and especially in our busy and increasingly crammed world is very different from doing nothing. The communication theorist Paul Watzlawick (1968) coined the

R. Zwahlen, *Assessing the Environmental Impacts of Hydropower Projects*, Environmental Earth Sciences, https://doi.org/10.1007/978-3-030-91185-0_6

phrase “one cannot not communicate”. The same principle could be applied here: “one cannot not do anything”; not doing one thing will, most of the time, mean to do something else.

Coming back from philosophy to hydropower, just a short illustration of this, by taking the example of the Amayla Falls HPP in Guyana (see Sect. 15.2). This project—or its site—is located at quite a distance from almost everything, illustrated by the fact that 100 km of access road through (still mostly primary) rainforest would be required for building it. So avoidance is seemingly very simple: just don’t do it, and the impacts will not happen. But it is really not as simple as that: just the fact that in anticipation of actually building it about half of the access road was already opened in the preparatory stage of the project has changed the situation in a potentially decisive way (see the discussion on impacts of access roads in Chap. 15). It could even be argued that the fact that the prospective developer has pulled out of the project makes the situation worse in the sense that now, a part of the road is there, but the access control, which was supposed to be put in place and maintained as an integrating part of the project, is not there, opening the area to whatever secondary effects such a road can—and in most cases does—have. And there is one other thing: as emphasised in Sect. 5.5, once a project has been identified, it cannot be “dis-identified” again: it is still there, for somebody to step in and take over.

Not every case needs to be like that and there are examples where the decision not to implement a project, for whatever reason, left the site unchanged. One such case is e.g. Curciusa, a dam project in a small lateral valley of one of the two main headwaters of the river Rhine in the Swiss Alps (see Fig. 5.3). For carrying out field work for the EIA one had to follow a small footpath for about two hours for getting to the dam site. Near the site, there was, and presumably still is, a small hut used by hunters during the hunting season, where field staff was lodged, but all the food had to be carried there. In winter, the only access was by helicopter, since there was a considerable risk of avalanches, and some parts of the path were actually blocked by them. No infrastructure—and especially no access road—was built during this period, and when the project was finally dropped, the valley was left as it had been before. The same was true for the Ayago-Nile project which was studied in the late 1970s (see 26.1.1); however, in that case the fact that the project has been studied again in recent years may have changed the situation.

One other point to consider: HPPs have the main aim to generate electricity. If a project is not being implemented, this does not mean that the need for energy is no longer there, and in most cases it will mean that a different way—alternative HPP site, alternative energy source—will be sought for satisfying the demand.

6.1.2 *Minimising*

Avoidance of an impact, however, does not necessarily mean in every case that a project has to be abandoned. It is absolutely possible that a specific impact can be avoided by a modification of the project. In each of the two cases of Kukule

Ganga in Sri Lanka and Rusumo Falls in Rwanda and Tanzania (see Sect. 14.1.2) three alternatives were evaluated, and in both instances the ROR solution was given preference over the two dam options because in this way a massive resettlement could be avoided; in the case of Rusumo, this would have had its main impact in Burundi, a country which is also participating in the development of this project, and which now is not negatively affected by it at all. Yet, since in both these projects there was still some resettlement, this might better be labelled as minimising the impact rather than as avoiding it.

Minimisation of an impact, for some parameters, can be very straightforward, namely, for those items which can be measured and for which there are legally defined threshold values. One such example is noise; here, threshold immission values are usually defined, e.g. for inhabited areas, and these may be different for day and night. In such a case, noise measurements have to be made, and the expected incremental noise due to the project needs to be evaluated. Based on the results, measures can then be defined which will allow to keep noise within the limits. Likewise, if e.g. on a construction site there are some extremely noisy places, as breakers or other inevitably noisy machinery or processes, the requirements for hearing protection for people working in such places can be determined. In either case, the aim is to reach an exposure which is acceptable and legally defined as such. However, while undoubtedly there are cases of that type, for most of the main impacts one has to deal with in hydropower projects no such rules and regulations exist. There is, to mention just these points, no threshold value for impacts on biodiversity, or on the aesthetic value of a landscape. In such cases, the answer to the question whether or not an impact is acceptable will always depend on the more or less subjective opinion of those who decide and, as shortly described in the introductory part, people looking at the same do not necessarily see the same.

Taking this into consideration, in the preparation of an ESIA this clearly means that the importance or relevance of an impact has to be considered and evaluated carefully, and the conclusion reached has to be described just as carefully. The aim here has to be to described it in a way that readers of the report, and mainly those who will have to decide, understand the reasoning; obviously, understanding it will not mean necessarily agreeing with it. It is then the same with the proposed measures and the evaluation and description of their effectiveness, or looking at it the other way round, about the remaining impact after the application of the measure. A good example here may be fish ladders (or any other type of fish passes). The basic approach here seems to be straightforward and rather simple, but in reality this is often not the case, since a number of points need to be taken into consideration. The first question is whether or not, in a specific situation and in a specific place, there are any migrating fish. As simple as this question may seem, there is not always an easy answer to it, sometimes simply because of a lack in information and potentially major difficulties in getting it. In case there are such migrating species, the next question would be about the importance of this migration: will blocking the route lead to the fact that they can no longer reach their spawning places, which might lead to the disappearance of a species from a river system (and in the case of European or North American salmons often has done exactly this), or will it just mean that some not very relevant local movements will no longer take place? Especially in large tropical rivers (as e.g. the Mekong, which presently is probably the one which comes under pressure by hydropower more than any other river of comparable size), there are usually many different species with different types of behaviour, requiring different conditions, and in addition to that there is often just not enough information available for answering these questions; no ESIA, which always is a specific study done for a specific project, will ever be able to answer such basic questions. Assuming that the answer to these two questions is that yes, there is migration, and yes, blocking it would affect fish populations in a very negative way, the next question would be about what to do to minimise this impact, i.e. to reduce it to an acceptable level. There is certainly an agreement that no fish pass structure would ever completely eliminate every negative impact on fish movement (which, just as a reminder, needs to enable upstream as well as downstream movements). There is some considerable experience with such structures for salmons, hardy swimmers able to withstand strong currents and to jump over obstacles of considerable height. However, this type of information is not necessarily available for rivers like the Mekong, whose fish fauna is very diversified, including species of widely differing size, swimming capability and migrating patterns, all of which might not be really well known. The design of appropriate structures, in such cases, may be very difficult, and an accurate statement on their effectiveness and efficiency cannot be made before some such structures will have been in operation—and hopefully under a good monitoring—for quite some time.

6.1.3 Compensation

In cases where avoidance is not possible and minimisation is not sufficient, compensation can be the answer, basically meaning to provide replacement of something lost in one place (usually the reservoir) in a different place. The concept of compensation is especially important in dealing with socio-economic impacts, most obviously where compensation has to be made for lost assets used by the affected population. One major issue in such cases is compensation for lost agricultural land if a rural population with a livelihood based on agriculture is affected, which is as good as always the case in HPPs with a large reservoir; exceptions to this rule may be reservoirs in uninhabited mountainous areas (e.g. Mauvoisin, see Sect. 25.1) or in protected areas (as e.g. in the cases of Ayago-Nile (Sect. 26.1.1) or Ivindo (Sect. 26.1.3). The main issue for such compensation usually is the problem of finding enough land of suitable quality, since by now almost everywhere on earth where there are good soils there are people using them, and in any given site, people will concentrate in the areas with the most productive and fertile soils, and these are, as a rule, the alluvial soils along the rivers. Unfortunately, these are the ones most likely to be affected by reservoirs; concerning this issue, see e.g. the short discussion of the Three Gorges HPP in China (Sect. 21.1.2), or Scudder (2005) for a much more detailed analysis.

Rehabilitation may be considered as a special case of compensation, since there a replacement of a lost asset is provided on the same site as it was originally. Rehabilitation means the restoration of a temporarily used area into something similar to the original situation. This is an important measure, applicable especially in the case of large construction sites, as the site for the construction of a large dam usually is, where considerable areas are used (and changed to a more or less considerable degree) temporarily. This latter point makes it clear that rehabilitation, in many cases, cannot really be anything like "restoring the original condition" of such a site. The most obvious example for this are quarries, where usually huge quantities of rock have been removed to be used for the construction of the dam, and where in addition to that very often large volumes of overburden had to be removed to reach the rock in the first place; it will never be possible to restore the original visual aspect of the terrain. The second such type are the locations where spoil, i.e. unused or unusable excavation material, among it the overburden just mentioned, was disposed of. In both cases, the landscape was changed, and sometimes quite considerably, and rehabilitation in such cases will have to be understood as reintegration of the sites into the surrounding landscape. Depending on the situation (see examples in Sect. 10.1.3.2, Lom Pangar and Ilisu) this can be rather easy or quite demanding, and integrating quarries in a satisfactory way can be very difficult; just painting the remaining rock face in green, as was apparently done in one case in China (https://www.nbcnews.com/id/wbna17153855), can really not be the solution. The best way to do it is certainly to think about this already during exploitation, by planning it in a way as e.g. to leave berms in different elevations of the rock face, where vegetation will subsequently be able to develop (see Fig. 6.1). In any case, it will be important

Fig. 6.1 Quarry rock face at the end of the construction phase. Not one sheer rock face; berms were left where vegetation will be able to develop

to clearly define the necessity for rehabilitation in the ESMMP, and to outline the conditions there.

In cases of considerable impacts on biodiversity for which compensation is required, this is now being done usually under the term of biodiversity offset (see Sect. 6.3).

6.2 Conflicting Issues

HPPs can cause impacts of widely differing types, requiring mitigation and compensation measures that differ just as widely. Since different issues are being handled by different experts, there is a risk that, when it comes to compensation measures, these to some extent contradict each other, and this has to be avoided; this is one major reason why in Chap. 3 the importance of the ESIA team leader is repeatedly mentioned, since all measures have to be listed and compared with each other. The risk of such contradictory measures is greatest in projects where there are impacts on the local population as well as impacts on biodiversity, both requiring compensation measures. The specialists dealing with either of these issues may be looking for suitable land for compensation, and it is easy to imagine what happens in case both find the solution to their respective problem in the same place. Another risk could be that e.g. land is identified as a replacement for agricultural land, and then it turns out that implementing compensation in this site would considerably increase the overall impact of the project on biodiversity. One example of a conflict, albeit of a somewhat different nature, became apparent in the case of Lom Pangar HPP in Cameroon. During the work of the E&S POE (Environmental and Social Panel of Experts) it was discovered that one village, a short stretch downstream of the dam site, had been totally ignored during the planning phase. This may have been acceptable at that time, since there was no direct impact of the project on this village whatsoever. However, the project occupied a small part of a National Park, whose main objective was the protection of a population of gorillas. In order to make compensation for this, land adjacent to the park was added to the park and given protection. As it turned out, this land had formerly belonged to this village; they had used it for collecting NTFPs (non-timber forest products), and for them it was a reserve of land that they could also have converted into agricultural land, should the need arise, which now was no longer possible. In a discussion of this case, the biodiversity specialist observed that in any case they had not used this land, to which the social specialist, and rightly so, answered that even so this was a case of expropriation. Quiet obviously, there is a big difference between having a resource that is not—or rather not yet—used, and not having the option to use it any longer.

6.3 Offset

Compensation for biodiversity losses of relevant dimensions are done by so-called biodiversity offsets. According to IFC (PS6 2012: 2) "Biodiversity offsets are measurable conservation outcomes resulting from actions designed to compensate for significant residual adverse biodiversity impacts arising from project development and persisting after appropriate avoidance, minimization and restoration measures have been taken."

6.3.1 The Critical Habitat Concept

IFC guidelines put special emphasis on so-called "critical habitats" when it comes to minimise or compensate impacts on biodiversity. The definition of "critical habitat" according to IFC (PS6, 2012) is the following: "Critical habitats are areas with high biodiversity value, including (i) habitat of significant importance to Critically Endangered and/or Endangered species; (ii) habitat of significant importance to endemic and/or restricted-range species; (iii) habitat supporting globally significant concentrations of migratory species and/or congregatory species; (iv) highly threatened and/or unique ecosystems; and/or (v) areas associated with key evolutionary processes."

Most of this is quite obvious and clear. When in a biological baseline investigation one finds one or more critically endangered or endangered species, it is only logical that such species will have to be taken into account, that a careful assessment of impacts has to be carried out, and that the potential for mitigation or compensation has to be examined. Examples of projects where this was the case and which are mentioned in this book are Lom Pangar in Cameroon (gorilla population in the National Park close to the dam site, see Sect. 24.1), Bumbuna in Sierra Leone (chimpanzees, see Sect. 16.1), and Palas Valley HPP in Pakistan (several species of endangered and restricted-range pheasants, see Sect. 26.2). There is also a clear reference point: a species is critically endangered or endangered if it is ranked in the corresponding categories of the IUCN Red List (https://www.iucnredlist.org/) and/or if it gets a similar ranking in the respective national legislation. The problem here may be to actually show that such a species is really present within the range that will be affected by the project. Not too seldom, in ESIA reports such endangered species are mentioned, based on information from the literature, which at a closer look may live within the catchment of the river in question, but certainly not in the part of it affected by the project, for the very simple reason that the habitat type required by this species is just not there. Such an enumeration is not only not helpful, but can actually lead to considerable confusion or even to the formulation of mitigation measures which are absolutely useless. On the other hand, it has also to be said that if suitable habitat types are affected, and the species in question may live in this area, it can be very difficult to actually prove that the species is not there. The second point, endemic species, is somewhat similar. As pointed out in Sect. 10.5.3, the term "endemic" has to be used with caution, and needs in any case to be defined in the sense of "endemic to what?". Information may not always be easy to come by for such species. One extreme case of an endemism is certainly the Kihansi spray toad (*Nectophrynoides asperginis*, Sect. 6.3.1). The habitats of importance for migratory or congregatory species as well can, in most cases, be identified rather easily. In the context with HPPs this may be e.g. spawning grounds of long-range migrating fish species which reproduce in the headwaters of major rivers; the best known in this respect are certainly the salmons, which have suffered from dams blocking their migration routes (see Rhine, Sect. 10.3.3). The definition of what is a highly threatened or unique ecosystem as well seems to be quite straightforward. Unlike in the case of endangered species, however, local conditions play probably a larger role. So e.g. one would certainly define any river in Switzerland not yet affected by some "correcting" measures, or even part of a river in that condition, as a threatened ecosystem, since there is hardly anything of that kind left, but one would not necessarily do so in Mongolia, where all rivers (still) flow freely (Fig. 6.2). Other examples of threatened as well as unique ecosystems are the case of Tigrovaya Balka in Tajikistan, a type of wetland called tugai (see Sect. 18.2.4.1) and the Tonle Sap-Great Lake system in Cambodia (see Sect. 22.5).

Fig. 6.2 River habitats. Freely meandering river (Egiin Gol, Northern Mongolia, top left), braiding mountain stream (Pamir, Tajikistan, right), and completely straightened and corrected river, banks fortified by boulders and squeezed in between a railway line and a road, all of the former floodplain having been taken over by agriculture, in this case pasture and hay meadows (Simme, Switzerland)

However, there is a considerable problem with the concept of "key evolutionary processes". Kolbert (2014:90), when talking about why the ammonites did not survive the impact of a large meteorite about 65 million years ago, the same that also eradicated the dinosaurs, states "In times of extreme stress, the whole concept of fitness at least in the Darwinian sense, loses its meaning: how could a creature be adapted, either well or ill, for conditions it has never before encountered in its entire evolutionary history?" Rather than a valid statement, this might well be a misunderstanding of the concept of fitness: fitness is not an active adaptation to a situation, but it has many similarities with a lottery. As long as the situation remains stable, the "fitness" requirements stay the same, and evolution under such circumstances will select against change. Under such circumstances, species can exist, without any or with very little change, for long periods. However, when conditions abruptly change, everything changes. And suddenly, species which were the most abundant and widespread, and which therefore can be considered as the "fittest" under these conditions, have, in a way, their fitness taken away from them. If in such cases, species (or somehow "aberrant" individuals or populations of some species), which happen to be able to cope—or maybe only to cope a little bit better—with the new situation, will suddenly be the fitter ones, and they may expand their range and increase in numbers quite rapidly. One of the best known, although very simple examples of such rapid change is the case of the birch moth *Biston betularius*. This moth, widespread in Europe, is normally white with black speckles, which makes it almost invisible when during the day it rests on the stems of birch trees and other tree trunks covered with lichens. When, most noticeably in England in the nineteenth century, lichens disappeared and tree trunks turned increasingly black due to emissions from coal burning, suddenly there were black specimens of this moth in increasing quantities. This black variety had been around before, but they were not "fit" in an evolutionary sense: too visible and therefore a too easy prey for birds. This had now changed, and fitness had passed over from the white to the black ones.

So, what is the "key evolutionary process": is it the stabilising selective pressure eliminating deviations from the "standard aspect" under stable conditions, or is it rather the potentially very strong selection of hitherto rare variations under changing conditions? I frankly do not know. Bonada et al. (2017) define evolutionary processes, and they define and discuss the importance of the ones they identified as key evolutionary processes, which in this case were mutation, gene flow, genetic drift, natural selection and sexual selection. These are undoubtedly very relevant drivers of evolution. However, here the effect of these processes were analysed on the very specific case of intermittent and ephemeral streams. According to Tatum-Hume et al. (2017:36), key evolutionary processes are "defined by the physical features of a landscape that might be associated with particular evolutionary processes, and/or subpopulations of species that are phylogenetically or morpho-genetically distinct and may be of special conservation concern given their distinct evolutionary history (IFC 2012, paragraph GN95)". This study, carried out for Bumbuna II HPP, Sierra Leone, shortly discusses the issue and comes to the conclusion that the situation there "is not considered to qualify under Criterion 5", mainly also because "no species of conservation concern have been recorded" (which then would have had to be considered under criterion (i) anyway). Other studies may come to similar conclusions.

Sites which could be considered as places harbouring such a process might be Lake Tanganyika, Lake Victoria and some others of the great African lakes (Worthington 1989; Worthington and Lowe-McConnell 1994; Huntley 1988; McGee et al. 2020). In Lake Tanganyika, over a period of only about 10 million years since this lake came into being due to tectonic events forming the African rift valley, from a small stock of cichlid fish 260 species have evolved (Ronco et al. 2021). While this is a short time in geological or evolutionary terms, it is still a very long time in terms of hydropower and impact assessment. In case such an extraordinary habitat would be affected by any type of project, the corresponding ESIA would probably deal with this under the heading of endemic - and in some cases potentially endangered—species rather than under "evolutionary processes". Still, evolution is going on, although mostly unnoticed, and some recent studies have shown that adaptive processes can happen surprisingly fast, as e.g. changes in the length of limbs in anoles lizards on small Caribbean islands as an adaption to the risk of being blown away by strong winds (Losos 2017: 155 ff.). One other evolutionary process presently under way—and hopefully fast enough—is the resistance of amphibians against the parasitic fungus *Batrachochytrium dendrobatidis* which is spreading almost worldwide, diminishing amphibian populations around the globe at an alarming rate (see Longcore 1999; Weldon et al. 2004; Byrne et al. 2019; and Kolbert 2014: 4ff., for an account of the devastation this disease is causing). Studies in Australia and in Panama (McKnight et al. 2017; Voyles et al. 2018) have shown that at least in some species some populations seem to develop resistance against this disease, and these populations will hopefully multiply and spread in a way as to re-conquer the parts of their habitat from which they were eliminated.

But even so, it is difficult to see what the concept can actually contribute, in a practical way, to biodiversity offset. There is also one very real obstacle to the application of this criterion: it would require a very sophisticated, probably lengthy investigation to confirm that a specific population is "phylogenetically or morphogenetically distinct"—or to prove that it is not—and this in itself might pose a great risk to a small local population especially if it is of conservation concern and therefore vulnerable.

6.3.2 The No Net Loss Approach

According to IFC (PS 6, 2012) "A biodiversity offset should be designed and implemented to achieve measurable conservation outcomes that can reasonably be expected to result in no net loss and preferably a net gain of biodiversity; however, a net gain is required in critical habitats. The design of a biodiversity offset must adhere to the "like-for-like or better" principle and must be carried out in alignment with best available information and current practices." As for the critical habitat, this sounds good, but has a number of practical problems.

Let's start with an example: in the case of Nam Ngiep 1 HPP in Laos, the main environmental impact of the project, in addition to water quality and related GHG issues (see Sect. 24.2) is the fact that about 70 river km are impounded, changing the situation from lotic to lentic, while the downstream area is influenced by the peaking operation pattern of the power plant which cannot be completely mitigated by the re-regulation pond. Much of the submerged forest was gallery type forest along the banks of the river. The project includes a (very far-reaching and good) biodiversity offset program, which aims at improving the situation in some biodiversity core areas in the upper part of the catchment area of Nam Ngiep, close to the border with Viet Nam. The main aim of the program is to provide actual protection to these areas, which so far were subject to heavy poaching. Real protection will eliminate or at least substantially reduce this pressure, and this in turn will allow a number of heavily exploited species to recover, certainly a gain for biodiversity (see Fig. 5.2, which can be seen as an illustration of this approach and potential outcomes). However: can in such a case "no net loss" be achieved, and, just as importantly, how would this be verified? As good as this program is, it does certainly not follow the like-for-like principle, since the offset for a loss in a river system with accompanying gallery forests is the protection of upland forest (which, along the streams flowing in this area, may contain some habitats of a similar nature, but is for the most part a definitely different habitat type than the one lost to the project). So the next question would be whether this protected habitat would qualify as "better". It would seem very difficult to identify any objective criteria which would allow to make such an evaluation in an objective, "mathematical" kind of way. There are certainly many possibilities for comparing the biodiversity value of habitats of very different types, to then possibly come to a conclusion on what, in a specific situation, would be "better". Still, one thing seems clear: the creation of a reservoir of that size in a river of that type will inevitably lead to losses, and whether an offset in a different type of habitat can actually lead to a situation of no net loss, or, as required for impacts in critical habitats, to a net gain, seems very difficult to ascertain.

This is not meant to criticise the principle of a biodiversity offset in a case where a project undoubtedly has impacts on biodiversity. It just questions the value of the no net loss approach, which seems just too theoretical, and which might lead to conclusions about comparative "worth" of habitats and species which are at least questionable. A good overview of the no net loss principle is provided by Maron et al. (2018), who stress the importance of having a clearly defined reference scenario, without which a discussion of what exactly no net loss means in any specific case is basically meaningless.

A good practical example for the application of a no net loss policy, and of its development over time, is the Swiss legislation on forest protection. Towards the end of the nineteenth century, increasingly unsustainable exploitation of forest was showing clearly adverse results, and this especially in mountainous areas, where forests were not only—or not even mainly—sources for wood, but were important for protecting settlements from avalanches and landslides. This led to a forest legislation, put in force in 1902, which forbid any further reduction of the forested area in Switzerland (Leibundgut 1970: 186). In practice, this meant that forest had to be maintained as forest, and that in every case where, for whatever reason, forest needed to be cleared, e.g. for the development of infrastructure projects, an area of the same size hat to be reforested. This had the intended effect of halting the reduction of the forested area in the country. However, over time it became clear that while overall deforestation had been effectively stopped, the aspect of quality had not been included. So e.g. the proportion of conifer forest in some parts of the Alps kept increasing, while on the other hand riparian forests along rivers were disappearing at an alarming rate. This led to a modification of the law in the sense that now reforestation had to be done within the same region, and under comparable environmental conditions. But even this had to be modified after a certain time. One aspect that had not been taken into consideration was the spontaneous increase of the area covered with forest. Practice had been that all land where trees had been growing for a certain time and which therefore now had the characteristics of a forest were included in the forest cadastre, and were henceforth protected from clearfelling. This led to an increase in the total forested area: while Leibundgut (1970: 185) mentions 25% of the surface of the country as forest, in 2004 this was 31% (BUWAL 2004: 24). Therefore, there is now an option that spontaneous regrowth will no longer automatically be considered as forest in any case, and that in areas where there is a considerable increase in forest area the requirement for compensatory reforestation may be replaced by specific compensatory measures in the interest of nature or landscape protection. This is a clear case of a biodiversity offset which replaces the like-for-like approach with an equivalent-or-better philosophy; the request formulated above, however, remains valid: a clear reference scenario will have to be formulated in every case for determining what, under the given local conditions, can be considered as not the same, but equivalent or better.

References

Bonada, N., S.M. Carlson, T. Datry, et al. 2017. Evolutionary and biogeographical processes in intermittent rivers and ephemeral streams. Ch. 4.10. In *Intermittent Rivers and ephemeral streams: ecology and management*, eds. T. Datry, N. Bonada, and A.J. Boulton, 405–431. London: Academic Press. https://doi.org/10.1016/B978-0-12-803835-2.00015-2.

Brenner, S. 2002. Nature's gift to science. Nobel Lecture, December 8, 2002 https://www.nobelprize.org/uploads/2018/06/brenner-lecture.pdf.

BUWAL. 2004. Waldprogramm Schweiz (WAP-CH). Handlungsprogramm 2004–2015. Schriftenreihe Umwelt Nr. 363, 117 pp. Bundesamt für Umwelt, Wald und Landschaft, Bern.

Byrne, A.Q., V.T. Vredenburg, et al. 2019. Cryptic diversity of a widespread global pathogen reveals expanded threats to amphibian conservation. *PNAS* 116 (41): 20382–20387. https://doi.org/10.1073/pnas.1908289116.

Huntley, B.J. 1988. Conserving and monitoring biotic diversity. In *Biodiversity*, ed. E.O. Wilson, 248–260. Washington D.C.: National Academy Press.

IFC. 2012. Performance standards on environmental and social sustainability. Performance Stadards PS1-8, Guidance Notes GN1-8. http://www.ifc.org/performancestandards.

Kolbert, E. 2014. *The sixth extinction. An unnatural history*, 319 pp. London: Bloomsbury.

Leibundgut, H. 1970. Der Wald. Eine Lebensgemeinschaft. Huber, Frauenfeld, 197 pp.

Longcore, J.E., A.P. Pessier, and D.K. Nichols. 1999. Batrachochytrium dendrobatidis gen. et sp. nov., a chytrid pathogenic to amphibians. *Mycologia* 91 (2): 219–227. https://doi.org/10.2307/3761366.

Losos, J.B. 2017. *Improbable destinies. Fate, chance and the future of evolution*, 368 pp. New York: Riverhead Books.

Maron, M., S. Browlie, J.W. Bull, et al. 2018. The many meanings of no net loss in environmental policy. *Nature Sustainability* 1: 19–27. https://doi.org/10.1038/s41893-017-0007-7.

McGee, M.D., S.R. Borstein, K.L. Meier, et al. 2020. The ecological and genomic basis of explosive adaptive radiation. *Nature* 586 (7827): 75–79. https://doi.org/10.1038/s41586-020-2652-7.

McKnight, D.T., R.A. Alford, C.T. Hoskin, et al. 2017. Fighting an uphill battle: The recovery of frogs in Australia's wet tropics. *Ecology* 98 (12): 3221–3223. https://www.jstor.org/stable/26602199

Ronco, F., M. Matschiner, A. Böhne, et al. 2021. Drivers and dynamics of a massive adaptive radiation in cichlid fishes. *Nature* 589 (7840): 76–81. https://doi.org/10.1038/s41586-020-2930-4.

Scudder, T., 2005. *The future of large dams. Dealing with social, environmental, institutional and political costs*, 389 pp. London: Earthscan.

Tatum-Hume, E., C. Fletcher, E.C. Regan, and V. Katariya. 2017. Bumbuna Phase II Project: Critical Habitat Assessment. Report prepared on behalf of Joule Africa. The Biodiversity Consultancy Ltd, Cambridge, UK.

Voyles, J., D.C. Woodhams, V. Saenz, et al. 2018. Shifts in disease dynamics in a tropical amphibian assemblage are not due to pathogen attenuation. *Science* 359: 1517–1519. https://doi.org/10.1126/science.aao4806.

Watzlawick, P., J.H. Beavin, and D.D. Jackson. 1968. *Pragmatics of human communication. A study of interactional patterns, pathologies and paradoxes*, 296 pp. London: Faber and Faber.

Weldon, C., L.H. du Preez, A.D. Hyatt, R. Muller, and R. Speare. 2004. Origin of the amphibian chytrid fungus. *Emerging Infectious Diseases* 10 (12): 2100–2105. https://doi.org/10.3201/eid1012.030804.

Worthington, E.B. 1989. The lake Victoria Lates saga. *Environmental Conservation* 16 (3): 266–267.

Worthington, E.B., and R. Lowe-McConnell. 1994. African lakes reviewed: Creation and destruction of biodiversity. *Environmental Conservation* 21 (3): 199–213.

Part II
ESIA Preparation

This part contains shorter or longer—depending on the topic—comments on the different subjects that normally have to be dealt with in an ESIA. This can, at least to some degree, be taken as a chapter by chapter guide through such a report. However, as pointed out in Sect. 3.4.1, the structure can be different, depending on the project, the specific situation of a project or, more often, the structure imposed by the guidelines prepared by the environmental authority of the respective country. Therefore, while as a rule all the topics discussed need to be at least mentioned somehow in an ESIA, this does not necessarily mean that this will have to be done exactly in the form as presented here. There is also the possibility that under given circumstances a subject might have to be added which is not dealt with here. To give just one example: in an ESIA prepared for a project in Laos there will have to be, in most if not all cases, a section on UXO: unexploded ordnance, the difficult and unfortunately still very present legacy from the Indochinese wars. Luckily, in most other places this is not a requirement. Since this aspect is really very country-specific, it is not discussed here.

The discussion of the individual subjects will, as a rule, cover the three main steps baseline situation, impacts and mitigation. A number of topics, which either are considered as being of specific importance, and/or cross-cutting issues which cannot easily be attributed to one or another of these environmental compartments will be discussed in more detail in Part III.

Chapter 7
Introductory Chapters to an Esia

Et quand bien même je saurais ce que je veux dire, est-ce que mon interlocuteur le reçoit exactement comme je le pense? (And even if I knew what I want to say, would my interlocutor get it exactly as I think it?).
Binet (2015: 336)

This is, as the title says, an introduction. It will usually shortly state the aim and importance of the project, the purpose of the report, mention the project proponent and the entity preparing the report, and of course it will duly list and discuss the points asked for in the relevant guidelines. Very important items in this introductory part are an overview of the legal and administrative framework for ESIA and project development, a detailed, albeit non-technical description of the project, and a definition of the study area, sometimes called area of influence, i.e. the area which will have to be covered by the study. In this part, a short overview of potential impacts to be expected in the specific subdivisions of the study area is useful.

7.1 Introduction

The first chapter of an ESIA report will cover a number of points which are important to understand as background information to the project. Some of the items usually addressed here are the following:

- Rationale for the project: short overview of the energy supply situation of the country and the role the project will have to play in order to improve the situation.
- Project proponent: a short description and identification of the entity which aims at implementing this project.
- Present stage of project development: in order to understand the ESIA it is important to know whether this is a study for a project in its feasibility or even pre-feasibility stage, or else for one in the detailed design stage, planned for rapid implementation. This, however, is not yet a detailed project description, see Sect. 7.3. In the case of a project at an advanced stage it may be useful or required

R. Zwahlen, *Assessing the Environmental Impacts of Hydropower Projects*, Environmental Earth Sciences, https://doi.org/10.1007/978-3-030-91185-0_7

to provide a description of the project history, including main findings of any earlier studies that have already been carried out.
- Short presentation of the entity in charge of preparing the ESIA.

Other points may be included in this introductory chapter, depending on the situation and especially also on the requirements defined in the relevant ESIA procedures which have to be followed.

7.2 Legal and Administrative Framework

The main purpose of the ESIA is to verify whether the project in question is compliant with the relevant (national, local) environmental legislation, and if there is a problem with this compliance it has to either show that measures can be taken to make it compliant, or, in an extreme case, show where it cannot be compliant, why, and what the potential consequences of this non-compliance are (Zwahlen 1991). As mentioned in Sect. 5.3, this legal compliance requirement may be very easy to fulfil for hard-fact parameters where legally defined thresholds apply, but can be much more difficult for soft issues like biodiversity.

Having said this, it could be argued that this chapter could potentially contain just one single sentence: The preparation of the present report follows all the relevant national legislation. This is, quite obviously, what it needs to do, it cannot possibly be in violation of such legislation. However, it is clear that this would never be accepted. The legal texts which are specifically applicable need to be listed, and the most important ones have to be at least shortly commented and their relevance for the project at hand to be emphasised. The same has to be done for relevant international guidelines, of which usually the ones issued by WB or IFC are the most important, but depending on the financing institutions the guidelines of the regional development banks apply, or the Equator Principles in case of commercial banks which adhere to them; these guidelines and standards can be found on the homepages of the respective organisations. The TOR usually clearly stipulate which of these guidelines have to be followed. This is very clear in every project where one of the multilateral development banks are involved, they will insist on their guidelines as the most relevant benchmark for impact assessment. They are usually more stringent in a number of points than the respective national legislation; still, compliance with this latter will always be a requirement. In most cases it will be useful to insert a table which lists the specific guidelines, provides a short appreciation of their respective importance for the case at hand, and makes a statement on the applicability of the respective safeguard item. One additional listing that may be required is the one concerning international treaties and conventions, highlighting the ones to which the country is a signatory party. Furthermore, the ESIA procedures to be followed need to be described. This is usually quite clearly regulated in the relevant legal and procedural standards of the country, or of its environmental protection agency.

In this way, this chapter serves its two main purposes: (i) to show the competent authority that the consultant is aware of the relevant legal and administrative framework, and (ii) to provide other stakeholders, e.g. private persons included in the ESIA process as directly affected parties, with basic information on the applicable legal and procedural background.

Not surprisingly, this process can get considerably more challenging and demanding in the case of projects which straddle international boundaries, since then legal and procedural requirements of both—or, if more than two, all—countries involved have to be followed, and these requirements can be very different from one country to another. It is not unheard of that already the structure asked for ESIA reports in two neighbouring countries can be different to a point of making the preparation of two separate reports, albeit for the same project, unavoidable, and this even more so in cases where the official language is not the same in both countries. A preparation of the reports in one language and then a translation into the other may not be sufficient in such cases. Such problems may arise in HPPs on border rivers (examples mentioned in this book: Rheinfelden: Germany and Switzerland, Sect. 13.2; Rusumo: Rwanda and Tanzania, Sect. 4.4.2).

7.3 Project Description

A good description of the project at hand is important for being able to assess, and for the reader of the study to understand, the impacts of the project. It is also a standard chapter in every template for ESIA reports. Surprisingly, this is something that is often done rather badly.

The description will always be based on the engineering already done for the project, and this depends on the stage in project development; obviously, the technical basis available is not the same whether we deal with a project in its prefeasibility, feasibility or detailed design phase. It also has to be noted that some relevant project parameters, like e.g. exact dam location, dam height, installed capacity or planned operation pattern can change during the planning process.

The easiest way of describing a project—and one that is often chosen in ESIA reports—is just to copy the project description, including plans illustrating it, from the technical documents. In my view, this is not a good way of doing it for the following reasons:

- One important requirement for ESIA studies, which in most cases must be made available to the interested—or affected—public, is that they must be written in a way that they can be understood by a non-specialist. In this respect, such a report is fundamentally different from the technical documents: while the latter must be technically correct and complete, the ESIA of course has to be correct, but does not need to be complete where technical details of the project are concerned. It has to clearly show the relevant parts of the project, but "relevant", from an environmental and social point of view, is not necessarily the same as from an engineering point of view. To give just a few examples: while the length of a headrace tunnel is an important piece of information, the question of whether its cross section is circular or horseshoe shaped is totally irrelevant from an environmental point of view. The type of dam (RCC, rockfill, double arch concrete …) is certainly very important from an engineering point of view, but much less so for the environmental impact. One specific information which is very important for both aspects is the storage volume of a reservoir, usually split into dead and live storage; however, the surface area of a reservoir is not terribly important for engineering purposes and this is sometimes lacking in engineering documents, but it is one of the most important parameters for the environmental and social studies. Just copying project descriptions out of technical studies often leads to pages and pages of ultimately useless information, which does not help non-specialists in understanding the project, and which in the worst case omits relevant data. One extreme example I have come by in this respect is the EIA for the Pak Lay HPP in Laos (NCG 2018): a 270 pages main report with a 240 pages Appendix entitled "Project layout and main structures", but not showing a single map or drawing of the dam and powerhouse and of the reservoir.

- The same is true for illustrations, which are important and should be clear and to the point. Engineering plans—for all the good reasons mentioned above—contain a lot of details which are not of relevance for the environmental aspects; an additional problem is that such drawings in an ESIA are often reproduced in a much smaller scale, which makes it impossible to really recognise and understand the details, let alone to read the captions. In addition to that, while such a figure may provide a considerable amount of information for the specialist, it may not be easy to understand for the non-specialist, and may even lead to confusion.

For these reasons, the following is recommended for the purpose of project description:

1. Provide a map which shows the location of the project site within the country and/or the river basin.
2. Provide a table with the relevant parameters (see Table 7.1, Kafue Gorge Lower HPP). This is a standard table, which can—and has to—be adapted to the specific conditions presented by the project at hand, by adding relevant items or leaving out irrelevant ones. It also allows to easily and directly compare several project alternatives, e.g. with different dam heights, in cases where such alternatives are under discussion. A table which contains all the details provided in a technical study and which stretches over several pages will often create more confusion than clarity, and may make it difficult for the non-specialist to identify the relevant parameters.
3. Provide an outline plan of the structures, based on a good map, and drawn specifically for the ESIA, which clearly illustrates the relevant parts; Fig. 7.1 gives an example which may be acceptable, but not very good (NT1) and a good one (Luang Prabang). Depending on the geometry of the project, several drawings may be necessary, e.g. in the case of a very large reservoir, and/or in projects with very long headrace structures; in such cases, at least a map with

Table 7.1 Key characteristics of Kafue Gorge Lower HPP, Zambia

Parameter	Unit	Value	
1. Dam		Preferred option	Alternative
Dam type		RCC	CFRD
Height	m	140	140
Crest length	m	376	363.4
Crest elevation	m asl	581	583
2. Reservoir			
Full supply level (FSL)	m asl	579.0	
Minimum operation level (MOL)	m asl	530.0	
Area at FSL	km^2	1.85	
Area at MOL	km^2	0.76	
Volume at FSL	Mm^3	83.0	
Volume at MOL	Mm^3	21.8	
Active storage	Mm^3	61.2	
3. Water ways and power house			
Headrace tunnel and penstock	km	4.5	
Total installed capacity (tentative)	MW	$5 \times 150 = 750$	
Minimum hydraulic head (net-head)	m	132.5	
Turbine output	m^3/s	$5 \times 96.8 = 484$	
4. Hydrology			
Catchment area	km^2	153,570	
Mean annual rainfall	mm	1010	
Potential annual evapotranspiration	mm	1630	
Actual annual evapotranspiration	mm	930	
Average discharge at dam site (DSS 1961–1990)	m^3/s	315	
Average discharge at dam site (KGU 1973–2012)	m^3/s	276	
Flood (HQ_{1000})	m^3/s	3770	
Flood (PMF)	m^3/s	7228	
5. Transmission lines			
Nominal voltage	kV	330	
TL 1 KGL to KGU s/s	km	11	
TL 2 KGL to LSMFEZ s/s	km	60	
TL 3 KGL to LW s/s	km	100	
6. Permanent access roads			
Construction (new road)	km	22	
Track upgrade	km	24	
7. Affected population			
Affected villages	N	0	
HH to be resettled	N	25	

Source Pöyry and Zenith, 2016 (modified)

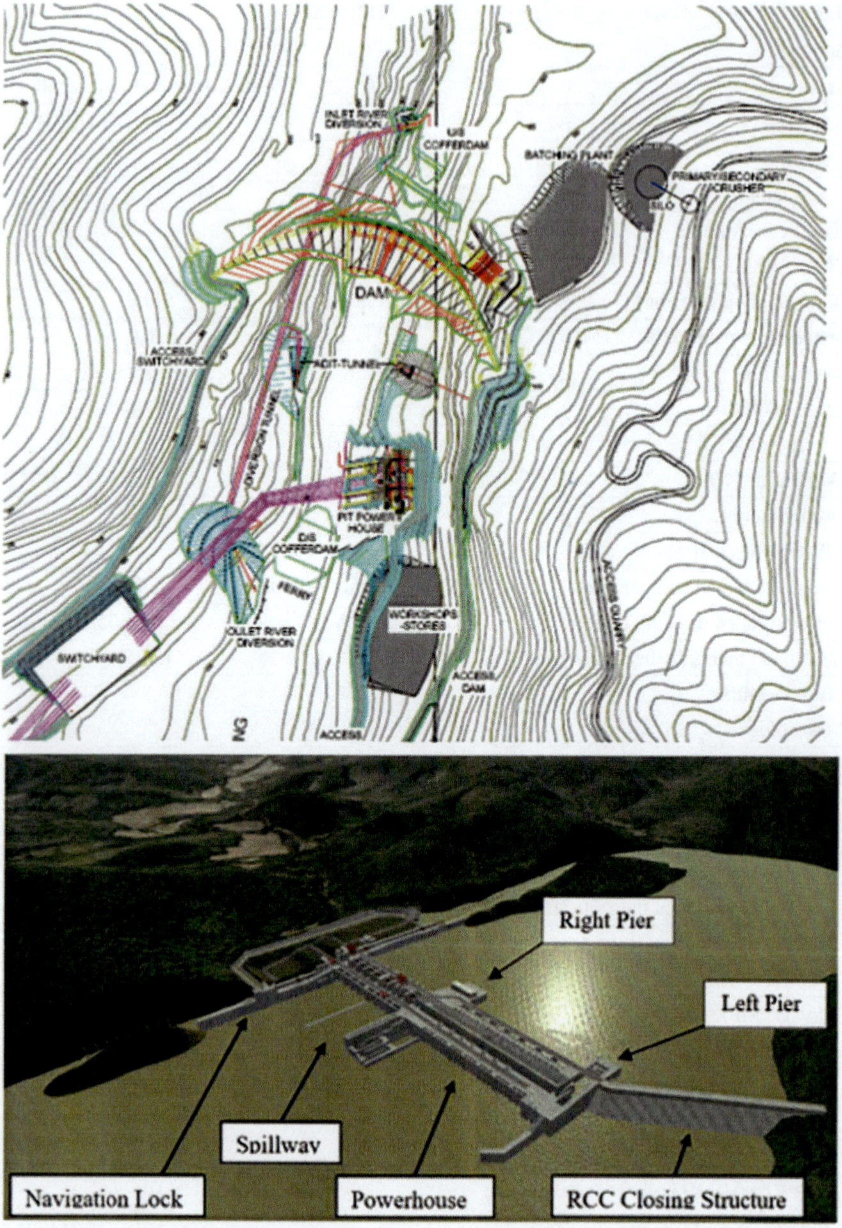

Fig. 7.1 Illustrations of projects in ESIA Reports. Upper: NT1 HPP (Electrowatt-Ekono 2005). Lower: Luang Prabang HPP (Pöyry 2020). Reproduced with permission

the entire project would be required, an additional one showing the dam, and possibly one showing the power house and related structures.

4. Provide a description of the relevant parts of the project in a non-technical language, which will allow the reader to understand it. It is also important to make this description according to the project stage and, in early stages, to mention structures or project parts which may not be known yet at this stage. As an example: in a prefeasibility phase, it will be clear that a construction site, a quarry and spoil dumping areas will be required, but there might not be any information yet about the exact location and dimensions of these important parameters.

In Fig. 7.1, two possibilities for illustrating the project in an ESIS are shown. The first is a layout map taken from a technical report, at the feasibility or early detailed design stage. It shows all the major parts of the project, with the exception of the reservoir (or at least the part of the reservoir that would be visible in this section of the map). The second shows a physical representation of the project as it will look once constructed.

Both have their good points and their shortcomings. The second one, Luang Prabang HPP, omits all temporary structures that will be required for construction, and since the landscape in which it will be placed is presented on the basis of a pre-construction satellite image, it very probably shows the vegetation in the surroundings of the project without the scars that construction will invariably leave behind. The NT1 example, on the other hand, shows much more detail, including at least part of the construction site structures. The main problem with this type of project representation (which can be found in a similar way in very many such studies), in addition to the fact that very often such figures, as is the case here, are reproduced in a scale that makes reading the integrated legends as good as impossible, is, as mentioned above, the fact that the average person, especially in the case of a local and potentially affected population, will most probably not really understand this type of illustration. The aim is clearly not to produce something that the colleagues from the civil engineering faculty will understand, but to provide information to the non-specialist. The Luang Prabang illustration does that very well, and the missing information can be added in a description or in a layout plan. In any case, it is often preferable not to take illustrations directly from technical documents, but to prepare them specifically for the purpose of the ESIA.

7.4 The Study Area

The study area will have to be defined at an early stage of the work to be done for an ESIA. Its exact delimitation will depend on project specific parameters and the situation of the site. Since effects of the project are not identical for all parts of this study area, the focus of the work will also be different, as described shortly below.

Basically, the study area for an ESIA for a hydropower project is composed of the following main parts.

- Dam and power house site, appurtenant structures and immediate surroundings: this will be a focal area for the impact assessment, since it will be affected in a relevant and permanent way by the project, and since project activities, mainly during the construction phase, will be concentrated in this area. Besides the location of the main structures (dam, water ways, power house, switchyard etc., including, depending on the specific conditions, cofferdams and other temporary or permanent structures), it includes the area affected by construction activities (construction sites, construction camps, quarries, stockpile, borrow and disposal areas, etc.). This will be a high intensity area for field investigations. Depending on the type and layout of the project, this part of the study area can be one compact area, or it can be split into two or more sites, especially in the case where dam and power house are separated by long, often subterraneous, waterways; this is a usual situation mainly in projects located in mountainous areas, where long headrace tunnels and penstocks are built in order to make use of a large head. Examples mentioned in the text where this is the case are e.g. Mauvoisin (Sect. 25.1) and Shnog (Sect. 9.1.1.1).
- Reservoir area: the future reservoir, i.e. the area which will be covered by water, is also an area in which the effects of the project will be very apparent and which therefore needs to be considered in detail, since it will be permanently and fundamentally changed by reservoir impoundment. Here, one specific point concerning this: in the literature on dam impacts, one usually reads that the reservoir area "will be flooded"; instead of using the term "flooding" in this context, it seems more appropriate to speak of "submerging", in the sense that "the reservoir area will be submerged". The process of reservoir filling is also addressed as "impoundment" instead of "flooding". With this, one expresses clearly that submerging an area during reservoir impoundment is something final and permanent: it will remain covered by water, at least during the life span of the dam. The term "flooding" should be used for what it is: a temporary, reversible—albeit often catastrophic—event that happens during floods, i.e. during times when a river brings exceptionally high amounts of water.
- Immediate reservoir catchment: this is the area directly surrounding the reservoir, which can be influenced by the project in different ways (e.g. change in groundwater regime, triggering of landsides, increasing pressure on habitats, etc.), and from where the water drains directly into the reservoir.
- River downstream of the dam: this area often has to be subdivided, depending on the project layout, into the two following parts: (i) between dam and powerhouse outlet; here, river discharge will be reduced, in an extreme case to zero, but now usually to the residual flow defined for the project, plus water from spilling; and (ii) area below the powerhouse outlet, where the river discharge pattern will be influenced by turbine operation (impact small in the case of run-of-river schemes, and potentially very important in the case of storage schemes with intermittent (e.g. peaking) power production, which can cause seasonal changes in river discharge

pattern and important short term fluctuations). The extent of the area to be taken into account depends on the specific situation like other power plants in the downstream area, confluence with other rivers, presence of floodplains etc. Potential issues are flow conditions as influenced by the project, cumulative impacts with existing power plants, water quality issues, effects on floodplain habitats, and effects on downstream water users. In the case of international rivers, this can include effects on other riparian countries.

- Catchment area: usually, the project as such does not influence the catchment area, but problems in this area (mainly: erosion influenced by human activities, water contamination) can have effects on the project. For this area, the situation will be described based on available information, problems, if any, will be identified, and proposals will be made how to address them. However, the ESIA does not include a full-fledged watershed management plan. It has to be seen very clearly that an entity operating a dam and power plant, certainly if it is a private company, but even in the case of state-owned projects, does not have any jurisdiction over the catchment area, and therefore it cannot, e.g., forbid logging in forests even if logging leads to higher erosion and therefore to a faster siltation of the reservoir, reducing its productivity and its life span.
- Access roads: these can have major direct or indirect environmental impacts and will therefore have to be addressed; see more details in Chap. 15.
- Other areas, as might be relevant. This can comprise, e.g., transmission line corridors, resettlement areas, or areas to be occupied for the necessary relocation of infrastructure (e.g. existing roads which will be submerged), if this is defined as part of the TOR for the ESIA. In such cases the question is often more an administrative one than anything else. The World Bank, e.g. rightly considers transmission lines as integrating parts of a hydropower project and therefore requests that their impacts are analysed together with those of the HPP as such. However, in practice there is often one department or ministry in charge of electricity generation, while there is another one in charge of transmission and distribution. In such cases, there are often two separate contracts, technical as well as environmental and social, and the two parts are considered as two different projects.

The definition of the study area will have to be made at the very beginning of the assignment, since other tasks will have to be carried out accordingly. The screening and scoping exercise will (or should) then define what exactly needs to be done in which part of the study area (which is often also called "area of influence").

The situation can be more complex than outlined here; it was mentioned that for dam, powerhouse and construction more than one site may have to be defined. The same is true in the case of interbasin projects, i.e. when either water from other rivers is diverted into the reservoir in order to increase the amount of water available, or when water from the reservoir is released into another river (as e.g. in the cases of NT2 and TH, see Sect. 22.3). Here, the other affected rivers will have to be included in the ESIA.

References

Binet, L. 2015. La septième fonction du langage. Grasset, Le Livre de Poche, 478 pp.

Electrowatt-Ekono. 2005. Nam Theun 1 Hydropower Project, Laos: Environmental Impact Assessment. GAMUDA Berhad, Malaysia.

NCG. 2018. Sanakham Hydropower Project. Environment Impact Assessment. Report prepared by National Consulting Group, Vientiane, for Datang (Lao) Sanakham Hydropower Co. Ltd. http://www.mrcmekong.org/topics/pnpca-prior-consultation/sanakham-hydropower-project/.

Pöyry and Zenith. 2016. Kafue Gorge lower HPP ESIA: Environmental and social impact assessment. Report prepared for ZESCO and Kafue Gorge Lower Power Development Corporation, Ltd., Zambia: 287 pp.

Pöyry. 2020. Luang Prabang HPP: Environmental impact assessment. Vol. 4.1 of the feasibility study. Report prepared for Luang Prabang Power Company Limited.

Zwahlen, R. 1991. Die Umweltverträglichkeitsprüfung (UVP). In *Öko-Brevier. Ein praktisches Handbuch für kleinere und mittlere Unternehmen*, ed. T. Dyllick. Bülach, Switzerland: Vetropack.

Chapter 8
Physical Environment

Continued global warming is projected to further intensify the global water cycle, including its variability, global monsoon precipitation and the severity of wet and dry events.
IPCC (2021: 25)

This chapter and Chap. 9 deal with the main aspects of the physical or abiotic (non-living) environment, which can be subdivided into three compartments, namely, (i) atmosphere (with the main topics of climate and related aspects as well as air quality), (ii) hydrosphere (issues related to water) and (iii) lithosphere (soils and geology). The subject of this chapter are the first and the third of these three groups. The main climate parameters are addressed, namely temperature, precipitation and evaporation, and two additional parameters, which in the context of hydropower are usually of less importance, namely wind and ice conditions, are mentioned shortly. Climate change as a rather recent, but increasingly important phenomenon, is discussed. Two specific issues, air quality and noise, are mentioned, whereby their usually less prominent role in the context of hydropower projects is emphasised. Geology, in most cases, is an issue of high relevance for technical and engineering purposes, but is usually of less concern for the ESIA as such, as are very often soil conditions. Finally, an aspect with potentially far-reaching consequences is sediment transport and sedimentation.

The second of the three compartments, namely water, will be discussed in detail in Chap. 9, since this is one of the most central issues in addressing impacts of HPPs.

8.1 Climate

A description of the main climatic conditions is required in order to understand the situation in the project area. The two most important parameters are temperature and precipitation. Evaporation can have some importance, and especially for very large reservoirs wind may have to be considered as well.

R. Zwahlen, *Assessing the Environmental Impacts of Hydropower Projects*, Environmental Earth Sciences, https://doi.org/10.1007/978-3-030-91185-0_8

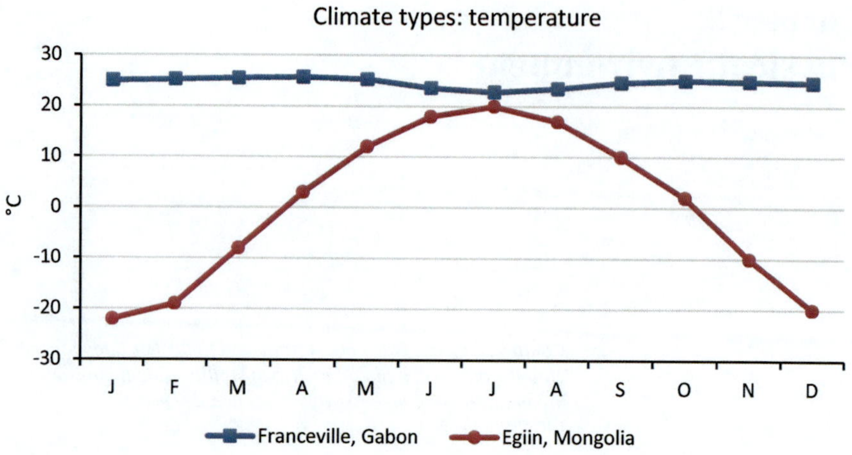

Fig. 8.1 Monthly mean temperature depending on latitude

8.1.1 Temperature

8.1.1.1 Baseline

Seasonal patterns of temperature can be very different, mainly depending on latitude and elevation of the sites chosen, as Fig. 8.1 shows.

The graph shows two patterns from locations of different latitude. Franceville is located in Gabon, close to the equator; temperatures are almost the same throughout the year, April, with an average of 25.7 °C being the warmest, July with 22.8 °C the coolest month. Egiin, however, located in northern Mongolia, shows a very marked difference between winter and summer, with − 22 °C in January and 20 °C in July; these are average values, minima and maxima can vary considerably, and here as well much more so in the boreal than in the tropical climate.

A seasonal pattern like the one shown for the site in Mongolia has important consequences for hydropower projects. It means that during winter, precipitation will fall as snow and that therefore river discharge in winter will be low to very low. In addition to that, when temperatures are that low—with even maximum values between November and March being rarely if ever above 0 °C—this also means that the reservoir, and in such a case probably even rivers, will be frozen every winter. This can have consequences for operating the power plant (see Sect. 25.2), and it may be of importance for planning and timing the construction process.

8.1.1.2 Impacts and Mitigation

Large water bodies influence the climate of their surroundings, especially temperature and humidity. In temperate and cold climates (here as long as the lake is not covered

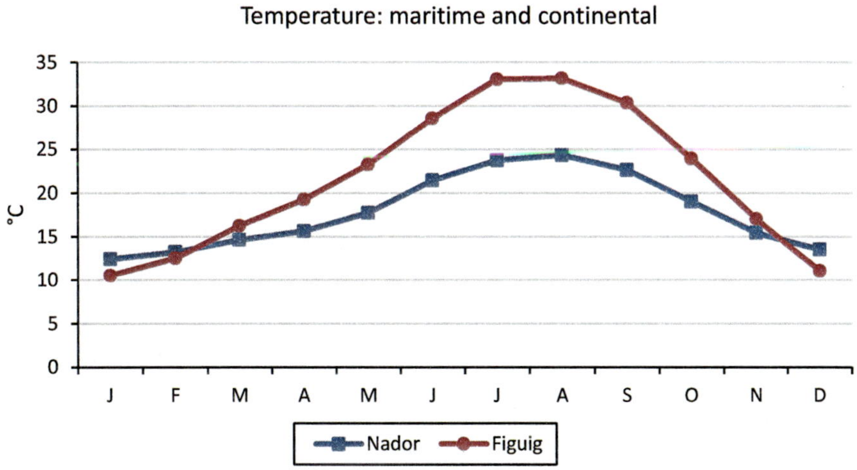

Fig. 8.2 Mean monthly temperature for two stations with different distance from the sea

with ice), the most noticeable effects are a general cooling in summer, a warming in winter, and a reduction of the daily and seasonal temperature variation. This effect can be clearly seen when the climate of a place on the seashore, with maritime climate, is compared to the one of a place far away from sea influence, with a continental climate type. Direct measurements have documented this effect. So it has been demonstrated that for instance a small island in the Finnish Bay exhibited a January temperature that was 1.3 °C higher than that of a station in the nearby land some distance away from the coast, while the average May temperature was 2.5 °C lower. Similar effects have been demonstrated in the vicinity of two lakes (Lake Peipus, Lake Chelkap), although in these cases the differences were smaller, reaching only + 0.3° in winter, – 0.7° and – 1.8°, resp., in summer (Alissow et al. 1956). Figure 8.2 shows average monthly temperatures for two stations in Morocco, where this effect can be seen clearly.

Nador is located on the Mediterranean coast. The smoothing effect of the sea on temperatures is clearly visible in comparison with the second station, Figuig, which is located at the edge of the Sahara desert, at a distance of almost 400 km from the sea, and where the differences between coldest and hottest months are much more marked than on the coast. A similar effect can be seen in the daily variations of temperature, as illustrated for Lake Aral in Fig. 8.3 showing measurements from August 1902—when Lake Aral was still a lake, see Sect. 23.5; the daily amplitude of surface water temperature is merely 0.6 °C, of the air 1 m above the water 1.5 °C, at the shore 6.9 °C and in greater distance from the lake 9 °C. Measurements from a very large reservoir, Lake Nasser (Aswan high dam, Egypt) showed July temperatures of 35 °C (land, 4 km away from lake), 32 °C (on a raft, centre of the lake), and water surface temperature of 27 °C, and January temperatures of 15°, 16° and 17 °C, respectively (El Bakry 1993). A similar effect of reducing the day-night temperature amplitude

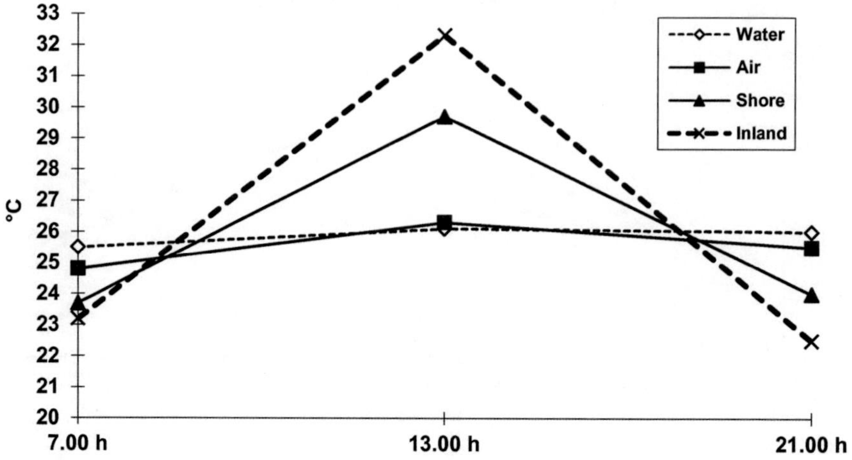

Fig. 8.3 Daily temperature variations as influenced by a lake. *Source* Based on data from Alissow et al. (1956)

was observed for the 1460 km^2 Itaipu reservoir on the Brazil-Paraguay border (Stivari et al. 2005).

This is mainly due to the fact that the water is able to store a considerable amount of heat, and that it reacts very slowly to changing temperatures. While on land the temperature (soil and air layers close to it) can exhibit large daily fluctuations, this is not the case for water. In Irkutsk, the daily variation in summer is 13.5–21.7 °C, in winter 5.7–14.5 °C (air 2 m above ground), the soil surface variation is even greater (29.8 °C in summer, 6.2 °C in winter; measurements for a day with rather low air temperature variation). It will be noted that Irkutsk is exhibiting a continental climate in spite of the proximity of Lake Baikal, which is one of the largest inland water bodies of the world. The biggest daily variations in air and soil temperature are recorded from deserts, which lack the cooling effect of evapotranspiration. In African deserts, daily surface temperature differences of up to 43 °C have been recorded, in middle Asian deserts up to 50 °C. In larger water bodies (seas and very large inland waters), normal daily fluctuations of temperature close to the surface are usually below 1 °C (and smaller for deeper layers), and this is also the case for the air layer of approximately one meter above the water surface, the difference getting more pronounced in higher air layers (Alissow et al. 1956; Geiger 1950).

In the case of smaller lakes, it is very difficult to clearly define their effect on ambient temperatures in their surroundings, as the complex mixing processes due to local winds tend to blur this influence. Nevertheless, it can be said from experience that, as a general rule, lakes have a beneficial effect on the local microclimate. This is shown by the fact that, in temperate regions, some plants normally limited to a warmer climate grow exclusively or at least much better in the vicinity of lakes (e.g. vineyards in central Europe). Furthermore, settlements tend to concentrate around lakes. While this has certainly different reasons, historical and scenic ones among others, the microclimate certainly contributes to it. In temperate and cold regions, this effect of a lake on the climate manifests only while it is not covered with ice. A compact ice layer covering the water body effectively blocks temperature exchange of the lake with its surroundings.

A few publications are concerned with the question of evaporation from lake surfaces (Kuhn 1977; Hoy and Stephens 1979). However, there does not seem to exist, to date, a thorough climatic study in relation with a man-made lake, which would compare temperature and humidity values before and after dam construction. The studies that at least mention potential climatic effects of artificial lakes attribute to them a minor or almost negligible effect (e.g. Odingo 1979), with the potential exception of the two studies mentioned above, namely Lake Nasser and Itaipu. This seems perfectly understandable given the fact that, while artificial lakes certainly create considerable impacts on the environment, climatic changes can, in the light of the details given above, be considered as of minor importance and, if at all noticeable, then rather beneficial, and if at all can only be detected in very large reservoirs.

For an ESIA for a dam project, therefore, impacts on local climate are usually on the list, but of very minor importance. Still, sometimes one can see examples of, in my opinion, misinterpretations or exaggerations in this respect, as shown by the following examples.

- When, in the early 1990s, we made the feasibility study for Egiin HPP in Mongolia (EWE 1994), we received an request for collaboration in the ESIA, and specifically for impacts on local climate, from a university institute in Ulaan Baatar. The project would have been to build an approximately 80 m high dam on the Egiin River immediately upstream of its confluence with Selenge River, in Northern Mongolia, and it would have created a reservoir of approximately 120 km^2. In order to show how important their contribution would be, the authors of this request described what would probably happen in this area once the reservoir would be there: a generally bad weather, with much more clouds and much higher precipitation, which would make life for the inhabitants of this area (of which there were very few) really difficult. Our answer to that was to point out that just about 200 km north of the site, on the same river, there was Lake Huvsgul, a natural lake about three times the size of the planned reservoir, that this area had a higher population density than the project area, that summer tourism was one of the economic activities in this area, and that all this presumably was not due to an especially unpleasant local climate caused by the lake. The project, as said, was located in the northernmost part of Mongolia, which is the part of the country receiving the highest amount of precipitation, but even there lack of rain is more likely to be a problem than an excess of it.
- One thing that can sometimes be seen, when it comes to a discussion of environmental impacts of reservoirs on local climate, are statements of a general nature, like e.g. the following one: "The reservoir lake will change the micro-climate of the Göksu Basin, adversely affecting the flora and fauna of the region, including some rare species." (Biro 1998). This refers to a hydropower project in Turkey;

the paper does not provide any data on climate, the species allegedly affected and the predicted effects, but the wording clearly shows that they must be adverse.

- An ESIA Report for a dam project in Lebanon (Gicome 2015) had identified the accumulation of an "enormous quantity of water" as leading to very serious impacts on the vegetation as well as on the agriculture in the entire region, since the increase in humidity, which allegedly would accumulate in the valley, would lead to a proliferation in parasitic fungi and insects. The project in question aimed at creating a reservoir of only about 1 km^2, in a climate with cool winters and hot, dry summers. An increase in humidity would, if at all detectable, be limited to the area in its immediate surroundings, and this only in conditions without wind, and this effect would be diminished or even turned into its contrary by the drawdown of the reservoir in summer, which would create a hot, dry zone of bare rock around it.

Only very large reservoirs would be big enough to have a measurable effect on the microclimate of their surroundings, and in most cases these effects would probably be rather beneficial. While dam projects do have adverse impacts, effects on local climate do certainly not count among them.

Since temperature is mainly a descriptive parameter and project impacts are observed only very exceptionally, mitigation is not required (and would not be possible!).

8.1.2 Precipitation

8.1.2.1 Baseline

For hydropower, precipitation is certainly the most important climatic factor, since the potential for generating electricity depends most of all on the availability of water. The relevant parameters are total yearly precipitation and seasonal distribution. This latter can show very different patterns, in some cases being more or less evenly distributed over the year, but in most places showing a marked seasonal pattern with often one peak, and sometimes, especially in proximity of the equator, with two yearly peaks (Walter and Breckle 1984: 20). Precipitation then translates more or less directly into river discharge, whereby modifications can be caused by the type of precipitation (rains versus snow) and the geometry of the catchment area (see Fig. 9.7). Figure 8.4 provides some examples of typical rainfall patterns in different climate zones.

Short comments to the data sets in Fig. 8.4:

- Grand Poubara, Gabon: close to the equator, rainfall pattern showing two rainy seasons separated by one long and marked (June–August) and one short and not very marked (December-January) dry season. Total annual precipitation 1855 mm. Tropical climate with very little variation in temperature.

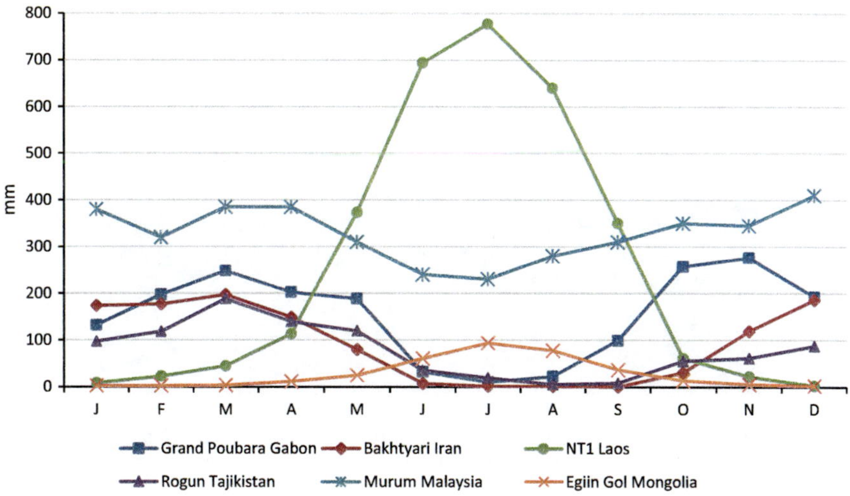

Fig. 8.4 Different annual rainfall patterns. *Sources* Grand Poubara: Pöyry 2008; Bakhtiari: DEZ 2006; NT1: Pöyry 2015; Rogun: Pöyry 2014; Murum: EWI 1998; Egiin: EWE 1994

- Bakhtiari, Iran: precipitation almost exclusively in winter, very dry summer (June–September). Total annual precipitation 1117 mm. Temperate climate with hot summers and cool winters, precipitation at higher elevations partly falling as snow.
- NT1, Laos: clearly seasonal pattern with one long, marked rainy season (monsoon). Total annual precipitation 3100 mm. Seasonal temperature differences, but no frost.
- Rogun, Tajikistan: similar pattern as mentioned for Bakhtiari, but most of the winter precipitation falling as snow; dry summer. Total annual precipitation 930 mm. Continental climate with marked seasonal temperature differences, hot summers and cold winters.
- Murum, Malaysia: site located in Sarawak, Borneo. Large amount of precipitation year-round, no dry season. Total annual precipitation 3945 mm. No marked seasonal variation in temperature.
- Egiin Gol, Mongolia: Very continental climate with warm summers and very cold winters, with marked daily variations in temperature. Precipitation mainly in summer, winter precipitation falling as snow. Total annual precipitation 329 mm; the location is in Northern Mongolia, precipitation there is just enough to allow tree growth under suitable topographic conditions, i.e. on slopes with a generally northerly exposition (Fig. 8.5). Further South, where it gets dryer, the vegetation changes to treeless steppe and finally to desert.

One important issue, in addition to total precipitation and its seasonal distribution, is the long-term trend in relation to climate change. In some cases, where sufficiently long data series are available, clear trends can be demonstrated (see Sect. 8.2).

Fig. 8.5 Forest-steppe transition zone in Northern Mongolia. North facing slopes covered with larch (*Larix sibirica*) forest, south-facing with steppe vegetation (see also short description in Hilbig et al. 1984: 33; 1989: 12)

8.1.2.2 Impacts and Mitigation

Hydropower projects, while strongly depending on precipitation, have no impact on it. No mitigation.

8.1.3 Evaporation

8.1.3.1 Baseline

Evaporation is the "loss" of water from soils or a water surface; its magnitude depends mainly on temperature and water availability. Three values have to be considered:

1. Potential evaporation: the evaporation that would occur if water were freely available.
2. Actual evaporation: the site-specific evaporation; if water availability is restricted, this can be much lower as potential evaporation.
3. Evapotranspiration: the sum of evaporation of water from soil or water surfaces plus the transpiration from plants (DeSanto 1978: 248).

Table 8.1 provides data on monthly and yearly averages of daily evaporation and on total yearly evaporation from locations with different climate. For the example from Franceville, yearly evaporation from an open water surface was estimated at 1140 mm and evapotranspiration at approximately 1000 mm. Table 8.2 summarises evapotranspiration values for a range of vegetation types. The four series from the USA are evaporation form open water surface. The values show the large variation in total yearly evaporation, but also the seasonal variation, which is large in climates with

Table 8.1 Mean daily and yearly evaporation

Site	J	F	M	A	M	J	J	A	S	O	N	D	Y	Year
Franceville (Gabon)[a]	3.4	2.5	3.8	4.0	3.6	3.2	3.2	3.4	3.7	4.3	3.4	3.2	3.48	1270
Germany[h]	0.2	0.3	0.7	1.6	2.3	2.5	2.5	2.3	1.8	0.7	0.3	0.2	1.3	480
N. Carolina, Bee Tree[c]	0.9	1.2	2.3	3.6	4.7	4.8	4.5	4.2	3.3	2.6	1.7	1.0	2.9	1046
Alabama, Fairhope[c]	1.7	2.0	3.1	4.2	5.2	5.2	5.0	4.7	3.8	3.1	2.0	1.4	3.4	1242
Hawaii, Waianae[c]	3.5	3.7	5.1	5.6	6.2	6.5	7.0	6.8	6.0	5.0	3.9	3.4	5.2	1883
Arizona, Bartlett Dam[c]	3.5	3.8	6.0	8.8	12.3	14.4	14.6	a2.3	10.8	7.8	5.1	3.5	8.5	3089

Source
[a] Pöyry (2008) potential evaporation
[b] Blüthgen (1980), actual evaporation
[c] US Dept. of Commerce (Anonymous, 1950), evaporation from open water surface
J–D and Y in mm/day, Year in mm

marked seasonal differences in temperature, and almost non-existent in Franceville, Gabon, where temperatures are almost the same year round.

Values from other sites are in the same range. Tanaka et al. (2008) discuss seasonal variations in evapotranspiration in seasonally dry forests (monsoon forests) in Thailand, which varies between 0.6 and 5.9 mm/day; they do not provide an average, but it would be about 2.5–3 mm/day, resulting in a yearly evapotranspiration of around 1000 mm. For Murum HPP in Sarawak, Malaysia, a region with year-round rain and a total yearly precipitation of 4200 mm, yearly and daily evapotranspiration were determined at 1368 mm and 3.7 mm, respectively (EWI 1998). Data on evaporation from water surface were collected for a large number of stations in the United States (Anonymous 1950); for 24 stations with complete yearly data sets, predominantly from Southern and South-Western states, they range from 1046 to 3089 mm/year, with an average of 1887 mm/year. As a summary, it can be said that evaporation from an open water surface is highest in a hot and dry climate. This is also true for PET, while AET largely depends on the availability of water; it is lowest in cold and dry climates. PET as well as AET are usually somewhat lower than EWS.

8.1.3.2 Impacts

Two different types of impacts have to be considered in relation with evaporation, namely, (i) increase in local humidity and therefore change of the local microclimate, and (ii) loss of water.

Table 8.2 Evapotranspiration from different vegetation types (mm/year)

Vegetation	Precipitation	PET	AET	EWS	Source
Montane rainforest, Papua	3500		860	1,380	Walter (1984: 4)
Lowland rainforest, Malaysia	2227		1.203	1,254	
Lowland rainforest, Malaysia	3053		1310	1,658	
Lowland rainforest, Sierra Leone	5795		1159	1,010	
Lowland rainforest, Australia	4037		1421		
Tree plantation, Java	4768		1075	1,344	
Hot desert	164	1022	164		Frank and Inouye (1994)
Tundra	239	278	202		
Cold desert	264	682	256		
Shortgrass prairie	440	851	413		
Taiga	506	481	380		
Tallgrass prairie	641	735	571		
Broadleaf-needle forest	884	651	543		
Broadleaf forest	965	665	588		
Conifer-deciduous forest	1185	941	821		
Savannah	1350	1245	894		
Wet tropical forest	2301	1510	1363		
Burkea savanna, South Africa	630		574		Bate et al. (1982)

PET potential evapotranspiration; *AET* actual evapotranspiration; *EWS* evaporation from water surface

Concerning the change in microclimate, this was discussed in Sect. 8.1.1.2. Water loss due to evaporation from the reservoir surface is a potentially more serious issue. Mainly large and shallow reservoirs in hot and dry conditions can lead to considerable losses of water, which can reduce water availability e.g. for irrigation. One example that is often mentioned in this context is Lake Nasser, created by the Aswan High Dam, with a surface of 6000 km^2 one of the largest reservoirs, located in a hot desert environment (e.g. Lavergne 1986). Annual losses due to evaporation from this reservoir were estimated at between 10 and 20 km^3 and were then determined to be between 12.5 and 16.3 km^3, on average 7.4 mm/day, and this from a total inflow into the reservoir of 55 km^3 per year (El-Magd and Ali 2012). However, while the reservoir certainly leads to a “loss” of water, without the dam most of this water would just flow through the country in the high flow season, leaving nothing for irrigation during the low flow season; overall water availability for irrigation would definitely be much lower without the dam.

The Rogun dam on Vakhsh River in Tajikistan, presently (2020) under construction, will create a reservoir with a surface of 170 km^2. In the ESIA (Pöyry 2014) it was calculated that evaporation from the reservoir would be in the order of magnitude of 800–1200 mm per year, which combined with the Nurek reservoir, further downstream on the same river, would translate into a yearly loss of about 0.2 km^3 of water, 1% of the average yearly inflow into the Rogun reservoir. Whether this is really a loss, after all, is not so clear. There is certainly evaporation from the reservoir surface, but since evapotranspiration in this area is in the same order of magnitude as evaporation it could be argued that most of this "lost" water would not even reach the river in the situation without a dam. Still, the calculated difference could be considered as a net loss of water availability in the downstream area, i.e. mostly for Uzbekistan, where the water is needed for irrigation. However, water losses from the Tuyamuyun reservoir (25 m high dam, reservoir area 303 km^2) on the Amu Darya River on the border between Uzbekistan and Turkmenistan, one of the major storage facilities used for alimenting the irrigated fields of this region, are estimated to be 2.6–10 km^3 per year (Dukhovny and de Schutter 2011: 259). Tuyamuyun is a shallow reservoir in a very hot and arid environment, and evaporation loss is much higher there than it is in the narrow and deep reservoirs in Tajikistan, located in a much cooler climate zone. It could therefore be argued that, when we look at the entire system, storing as much water as possible in the mountains of Tajikistan would be better for the irrigation system in Uzbekistan than storing it there in the desert. Obviously, this would require a good and reliable organisation for the entire system (see Chap. 23 for more details).

In a moist climate, the differences will be smaller. Evaporation from an open water surface is lower if the relative humidity of the atmosphere is already almost at saturation, and evapotranspiration from the surrounding vegetation (or, for that matter, the vegetation in the reservoir area prior to impoundment) is higher and can be almost as high as that from an open water surface. There is conflicting information on evapotranspiration from floating water plants like the water hyacinth (*Eichhornia crassipes*): While Walter and Breckle (1984: 189) state that evapotranspiration from a dense mat of this plant is three times as high as that from an open water surface, Doorenbos and Pruitt (1977) report that when a water surface is covered by aquatic weeds the water loss into the atmosphere will be lower than that from a free water surface; according to them this is due to a combination of the sheltering of the water surface by the weeds and a higher reflectance of the green plants and their internal resistance to transpiration.

8.1.3.3 Mitigation

Mitigation is not possible. For very small reservoirs used uniquely for irrigation purposes covering the surface with plastic sheets is sometimes recommended. This is obviously not feasible for larger reservoirs, and definitely not for multi-purpose reservoirs which are also used e.g. for fishing, transport or leisure. However, given the insignificance of the impacts mitigation is certainly not required.

8.1.4 Other Climatic Conditions

8.1.4.1 Wind

Wind, i.e. seasonal distribution of wind directions and wind velocity, is sometimes mentioned in ESIAs for dam projects, but it is usually not a major issue. Only in the case of very large reservoirs wind velocities can be influenced locally, when roughness of the surface can be reduced, which may lead to an increase in wind speed. I have not come across any case where this would have been a problem, and mitigation, other than not implementing the project, would not be possible.

8.1.4.2 Ice

Under special conditions, ice can be an issue. This will be the case in places with cold winters, i.e. in high mountain areas and in Northern areas like Scandinavia, Russia and Canada. Prowse and Gridley (1993) give an overview of environmental issues related to river ice, and Gebre et al. (2013) review ice effects on hydropower in Norway.

According to Gebre et al. (2014), the main issues for hydropower related to ice are the following:

- Intake blockages with frazil ice and anchor ice causing head losses and even complete shutdowns;
- Flow reductions to the intakes in case of run-of-river intakes causing reduced output and even complete shutdowns;
- Upstream and downstream flooding caused by ice jamming;
- Icing of structures, especially gates (intake gates and spillway gates) that causes operational and safety concerns in case of spillway gates;
- Creating open water reaches downstream of power plant outlets which may lead to extensive frazil ice formation and jamming affecting downstream facilities;
- Operational restrictions on hydro-electric operators to avoid ice problems.

Most of these issues are mainly of a technical nature, concerning operation of hydropower plants under conditions with ice formation. My experience with such cases is limited, therefore I shall not go into much detail here.

In cases like Mauvoisin, located at over 2000 m asl in the Swiss Alps (Sect. 25.1), ice will cover the reservoir in the winter months. However, since the intake is far below the surface of the water, this will not be a problem for the operation of the plant. Drawdown during winter is considerable, and the ice cover will sink with it, often cracking, since the reservoir gets narrower with decreasing water level. However, since the reservoir is not accessible in winter, not least due to the high risk of avalanches in the valley leading to it as well as on the mountain slopes along the reservoir, this is also not a safety risk. The power plant is at much lower elevations, where the temperatures are not as low as to let rivers freeze. The issue was different for Egiin Gol HPP in Mongolia, which would have used the Egiin Gol (Egiin River), the outflow of the large Lake Huvsgul in Northern Mongolia, shortly upstream of its confluence with Selenge River, the main tributary of Lake Baikal in Russia. There, the rivers as well as Lake Huvsgul freeze in winter, the latter to the extent that it is used as a shortcut for lorries transporting goods from Russia, which drive then across the lake. The risks that were anticipated were that if water to be released from the reservoir by peaking operation of the power plant would flow on top of the ice covering the river, this might lead to an accumulation of ice at this place, or if the ice cover on the river would be broken up by strong fluctuations in discharge ice blocks might be swept downstream; in both cases this could lead to ice jamming as listed above, which might be a risk for the downstream area and could negatively affect operation of the power plant. A different problem was encountered in the case of Pamir 1 HPP in Tajikistan (see Sect. 25.2, Fig. 25.2). There was a plan to increase water available for power generation in winter by using an additional small high mountain lake, Lake Turumtaikul. The idea was that a regulating structure at the outlet of the lake would have allowed to retain some additional water during summer, to release it at the rate of approximately 0.5–1 m^3/s during January and February, the two winter months where water availability is lowest and electricity demand highest, to be used in the power plant which is located approximately 100 km further downstream. However, given the fact that the lake is located at an altitude of 4200 m asl, with winter temperatures of – 30 °C or lower, there seemed to be a high risk that the intake structure, after having been closed for several months, might be frozen, and mainly that the relatively small amount of water to be released to a then completely dry and rather wide river bed would just freeze and that no water from this source would actually reach the power plant.

8.2 Climate Change

8.2.1 Baseline

I smoked my first cigarette at about the age of 12, i.e. in the late 1950s. I am quite convinced that already at that time we knew that smoking was bad for one's health, and we knew then—or shortly afterwards, that it can cause lung cancer. For this reason I find it quite astonishing that still rather recently people who had been smoking for 30 or 40 years and then developed lung cancer could successfully sue the tobacco industry on grounds that they had not been warned. A similar phenomenon can be observed now where climate change is concerned: in the late 1960s, students were shown the graph with the CO_2 concentrations measured on Mauna Loa in Hawaii, which showed a very clear, not to say dramatic, increase in CO_2 concentrations in the atmosphere; this was also mentioned in the 1972 Club of Rome Report (Meadows 1972: 60) which, interestingly, predicted a CO_2 concentration of 380 ppm for the

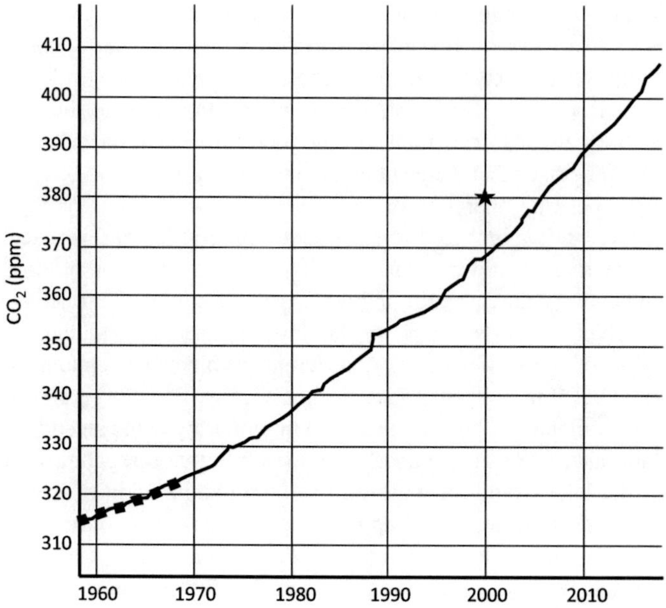

Fig. 8.6 CO_2 concentrations on Mauna Loa. *Source* Redrawn after Hagens (2020: 10). Fat dotted line 1958–1968: values shown in Meadows (1972: 60). Asterisk: Estimated concentration in 2000 (Meadows 1972: 60)

year 2000; the most recent IPCC Report shows 360 ppm for the year 2000 (IPCC 2014: 3), and this is quite a remarkable fit (see Fig. 8.6).

And if not right then yet, it became clear shortly afterwards that a rise in CO_2 content in the atmosphere would lead to a rise in temperature. Graphs that appear regularly in the press show a dramatic rise in average global temperature, among other observations the fact that the ten or so hottest years since start of measurements were observed since the year 2000. Consequences of this are visible everywhere, most clearly in the melting of glaciers and permafrost. There was even a warning that global warming could lead to a reappearance of dragons, which reportedly have been in hibernation since the Little Ice Age which started in the fifteenth century (Hamilton et al. 2015); it remains to be seen if this will actually materialise. In any case it is difficult to understand how global warming can still be ignored or negated today. Voices which postulate alternative causes for climate change (e.g. Calder 1999, who saw raising CO_2 concentrations as a consequence rather than a cause of global warming) seem to be in decline. Publications proposing solutions for the problem are numerous and differ widely (to list just a few: Bekkering 1992; Rosen 2018; Gates 2021).

In the context with hydropower, one has to consider two main aspects concerning climate change, namely:

- Effects of climate change on hydropower projects, mainly through changes in precipitation and river flow patterns; this also includes role of water storage as possible contributions of dams to mitigate climate change effects.
- Contributions of energy sources, including hydropower, to greenhouse gas emissions and therefore to climate change, or the potential for specific energy sources to reduce GHG emissions.

8.2.2 Effects of Climate Change on Hydropower Projects

Climate change can have very important consequences for hydropower projects, and this in three major ways:

- Change in precipitation: the IPCC models predict changes in the amount of precipitation which can be very different in different parts of the world; in some areas precipitation is likely to increase, while in others it will decrease. This, in the middle and long term, will affect energy production and thus the economic performance of affected projects.
- Glacier melt: presently, glaciers are melting at an unprecedented rate. This increases the amount of water from melting and therefore can make more water available for hydropower (see the case of Rogun HPP discussed in Sect. 23.6). It also creates a risks for valleys receiving this water, namely by GLOFs (Glacier Lake Outburst Floods; Cook et al. 2018; see also Kraaijenbrink et al. 2017). Once the glaciers gone, however, flow in these areas will diminish, and so will energy generation. Increasing risks in high mountain areas, and specifically to hydropower structures in such areas, related to or enhanced by glacier and permafrost melt due to raising temperatures are pointed out by Shugar et al. (2021), who analysed the 2021 disaster at Chamoli, Indian Himalaya, which killed 200 people and destroyed two hydropower plants.
- Increase in extreme weather situations: this is an aspect that often seems to be forgotten in the public discussion on climate change: as soon as there is an extreme snowfall or a short period of extremely low temperatures (both has, e.g., been the case in the north-central part of the USA in January 2019), there are immediate reactions of the type of "so much for global warming!". However, the models predict not only a rise in average temperature, but also a rise in frequency of weather extremes. This, on the one hand, could make reservoirs more important: if rainfall is less reliable, with more pronounced spells of droughts, storage of water during more severe rainfall events could get a higher priority (Muller 2018). On the other hand, this same effect - extremely dry periods and very heavy rainfalls—could lead to an increase in erosion, and this in turn could speed up the siltation of reservoirs, in this way reducing the life span of dams.

Climate change is, for the reasons indicated, certainly a topic which needs to be addressed in ESIAs for dam projects. The two graphs in Fig. 8.7 show a development which is not anticipated, but already takes place: diminishing rainfall. The

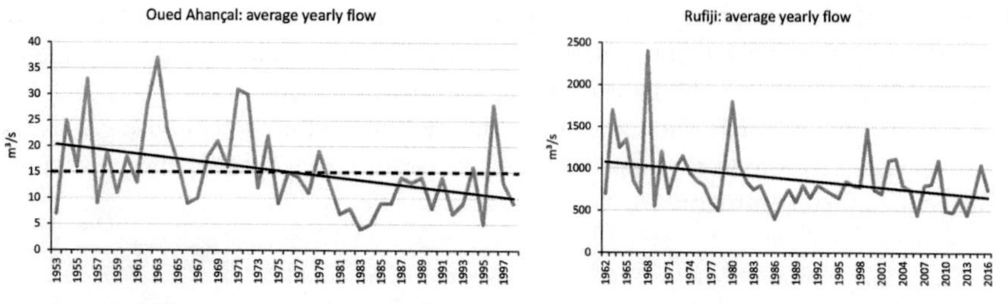

Fig. 8.7 Long-term trend in annual discharge of two African rivers. Based on data from Pöyry (2007) for Oued Ahançal and Powerchina (n.d.) for Rufiji

first example is the Oued Ahançal, a small river in Morocco, the second one is the Rufiji in Tanzania: two rivers from almost opposite places in Africa, which show a very similar pattern of river discharge: high year-to-year variations, and an overall decreasing trend, indicated by the solid black line. This can mean two things: on the one hand, there might be an increasing risk for the economic viability of the HPP, on the other hand this could also indicate that dams (and reservoirs for storing water which gets ever scarcer and more unpredictable) are even more important under these changing conditions. It is not the duty of the ESIA to make this decision, but it is certainly its duty to point out the problem.

The curve for the Oued Ahançal also shows something else, namely, an extended dry period between 1980 and 1993, with 13 consecutive years showing a mean annual flow below the long-time average of 15.2 m^3/s (dotted line); this dry period is also referred to in the discussion of the Aswan High Dam project (see Sect. 21.1.1). Interestingly, Rufiji shows a similar pattern in the same period.

Depending on the site, the situation can be different from what is shown above. Rainfall data from three stations around a project site in the upper Indus valley in Pakistan indicated, along with higher irregularity, increasing rainfall (ILF 2008: 34). In the case of the Central Asian Rogun HPP, Tajikistan, recent climate models led to the following conclusions for future climate development (Pöyry 2014):

- Mean air temperature will rise by about 1.7 °C till 2050; this will increase evapotranspiration, which in turn means less surface runoff.
- Mean annual precipitation will remain unchanged, but variability from year to year is likely to increase.
- Runoff of many rivers will increase due to the melting of glaciers until about 2080, and will then decrease.
- Higher temperatures and the retreat of glaciers will lead to an upward shift of permafrost, which can mobilise large amount of glacier sediments, increasing the sediment load in rivers, and which could increase the risk of landslides. This also could mean a higher risk of GLOFs (glacier lakes outburst floods); effects of such events on the valleys below these glaciers have been emphasised by Cook et al. (2018).

- Peak flows in rivers will tend to occur earlier in the year due to more precipitation falling as rain instead of as snow, and due to earlier snow melt (Blöschl et al. 2017).
- Variability of flows from year to year, and therefore cases of dryer and wetter than average years, is likely to increase.

8.2.3 GHG Emissions from Reservoirs

In comparison to other, and obviously mainly fossil fuel fired thermal power plants, hydropower is a very clean form of energy generation. So says Muller (2019), and proves it with the data shown in Fig. 8.8. Similar figures are presented by Berga (2016).

Figure 8.8 speaks for itself: while electricity based on coal causes a greenhouse gas (GHG) emission of over 800 g CO_{2eq}/kWh, it is only 24 g in the case of hydropower. What Muller does not mention (but his source, Schlömer et al. 2014 IPCC Report, does) is that 24 g is the median value for hydropower, while the maximum for a specific project can be as high as 2000 g CO_{2eq}/kWh, i.e. almost three times as much as that from a coal fired power plant!

This just reinforces one of the main points made several times in this book, and which is very important: when talking about environmental impacts of hydropower, one has to look at every particular project very carefully, and should not try to assess impacts based on any average values; these are, of course, important and valuable, and the graph above still provides an overall picture of various energy sources which is realistic and good to know. However, it does not tell anything about an individual project, where the situation can be completely different.

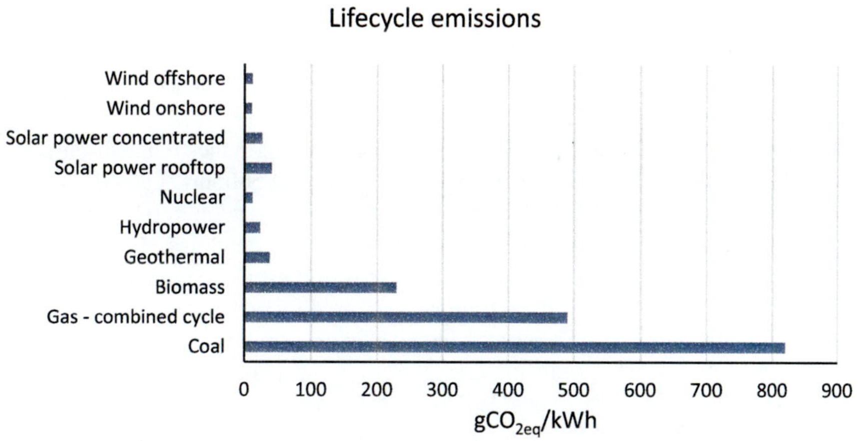

Fig. 8.8 Life cycle GHG emissions for different energy sources. Based on data from Schlömer et al. (IPCC 2014)

One of the main critics of hydropower in terms of GHG emissions is Fearnside (1989, 1995, 1997, 2015); he, by the way, also objected to Muller's overall statement in a letter published in Nature 568 (7750: 33, 2019). Muller mentions that the biomass in question, which is the source of GHG emitted from reservoirs, would decompose anyway, dam or not. This is true, but there is one very important difference: in a river (and, at least to some degree, also in natural wetlands, see Pangala et al. 2017) biomass degrades usually under aerobic conditions, and is therefore decomposed to CO_2. However, in a reservoir, unless it is a very shallow one, biomass will very often, and as a rule in reservoirs containing large amounts of biomass, decompose under anaerobic conditions, and therefore be transformed to methane (CH_4), which is a GHG about 20 times as potent as CO_2. That the issue over GHG emissions from reservoirs it not easily resolved is shown by its discussion in the literature, which has been going on for some time. Rudd et al. (1993), investigating a reservoir in Canada which submerged significant quantities of peat in addition to fresh biomass, came to the conclusion that in that case GHG emissions reached values which were comparable to a coal fired power plant of the same capacity. This was objected on the grounds that the case analysed was very untypical for the situation in Canada (Gagnon and Chamberland 1993) and that as a rule emissions from gas and oil fired thermal power plants would be higher than those caused by hydropower by a factor of 25–30. It was also pointed out that GHG problems would arise mainly in hydropower plants with large reservoirs in relation to the installed capacity and that therefore the results from both Rudd and Fearnside could not be generalised (Gagnon 1998). Rosa et al. (2002) found large differences in actual GHG emissions from a number of Brazilian HPPs, without really finding an explanation for this observation. Bauer and Hirschberg (2017) found that estimated life cycle emissions of hydropower in Switzerland are 6–7 g CO_{2eq}/kWh "predominantly caused (directly and indirectly via material supply chains) by infrastructure construction of the hydropower plants", and that mainly due to the sparse vegetation in alpine regions, where most reservoirs are located, this is not an issue in large parts of Europe. This mainly shows, once more, the importance of closely looking at the project- and site-specific conditions in each case. This is discussed in more detail in Chap. 17, which deals with the different aspects of reservoir stratification.

Another issue, not related to GHGs, but of potential importance in the discussion on energy and climate change, was brought up recently (Wohlfahrt et al. 2021). In a study including over 700 HPPs distributed worldwide they came to the conclusion that the fact that reservoirs are often darker than their surroundings, this leads to a storage of heat in the reservoirs which otherwise would have been reflected back to space by the higher albedo of the terrain now covered by water. According to their calculations, about half of the reservoirs included in the sample produced net climate benefits, in comparison to producing energy from fossil fuels, within four years, but due to this additional storage of heat 19% of them needed more than 40 years to reach this stage, and 12% would fail to do so over an assumed lifetime of 80 years. This might be an additional topic to be reflected in future ESIAs for hydropower projects. However, this study does not answer another question: how much is the albedo loss due to climate-change induced loss in glaciers and the extent of snow cover, and how much of this could be prevented by replacing a fossil fuel based thermal power plant with a reservoir and hydropower plant of the same capacity? The answer to that question might well be in favour of hydropower.

8.3 Air Quality and Noise

These are two classes of environmental impacts which can present a considerable nuisance to affected people, and can result in serious health issues, especially in urban and other densely populated areas. They rank prominently in ESIAs on road projects. Both of them have some advantages when it comes to baseline data collection and to impact assessment, namely:

- As good as every country, and a number of various organisations dealing with environmental and health issues, have well-defined thresholds for noise as well as for air pollution.
- Measuring equipment is available which allows for fast and reliable data collection.
- There are well-established and widely accepted models for calculating future noise and air contamination levels, e.g. based on forecasts of traffic development in the future.

While, as said, these are very important parameters e.g. for road projects, very often for hydropower projects this is not the case. Traffic due to construction activities can locally be a source of impacts (security, noise, air pollution and dust), and noise can be a serious issue on the construction site, mainly at specific places and/or related to specific activities, as e.g. breakers for aggregate preparation, and these points have to be adequately covered, especially in the respective sections of the ESMMP (see Chap. 12). Still, and once again in my opinion, in ESIA studies there is often an effort dedicated to these issues which is not justified.

One example: for an HPP ESIA, an assessment of the noise situation was done by measuring noise levels for 24 h in four places. Approach, work carried out and results were described on 4 pages of the report, and in annex there was a table that took 34 pages providing minute-by-minute results of noise measurement on the four sites. It did not come as a surprise that the result showed that in the nearest town there were noise levels which were higher than the accepted threshold values; this was not astonishing, since this town is located directly by the road leading up the main valley, and there is heavy traffic almost day and night. However, even measures at the project site indicated higher than threshold noise levels during the night, and this in spite of the fact that this place is located in a lateral valley way out of hearing distance from the main road, and that there was no road leading to this site, and therefore no traffic which could cause noise. As it turned out, the river at the project site was the only relevant, permanent source of noise.

To say that very clearly: it was not the consultant who prepared this study who had done anything wrong (maybe with the exception that placing the recording device at the project site in a more suitable place somewhat away from the river might have produced a more relevant result): the TOR for this study, which were in compliance with the national guidelines, clearly asked for such an analysis. While, as mentioned, this may often be required and fully justified, it was certainly not in this case. It would have been sufficient to mention that noise levels are rather high in the town, and that here the project might lead to a somewhat worse situation due to traffic generated during the construction phase, and that the silence in the project site would also be disturbed during this period, that some specific places would likely be exposed to higher than tolerable noise levels, and that specific measures would be required for such situations. However, such an approach would be acceptable only if in an earlier stage of the study, during the screening and scoping exercise, this would clearly be defined in this way, and would be accepted by the competent authority in charge of examining and ultimately accepting the report. This, unfortunately, is not done very often, a situation that leads to much unnecessary effort.

8.4 Geology

8.4.1 Geology of the Dam Site

The geology of the dam site is of crucial importance for the engineering part of the project: choice of the exact dam location, the dam type, questions of stability of the dam, risk of seepage (and therefore the need for grouting), seismic situation of the area of dam and reservoir including the risk of reservoir triggered seismicity, etc. For the ESIA as such however, geology is very often not of high importance, given the fact that the project will not change the geology of the site, and therefore a summary of the relevant information provided in the technical documents will be sufficient. This summary should, as other parts of the ESIA, be done in a "not too technical" language, since it is not directed at experts in geology and engineering (as the corresponding part of the technical feasibility study certainly is), but at a general public.

8.4.2 Geology of the Reservoir Area

One aspect of surface geology in the reservoir area may be of importance, namely, slope stability, as it was e.g. in the case of Rogun HPP, Tajikistan. The reservoir to be created by this dam is located in an area which is prone to landslides, and this is can be exacerbated by the presence of a reservoir, mainly by a reservoir with a considerable seasonal drawdown, which can increase the risk of such landslides due to the rather rapid change in water level. Figure 8.9 illustrates this situation.

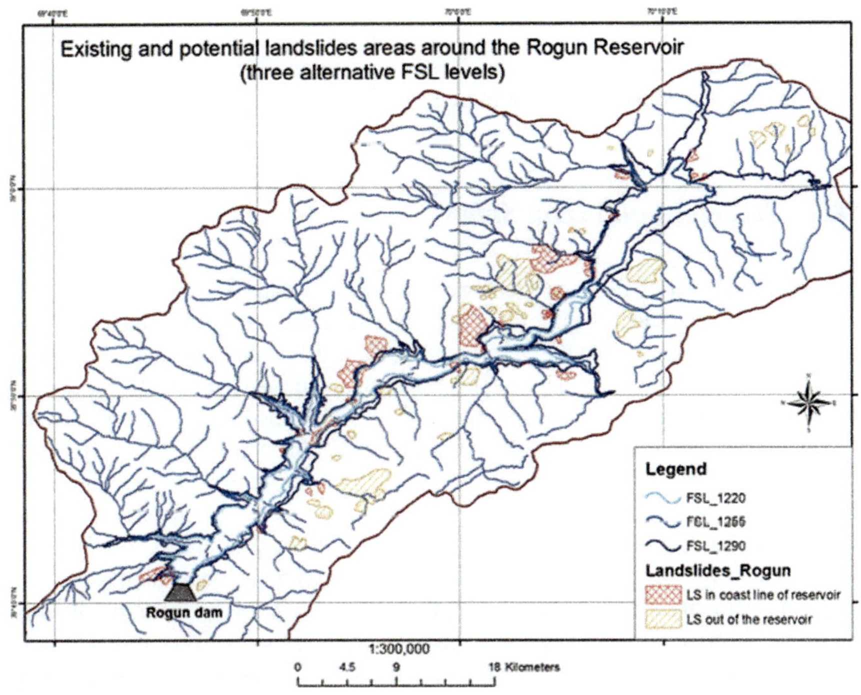

Fig. 8.9 Map of landslide prone areas around the future Rogun reservoir. *Source* Pöyry (2014). Reproduced with permission

The figure indicates the reservoir areas of the three alternatives which were studied (FSL 1220, FSL 1255 and FSL 1290 alternatives) in addition to the landslide prone areas, where landslide processes have been observed and are expected in the future. All these areas are located in the soft deposits of deluvial and proluvial genesis and some crushed rock outcrops, characterised by very active erosion processes.

These areas are subdivided into two groups, depending on location:

- on the shoreline of the future reservoir (crosshatched in the figure), which can be affected by the oscillation of the reservoir water level;
- out of the reservoir area (hatched in the figure), which are not directly influenced by the reservoir, but could block tributaries of the Vakhsh river.

No large active landslide blocks have been identified in immediate proximity to the shoreline of the reservoir which could trigger large impulse waves. Rockslides could be triggered by the dissolution of salt deposits in the foot of slopes. Oscillations of the reservoir water level can accelerate landslide processes, but none were identified which could create a problem for dam safety. Preventive measures were proposed for cases where such areas are in close proximity to infrastructure and settlement areas. Even if potential landslides were not expected to present a risk of impulse waves overtopping or damaging the dam and the hydraulic facilities, they still present a risk for settlements located nearby. Therefore it was recommended to study these issues in more detail, and to install a monitoring system in areas to be found as presenting a risk. For dealing with these issues, a close cooperation between the technical and the environmental consultant was required.

8.4.3 Seismicity and Dam Safety

One specific aspect of the local geology in relation to dam construction and HPP operation, and specifically to dam safety, is seismicity, i.e. the regional and site specific situation concerning the risk of earthquakes. While this, especially when the site is located in a seismically active zone, has to be mentioned in the ESIA report, the specific studies required for assessing the situation, drawing the relevant conclusions and taking the required measures are part of the technical studies for the project, geology first of all, and then structural engineering and dam instrumentation for monitoring the situation. The relevant parts of these studies will be taken over into the ESIA report, but additional evaluations of the situation are not a subject of the ESIA as such. One issue of special importance in this context is the question of reservoir triggered seismicity, i.e. the question whether the presence of the reservoir, mainly due to the sheer weight of the accumulated water, might or is likely to exacerbate the risk of earthquakes in the region of the project. This is just an example that illustrates the importance of communication and exchange of information between the technical and the E&S teams (Table 8.3).

8.5 Soils

In an EISA report, under the heading "Soils", one can read the following: "The soils are identified as Luvisols, Ferrasols, Arenosols, Acrisols, Nitosols, Cambisols and Lithosols. The upland soils are mostly well to moderately well drained, gravelly with a light-textured matrix which in some areas overlies an iron pan developed in situ at shallow depth. … The lowland soils within the zone include Fluvisols, Gleysols, Vertisols and are mostly non-gravelly with a reasonably moderate to high inherent

Table 8.3 Some earthquakes with damage to dams

Year	Place	Country	Magnitude	Damage to dams
1971	San Fernando	USA (Calif.)	6.5	Lower San Fernando, Pacoima
1990	Manjil	Iran	7.5	Sefid Rud
1990	Philippine	Philippines	7.7	Ambuklao, Binga, Magat, Pantabangan
2001	Bhuj	India	7.7	~ 245 earth dams
2010	Maule	Chile	8.8	Several

Source Martin Wieland (oral comm., 2020)

fertility status, making them more productive than their upland counterparts. … Other lowland soils include Plinthic Luvisols, Regosols and Histosols." A soil expert may be able to draw some conclusions from that, but otherwise it is probably not of much practical use. Still, descriptions of that type can be found rather often, and in most cases one can argue that this is sufficient, although it might not exactly follow the general ESIA rule that the report should be written in a way that it can be understood by non-specialists. But since very often such a text is not followed by any specific information on impacts, let alone compensation measures, it is usually accepted as sufficient.

More detailed information on soils may be required if:

- considerable agricultural areas will be submerged and need to be replaced/compensated;
- slopes of the future reservoir may be unstable, presenting a risk of landslides, although in such cases there might be an argument whether this is an issue to be dealt with as an aspect of local geology (see previous section) or under the heading "soils";
- soils with high content of organic material (peat) will be submerged, which could contribute to the GHG issue.

In most cases, however, this will be one of the topics which are required by the specific guidelines (sometimes called "terms of reference", although in reality these are basically just checklists), and therefore texts like the one provided above, copied from some pedological publication, or maybe just from another ESIA which had been produced for a project in the same area, will be included. No harm done, but not of much use either. Here, as in some other cases, it should be decided in the screening and scoping phase of the study whether or not the topic is of importance; if yes, it needs to get more attention, if not, it is better left away.

8.6 Sediments and Sedimentation

8.6.1 Prevailing Situation

Rivers carry large amounts of sediments, partly as bedload (coarse material transported along the bottom of the river bed) and partly as suspended solids, fine material which, at least as long as the movement of the water is strong enough, remains in suspension. The capacity for sediment transport depends on the gradient of the river, i.e. flow velocity, and on discharge. The amount transported, however depends largely on the erodibility of the soils in the area through which the river is flowing. Erosion and sediment transport is highest in steep areas with strong surface runoff after rains and high water velocity in the rivers. In areas with less or almost zero slope, sediment of different granulometry is being deposited, and this can lead to a status of equilibrium, where gravel or sand is deposited during receding and low flows, part of it

Table 8.4 Discharge and sediment loads of selected rivers

River	Drainage area 10^6 km^2	Water discharge km^3 year^{-1}	Sediment load 10^6 t year^{-1}	Sediment yield t km^{-2} year^{-1} [a]
Mississippi	3.27	580	210	64
Orinoco	0.99	1100	210	152
Amazon	6.15	6300	900	195
La Plata	2.83	470	92	35
Yellow (Huangho)	0.77	49	1080	1400
Yangtze	1.94	900	478	246
Mekong	0.79	470	160	202
Ganges/Brahmaputra	1.48	971	1670	716
Indus	0.97	238	100	61
Tigris-Euphrates	1.05	46	(53–105)[b]	(50–100)
Nile [c]	2.96	30	0	0
Niger	1.21	192	40	33
Congo	3.82	1250	43	12
Zambezi	1.20	223	20	17

Source Milliman and Meade (1983)
[a] figures mainly from Bukaveckas (2009: 33)
[b] cites figures from other studies, no own result
[c] situation after the construction of the Aswan Dam; before, sediment load was 100×10^6 t year^{-1}

being mobilised again when discharge is rising. In the end, sediments will eventually reach the sea, where depending on coast morphology, tidal and wave conditions vast deltas can be formed. On a smaller scale, this also happens in lakes, which over time will be completely filled with sediment.

High sediment yield can be attributed to natural factors leading to high soil erosion, like soil composition, steepness of slopes and high rainfall, but it can also be caused, or at least strongly influenced, by human activities like e.g. deforestation or other activities leading to the destruction of natural vegetation and exposure of the soils to erosion. The by far highest yield listed in Table 8.4 is the one from the Yellow River in China, which mainly flows through large areas with easily erodible loess soils, where human activities tend to increase erosion (Pearce 2018:118).

8.6.2 Impacts of Dams

Dams retain sediments. Since flow velocity in reservoirs is very much reduced, in large ones to the point of resulting in stagnant water, as in a large natural lake, bedload will no longer be carried along and suspended solids will sink to the bottom of the reservoir. This is illustrated here for the Nurek HPP, Tajikistan, for which the situation

Fig. 8.10 Inflow of Vakhsh River into Nurek reservoir, showing sedimentation process

was analysed in the ESIA for Rogun HPP, to be located a short stretch upstream of it on the same river.

Since the construction of Nurek HPP, the majority of the sediments of the upper Vakhsh catchment does no longer reach the lower catchment but instead is deposited in the reservoir of Nurek. Figure 8.10 shows the inflow of Vakhsh River into Nurek reservoir. It demonstrates that suspended solids brought by the river sediment quickly and are trapped in the reservoir.

In the case of Nurek, ongoing sedimentation would not only shorten the lifespan of the dam and HPP, it would also require a redesign and replacement of several hydraulic structures, and dam safety would be questioned. The result of sediment retention in the reservoir is the formation of a delta at the inflow and a gradual filling in of the reservoir. As long as sedimentation is restricted to the dead storage volume of a reservoir, this will not have negative consequences on the performance of the hydropower plant. However, bottom outlets can be blocked, which may present a problem in cases where these structures should be available for flood evacuation, and eventually the live storage area will also be filled in, to the point where the reservoir has no storage volume left and can therefore no longer be used. Without any regulation capacity in the reservoir, winter energy output would drastically decrease. Construction of Rogun dam will largely decrease the Nurek filling rate, ensuring the river regulation for a significant additional time period, and delay the need of rehabilitation of the flood evacuation system with respect to the sedimentation issue.

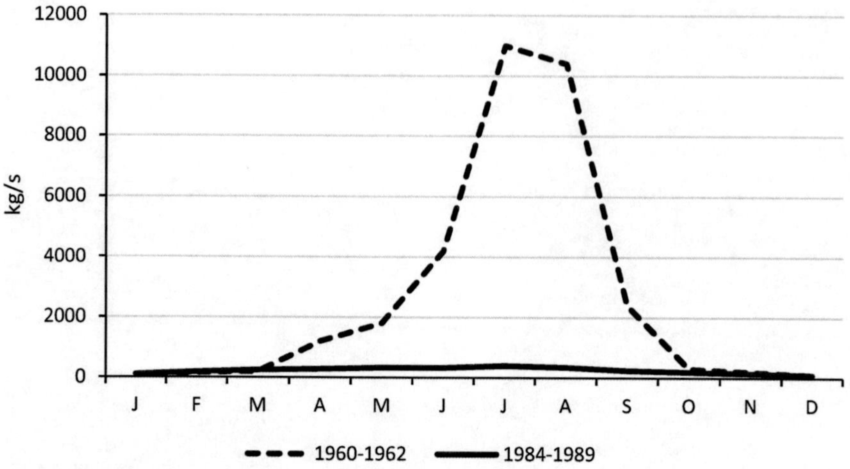

Fig. 8.11 Suspended solid transport below Nurek dam before and after commissioning. Based on data from Pöyry (2014)

This, in short, is the effect of sediment and sedimentation on reservoirs and hydropower. However, the main direct impact of sediment retention is downstream of the dam. One point is the fact that water released from the reservoir does not contain any relevant amounts of sediment, neither as bedload nor as suspended solids. The graph in Fig. 8.11 shows the amount of suspended solids at Tigrovaya Balka, downstream Nurek HPP, before and after commissioning of the HPP. Such a massive reduction in sediment transport has a considerable impact on river dynamics in the downstream area, creating a situation of "hungry water", with a strongly increased erosive capacity, and this effect can highly increase the erosion in the river bed downstream of a dam.

One important threat caused by this effect of sediment retention, however, is the lack of sediment supply in the river deltas. As Giosan et al. (2014) put it: "Most river deltas, and especially the large ones, are under a double threat: rising sea level due to climate change, and reduction of sediment inflow due to dams." Lack of sediment inflow from the river must invariably lead to delta erosion, which means a loss of (often very productive) land and loss of specialised and valuable habitats and is considered as one main impact of dams (Grill et al. 2019; Stone 2016; Nicholls et al. 2020a, b, c) and is usually listed as such in ESIAs mainly in the case of large dam projects on large rivers, as e.g. the Mekong. A well-known case of such an impact is the effect of the Aswan High Dam on the Nile in Egypt; see discussion of this situation in Sect. 21.1.1. Considering the information on sediment yield in the Nile shown in Table 8.4, one can easily imagine that this is a problem for the Nile delta. However, direct impacts of this type, and the indirect effect on the delta, are difficult to assess in detail, and might often require extensive studies lasting several years for actually showing whether such an impact is happening and what the extent is.

Still, I have to put a question mark on the generalised assumption that retention of sediments in reservoirs invariably leads to an increased erosion in the deltas, and this mainly for the following reason: it is certainly the case that sediments are being retained in reservoirs. On the other hand, in very many places, and especially in tropical areas, erosion has increased tremendously mainly due to deforestation, i.e. slash-and-burn cultivation far beyond a sustainable level, logging and clearing of land for agriculture or other use (Wilkinson and McElroy 2007). So for instance, between 1940 and 1990 forest cover in in Costa Rica has diminished from 80% to less than 30% (Harcourt and Sayer 1996: 166) and in Laos from70% to about 40% (Azimi et al. 2000), and has probably been shrinking again since then. For this reason, it could even be that at least some deltas would have received a much higher quantity of sediments in comparison to an earlier, less disturbed situation in their catchment areas, a development that according to Malakoff et al. (2020) in the Mediterranean region has started about 4000 years ago, at about the time that marks the onset of agriculture in this region. Building dams in such places might then reverse this development, but in any case the question of reduced sediment input seems much less clear-cut than it was often described.

This was at least partly confirmed recently. In a study which examined about 11,000 coastal deltas worldwide, the main conclusion was: "In total, 970 deltas have had their fluvial sediment supply reduced by > 50%, collectively from ~ 9×10^4 kg s^{-1} to ~ 2×10^4 kg s^{-1}, resulting in a shift towards wave or tide dominance. On the other hand, human-driven soil erosion, mostly through deforestation, is predicted to have caused a > 50% increase in sediment flux, or ~ 5×10^4 kg s^{-1}, to ~ 1,500 deltas. We find that over the past 30 years, deltas globally have gained 181 ± 8.3 km^2 $year^{-1}$ and lost 127 ± 8.3 km^2 $year^{-1}$, resulting in a net gain of 54 ± 11.8 km^2 $year^{-1}$ (2 s.d.). With a combined ~ 9×10^9 m^3 $year^{-1}$ fluvial sediment flux to the global ocean, deltas on average require 150 m^3 of sediment delivered to the coast for every square metre of land gain." (Nienhuis et al. 2020: 514).

A new book (Nicholls et al. 2020a; b, c) focusses on the development of a number of major deltas mentioning, in addition to raising sea levels due to climate change, sediment retention by dams as one of the major problems deltas are presently facing. However, without going into more details on this issue, it is also mentioned that "prior to these changes of the last few centuries, many of the world's deltas experienced increased rates of fluvial sediment supply associated with major land use changes driven by the onset and subsequent intensification of agriculture" (Darby et al. 2020: 107). These authors also make the following interesting observations in a detailed discussion of the situation of the Mahanati delta in India. According to them, this delta expanded by about 4 m per year since about 6000 years BP, and this increased to about 9 m per year in the period from 1700 to 1950, while since then the shoreline has retreated significantly; this latter effect is supposed to be caused by a dam which was built 300 km upstream of the delta in 1957, which reduced sediment input by about 70%. No explanation is given for the strong increase in the "pre-dam" phase, but it might well be that this was caused by intensification of agriculture in the colonial period.

For the case of the Mekong, whose delta seems to be one of those that are presently loosing (Giosan 2014), Hackney et al. (2020) point out another factor: according to their figures, sand mining is practiced there to an extent which is beyond the replacement by sediment inflow, and can therefore no longer be considered as sustainable.

All this is not meant to say that sediment retention is not an impact caused by dams which has to be taken seriously. The aim is just to point out that very often, the situation might be less uni-directional and clear-cut as it may seem at first.

8.6.3 Measures

Undoubtedly, there are two main negative effects due to sediment retention in reservoirs, as mentioned above, namely:

- Filling in of reservoirs, which reduces the life span of the reservoir and the associated hydropower plant.
- Lack of sediment in the downstream area, with negative effects on the river bed, floodplains and deltas.

It will therefore often be required to do something about reservoir sedimentation. Mechanical removal of sediments may be a solution in specific cases, but probably only for small reservoirs, just given the vast amount of material that would have to be moved and transported, and the problem of its disposal. Still, doing nothing may be a bad idea, or, as Prof. Anton Schleiss, an expert in reservoir and sediment management, put it in a 2019 conference: "Not to flush sediments is a waste of money, since this reduces the life span of HPPs. Increasing the dam height to increase dead storage (for trapping sediments) is also a waste of money."

Sediment flushing is a method which is sometimes used, especially in the case of high dams with periodic substantial drawdown. This procedure is used in the case of Mauvoisin dam, a 250 m high double arch dam in the Swiss Alps. Here, the reservoir was periodically drawn down to a very low level, and when the water level came close to the level of the silt (which is very fine material resulting from the glaciers which feed this reservoir) large quantities of this silt were mobilised and carried downstream. This procedure has two serious disadvantages. First, it requires a very substantial drawdown of the reservoir, since silt will not be mobilised as long as the water column above it is too high; this means a "waste" of water from the point of view of the operator of the power plant, this water not being available for generating electricity. Secondly, this flushing will result in water with an extremely high silt content, in large quantities over a short period, in the river downstream of the dam, a river which at least in this case is normally small and contains very low quantities of suspended solids. Much of this very fine material sediments quickly in the entire stretch of the affected stream, and since it is so fine it will block all the interstices in the gravel ground of the stream as well as the gills of fish and other water organisms, leading to an almost complete eradication of the fauna potentially over several river kilometres. Although it has been shown that recolonization of the affected stream took place astonishingly fast, this is still a serious impact, especially when large populations of fish are negatively affected.

New developments in this question promote routing of sediments through the reservoir; Schleiss (2019) provides an overview of recent developments in this sense, which is already being applied in a number of projects. In any case, this will clearly be an engineering solution, which has to be considered already in the planning of the dam project. In order to be effective, it requires careful hydrological modelling as well as suitable arrangement of bottom outlets. In the case of the dams on the Mekong mainstream, the recently completed Xayaburi dam and the Luang Prabang dam, which is under preparation, the aim is to reach a new equilibrium, from a short period during which sediment may be accumulated in the reservoir to some extent, to a situation when an amount of sediment similar to the one being brought in is constantly washed out and delivered to the downstream area. As Staehly et al. (2019) have shown, sediment replenishment combined with artificially created flood peaks can also help in restoring some of the floodplain dynamics downstream of dams.

References

Alissow, B.P., O.A. Drosdow, and E.S. Rubinstein. 1956. *Lehrbuch der Klimatologie*. Berlin: Deutscher Verlag der Wissenschaften.

Anonymous. 1950. Mean monthly and annual evaporation from free water surface for the United States, Alaska, Hawaii and West Indies. Technical Paper No. 13. Hydrologic Branch, Weather Bureau, U.S. Department of Commerce. 10 pp. https://www.nws.noaa.gov/oh/hdsc/Technical_papers/TP13.pdf.

Azimi, A., W. Knowland, J. Carew-Reid, I. Ruzicka, A. Zola. 2000. Environments in transition: Cambodia, Lao PDR, Thailand, Viet Nam. Asian Development Bank, Manila, 114 pp.

Bate, G.C., P.R. Furniss, B.G. Pendle. 1982: Water relations of southern African savannas. In Ecology of tropical savannas, eds. B.J. Huntley, and B.H. Walker, vol. 42, 336–358. Berlin: Springer.

Bauer, C., and S. Hirschberg (eds.). 2017. *Potentials, costs and environmental assessment of electricity generation technologies. PSI, WSL, ETHZ, EPFL*, 783. Switzerland: Paul Scherrer Institut, Villigen PSI. https://www.psi.ch/sites/default/files/import/ta/HomeEN/Final-Report-BFE-Project.pdf.

Bekkering, T.D. 1992. Using tropical forests to fix atmospheric carbon: The potential in theory and practice. *Ambio* 21 (6): 414–419.

Berga, L. 2016. The role of hydropower in climate change mitigation and adaptation: A review. *Engineering* 2 (3): 313–318. https://doi.org/10.1016/J.ENG.2016.03.004.

Biro, Y.E.K. 1998. Valuation of the environmental impacts of the Kayraktepe Dam/Hydroelectric Project, Turkey: An exercise in contingent valuation. *Ambio* 27 (3): 224–229.

Blöschl, G., J. Hall, J. Parajka, R.A.P. Perdigao, B. Merz, et al. 2017. Changing climate shifts timing of European floods. *Science* 357 (6351): 588–590.

Blüthgen, J. 1980. *Allgemeine Klimageographie*, 720 pp. Berlin: Gruyter.

Bukaveckas, P.A. 2009. Hydrology: rivers; Likens G.E. (ed.). 2010a. *River ecosystem ecology. Encyclopedia of inland waters*, 32–43. Oxford: Elsevier.

Calder, N. 1999. The carbon dioxide thermometer and the cause of global warming. *Energy and Environment* 10 (1): 1–18.

Cook, K.L., C. Andermann, F. Gimbert, B.R. Adhikari, and N. Hovius. 2018. Glacial lake outburst floods as drivers of fluvial erosion in the Himalaya. *Science* 362 (6410): 53–57.

Darby, S.E., K.A. Addo, S. Hazra, M.M. Rahman, and R.J. Nicholls. 2020. Fluvial sediment supply and relative sea-level rise. In: *Deltas in the anthropocene*, eds. R.J. Nicholls et al., 103–126. Palgrave Macmillan/Springer Nature. https://doi.org/10.1007/978-3-030-23517-8_5.

De Santo, R. 1978. *Concepts of applied ecology*, 310 pp. New York: Springer.

DEZ, Electrowatt-Ekono, Moshanir and Stucky-Pars. 2006. *Bakthyari dam and hydroelectric power project. Environmental Impact Assessment*. Iran: Ministry of Energy.

Doorenbos, J., and W.O. Pruitt. 1977. Guidelines for predicting crop water requirements. FAO Irrigation and Drainage Paper No. 24, 144 pp. Rome.

Dukhovny, V.A., and J. de Schutter. 2011. *Water in Central Asia. Past, present and future*, 410 pp. Taylor and Francis, London: CRC Press.

El Bakry, M.M. 1993. Comparison between some meteorological parameters over land at Asswan and high dam lake. Internat. Symposium on High Asswan Dam vital achievements, fully controlled. ICOLD, 240–250.

El-Magd, I.H.A., and E.M. Ali. 2012. Estimation of the evaporative losses from Lake Nasser, Egypt using optical satellite imagery. *International Journal of Digital Earth* 5 (2). https://www.tandfonline.com/doi/abs/10.1080/17538947.2011.586442?src=recsys&journalCode=tjde20.

EWE. 1994. Egiin hydroelectric project. Environmental impact study. Vol. I: Text, 187 pp. Vol. II: Tab. and Fig. Vol. III: Executive summary, 11 pp. Electrowatt engineering with electroconsult and Usni Erchim for Asian Development Bank and Central Energy System, Mongolia. TA No. 1653-MON.

EWI. 1998. EIA for Murum HPP, Sarawak, Malaysia. Report prepared for Sarawak Electricity Board.

Fearnside, P.M. 1989. Brazil's Balbina Dam: Environment versus the legacy of the Pharaohs in Amazonia. *Environmental Management* 13: 401–423. https://doi.org/10.1007/BF01867675.

Fearnside, P.M. 1995. Hydroelectric dams in the Brazilian Amazon as sources of "greenhouse" gases. *Environmental Conservation* 22 (1): 7–19.

Fearnside, P.M. 1997. Greenhouse-gas emissions from Amazonian hydroelectric reservoir: The example of Brazil's Tucuruí dam as compared to fossil fuel alternatives. *Environmental Conservation* 24 (1): 64–75.

Fearnside, P.M. 2015. Emissions from tropical hydropower and the IPCC. *Environmental Science and Policy* 50: 225–239. https://doi.org/10.1016/j.envsci.2015.03.002.

Frank, D.A., and R.S. Inouye. 1994. Temporal variation in actual evapotranspiration of terrestrial ecosystem patterns and ecological implications. *Journal of Biogeoraphy* 21 (4): 401–411.

Gagnon, L. 1998. State of research in 1998 confirms that hydropower is a tool to reduce greenhouse gas emissions. Modelling, Testing and Monitoring for Hydro Powerplants—III, Aix-en-Provence 1998; Proceedings, 95–105.

Gagnon, L., and A. Chamberland. 1993. Emissions form hydroelectric reservoirs and comparison of hydroelectricity, natural gas and oil. *Ambio* 22 (8): 568–569.

Gates, B. 2021. *How to avoid a climate disaster*, 257 pp. New York and Toronto: Knopf.

Gebre, S., N. Timalsina, and K. Alfredsen. 2014. Some aspects of ice-hydropower interaction in a changing climate. *Energies* 7: 1641–1655. https://doi.org/10.3390/en7031641.

Gebre, S.B., K. Alfredsen, L. Lia, et al. 2013. Review of ice effects on hydropower systems. *Journal of Cold Regions Engineering* 27 (4): 196–222. https://doi.org/10.1061/(ASCE)CR.1943-5495.0000059.

Geiger, R. 1950. *Das Klima der bodennahen Luftschicht*, 460 pp. Braunschweig: Vieweg.

GICOME. 2015. Janneh dam. Specific EIA Study. Republic of Lebanon, Ministry of Water and Energy. Mount Lebanon and Beirut Water Establishment. 418 pp.

Giosan, L., et al. 2014. Protect the world's deltas. *Nature* 516 (7529): 31–33.

Grill, G., B. Lehner, M. Thieme, et al. 2019. Mapping the world's free-flowing rivers. *Nature* 569 (7755): 215–221. https://doi.org/10.1038/s441586-019-1111-9.

Hackney, C.R., S.E. Darby, and D.R. Parsons. 2020. River bank instability from unsustainable sand mining in the lower Mekong River. *Nature Sustainability* 3 (3): 217–225. https://doi.org/10.1038/s41893-019-0455-3.

Hagens, N.G. 2020. Economics for the future—beyond the superorganism. *Ecological EconOmics* 167: 106520. https://doi.org/10.1016/j.ecolecon.2019.106520.

Hamilton, A.J., R.M. May, and E.K. Waters. 2015. Here be dragons. *Nature* 520: 42–43.

Harcourt, C.S., and J.A. Sayer (eds.). 1996. *The Conservation atlas of tropical forests: The Americas*, 335 pp. New York: Simon and Schuster.

Hilbig, W., K. Helmecke, Z. Shamsaran, and D. Bumzaa. 1989. Mikro-klimatische Untersuchungne in Plfanzengesellschaften verschiedener Höhenstufen in Hochgebirgen der Nodwest- und Südmongolei. *Biologische Ressourcen Der Mongolischen Volksrepublik* 7: 5–79.

Hilbig, W., M. Stubbe, N. Dawaa, Z. Shamsran, M. Dorn, K. Helmecke, D. Bumzaa, and K. Ulykpan. 1984. Vergleichend biologisch-ökologische Untersuchungen in Hochgebirgen der Nordwest- und Südmongolei. *Biologische Ressourcen Der Mongolischen Volksrepublik* 4: 5–59.

Hoy, R.D., and S.K. Stephens. 1979. *Field study of lake evaporation*, 177 pp. Canberra: Australian Government Publishing Service.

ILF. 2008. Palas valley/spat gah hydropower complex. Prefeasibility study report, Vol. 3: Initial environmental examination. Report prepared by ILF, Pöyry and ACE for WAPDA (Water and Power Development Authority), Pakistan.

IPCC. 2014. *Climate change 2014: Synthesis report. Contribution of working groups I, II and III to the fifth assessment report of the intergovernmental panel on climate change*. In eds. Core Writing Team, R.K. Pachauri, and L.A. Meyer, 151. Geneva, Switzerland: IPCC.

IPCC. 2021. Summary for Policymakers. In *Climate change 2021: The physical science basis. Contribution of working group I to the Sixth assessment report of the intergovernmental panel on climate change*, eds. V. Masson-Delmotte, P. Zhai, A. Pirani, S.L. Connors, C. Péan, S. Berger, N. Caud, Y. Chen, L. Goldfarb, M.I. Gomis, M. Huang, K. Leitzell, E. Lonnoy, J.B.R. Matthews, T.K. Maycock, T. Waterfield, O. Yelekçi, R. Yu, and B. Zhou. Cambridge University Press. In Press. https://www.ipcc.ch/report/ar6/wg1/downloads/report/IPCC_AR6_WGI_SPM.pdf.

Kraaijenbrink, P.D.A., M.F.P. Bierkens, A.F. Lutz, and W.M. Immerzeel. 2017. Impact of a global temperature rise of 1.5 °C on Asia's glaciers. *Nature* 549 (7671): 257–260.

Kuhn, W. 1977. Berechnung der Temperatur und Verdunstung alpiner Seen auf klimatologisch-thermodynamischer Grundlage. Arbeitsbericht der Schweizerischen Meteorologischen Zentralanstalt Nr. 70. Zürich, 34 pp.

Lavergne, M. 1986. The seven deadly sins of Egypt's Aswan High Dam. In *The social and environmental effects of large dams*, .eds. E. Goldsmith, and N. Hildyard, Vol. 2, 181–183. Case Studies.

Malakoff, D., N. Desai, and X. Liu. 2020. Mud on the move. *Science* 369 (6506): 896–897.

Meadows, D. 1972. Die Grenzen des Wachstums. Bericht des Club of Rome zur Lage der Menschheit. DVA, Stuttgart, 180 S.

Milliman, J.C., and R.H. Meade. 1983. World-wide delivery of river sediment to the oceans. *Journal of Geology* 91 (1): 1–21.

Muller, M. 2018. Lessons from Cape Town's drought. *Nature* 559 (7713): 174–176.

Muller, M. 2019. Dams have the power to slow climate change. *Nature* 566: 315–317.

Nicholls, R.J., N.W. Adger, C.W. Hutton, et al. 2020a. Sustainable deltas in the anthropocene. In *Deltas in the anthropocene*, eds. R.J. Nicholls et al., 247–279. Palgrave Macmillan/Springer Nature. https://doi.org/10.1007/978-3-030-23517-8_11.

Nicholls, R.J., W.N. Adger, C.W. Hutton, and S.E. Hanson (eds.). 2020b. *Deltas in the anthropocene*, 282 pp. Palgrave Macmillan/Springer Nature. https://doi.org/10.1007/978-3-030-23517-8.

Nienhuis, J.H., A.D. Ashton, D.A. Edmonds, et al. 2020c. Global-scale human impact on delta morphology has led to net land area gain. *Nature* 577 (7791): 514–518. https://doi.org/10.1038/s41586-019-1905-9.

Odingo, R.S. 1979. An African dam. Ecological survey of the Kamburu/Gtaru hydro-electric dam area, Kenya. *Ecological Bulletins* 29: 183 pp.

Pangala, S., R., Enrich-Prast, A., Basso, L.S., et al. 2017. Large emissions from floodplain trees close the Amazon methane budget. *Nature* 552 (7684): 230–234

Pearce, F. 2018. *When the rivers run dry. The global water crisis and how to solve it. Updated edition*, 312 pp. London: Granta Publications.

Powerchina. (n.d.). Rufiji Hydropower Project (RHPP). Bidding document Vol. III: Technical part. RCC gravity dam design report. Ministry of Energy and Minerals, Tanzania.

Pöyry. 2007. Projet hydroélectrique Oued Ahançal. Etude d'Impact Environnenemtal et Social. ONE, Maroc.

Pöyry. 2008. Projet Hydroélectrique du Grand Poubara. Volume I: Etude d'Impact sur l'Environnement et Etude d'Impact Social. Rapport préparé pour MMPHERHPEN (Ministère des Mines, du Pétrole, des Hydrocarbures, de l'Energie, des Ressources Hydrauliques et de la Promotion des Energies Nouvelles), Gabon.

Pöyry. 2014. Environmental and social impact assessment for Rogun hydropower plant. ESIA Vol. I: Text. Vol. II: Annexes. Report prepared for OSHPC "Barki Tojik", Republic of Tajikistan.

Pöyry. 2015. Biomass clearance plan. Environmental and social studies, Nam Theun 1 Hydropower Project. Report prepared for Phonesack Group, Lao PDR: 65 pp.

Prowse, T.D., and Gridley, N.C., 1993: Environmental aspects of river ice. National Hydrology Research Institute, Saskatoon, Canada. NHRI Scientific Report No. 5:155 p.

Rosa, L.P, M.A. dos Santos, B. Matvienko, and E. Sikar. 2002. Hydroelectric reservoirs and global warming. Rio 02—World Climate and Energy Event, January 6–11, 2002, 7 pp. https://www.researchgate.net/publication/228816992_Hydroelectric_reservoirs_and_global_warming.

Rosen, J. 2018. The carbon harvest. *Science* 359 (6377): 733–737.

Rudd, J.W.M., et al. 1993. Are hydroelectrical reservoirs significant sources of greenhouse gases? *Ambio* 22 (4): 246–248.

Schleiss, A. 2019. Recent developments on management and release techniques of fine sediments in reservoirs. Paper presented at the Africa 2019 Hydropower Conference, Windhoek, Namibia, April 2019.

Schlömer S., T. Bruckner, L. Fulton, E. Hertwich, A. McKinnon, D. Perczyk, J. Roy, R. Schaeffer, R. Sims, P. Smith, and R. Wiser, 2014. Annex III: Technology-specific cost and performance parameters. In *Climate change 2014: Mitigation of climate change. Contribution of working group III to the fifth assessment report of the intergovernmental panel on climate change*. New York, NY, USA: Cambridge University Press, Cambridge, United Kingdom. https://www.ipcc.ch/site/assets/uploads/2018/02/ipcc_wg3_ar5_annex-iii.pdf.

Shugar, D.H., M. Jaquemart, D. Shean, et al. 2021. A massive rock and ice avalanche caused the 2021 disaster at Chamoli. *Indian Himalaya. Science* 373 (6552): 300–306. https://doi.org/10.1126/science.abh4455.

Staehly, S., M.J. Franca, D. Tonolla, et al. 2019. Increase of hydraulic habitat diversity downstream of dams with sediment replenishment and artificial flooding. Paper presented at the Africa 2019 Hydropower Conference, Windhoek, Namibia, April 2019.

Stivari, S.M.S., A.P. de Oliveira, and J. Soarez. 2005. On the climate impact of the local circulation in the Itaipu Lake area. *Climatic Change* 72 (1–2): 103–121. https://doi.org/10.1007/s10584-005-5923-2.

Stone, R. 2016. Dam-building threatens Mekong fisheries. *Science* 354 (6316): 1084–1085.

Tanaka, N., T. Kume, N. Yoshifuji, et al. 2008. A review of evapotranspiration estimates from tropical forests in Thailand and adjacent regions. *Agricultural and Forest Meteorology* 148: 807–819. https://doi.org/10.1016/j.agrformet.2008.01.011.

Walter, H., und Breckle, S.-W., 1984. *Oekologie der Erde Band 2 : Spezielle Ökologie der Tropischen und Subtropischen Zonen*, 461 S. Stuttgart: UTB Grosse Reihe, Fischer.
Wilkinson, B.H., and B.J. McElroy. 2007. The impact of humans on continental erosion and sedimentation. *GSA Bulletin* 119 (1–2): 140–156. https://doi.org/10.1130/B25899.1.
Wohlfahrt, G., E. Tomelleri, and A. Hammerle. 2021. The albedo-climate penalty of hydropower reservoirs. *Nature Energy*https://doi.org/10.1038/s41560-021-00784-y.

Chapter 9
Water

Whatever there is that might be is simply not there: only the pooled water, only the calm sea, only it alone is pooled.
From the Popol Vuh, Maya creation myth
(cited after Leeming 1994: 188)

As mentioned at the beginning of this book, water is one of the three main topics when it comes to assessing impacts of hydropower structures. It is therefore important to deal with the issues related to this with some detail. In order to illustrate the major impacts to be looked into, two projects showing these points at least to some degree are used to start with. Additional examples will be provided in some of the case studies presented in Part III. The chapter is structured in a way as to deal with the three main sectors, the three water-related issues where impacts of HPPs have to be expected, namely, hydrology (importance of and influence on aspects of long-term, seasonal and short-term flow distributions), hydrogeology (changes in ground water table and potential consequences), and water quality. In addition to assessment of baseline conditions, identification of impacts and mitigation measures, a specific focus is put on monitoring.

9.1 Hydrology

9.1.1 First Approach

9.1.1.1 Shnogh HPP

Shnog HPP is a project for a ROR structure on Debed River in Armenia. The main features are illustrated in Fig. 9.1.

The project is planned to use the water and the head provided by Debed River in Armenia. It will have no storage with any regulating capacity, but it involves a considerable amount of tunnelling. A first tunnel of about 2 km leads the water from intake 1 on Debed River to intake 2, the next right bank tributary. From there the water

R. Zwahlen, *Assessing the Environmental Impacts of Hydropower Projects*, Environmental Earth Sciences, https://doi.org/10.1007/978-3-030-91185-0_9

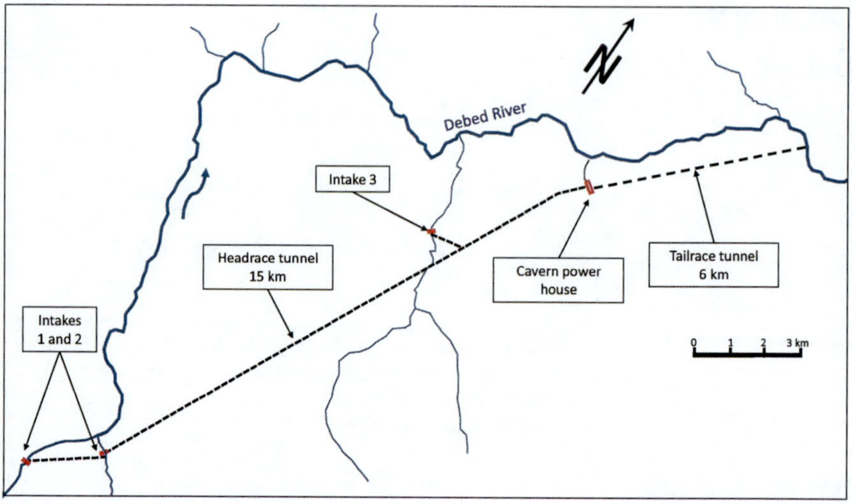

Fig. 9.1 Layout of Shnogh HPP

is conveyed through the headrace tunnel of about 15 km to the underground power house; on its way a second major tributary is captured at intake 3. A perpendicular 227 m high penstock brings the water to the turbines, from where it flows through the 6 km long tailrace tunnel to be released again to Debed River. This describes the project layout analysed during the feasibility study (Pöyry 2018), which is not necessarily as it will eventually be built.

Between the intake and the tailrace outlet, Debed River flows in a rather narrow valley. Some settlements are located in this valley along the river, while others are on the plateaus on either side of the river, at an elevation of roughly 250 m above the valley floor. The immediate surroundings of the river are characterised by metallurgy industry (copper factory) and mining, although these activities were much reduced after the end of the Soviet Union.

The project would reduce the flow in the river on a stretch of 37 km where, in addition to water from some minor tributaries, only the residual flow and water from spilling during the high flow season would remain. One major problem is the severe contamination of the water in the river. There was a least one large waste water treatment plant in the valley, which presumably also treated the waste water from the factory, but this is no longer in operation. A very considerable part of the contamination stems from a mining area close to the lower end of the deviation, drained by the last left bank tributary shown in the Figure. Contamination with heavy metals (molybdenum, zinc, manganese, copper and cadmium), stemming from the copper smelter, but to a large extent from the tailings of the mine, is high and far above environmental standards; water contamination from such sources is well documented, e.g. for a case in Peru (Balvín Díaz 1995: 90 ff.). The HPP will not change anything in the amount of polluted water flowing into the river, but reducing the flow will increase concentrations, which in turn will mean a further deterioration of water quality upstream of the tailrace outlet.

9.1.1.2 The Kafue Cascade

The Kafue cascade as such consists of three dams and power plants, namely (from up- to downstream) the two existing dams Itezhi-Tezhi (ITT) and Kafue Gorge Upper (KGU) and finally Kafue Gorge Lower (KGL), presently under construction. However, in order to understand the situation it is necessary to take two other dams into consideration as well, namely Kariba (shared between Zambia and Zimbabwe) and Cahora Bassa (Mozambique), the two major dams and reservoirs on the Zambezi River. This entire system is shown in Fig. 9.2; note that in this Figure KGU and KGL are indicated only as dam location, both of these reservoirs are too small as to be shown at this scale. The main features of the five dams in the system are summarised in Table 9.1.

The three dams on the Kafue River can be used for illustrating three typical downstream impacts of hydropower plants:

- Change of seasonal discharge pattern: Fig. 9.2 also shows the Kafue Flats, a large and ecologically important wetland (river floodplain); the regulation of the flow, reducing wet season high flows and increasing dry season low flows, had considerable negative impacts on this wetland (Godet and Pfister 2007; WWF 2003). The difference in seasonal discharge is illustrated in Fig. 9.3.
- River deviation: both dams in the Kafue Gorge operate with river deviation of about 5 km each. In this case, the impact of the deviation can be considered as rather small (e.g. in comparison with Shnogh HPP mentioned in the previous section), since only a limited stretch of river is affected, and the river was changed anyway by the presence of the dams.

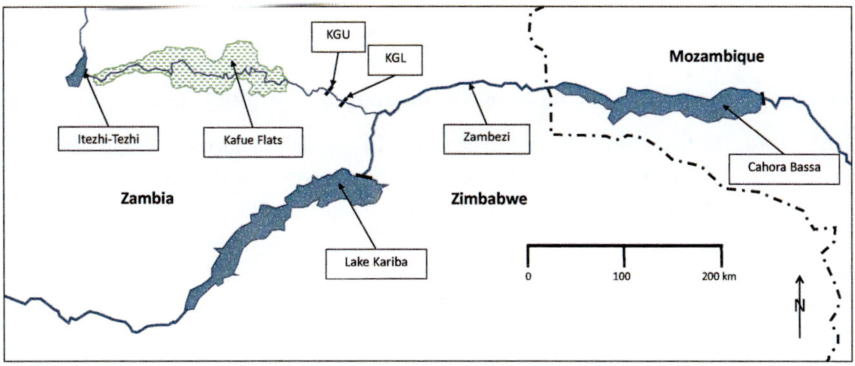

Fig. 9.2 The Zambezi-Kafue cascade. *Note* Zambezi and Lake Kariba form the border between Zambia and Zimbabwe (not shown in the graph); based on Zwahlen et al. (2019)

Table 9.1 Major reservoirs on the Kafue and Zambezi rivers

Name	Abbr.	River	Storage capacity (Mm^3)	Water surface (km^2)	Start year	Dam height (m)	Inst. capacity (MW)
Itezhi Tezhi	ITT	Kafue	7000	375	1978	62	120
Kafue Gorge Upper	KGU	Kafue	950	13	1971	50	900
Kafue Gorge Lower	KGL	Kafue	83	1.85	uc (2017)	140	750
Kariba	/	Zambezi	200,000	5500	1959	128	1626
Cahora Bassa	/	Zambezi	85,000	2700	1975	171	2075

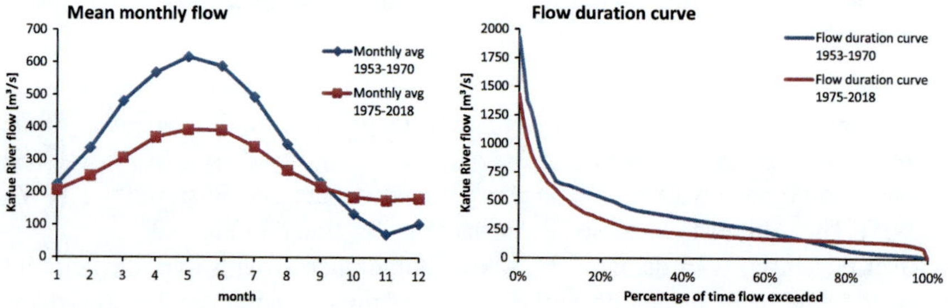

Fig. 9.3 Mean monthly flow at KGU before and after construction of ITT

- Peaking regime: originally, KGL had been conceived as a ROR structure, which basically would just turbine the water released from KGU; in this case, its effect downstream of the tailrace would have been negligible. It was then decided to operate it as peaking plant, which brings about a considerable change in the downstream area due to short-term fluctuations in water discharge.

KGU was the first dam built on Kafue River. Like many dams on strongly seasonal rivers, it faced the problem of low flows in the dry season, insufficient for generating energy at full capacity, while in the wet season a considerable amount of water had to be evacuated over the spillway. In order to improve this situation, Itezhi-Tezhi dam was built, which originally had the only task of regulating flows for KGU; only recently, a power plant was integrated in this dam which allows to turbine the water that is passed through the dam. Figure 9.3 shows the discharge pattern in the Kafue River.

In Fig. 9.3 the same information is given in two different ways, once as mean monthly flow, and once as flow duration curve (see Sect. 9.1.2 for additional explanation of the latter). It shows, first of all, the considerable shift from high flow to low

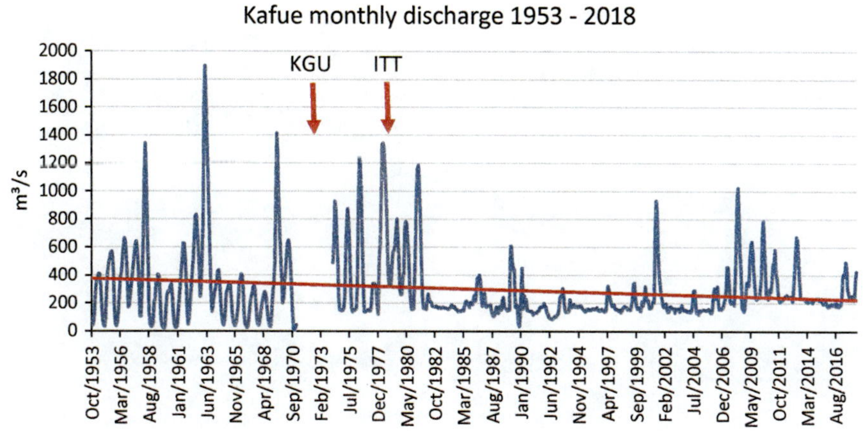

Fig. 9.4 Kafue monthly discharge 1953–2018

flow season, which is the effect of Itezhi-Tezhi. But at a second look it also shows that apparently in the period after the construction of the dam total flow was reduced in comparison to the earlier period. In fact, in the period 1953–1970 the average flow was 343.8 m^3/s, while in the period 1975–2018 it was only 277.4 m^3/s. This is also shown in Fig. 9.4, which indicates the monthly flow over the entire period. Most of the years between 1982 and 2006 showed lower that average total flows, an indication of an extended dry period (see Sect. 8.2 for other examples of this effect).

The data for the period 1953–1970 stem from a hydrograph which was located downstream of the Kafue flats, a short stretch upstream of KGU dam site, this station was no longer available with KGU, and there are no data for the construction period of KGU. The two red arrows indicate the start of operation of the two dams, the figures for the period 1975–2018 are outflow from KGU (turbines and spilling). Whether the flow reduction indicated by the red trend line is an ongoing development, or whether it just reflects the rather marked dry spell in the 1980s and 1990s, remains to be seen.

River deviation in this case, as indicated above and unlike what was described for Shnogh HPP, has a minor impact. However, the decision to operate KGL as a peaking plant changed the nature of the downstream impact in a very decisive way: the plan is to operate it only a few hours per day at full capacity, depending on demand in one or several episodes, whereby the amount of water released to the river will change from 310 to 484 m^3/s and back again in the proposed operation regime; it would however also be possible to increase peaking by turbining only 97 m^3/s on a continuous basis. This, whether the difference is 174 or 397 m^3/s, will cause a surge wave which will flush out the remaining part of the gorge downstream of the tailrace structure (5 km), and it will be a risk to persons who may happen to be in proximity of the river. This wave will be smoothed out to some extent by the time the river leaves the gorge for entering the remaining 45 km across the Zambezi floodplain, but the daily fluctuations in water level will still be sufficient for at least partially flooding the river banks, an effect that will make the use of these banks as river bank gardens

Fig. 9.5 River bank cultivated during the low-flow season. If due to peaking such river banks will be flooded repeatedly during the low-flow season, cultivation will no longer be possible

during the low-flow season impossible. Interesting detail: the about last 10 km of the Kafue River before its confluence with the Zambezi River are affected in the same way by peaking at the Kariba HPP, which causes a backwater effect into Kafue River (Fig. 9.5).

9.1.2 Prevailing Conditions

A dam project is all about water, or more precisely about quantities available, about temporal distribution of availability, and about the possibility for storage. This means that in most cases there is a large amount of reliable information available on this topic from hydraulic stations which have been operated often over several decades, and from previous studies, where hydrological information was of special importance. Usually in a study team for a dam and hydropower project there is a hydrologist who can prepare and analyse these data in a competent way. In case no or insufficient hydrological data are available, the hydrologist will use a suitable model which allows him, e.g. based on data stemming from gauging stations in the wider area, and/or on an incomplete data set, to come to a reasonably accurate simulation of flow data. This means that usually, where it comes to hydrology, no specific field work or data analysis will have to be done in the framework of the ESIA for getting sound basic data. It is even important that the technical and the environmental side do their work based on the same data. The way to look at the data however, as mentioned already, can be rather different.

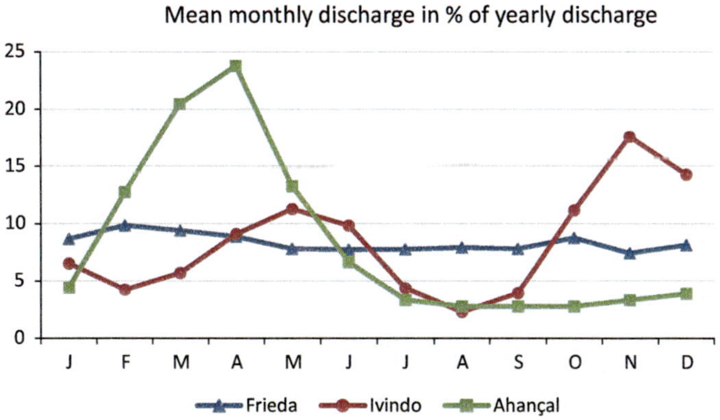

Fig. 9.6 Three characteristic patterns of seasonal flow in rivers. Based on data from the following sources: Frieda River, Papua New Guinea (SKM, Pöyry and SMEC 2010).Ivindo River at Kongou Falls, Gabon (Pöyry 2009). Oued Ahançal, Morocco (Pöyry 2007)

From an environmental point of view, three characteristics of river flow are of importance, namely:

- quantity and seasonal distribution of discharge;
- year to year variability of discharge; and
- long term trend.

The first of these three is the seasonal pattern of discharge. In rather rare cases—restricted to regions with rainfalls evenly distributed over the entire year—river discharge does not show much month-to-month variation. In most cases, however, there are marked seasonal differences in average monthly discharge. Characteristic examples are shown in Fig. 9.6.

Since the figure includes data on three rivers which do not only have different seasonal flow patterns, but are also of very different sizes, mean annual discharge being 220 m^3/s for Frieda, 800 for Ivindo and 15 for Ahançal, the graph does not give direct flow data, but shows the monthly values as percentage of annual flow. The first example is from Frieda River in Papua New Guinea. This is a region characterised by high rainfalls throughout the year and correspondingly a river discharge pattern without marked seasonal differences. The second one shows the Ivindo, a major tributary of the Ougoué, the main river in Gabon. Here, in this region close to the equator two distinct rainy seasons lead to two yearly peak flow seasons, one in May and one in November. Finally the third example is the Oued Ahançal, a small stream in Morocco, with a very marked seasonal difference between wet and dry season; a pattern of this type can often be seen where in winter precipitation falls as snow, and where then peak flow is due to snow melt in spring and early summer.

This is an example of the different way of looking at things mentioned in Sect. 2.2, and an important one: strong seasonal differences in river discharge will be seen by the hydraulic engineer as shortage in the low-flow season, while the peak flows in the high-flow season are a surplus, and even a waste, since they cannot be used, at least as long as they cannot be stored in a reservoir, to then be productively used during the dry season. The surplus stored in the reservoir will then fill the gap in the dry season, reducing loss (spilling) and increasing production of electricity, as was shown above for the case of the Kafue cascade. From an environmental point of view, however, such a marked seasonal variation means that in all likelihood there are species, habitats and ecosystems that depend on this marked seasonal change, as, most importantly, floodplains, which will be fundamentally affected if the system is changed in a way that high flows are reduced and low flows increased. An extreme such case are intermittent rivers, i.e. rivers which usually run dry during certain periods of the year. A large number of plant and animal species live in exactly such habitats, and many of these species depend on them (Datry et al. 2017, 2018); managing the water resources in a way as to have flowing water the whole year round, while it may be beneficial for human users and potentially for some species of plants and animals as well, may well lead to the disappearance or species and entire ecosystems adapted to such situations. This is one of the crucial conflicts between hydropower and ecology. One prominent example of a large floodplain functioning under a regime of distinct seasonal flow, and now under pressure due to the reduction of seasonal differences by dam projects is the Tonle Sap in Cambodia (see Sect. 22.5).

Seasons—at least in tropical areas, where "winter" might be a misleading term—are often referred to as dry and rainy seasons. However, for describing the situation in the river, it might be better and more accurate to use the somewhat clumsier, but more precise terms "high flow" and "low flow" season. The reason for this is illustrated in Fig. 9.7.

As can be seen here, discharge and precipitation pattern do not match in this case: highest river discharge is from April to July, while highest precipitation is from

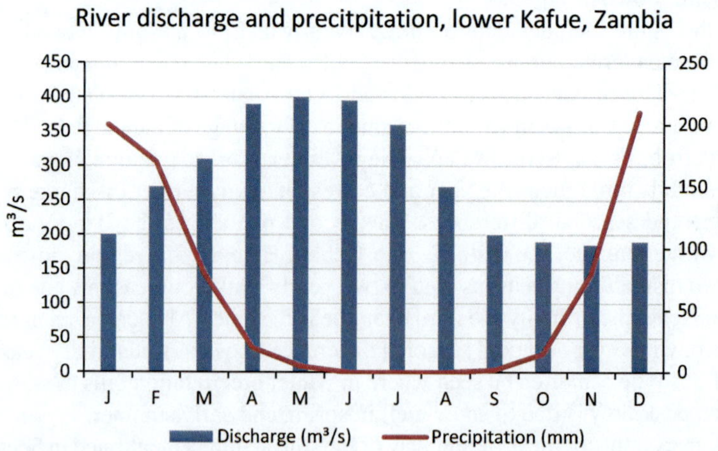

Fig. 9.7 River discharge and precipitation, lower Kafue, Zambia. Based on data from Pöyry and Zenith (2016)

December to February. The reason for this difference is simply the fact that both values are the ones recorded in the lower part of the catchment of the river. The by far largest part of the catchment, from where the bulk of the water in the river comes, is located several hundred kilometres away and further north, and by the time peak flows reach this part of the river, the local rainy season is over. A similar shift can be seen in cases where most of the precipitation falls in winter in the form of snow, while the peak discharge can be observed at the time of snow melt, several months later.

As was shown in Fig. 9.3, there are two ways for showing differences in flow during the year. One is the average monthly flow, the other the flow duration curve; this latter means to show the percentage of a year (sometimes also expressed in days) during which a certain flow will on average be exceeded. For the case of the Kafue HPPs illustrated there, these curves show e.g. that under natural conditions, there will be 23% of the year or 84 days with a discharge of over 480 m^3/s; this is the turbining capacity of the power house, whatever flow exceeds this value has to be spilled and is, in engineering terms, lost. With regulation, such a loss will be observed only during 12% of the time or 44 days, the difference in water, i.e. the amount otherwise lost to spilling, is retained in the reservoir and saved for producing energy during the low-flow season. This shows clearly that from an engineering point of view the flow duration curve provides very relevant information. It can also be used for determining the requirements for ecological flow (see Chap. 20).

However, from a biological, i.e. strictly environmental point of view, this curve does not provide much information, while the seasonal or actually monthly distribution is very relevant, since it has a direct effect on all types of organisms living in or near the river, and on all habitats associated in one way or another with the river. These points are discussed below, e.g. in Sect. 10.3 for fish.

There is one other relevant point which makes this way of providing information very important: public participation. One general rule for ESIA reports is that they should be scientifically correct, but at the same time written in a way as to be understood by a non-specialist. Any person who will read such a report will understand that in a certain case there will be much water in the river, e.g., in late spring/early summer, and very little water in winter; the flow duration curve does not provide this simple, but essential information.

Wherever long term data are available, they can be used not only for calculating the yearly pattern, but also for showing the year-to-year irregularity in river flows, which can be very considerable, wet years alternating with dry years. It is then also possible to see whether during the period of monitoring longer dry or wet spells occurred, in the way that Fig. 9.4 shows a prolonged drought period between 1980 and 2000 for the Kafue basin in Zambia. Such a time series also allows to see whether there is a trend over the years, which suggests either increasing or decreasing rainfall. This is an important piece of information for the discussion of potential effects related to climate change (see Sect. 8.2).

9.1.3 Project Impacts

9.1.3.1 Main Types of Impacts

A dam can lead to a considerable change in river discharge in the downstream area. While a ROR will have no or a very limited effect in this respect, in the case of storage dams it can be very considerable. The following main types can be distinguished:

- Run-of-river (ROR) structures: in this case, the dam or weir just serves as an intake structure and possibly for providing additional head, the reservoir is usually small, has no or very limited storage capacity, and basically the water just flows through the structure, with no or very little change in downstream discharge. However, a ROR structure can be combined with river deviation (see below).
- Dams with daily storage capacity provide the possibility for peaking operation, meaning that the power plant will operate during the time(s) of day with the highest energy demand. During this time, the water level in the reservoir will be drawn down, and the discharge in the river downstream will rise. If the turbines are started very quickly, in extreme cases within a few seconds, this will result in a surge wave, which can be very destructive and which can pose a serious danger for people using the river, and this can happen several times a day. When the turbines are shut down, the water level in the river will drop very fast again, and the reservoir will be filled during this period. KGL in Zambia (Fig. 9.2) is a scheme which is working in this way.
- Dams with seasonal and overannual storage can completely change the seasonal distribution of river discharge in the downstream area, and this is very often their main purpose: the aim here is to store water in the high-flow season in order to make it available in the low-flow season, be it for irrigation or for generating electricity during this latter season. In addition, they can also be used, if required, for peaking production. This means that in such a case every possibility for influencing the downstream discharge is given, from short-term variations to very marked seasonal changes. Examples discussed in other chapters include e.g. Mauvoisin in Switzerland (Sect. 25.1), Rogun in Tajikistan (Chap. 23) and Ilisu in Turkey (Sect. 26.3).
- Deviation projects, i.e. projects where the water is taken from the reservoir into headrace structures (open channels or tunnels) to be directed to the power house further downstream, often over considerable distances, have an additional effect: they permanently reduce the discharge in the river stretch between dam and power house (or tailrace outlet, which again can be further downstream). In an extreme case this can mean zero flow, at least down to the next tributary. Such a structure can be combined with any of the three types described above, and in this way also a ROR can cause this type of impact; an example, Shnogh in Armenia, is described in Sect. 9.1.1.1.
- Transbasin projects, which create a reservoir on one river and then divert the water into another river, can be considered as extreme cases of deviation projects. Here, the entire river downstream of the dam will lose the diverted quantity of

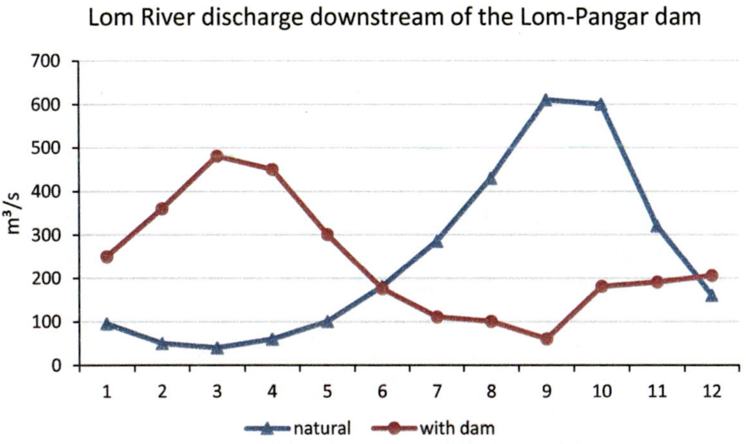

Fig. 9.8 Shift in seasonal discharge, Lom-Pangar HPP. Based on data from EDF (2016)

water. In a way, irrigation projects have the same effect: since irrigation, unlike power generation, is a consumptive use of water, the quantity of water that may reach the river again, if any, will be usually much smaller than the intake quantity (and in addition to that, drainage water from irrigation schemes will usually be contaminated). An additional impact of a deviation HPP is obviously the fact that the water abstracted from one river will be released into another one, in this way changing the conditions there as well. Examples of such types are shown in Fig. 5.3 (Curciusa) and Fig. 22.5 (Theun-Hinboun and Nam Theun 2).

In the following sections, the three main impacts—seasonal shift, peaking and river diversion—will be described and discussed shortly. However, it has to be kept in mind that depending on the situation, these effects can be combined. An example for this may be the case of the Kafue cascade presented above: Itezhi-Tezhi has a regulating capacity, and the ensuing shift in seasonal discharge had a definitely negative impact on the large floodplain, the Kafue Flats. The two other dams, KGU and KGL, both have a river diversion, leaving a part of the river with reduced flow, and KGL will be operated as a peaking plant causing a surge wave downstream of its tailrace outlet.

9.1.3.2 Shifting Discharge

Impacts

Figure 9.8 sshows an extreme example of shift in seasonal discharge, the case of Lom-Pangar dam in Cameroon.

As the figure clearly shows, the River, Lom has a very seasonal discharge pattern, which is completely changed by the operation of the dam: the high flow is now during the dry season, and discharge during the rainy season is reduced almost to the minimum the dry season had before. In this case, this was the main purpose of the dam, which is called Lom-Pangar since it was constructed immediately downstream of the confluence of these two rivers (see shape of the reservoir in Fig. 16.2). The major river in Cameroon, the Sanaga, of which the Lom is one of the main tributaries, has the same flow regime as the Lom, with a very high discharge in September and October and a minimum from February to March. The main objective of the Lom-Pangar dam, together with a number of additional reservoirs on other tributaries, is to retain water during the rainy season and to make it available in the dry season so that the cascade of ROR (some with small storage capacity) hydropower projects along the Sanaga (Song Loulou and Edéa, existing, and Nachtigall and Song-Mbengué, planned), have enough water for generating electricity at or near full capacity year round (see Fig. 24.1).

A change of this type in the seasonal discharge pattern will inevitably have negative impacts on wetlands along the downstream area of the river. The natural high water level in such wetlands will no longer be reached, or not at the usual time, and the low water level will be much higher than it was under natural conditions. This will lead to a reduction in the area hitherto influenced by the flood regime of the river, and a shift of the zonal vegetation within the remaining floodplain area. A study on downstream effects of Lom Pangar should have been prepared before commissioning the dam, unfortunately, this was not done. A short introduction to floodplain dynamics is provided in Sect. 18.2, and examples of such impacts are described for the Tonle Sap in Cambodia (Sect. 22.5), and Tigrovaya Balka in Tajikistan (Sect. 18.2.4.1).

One special point may have to be considered especially for large reservoirs: the filling phase. During this time, water will actually be retained "for good", and not just for being released again a few months later. This means that during this phase, overall flow in the downstream area will be reduced by the amount required for filling the reservoir. This can be substantial, e.g. in the case of Rogun HPP in Tajikistan, a large reservoir with a storage capacity of 13.3 km^3, corresponding to 82% of the average yearly flow of the river; in order to reduce the main negative impact of this filling phase, a reduction of water in the downstream area, to a minimum, it is planned to stretch filling the reservoir over a duration of 16 years, and water retention will be further reduced in exceptionally dry years (see Sect. 23.2).

Figure 9.9 illustrates another case of a large reservoir, Ilisu in Turkey, where the downstream discharge will be influenced during the filling phase. It shows a situation as the one described above: a seasonal river where the discharge pattern is strongly changed by the dam; in addition, it shows a change, but mainly a very marked reduction in flow, during the filling phase. The considerable amount of water released in April and May is conditioned by the need to operate the irrigation systems downstream of the dam, but may also be related to the fact that the water level, during first filling of a reservoir, is not allowed to raise too fast. In such a case it might be helpful to check the time series available for the river in order to see whether under natural conditions something like the situation expected for the dry season during the period of reservoir filling may occur naturally in an exceptionally dry year. On the other hand, especially in cases where drought conditions are not too rare, such

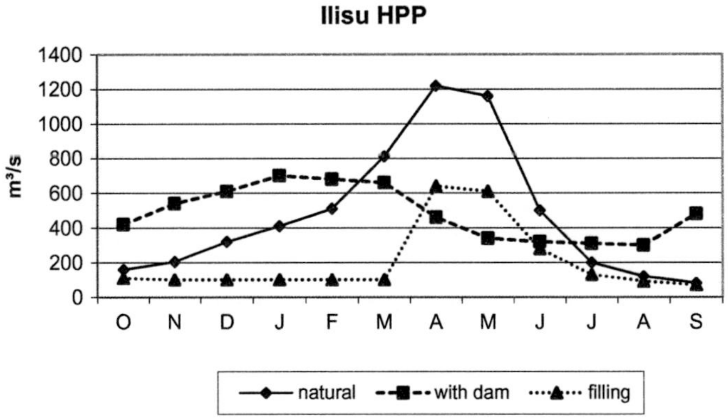

Fig. 9.9 Tigris River discharge as influenced by Ilisu HPP

an analysis might also help in the preparation of an alternative filling plan if such a dry spell should occur during the time foreseen for filling the reservoir.

Mitigation

As for any case of impact, the first question that needs to be answered is whether mitigation of the impact is required and possible, and what its relevance would be. In case of a seasonal shift of discharge into a river bed which is essentially narrow and limited in lateral extent, where therefore the impact is only to have a higher than natural water level part of the year, and a lower level another part of the year, both probably associated with changes in flow speed, mitigation may not be really important, since the general characteristics of the habitat may not have changed very much. Still, organisms adapted to and depending on seasonally changing conditions may be affected.

On the other hand, changes will be very relevant where extensive floodplains are affected. As explained in more detail in Sect. 18.2, this can have very far-reaching effects both on natural habitats and on human use of such areas. And here we have a classical conflict of interest: as described above, increase in low season flow is a requirement for generating electricity during this period, and we should not forget that this often happens during a time of the year when natural flows in rivers are at their minimum, but electricity demand is at its maximum, as is the case e.g. in Central and Northern Europe. Increasing flow in this period is not possible without decreasing flow during the high flow season. A real mitigation in such a case is not possible, since from an environmental point of view this would require to restore the natural discharge pattern, which would mean to abandon the purpose of the HPP. The only way of at least somewhat reducing the impact can consist of specific floodplain management measures. One measure which may be feasible is to periodically release an amount of water from the dam which corresponds to a normal high flow season peak, in this way mimicking a normal (average) flood event. Even if this is shorter and occurs with less frequency as would be the case under natural conditions, it may still be sufficient for preventing plants from higher elevation, which cannot sustain periodical flooding, from invading the floodplain, changing the habitat characteristics in a fundamental way. The measures will have to be defined on a case to case basis, where the value of the affected floodplain—value specifically in terms of biodiversity conservation, but in certain cases certainly also in terms of flood protection for areas further downstream—have to be considered carefully. A monitoring program will be required in such cases, and other measures, as e.g. direct interference in the physical or biological conditions of the floodplain or parts of it, may be required.

9.1.3.3 Peaking

Impacts

Storage hydropower structures have one major advantage over any other way of generating electricity: as long as there is enough water in the reservoir, turbines can be switched on and off any time, and this on very short notice. It is a fact that electricity demand varies seasonally, but also, and sometimes even more so, during the day, with hours with peak demand alternating with hours with considerably lower demand. No other way of electricity generation has this flexibility: not thermal plants (whether fossil fuel or nuclear), and certainly not the alternative sources wind or solar. This means that even in cases where electricity demand would be met mainly by one of these other sources, hydro may still play an important role for adjusting production to short-term variations in demand. This is being done by peaking, which means that if demand is raising, turbines are switched on, only to be turned off again as soon as demand decreases. This great advantage of hydropower, however, comes at an environmental cost, which can be considerable.

Figure 9.10 shows such a peaking pattern, in a case where two different operating patterns were envisaged, one for summer, and one for winter. It has to be pointed out clearly that this is a pattern that was envisaged before the start of operation of Atatürk HPP, and is not necessarily the pattern which was then actually followed during operation, and that it might have changed in the meantime. The objective here is not to show a pattern of a specific power plant, but to illustrate the issue of peaking.

From an environmental point of view, in this case there are two main problems. The first is the very rapid shift in the water volume flowing out of the tailrace structure, which can increase from 400 to 1200 m^3/s in summer, and from 0 to 1600 m^3/s in winter, and then drop again, in a very short amount of time. This leads to an enormous surge wave every time the turbines are switched on, and to an extremely fast reduction in flow when they are turned off again, and this every day; in other cases, where peaks are shorter but more frequent, this can even happen several times a day. First of all, this can put people at risk who, for whatever reason, stay on the river or within the river bed at the time when the surge wave arrives. The people most at risk are usually fishermen, women fetching water or washing, and children playing on or near the

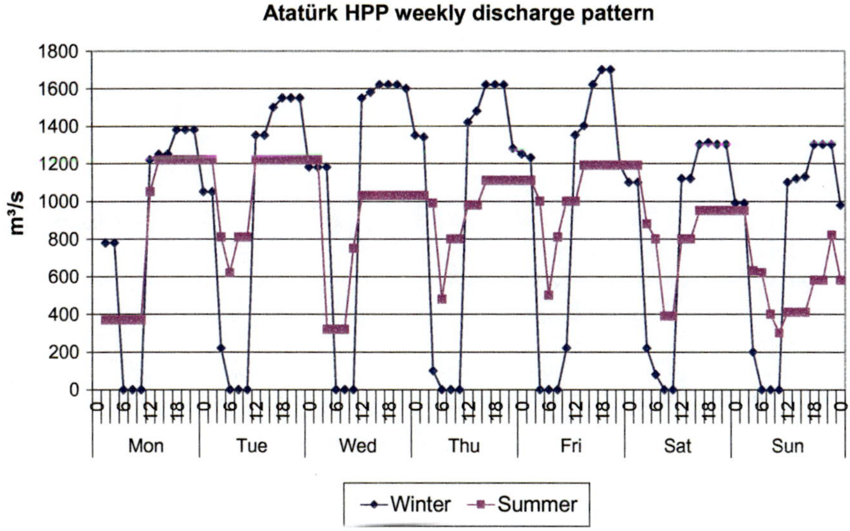

Fig. 9.10 Operating regime of Atatürk HPP, Turkey. Based on date from IEG (2001) and Zwahlen (2003)

river. The second impact is the one on aquatic organisms, since such a wave will flush out the river every time, washing out sediments as well as anything living in the affected part of the river. The surface covered with water will greatly expand and then shrink again, and water depths and flow velocity will change abruptly. In addition, especially in the case of large and deep reservoirs exhibiting thermal stratification (see Sect. 18.1) the temperature of the water delivered during turbine operation can be significantly colder than the residual flow of the river, and in extreme cases it can also be oxygen deficient. The development of a normal aquatic vegetation and fauna is not possible under such conditions. The surge wave will smooth out gradually in the downstream area, but depending on size and frequency of the wave and on morphology of the river bed the affected part of the river can still be long, the effect being felt over dozens of kilometres.

An additional issue is the fact that in a peaking regime the discharge between two peaks can actually go down to zero, as shown in the figure.

Mitigation

Here again, mitigation in the sense of eliminating the impact is not possible as long as a peaking regime is being maintained. However, a number of mitigation measures are available which aim at reducing the impact.

The first is to slow down the raise in discharge by gradually starting the turbines (as far as this is technically feasible), and, in the case of structures with several turbines, not to start all of them at once. As Fig. 9.10 shows for the summer pattern, this was the

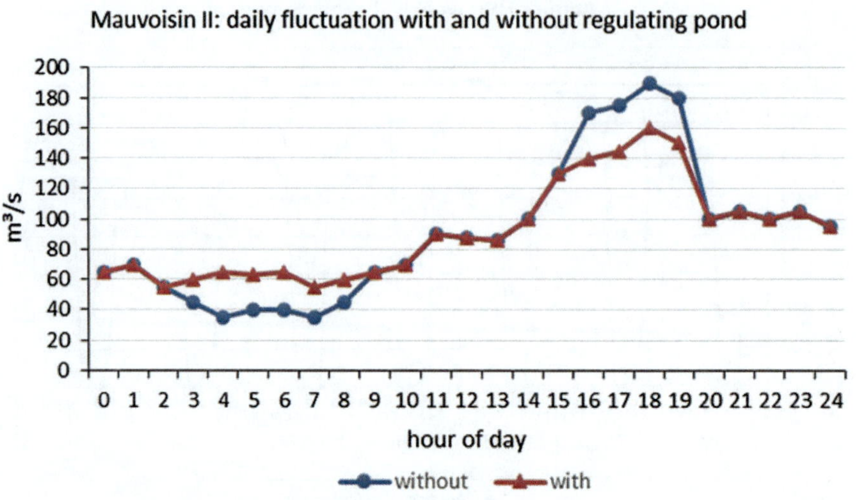

Fig. 9.11 Effect of a re-regulation pond (model calculation)

plan for the summer weekly pattern, where the turbines are put in march one after the other, and this over a stretch of several hours. This will enable water discharge to raise gradually, which will make the surge wave less abrupt and therefore less dangerous. If this is coupled with a baseline flow of a considerable amount, as also shown in the Atatürk summer pattern, this already reduces the impact to a considerable extent. This is obviously only possible in close concertation with the operator of the power plant and will have to be coordinated with the management of the entire power grid into which the electricity is fed.

Another possibility is to integrate a re-regulation pond into the structures. This is a pond into which the water from the turbines is flowing, to be released from there gradually with the aim of smoothing the raise of the surge wave, reducing the peak flow and raising the minimum flow. Examples for HPPs with such a pond are Nam Theun 2 and Nam Ngiep 1 in Laos. Considerable efforts in that direction were made for a project called Mauvoisin II, in Switzerland (see Sect. 25.1 for additional information on this project). Its aim was to increase the capacity for peaking without adding storage capacity to the system. This would have meant to enable a higher peak (for providing electricity during periods of high demand), but of a shorter duration than what is possible with the existing scheme. The negative effect would have been a very considerable surge wave in the River Rhone, which receives the water from this power plant. Figure 9.11 shows the anticipated effect of this regulation pond.

The figure shows the situation in the River Rhone downstream of the tailrace outlet from Mauvoisin HPP with and without this mitigation. Two main effects are clearly visible: the trough, i.e. the lowest flow during the night, is less marked, and the peak, the maximum flow in late afternoon and evening is reduced; with some fine-tuning it might ultimately have been possible to smoothen the daily raise and the drop in late afternoon, somewhat more. Such a structure, however, does not come without

Fig. 9.12 Water intake structure providing residual flow

its own impact, not even mentioning the costs: it would have required a pond with a surface area of 9.95 ha (irregular shape, 450 × 200 m), surrounded by a 10 m high dam, providing a retention volume of 470,000 m^3. The figure also shows that, while the pond can reduce the impact, it is not able to completely avoid it. Ultimately, for strictly economic reasons, Mauvoisin II was not implemented.

One measure that will often have to be taken along rivers affected by peaking power is to inform the population potentially at risk and to position warning signs in places where people might access the river.

Finally, a residual flow can contribute to mitigating the effect, since it will at least guarantee that flow will never be reduced to zero. Still, in the case of Atatürk, with a potential surge from zero to 1600 m^3/s, a residual flow of a few dozens of m^3/s would not really change the situation. See Chap. 20 for a more in-depth discussion of the residual flow issue.

9.1.3.4 River Deviation

Impacts

A river deviation will, as mentioned above, reduce the flow in the affected part of the river by the quantity of water taken in for the power plant. This reduction can be—and usually is—very substantial, as shown in Fig. 9.12, although a reduction to zero flow today would be accepted only very exceptionally, since most regulations in this respect foresee a minimum flow to be left in the river at any time.

Such a reduction in flow volume will definitely change the nature of the affected part of the river. It may still be a habitat for a number of aquatic organisms, but in most cases it will not be able to sustain the entire set of organisms that lived in the original river, and it could also be that some species which were considered as being of special importance will no longer be able to live there; as always, the impact in a specific case will have to be assessed carefully based on the local conditions.

A number of indirect impacts may follow. One point to consider is the use of the affected part of the river by the local population, e.g. for irrigation or the provision of drinking water, or for dilution and elimination of waste water; a reduction in flow will have a fundamental impact on these services of the natural river. Another point to potentially consider is the fact that the river bed, given the reduced flow, will also be reduced, e.g. by plants starting to move in and growing in the part of the river bed which is now dry. This can reduce the capacity of the river bed to deal with exceptionally high flows, i.e. with flood events. An intake is usually structured in a way as to capture the amount of water that can be used in the power plant, whereas peak flows will be passed over or through the spillways. Flow in the old river bed, during a flood event, will often be as high as it was under natural conditions, but the difference between average flow, i.e. residual flow, and flood discharge will be much greater than it was before. If the width of the river bed is now reduced, it is possible that the flood wave can no longer be passed, and that more flood damage occurs than what was the case earlier.

Mitigation

Here as well, avoidance of the impacts is not possible, it is only possible to reduce them to some degree. The most important mitigation measure here is certainly the definition of a residual flow, which has to be done based on the specific circumstances of the site; see Chap. 20 for a more detailed discussion of this topic. The overgrowing and ensuing width reduction of the river bed can be controlled by specific measures on the ground, i.e. by constantly eliminating undesired plants that start to grow there. This, however, is not a one-time measure to be taken, but requires a continuous effort throughout the operation period. Releasing flood pulses from the weir that imitate natural high flow conditions, albeit much shorter, may be a measure supporting such activities.

9.2 Hydrogeology

A dam and reservoir project will have influence on groundwater mainly around the reservoir and potentially in the downstream area. The groundwater level around the reservoir will usually stabilise approximately at the water level in the reservoir, which means that it will be considerably higher than it was under natural conditions. This development was of some concern in the case of Rogun HPP in Tajikistan, since there drawdown of the reservoir and accelerated movement of the ground water level along slopes covered with large amounts of lose material, some of it stemming from earlier rock falls or mudslides, might trigger landslides into the reservoir (see Fig. 8.9). In that case it was concluded that there was no risk that waves which might overtop the dam might be caused by such landslides. However, movements of this material might accelerate reservoir siltation. It was recommended to install a monitoring system which would allow to recognise any such movements at an early stage (Pöyry 2014).

The groundwater situation also was of high relevance in the case of decommissioning or replacing the Rheinfelden HPP (see Sect. 13.2). There, the groundwater level had stabilised at about the water level in the reservoir, which close to the dam was roughly 10 m higher than the former natural water level in the river. In the about 100 years of existence of the dam the entire situation had adapted to that water level, and there was a risk that if the groundwater level would go down to the original elevation, a large number of buildings in the area might be damaged as a consequence of this. While this was certainly not the one deciding issue, it was definitely one point in favour of replacing the dam with a new structure, keeping the reservoir level constant, as opposed to removing the dam. During the studies, a sophisticated groundwater model was applied to the situation, which was fed by data obtained from a network of piezometers installed specifically for that purpose (EWI 1987; EWI and Simultech 1988).

If the water discharge pattern in the area downstream of the dam changes, a groundwater system linked to the river can be influenced by these changes, e.g. in a floodplain (Grill et al. 2019). In such a situation, a system for surveying the groundwater may be required, as e.g. proposed by Kummu et al. (2014) for the Tonle Sap.

Projects in an area characterised by karstic formations can pose some very specific challenges, since in such a case it might not be clear which ways the water takes in the natural situation, and what alternative ways it may find if e.g. a reservoir is created. The river flowing out of Lake Ira Lalaro, East Timor, disappears into a sinkhole about 5 km away from its origin in the lake. There is (or was) a project to use this outflow from the lake for a hydropower plant. However, it was not known which way the water actually was taking after its disappearance from the surface, and what therefore the impacts would be if the project should be implemented. It was therefore recommended that this situation would have to be closely examined before any decision on the project would be taken (Electrowatt-Ekono and SKM 2003). Another case of a project in a karstic area, characterised by fractured limestone cliffs with a considerable number of caves, is Jannah in Lebanon. Before construction started, a number of hydrogeological investigations were made with the aim of clarifying whether some of the water to be retained in the reservoir would actually leak out and in this way be lost for power generation. It was, however, not possible to reach a complete agreement on these questions (Pöyry 2017).

9.3 Water Quality

9.3.1 The Parameters: What, How Much and Why?

9.3.1.1 The Full Range

Depending on the objective and the exact resource to be addressed (surface, ground, drinking, or waste water), the list of water quality parameters to be analysed can be quite different, and there are certainly also differences in the respective regulations from one country to another. Just two examples: the Irish EPA (2001) list comprises 101 parameters, and the Concession Agreement for a hydropower project in Laos lists 51; the latter contains, e.g., 8 heavy metals and 5 different organochlorine pesticides.

9.3.1.2 The Essential Parameters

A rather limited number of water quality parameters are of utmost importance in relation with hydropower projects, and therefore for impact assessment for such projects. Of them, two are the most relevant, namely:

- Oxygen: the single most important parameter, at least in projects which include a reservoir with a storage capacity of more than just a few days or weeks, but also in high head power plants. The importance of this parameter is given by the simple fact that oxygen is an important requirement for all fish (although to very varying degree, from species like salmons which require very high oxygen content of the water to some fish species which are able to take in air directly from the surface or which are otherwise very tolerant of low oxygen content) as well as for many other aquatic organisms, and by the risk of oxygen depletion in the reservoir after impoundment, especially in cases where large amounts of biomass are submerged (see Chap. 17). On the other hand, fast and large differences in pressure, sometimes combined with gas oversaturation in very turbulent water (oxygen, but probably more often nitrogen), can lead to fish deaths due to the so-called bubble disease, i.e. the formation of gas bubbles in the blood vessels of fish when pressure drops very fast (Reichenbach-Klinke 1980: 358; Zhong and Power 1996).
- Temperature: in addition to what can be considered as a normal baseline description there are two main reasons why temperature is or can be of importance: (i) it determines the amount of oxygen that can be dissolved in water, and (ii) in deep reservoirs with a thermal stratification (see Chap. 17) water taken from deeper zones of the reservoir can be significantly colder than the river water, and this can have consequences in the area downstream of the tailrace structure.

  Monitoring of these two parameters will usually be required during the construction period and after impoundment as long as a new equilibrium in the reservoir has not yet been reached.

A number of water quality parameters are of specific importance during the construction period. These are:

- pH: the pH value of river water varies around 7 (neutral), usually between 6.5 and 8.5, but can go down as far as 5, in the case of "black water rivers", i.e. rivers flowing through wetlands or wet tropical areas with high amounts of organic material and therefore containing high concentrations of humic acids. In the context of a hydropower project, this value can be influenced mainly during the construction phase, in cases where water contaminated with large amounts of concrete reaches the river. Such a contamination has the potential to strongly increase the pH value in the water, to values of 9 or higher, and this in turn can be harmful for aquatic organisms, mainly for fish. Acidic water, i.e. water with a natural pH of lower than 6 will react most strongly, or turning it the other way round, fish adapted to such conditions will be most sensitive to such changes. The pH value must be monitored throughout the entire construction period and, depending on the situation, potentially into the operation period.
- TSS (total suspended solids): this is mainly fine silt which is kept in suspension by movement of the water. Total amount and granulometry will be decided mainly by turbulence; as soon as flow velocity and with it turbulence drop, different fractions of TSS will start to sediment (see Fig. 8.10, Nurek, as an example). The main source of TSS is erosion, and runoff from bare surfaces can contain very high concentrations. This means that a dam construction site, with usually very large surfaces from which vegetation was removed, and on which in addition to that soil structure is constantly being disturbed by movements of vehicles and machines, is an obvious source of TSS.
- Oil and grease: such substances can stem from water contamination e.g. on a construction site, whereby two main types of such a contamination can be distinguished, namely (i) accidental spill of potentially large amounts of oil, e.g. stemming from the rupture of a large container, and (ii) more or less continuous losses of fuel and lubricants from all kinds of machines and vehicles. The first type is rare on a reasonably well managed construction site and can rather easily be prevented, most efficiently by storage and handling of these substances according to well-established rules and principles, but if it happens the results can be rather dramatic, putting the entire river downstream of the site at risk. The second type is very difficult to prevent and will often go unnoticed, since there are usually small quantities involved, which will quickly be swept away.
- Coliform bacteria: these are of importance in the case of large construction sites with a large work force living on site, which therefore produce considerable quantities of domestic waste water. Here, several categories are usually determined: total coliforms, *Escherichia coli* (the species most frequently associated with all forms of domestic waste water) and faecal coliforms (indicating, if present, a potentially serious contamination of the water with insufficiently treated domestic waste water).

9.3.1.3 Special Conditions

Concerning water quality, as for other items, it is important to check the specific situation of the project, since depending on local conditions other parameters can be of importance. One of these is mercury (Hg), which is usually not of concern, but can be so under specific circumstances, mainly if there is gold mining, including artisanal gold mining, in the upstream area, or if soils to be submerged contain mercury, which then can be transformed to methylmercury, a soluble form which can enter the food chain. Accumulation of mercury in fish can lead to toxic concentrations in humans consuming contaminated fish. Most publications dealing with this problem investigated the situation in the Amazon basin, which seems to be especially exposed to this type of water contamination (Porvari 1995; Zhong and Power 1996; Dorcey et al. 1997: 98; Meech et al. 1998; Guimaraes et al. 1999; Fearnside 2016). It cannot be excluded that other heavy metals could, in very specific cases, become an issue as well. This would seem to be a risk especially if there are any mining activities in the catchment area of the reservoir, from where sludge containing such substances could reach the reservoir; an example of this type was mentioned in Sect. 9.1.1.1.

9.3.2 Baseline

Water quality is undoubtedly an important aspect of the baseline situation description for any HPP. It is essential to know what the water quality was like before any work for the project started, and it is just as essential to know whether at a specific site there is a special situation, like e.g. higher concentrations than allowed by relevant guidelines of any of the parameters checked, which might indicate a source of contamination in the upstream area, be it of natural origin or caused by human activities. Here, it is certainly good to determine a wide range of parameters, e.g. the 51 mentioned above for a project in Laos, in order to have a good picture of the situation, in order to identify any specific actual or potential problem, or to accurately exclude such a situation. The problem here can be timing of the sampling, since water contamination can show important seasonal variations, e.g. through dilution during the high flow season, or through periodicity in some relevant human activities like e.g. use of fertilisers or pesticides. Concentrations of some pollutants can go up greatly at the onset of the rainy season, where during the first rains on dry soils there is a considerable surface runoff which sweeps large amounts of soil particles, often contaminated with agrochemicals, into the rivers. It is therefore recommended, if the schedule of the study allows it, to take and analyse a number of samples during different periods of the hydraulic cycle in order to have a better picture of the situation. In an ideal situation—and this is not always the case—there are water quality data available from third parties which can be used to better reflect the situation.

9.3.3 Impacts

Impacts due to HPPs can roughly be grouped into the following three categories:

1. Impacts during construction: this is the period where some impacts are inevitable (e.g. removing vegetation from large surfaces, exposing the soils to erosion, which will increase the TSS value; see Fig. 9.13) and where there is a high risk for a number of other impacts, as e.g. water contamination with coliform

Fig. 9.13 NT1 dam site, Laos. Dam site twice seen from approximately the same place, once before and once shortly after start of construction work

Fig. 9.14 Fuel oil leaking out of machine or storage tank. A serious case of leakage, oil flowing directly into the river. Bakhtiari dam site, Iran

germs, increase in pH due to inflow of water contaminated with concrete, or water contamination with fuels and lubricants (Fig. 9.14).

2. Impacts during impoundment: this is a very relevant period in relation to water quality, in the reservoir as well as in the downstream area. The effects of impoundment are greatest in projects with a large and deep reservoir submerging large amounts of biomass (see discussion of this issue in Sect. 10.2.3 and Chap. 17). The overall impact of impoundment is determined not only by the fact that water is retained in the reservoir, but also by the way in which this is done. So e.g., in the case of Lom Pangar HPP in Cameroon, where a large area of dense forest was submerged, the reservoir was partly filled twice and then emptied again during the impoundment period. In this way considerable amounts of biomass were flushed out of the reservoir and swept downstream. This may have reduced the water quality problem in the reservoir to some extent, but has certainly lead to problems in the downstream area; unfortunately, this procedure was not accompanied by a suitable monitoring.
3. Impacts during operation: this period officially starts once the power plant has taken up its regular performance. The transition from impoundment to operation, however, is not necessarily very well defined. In projects with a large reservoir and an important installed capacity it can be that commissioning of turbines is being done in stages, and that some units start operating normally before the reservoir has reached its final water level. In addition to that, there can be a long transition period from the start of impoundment, with deteriorating water quality in the reservoir mainly due to lack of oxygen in the deeper parts of it, till the moment when a new steady state, or a new equilibrium, has been

reached. As far as possible, this development of the situation in the reservoir needs to be taken into account in the ESIA, and a monitoring programme has to be designed accordingly. The situation can change to a very considerable extent during the operation period, which can last up to a century. Drivers of such changes are usually developments in the catchment area, like deforestation, expanding agriculture, growth of settlements or industrialisation. As far as such developments can be identified in the planning phase, the ESIA will have to point out the risks and potential solutions, but it is impossible to make a reliable forecast of future developments over a longer time span.

In addition, or possibly as a fourth category, one could also mention impacts on water quality during decommissioning. If a dam is removed, the work to be done can lead to a number of water quality problems which are similar to the ones experienced during construction; in addition, mobilisation of the sediment accumulated in the reservoir during the operation period can lead to problems in the downstream area (see example in Sect. 13.2).

9.3.4 Mitigation

Since most impacts on water quality become manifest during construction, the identification of mitigation measures has to concentrate on this phase. Mitigation has to be done largely by taking precautions which are likely to reduce the risk of accidents or effects of normal construction activities. These measures are described in the ESMMP (Environmental and Social Management and Monitoring Plan), where the most important risks are identified and measures as well as responsibilities are defined; an outline of such a plan is provided in Chap. 12.

9.3.5 Monitoring

Monitoring of water quality is essential for the proper environmental management of a dam and hydropower project. Therefore it is important that as part of the ESIA—or of the ESMMP mentioned above—a water quality programme is developed, or at least outlined, for being developed further in the following stages of project development. As s first step, it has to be clear what the purposes of such a monitoring are. There are three main objectives:

1. Description of the prevailing situation before the start of project activities on the ground, and evolution of the situation over time.
2. Identification of problems that arise due to the project or project related activities (which are mostly construction activities) with the aim of being able to take appropriate mitigating measures where necessary.

3. Identification of situations which could lead to problems and complaints in the downstream area, but which do not originate from the project.

These three points already define where in principle the monitoring points have to be placed namely (i) upstream of the area of influence, (ii) within the area, and (iii) downstream of it. It has to be pointed out clearly that "monitoring point", in a simple situation, can actually mean one point in these three locations, while very often quite a series of points will be required; "upstream" may be one site located in the river upstream of the future reservoir, but in cases like e.g. Lom Pangar (see Sect. 17.2) or Rusumo Falls (see Sect. 9.3.6), where the dam site is located downstream of the confluence of two rivers, at least one point in each of these will be required. If the reservoir is very long, several points may be required within the reservoir area (to be maintained later on, after impoundment). In addition, it is certainly required to have such a point (or several in the case of more than one main tributary) upstream of the future reservoir; this will provide a data base for future monitoring during the operation period. Several points may also be necessary within the construction site as such, and at least one point downstream in a sufficient distance as to make sure that no direct project impacts can become manifest only downstream of this point. Once again, the project specific situation may require specific approaches. In the case of Nam Ngum 2 in Laos, e.g., the water flowing into the existing Nam Ngum 1 reservoir a short stretch downstream of the construction site was often heavily loaded with suspended solids, obviously stemming from the construction site. Only monitoring of TSS could demonstrate that during the rainy season a considerable part of this silt actually originated from the area upstream of the construction site, and that another very considerable part was contributed by a tributary which joined the river about half way between construction site and inflow into the existing reservoir. This should not be understood as an excuse for not taking appropriate measures for erosion protection on the construction site. The example should just illustrate that there may be sources of water contamination on which the project has no influence. Still, in cases of problems with water contamination in the downstream area the obvious culprit will always be the large construction site, and only good monitoring data will be able to show whether or not this assumption is really justified.

Monitoring, and with it the selection of parameters to be monitored and frequency of sampling, will be different for different purposes of the monitoring. The most important categories are the following, whereby the parameters mentioned are not meant to represent a comprehensive list for every case:

- Baseline situation: as mentioned above, the entire range of parameters can or may have to be monitored here, in order to get a good comprehensive impression of the situation. It is certainly justified to carry out such a program throughout the process, i.e. from the planning phase into operation, however, with a low frequency, depending on the situation possibly in intervals of 3 to 6 months. Parameters which have never been identified during the first few tests, like e.g. certain heavy metals or pesticides, should then gradually be eliminated from further testing.

- During the construction phase, a number of different types of water may have to be monitored:
  - River water quality: basic parameters, i.e. pH, temperature, oxygen, TSS; some other parameters may have to be included, like oil and grease, coliform germs, or conductivity (as a proxy for total dissolved solids). Here frequency should be as high as possible. Once a month, as is often being done, seems to be insufficient, since then there is a low probability of detecting, e.g., a short time massive raise in pH due to insufficient retention of sludges from concrete batching plants.
  - Domestic waste water: this will have to be treated before being released to the river, and monitoring of the relevant parameters, most of all coliform germs, will have to be made in order to verify whether legal standards for waste water flow into surface waters can be met.
  - Waste water from retention structures, sedimentation ponds etc.: here, the relevant parameters are pH (for water from sedimentation ponds of crushing and concrete plants), and TSS.
  - Drinking water: where drinking water is taken from a natural source and distributed through an internal piping system, the quality will have to be checked regularly, again with coliform germs as the most important parameter; residual chlorine may have to be checked where this is used for water treatment before distribution.

- Throughout the construction phase, impoundment, commissioning and into the operation phase, the two most important parameters to be closely monitored are temperature and oxygen. This must be done in the reservoir as well as in the downstream area, and an elevated number of sampling points may be required. Here it is essential that these two parameters are analysed over the entire water column within the reservoir, i.e. at different depths from the surface to the bottom of the deepest part of the reservoir; see Chap. 17 for a discussion of the issues at hand and a number of examples.
- Monitoring during operation, i.e. once the situation in the reservoir has stabilised, will have to be done mainly with respect to the situation around the reservoir and further upstream in the catchment. If this includes large parts of agricultural areas, main nutrients like nitrates and phosphates should be monitored in order to be able to judge the risk of eutrophication, i.e. the accumulation of nutrients, in the reservoir.

This is not meant to be understood as a complete overview of requirements for water quality monitoring, as always the specific situation of the project (which may include specific legal requirements) have to be observed.

9.3.6 *Importance of Qualification*

The importance of monitoring was pointed out several times above. Here, an additional point of relevance has to be made, illustrated with the results from two monitoring programmes, one for the Bujagali HPP, Uganda, monitoring the pH in the Victoria Nile up- and downstream of the construction site (Fig. 9.15) and the other monitoring oxygen content in the Akagera River, up- and downstream of Rusumo Falls HPP construction site, on the border between Rwanda and Tanzania (Fig. 9.16). Please take a moment to look at the two graphs before reading on.

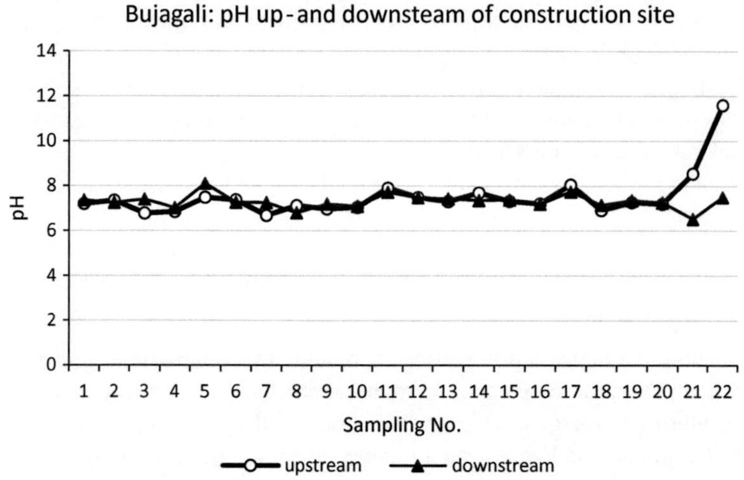

Fig. 9.15 pH values, Bujagali HPP, Uganda. Results of weekly monitoring. Further explanations in the text

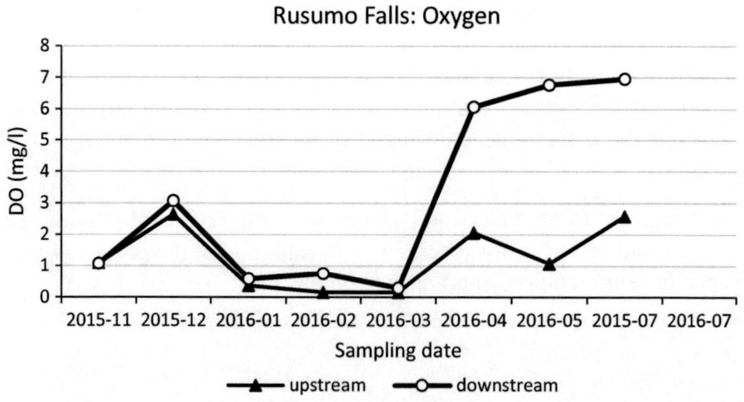

Fig. 9.16 Monitoring of oxygen content, Rusumo Falls HPP

In both cases, there had been monthly reports in which the data recorded during the reporting period were mentioned without further comments, and without provoking any reaction from anybody. Finally, the POE insisted that the date were presented as shown above, not as isolated monitoring values, but as a graph showing the entire data set, and just adding the new values to the graph in each of the subsequent reports. This does not only give a better picture of the situation, and of possible changes in the situation, it also allows to identify any problems that might arise with the data much easier, and the two data sets shown in the graphs clearly have such problems.

In the case of Bujagali, during months pH had constantly been between about 6.5 and 8, values which would be expected in a natural river, and with small differences between the upstream and the downstream data set. And then suddenly there was a pH value of 11.6 in the upstream area; this value had been reported, like all other values before, without any comment. A similar situation had been observed in the reporting for Rusumo HPP. Here as well, the reports had listed the measured values and left them uncommented.

A sudden raise of the pH to a value of more than 11, in a river which might not be completely free of contamination, but which is by no means heavily contaminated, is not only highly unlikely, but is almost impossible. The only "natural" explanation would have been if the sampling point had been located directly at the mouth of a drainage pipe bringing highly charged waste water, e.g. from a concrete plant, which was certainly not the case. The Rusumo example shows something similar: the upstream values may seem not too much out of the range that one would expect, since at this measuring point the rivers flows very slowly and without any turbulence out of a swamp area of a length of several dozens of kilometres, where the amount of organic material in the form of rotting plants is extremely high, and where therefore oxygen consumption in the water is high as well. Even so, the values of close to 0, recorded from January to March, seem to be a bit on the low side. It is completely different for the downstream values: this sampling point is located immediately downstream of Rusumo Falls Fig. 18.5a gives an impression of how these falls look like, and what is inevitably happening there. It is physically just not possible to have water with close to 0 oxygen just downstream of a water fall with such a turbulence. This is confirmed by the values recorded from April onwards, which are about at saturation, i.e. at the maximum oxygen content possible under the prevailing temperatures. A discussion with the person in charge of this monitoring then revealed that at some point in time they had realised that their measuring device had not been working properly; it can be assumed that the same had happened at Bujagali.

The relevant point here: when a monitoring is being done, it is not sufficient to have a technician in the field who knows how operate the device used; this person must also have an understanding of what it is all about. In both cases the measuring was done by using one of these very efficient devices (if handled and calibrated correctly) which show the measured results onsite immediately. In both cases, it should have been obvious for the person taking the measurements that the recorded values could not possibly have been correct. One obvious reaction to that would certainly be to immediately take another measurement, preferably not in exactly the same place in the case of the pH value. In no case should these values have gone uncommented into monitoring reports.

Finally, this means that the person in charge of the overall monitoring program needs to be sure that the specialists working in this program know what they are doing. Training is sometimes included in programmes for the preparation of an ESIA, and this is certainly a valid point. However, training is also required at other levels or points in time during the work for such a project.

References

Balvín Díaz, D. 1995. Agua, Minería y Contaminación: El caso Southern Perú. Labor, Ilo, Peru: 346 pp. + anexos.

Datry, T., N. Bonada, and A. Boulton. 2017. *Intermittent Rivers and ephemeral streams. Ecology and management*, 597 pp. London: Academic Press, Elsevier.

Datry, T., A.J. Boulton, N. Bonada, et al. 2018. Flow intermittence and ecosystem services in rivers of the Anthropocene. *Journal of Applied Ecology* 55 (1): 353–364. https://doi.org/10.1111/1365-2664.12941.

Dorcey, T., Steiner, A., Acreman, M., Orlando, B. (eds.), 1997. Large Dams: Learning from the past, looking at the future. In *Workshop Proceedings*, Gland, Switzerland, 145 pp, April 11–12, 1997. IUCN / The World Bank, Gland and Washington.

EDF. 2016. Lom Pangar – Suivi de la qualité d'eau et des émissions de gaz à effet de serre. Rapport de référence. EDC, 52 pp.

Electrowatt-Ekono and SKM. 2003. Power Sector Development Plan for East Timor. ADB and Government of East Timor.

EPA. 2001. *Parameters of water quality. Interpretation and Standards*. Wexford, Ireland: Environmental Protection Agency. https://www.epa.ie/pubs/advice/water/quality/Water_Quality.pdf.

EWI and Simultech. 1988. Neubau Kraftwerk Rheinfelden. Ergänzungsbericht zum UVP-Bericht 1987. Kraftübertragungswerke Rheinfelden AG.

EWI. 1987. Neubau Kraftwerk Rheinfelden. UVP-Bericht. Kraftübertragungswerke Rheinfelden AG.

Fearnside, P.M. 2016. Environmental and social impacts of hydroelectric dams in Brazilian Amazonia: Implications for the aluminum industry. *World Development* 77: 48–65. https://doi.org/10.1016/j.worlddev.2015.08.015.

Godet, F., and S. Pfister. 2007. Case study on the Itezhi-Tezhi and the Kafue Gorge Dam: The science and politics of international water management. https://www.researchgate.net/publication/259533453.

Grill, G., B. Lehner, M. Thieme, et al. 2019. Mapping the world's free-flowing rivers. *Nature* 569 (7755): 215–221. https://doi.org/10.1038/s441586-019-1111-9.

Guimaraes, J.R.D., J. Fostier, M. Forti, J. Melfi, H. Kehrig, J. Mauro, O. Malm, and J. Krug. 1999. Mercury in human and environmental samples from two lakes in Amapá, Brazilian Amazon. *Ambio* 28 (4): 296–301.

IEG. 2001. Ilisu Dam and HEPP. Environmental Impact Assessment. Ilisu Engineering Group, Consortia for Ilisu: Hydro Concepts Envineerign, Colenco, Hydro-Québec International, Dolsar.

Kummu, M., S. Tes, S. Yin, J. Adamson, J. Józsa, J. Koponen, J. Richey, and J. Sarkkula. 2014. Water balance analysis for the Tonle Sap Lake-floodplain system. *Hydrological Processes* 28: 1722–1733. https://doi.org/10.1002/hyp.9718.

Leeming, D.A. 1994. *A dictionary of creation myths*, 330 pp. New York: Oxford University Press.

Meech, J.A., M.M. Veiga, and D. Tromans. 1998. Reactivity of mercury form gold mining activities in darkwater ecosystems. *Ambio* 27 (2): 92–998.

Porvari, P. 1995. Mercury levels of fish in Tucuruí hydroelectric reservoir and in River Mojú in Amazonia, in the state of Pará, Brazil. *Science of the Total Environment* 175 (2): 109–117. https://doi.org/10.1016/0048-9697(95)04907-X.

Pöyry and Zenith. 2016. Kafue Gorge Lower HPP ESIA: Environmental and Social Impact Assessment. Report prepared for ZESCO and Kafue Gorge Lower Power Development Corporation, Ltd., Zambia: 287 pp.

Pöyry. 2007. Projet hydroélectriqueOued Ahançal. Etude d'Impact Environnenemtal et Social. ONE, Maroc.

Pöyry. 2009. Projet Minier de Belinga: Etude d'Impact Environnemental et Social pour le Projet Hydroélectrique de Kongou. Volume III: Projet Hydroélectrique de Kongou. Report prepared for Compagnie Minière de Belinga (COMIBEL).

Pöyry. 2014. Environmental and social impact assessment for Rogun hydropower plant. ESIA Vol. I: Text. Vol. II: Annexes. Report prepared for OSHPC "Barki Tojik", Republic of Tajikistan.

Pöyry. 2017. Jannah dam and lake project. Update of Environmental Imapct Assessment Reports. Eaux de Beyrouth et du Mont Liban.

Pöyry. 2018. Environmental and social screening. Shnogh Hydroelectric Project—Feasibility Study.

Reichenbach-Klinke, H.H. 1980. *Krankheiten und Schädigungen der Fische*, 472 pp. Stuttgart: Gustav Fischer.

SKM, Pöyry and SMEC. 2010. Determination of residual flow for the Frieda River hydro dam. Frieda River Feasibility Study. Report prepared for xstrata copper

WWF. 2003. Managing rivers wisely. Case Study: Kafue Flats http://d2ouvy59p0dg6k.cloudfront.net/downloads/managingriversintroeng.pdf.

Zhong, Y., and G. Power. 1996. Some environmental impacts of hydroelectric projects on fish in Canada. *Impact Assessment* 14 (3): 285–308.

Zwahlen, R. 2003. Identification, assessment and mitigation of environmental impacts of dam projects. In *Modern trends in applied aquatic ecology*, ed. R.S. Ambasht, 281–370. New York: Kluwer Academic Publishers.

Zwahlen, R., L.W. Kahongo, and V. de Genot de Nieukerken. 2019. Kafue Gorge Lower HPP—A low-impact addition to an existing HPP cascade, Zambia. Paper presented at the Africa 2019 hydropower conference, Windhoek, Namibia, April 2019.

Chapter 10
Biological Environment

No one definition [of the term 'species'] has as yet satisfied all naturalists; yet every naturalist knows vaguely what he means when he speaks of a species.
Charles Darwin, 1859 (Chapter II)

This subject includes all topics related to living organisms (with the exception of the human population) and their habitats. For the practical purpose of an ESIA, under this heading one has to deal with the species of plants and animals living in the affected area, and with their habitats. It is important to mention here already that no such investigation will ever be able to cover all the species; this point is discussed in Sect. 10.2.

A large number of species have become very rare or even extinct through the progressive shrinking of specific habitat types as e.g. primary forests or wetlands, sometimes in combination with other influences like hunting. Species that are rare by nature, as the large predators, especially attractive ones for one reason or another, as furbearers or "trophy animals", or species that depend entirely on a specific, threatened habitat type as amphibians (wetlands) are especially vulnerable.

Hydropower projects can affect biodiversity mainly in the following ways:

- direct destruction of habitats, mainly by submerging large areas through formation of a reservoir;
- indirect disturbance through improved access to hitherto inaccessible areas or through the resettlement of the people living in the reservoir area to formerly natural areas;
- disruption of a river by formation of a lake (reservoir), by impeding migrations of aquatic species through construction of a dam and by changing the downstream flow pattern in this river.

The first two points affect mainly terrestrial vegetation and fauna, while the third point refers to impacts on hydrobiology (fish and other aquatic life). In addition to these points, the construction phase can have a major impact on specific species, e.g. by increasing pressure on certain plants or animals, mainly by collecting or (usually

R. Zwahlen, *Assessing the Environmental Impacts of Hydropower Projects*,
Environmental Earth Sciences, https://doi.org/10.1007/978-3-030-91185-0_10

illegal) hunting carried out by the work force or by people moving into the area as so-called camp followers.

10.1 Terrestrial Vegetation

10.1.1 Baseline

In the case of plants, two aspects have to be considered, namely:

- Flora, i.e. the individual species of plants living in the area under study. For impact assessments, emphasis lies on the most important or frequent plants (which constitute a characteristic feature of the area, e.g. trees characterising a certain type of forest), on plants which are of special importance due to them being used by the local population (as food, construction material, fuel, medicine and/or for economic purposes), and on rare, endangered and/or endemic plants, which are important from a biodiversity conservation point of view.
- Vegetation: these are the plant associations to be found in an area, which determine to a large extent the visual aspect of this area (forests, grasslands, wetlands, steppes etc.) and which also constitute a main component of the habitat of animals. Today, in most places (natural) vegetation overlaps with and is influenced to a varying degree by all forms of human land use (agriculture, forestry, pasture, etc.), and in many places natural vegetation has largely or completely been replaced by human activities. Important and ongoing such changes mainly consist in replacing forests by agriculture, plantation or pastures. Such a transformation also has consequences for the fauna, which will change accordingly.

The description of vegetation and flora of the area to be affected by the project will focus on vegetation on the land used for construction and to be submerged at reservoir impoundment, on any other land that might be occupied by the project, on wetlands in the downstream area that might be affected by changes in river discharge pattern by the project, and on vegetation in the surroundings of the project, i.e. mainly of the reservoir, which might play a role for reservoir management and which might come under pressure due to the project, and certainly also due to new access roads; for this latter issue see Chap. 15. It is important that all types of potentially affected areas and their vegetation is adequately covered by the baseline investigations, whereas the detail of investigation needs to be commensurate to the expected or potential impact.

10.1.2 Impacts

Impact on vegetation can be roughly divided into two categories, direct and indirect. The most important cases of direct impact are the following ones:

Fig. 10.1 Transmission line corridors in different vegetation types. Left: dense, closed canopy forest: potentially considerable impact, ROW can be an insurmountable obstacle for canopy- or undergrowth-dwelling animals (Uganda). Middle: Miombo woodland, with naturally patchy, open-canopy vegetation; minor impact (Zambia). Right: land used for agriculture; very little if any impact (Switzerland)

- Destruction of the vegetation due to permanent occupation of land by project structures. These are mainly the reservoir (which, depending on the project, can occupy very large areas, going into the hundreds of square kilometres), dam and powerhouse with appurtenant structures like switchyard, housing and infrastructure for the operating staff, etc., and any permanent structures that may be required; in this respect, access roads can play a major role.
- Permanent change in the vegetation, although it might not be completely removed by structures occupying the surface. Important examples here are changes in vegetation in the transmission line corridor or right-of-way (TL ROW), which might be limited or almost nil in case such a line leads through areas with steppe or savanna vegetation or used as agriculture or pasture land, but which might be very considerable if forests are crossed and the ROW has to be kept permanently clear of tree growth (Fig. 10.1).
- Destruction of vegetation due to temporary occupation of land; this is the case mainly for land used for construction purposes and includes land used for infrastructure like workers' camps, storage and workshop areas, borrow areas, spoil disposal sites, quarries etc. Depending on the type of use, such areas of land can be changed fundamentally, e.g. in the case of quarries. Still, once the work is finalised and the areas are no longer being used, some vegetation can—and eventually will—re-develop.

Indirect impact on vegetation can also be very important. Here, the main categories are:

- Change in aquatic or riparian vegetation along the river in the downstream area due to change in river discharge. Several such hydrological changes have to be mentioned in this respect here, the most important being:

  - Reduction in flow in the stretch of river between dam and tailrace outlet; here, usually only the amount of water defined as residual flow will be left in the river bed, but during spilling, and especially during flood evacuation, this amount can increase enormously.
  - Rapid and short-term fluctuations in discharge due to peaking operation of the power plant; this will constantly flush out the affected stretch of river and in this way permanently change its quality as a habitat, not only for the vegetation.
  - Change in seasonal distribution of river discharge. This is very often a main purpose of a reservoir, shifting water from the high flow to the low flow season. Such a change has a decisive influence on floodplains downstream of the power plant and will induce profound changes of the floodplain vegetation in the mean and long term. Since these changes are slow in taking place, they often go almost unnoticed.

- Potential threat to vegetation due to the creation of access to formerly isolated and more or less inaccessible areas. Here, the access roads play a very important role (see Chap. 15). However, the creation of a reservoir, which is often much easier to navigate with all kinds of boats than the original river was, can also open access to areas surrounding the reservoir, from which pressure on vegetation can result.
- Finally, one more potential and sometimes very important type of indirect project effects on vegetation has to be mentioned: resettlement. If a large population with a livelihood based on subsistence agriculture and use of NTFPs (non-timber forest products) has to be relocated, new land will have to be found where this population can get an adequate compensation for the losses.

10.1.3 Mitigation

Mitigation has to respond to the specific impacts identified and is very project-and site-specific. A few general observations can be made.

10.1.3.1 Mitigation of Permanent Losses

One thing has to be stated very clearly: compensation for vegetation lost due to impounding of large reservoirs is not possible. Take as one example the case of Lom Pangar in Cameroon (see Sect. 24.1): a reservoir impounding 540 km^2, roughly half of it primary forest. There is simply no possibility for a “real” compensation for an impact of this type. The question in such a case, therefore, usually is (or should be) not as much about compensation but rather about the importance of the impact. If the affected—or, in such a case, destroyed—vegetation and the habitat it creates is of a type that is still widespread and abundant, the impact would usually be considered as acceptable. If, however, one or several vegetation types or habitats would be affected, which are considered as special for any reason—rare, limited in distribution, home to specific rare, protected or endemic species of plants and animals—this could then mean a no-go for the project, a reason for abandoning it and looking for an alternative.

Still, a number of mitigation measures are often proposed, and some of them may actually work under specific conditions. This sentence is formulated in this way on purpose: one can very often see proposed mitigation measures in ESIAs, of which it could be clear right from the start that they will not be implementable and that they will not serve the purpose, and which are usually forgotten as soon as the project gets its environmental licence (or however this document is called in the given setup). I'm definitely not in favour of proposing such measures, an ESIA should clearly describe the situation as it is, and this includes the feasibility of mitigation measures.

One such measure is the idea to go and recover specimens of rare and endangered plants in the area to be submerged, for planting them in a different place outside of the affected area. This is not to say that this never works, however, it is difficult to find an example where it actually did. While it might be possible to cultivate and reproduce such plants under artificial conditions—nurseries, greenhouses, botanical gardens etc.—which in specific cases may be a good thing to do, just replanting them outside of the affected area will often not lead to a success. It has to be kept in mind that rare species usually are rare for a reason, and this reason is very often that they require very specific conditions to be able to grow and propagate. These conditions can be highly different depending on the requirements of the species in question and may include things like soil conditions, availability of water, presence of specific pollinators, competition with other species etc. If these conditions cannot be met, there is little chance of success of such a relocation, and where the conditions are met, the species will very often already be present naturally. This just means that measures of this type have to be evaluated very carefully, and implemented just as carefully if considered as suitable. Only doing it for the sake of having a measure to show as a demonstration of goodwill is not useful, and is definitely not a sustainable solution.

If forest is affected, reforestation may be an option, and in some countries reforestation is a legal requirement. It can be a valuable compensation measure especially in places where forests have come under pressure, and where suitable land for reforestation is available. Here as well, for a reforestation to be successful it has to be done in an appropriate way, it is certainly more than just "planting trees". However, in most places there is ample experience with reforestation, and there are institutions and specialists able to do it.

One measure that I consider as a valuable compensation, especially if habitats of high biodiversity value are affected, is to provide an offset by means of providing effective protection, and possibly rehabilitation and upgrading, to similar habitats, or habitats of a similar value, in another place. This is now being done increasingly in the form of a specific biodiversity offset (see Sect. 6.3). Here again, the measure is only as good as its implementation. Just two occasions where this did not work: one was in Laos, where the ESIA for a dam project proposed to keep a parallel river, tributary to the same main river, in its natural state for providing protection for riparian habitats and short-range fish migrations which would be affected by the project. The ESIA was accepted, and seemingly so was the proposed offset measure.

Shortly afterwards a new dam project was being studied on exactly this river. The other example was in Malaysia, in a dam project in an area largely covered by natural primary forest. Logging concessions had recently been awarded for about half of the immediate catchment area of the future reservoir. In discussions with different authorities, the importance of keeping parts of these forest intact was pointed out, partly for biodiversity conservation, but also as an offset for the forest that would be lost due to the reservoir, and for preventing or at least reducing erosion in the surroundings of the reservoir, thus reducing sedimentation. As far as I recall, in all these discussions there never was any argument against this measure. Still, later on it turned out that while these discussions were ongoing, logging concessions had been awarded for the remaining forests in these areas. The problem here is that sometimes different authorities are dealing more or less in parallel with certain issues, and without much coordination among each other, and that as a rule the "stronger" authority wins; this in most cases is not the Ministry of Environmental Protection.

10.1.3.2 Mitigation of Indirect Impacts

An impact of potentially major importance can be caused by smoothing the seasonal differences of river discharge, by increasing discharge in the low-flow season and reducing it in the high-flow season. If important floodplains in the downstream area are affected by such a shift, this will mean that the difference between high and low water levels in these floodplains will be reduced. As mentioned, in the medium and long term this will change the vegetation in the floodplain, as plants which do not tolerate flooding will move into the areas which will no longer be influenced by seasonal floods. This invasion will invariably replace species which tolerate such seasonal flooding, or which even depend on them (see Sect. 18.2). Here, one potential mitigation measure consist in creating artificial flood peaks by releasing a sufficient amount of water from the dam, at appropriate times of the year, in order to mimic a natural regime. Such a measure, which can be rather costly in terms of lost energy production, will have to be planned carefully, on a case to case basis, in order to maximise the effect. It could be a matter of debate whether this issue should be treated under the heading of "impacts on terrestrial vegetation", since in this case wetland, i.e. semi-aquatic vegetation would be affected. As in other similar situations, however, it is more important that the subject is adequately covered at all.

One indirect effect, which can be very decisive and long lasting consists in creating access to a hitherto inaccessible area by the access roads required for a project, even if the impact of the road as such, at least in terms of direct land occupation, may be rather small. Mitigation of this type of impacts can be very difficult, since it would require constant very tight access control. Issues related to access roads are dealt with in more detail in Chap. 15.

10.1.3.3 Rehabilitation of Temporarily Used Areas

General Requirements

Surfaces of considerable extent are normally used temporarily for the construction of a large dam. This is required, e.g., for the construction site as such, which includes workers' camps, office areas, storage areas, workshops, crushing plant, concrete batching plant, temporary site roads, spoil disposal areas, borrow areas, quarries, etc. In addition of the removal of vegetation, most of these areas are physically changed to some extent (e.g. terrain levelling for camps, offices, workshops and storage areas), some in a quite considerable way, either by adding large quantities of material (spoil disposal areas) or by removing material (borrow areas, quarries). Once their use is no longer required, they have to be rehabilitated. This needs to be done at the end of the construction phase, but preparation for this has to start in the ESIA stage. It is important that this rehabilitation is planned carefully and in time; although the required measures might be implemented only several years after the start of the work, preparation for this has to commence much earlier. Some of the relevant points to be defined in the planning phase are the following:

- Planning of the sites: construction site layout has to be planned already with the aim of minimising the impact, i.e. of limiting temporarily used areas to the minimum required. This is of importance especially if the project is located in a sensitive area, e.g. within or close to a protected area.
- A potentially good way of minimising the impact is to place as much of the temporarily used areas within the future reservoir. Since this land will be submerged and permanently covered with water, rehabilitation other than removal of any substances that may pose a threat to water quality will not be required. One rehabilitation issue is often posed by quarries, and this is not a problem if the quarry is then impounded; obviously, this can be done only if sites offering suitable conditions, and especially suitable rock material, are available in such a place.
- Responsibility: it must be well defined who will be responsible for the rehabilitation, and what (in great lines) will be expected. The contractor must know the requirements on beforehand, and must include them in his planning, and mainly also in the budget. Responsibilities must also be defined e.g. in cases where parts of temporary structures (e.g. workshops, camps, etc.) are handed over from the civil contractor to the electro-mechanical contractor (or to any other entity for future use).
- Timing: as a general rule, each temporarily used site should be rehabilitated as soon as it is no longer in use. This means, that some of these areas can be rehabilitated long before the end of the works. However, much will remain to be done after the main construction work has ended. The planning must be made in a way as to ensure that the contractor can only move out (and remove heavy equipment) once the work has been done and was received by the project owner. Once the

contractor has moved out and machines are no longer available on site, it is very difficult to re-mobilise again.

- Definition of expected end results: at the moment of preparation of the ESIA or the ESMP it is not yet possible to define in detail and for every site to be used what should be done with it in the end. Still, it is important that the aims are defined in the great lines. One sentence, which can be found quite often in documents like the TOR or an ESIA, namely that the sites "have to be restored to their original state" is in general useless: in the case of, e.g., occupying areas originally covered with primary forest of any type, such areas can be rehabilitated by reforestation, and in the best case be prepared in a way that natural succession can take over rapidly, but it is impossible to revert the area to its original state; this process can take several hundred years, which may be considerably longer than the life span of the project itself. And there is no way that land which was used as spoil disposal areas or quarries can be reverted to their original state. The aims have to be realistic.
- Carry out the work correctly right from the start: two conditions that should be clearly formulated and implemented, and which in my experience are rarely respected, are the requirements to (i) primarily remove topsoil from all affected areas within the construction site (including the permanently used ones!) and store it separately and in a way that in the end it can be used to cover temporarily used areas, and (ii) to carefully place spoil material (in layers that are compacted, leaving berms every 10 m or so, and avoiding too long and too steep slopes which are exposed to erosion; see Fig. 10.3 for incorrectly put spoil material).
- In the end, there should be a plan which clearly identifies each site, giving its exact location, the type of temporary use, and the type of rehabilitation measures taken. This is especially important in view of a potential different use of such areas at a later point in time. It should not happen, e.g., that if 20 years after the end of the construction a new building is going to be erected, in this process one has to find out that the place chosen for this purpose had been used as a domestic waste landfill area during the construction period.

These steps are all required, but it always has to be kept in mind that this will work only if in addition to good planning there is also a permanent follow-up and control of the implementation. One case in point: if a contractor starts preparing the construction site and is eager to do this quickly, in order to be able to start with the real work, and if at that stage there is no proper supervision, the outcome will often be that topsoil has been removed and was then buried under large amounts of other spoil material, lost forever.

Rehabilitation Under Various Site Conditions

To put it simply: rehabilitation first of all mainly aims at integrating temporarily used sites, to the extent possible, into the landscape, and to protect the soils against erosion. The best protection against erosion is a good vegetation cover which stabilises the soil, reduces or prevents superficial runoff, and shields the soil from direct rainfall. In most cases, a natural vegetation type is usually preferable; however, the question of the definition of "natural" remains. In this case, there can be two such definitions: one is "what has been growing there before", and the other is "what will grow under the present conditions". In Sect. 3.5, the difference in approach between "the engineer" and "the ecologist" was mentioned, and this may be a good example for illustrating this difference. One approach, in the case of an area which originally was covered with forest, is to aim at re-creating just such a forest. In rehabilitation plans prepared for specific sites, one can often see lists of trees to be planted, and just as often one may wonder how and at what costs seedlings of such trees might have to be purchased, and whether they would grow under the conditions of the areas to be rehabilitated; and quite often one might come to the conclusion that many of them would not grow, since these were specific tree species of the dense, mature forest; this may happen especially if the list of trees to be planted was just basically copied from the list of trees which grew there under natural conditions. This would correspond to a "recreate the original situation" approach. However, conditions presented on such areas to be rehabilitated are very different, and not many plant species are able to grow under such conditions. Species that can cope with this are often called "pioneer species", since they are able to grow under conditions where they are fully exposed to sunlight, and may suffer from temporarily dry conditions (Fig. 10.2). Such species can also be planted, but many of them, and among them the ones best adapted to such conditions, will grow spontaneously and will not need any special care. Leaving this type of vegetation to grow, or encouraging it to grow by improving the situation as much as possible, would mean to choose the approach of "what can grow under present conditions". One barrier to this approach is often the view that these plants

Fig. 10.2 A typical pioneer species, the umbrella tree *Musanga cecropioides*. The umbrella tree (*Musanga cecropioides*), a fast growing tree, easily recognisable by its large, palmate leaves, gests established very quickly on newly opened areas in forests, e.g. along roads. It is widespread in moist tropical Africa (Savill and Fox 1967; here a group in Cameroon)

are considered as weeds by horticulturists, and measures are often taken to eliminate them. However, in many cases they will do better than planted trees. One anecdote to illustrate this; in a project in Italy (tourism, nothing to do with dams), efforts were made to develop the surroundings of the newly built houses. For this purpose, a large number of already quite big trees were planted with a lot of effort, and in order to prevent the sometimes rather strong wind in this area from uprooting the newly planted trees, they were given support by attaching them to a solid pole. Many of these young trees died, but on the other hand about half of the poles provided as support, mainly strong branches from poplar trees, started to grow: development of vegetation able to cope with the prevailing conditions! Over time (given the time), such pioneer species will then gradually be replaced by species growing in and dominating dense forest. That an approach for fighting weeds in forest regeneration plots can be counterproductive has also been shown in a test in Australia: on half of the area under observation efforts were made to eliminate or at least reduce the abundance of an invasive grass species; this did not only come at a considerable cost, in addition to that 6.5% of the tree saplings died due to the herbicides used to fight the grass (Radford-Smith et al. 2021), whence the very appropriate title of this paper: better left alone!

In a different context, the approach of protecting and letting grow spontaneously developing vegetation is the method called "Farmer Managed Natural Regeneration" developed by the Australian agronomist Tony Rinaudo (https://de.wikipedia.org/wiki/Farmer_Managed_Natural_Regeneration), which is based simply on protecting coppicing from old tree trunks instead of constantly removing them; this led to enormous re-greening of areas threatened by desertification in the Sahel (TA 2010-0-06:14) and proved to be much less costly and more effective than tree planting. On a smaller scale, I have seen two such "experiments": a safari park close to Dakar, Senegal, and a graveyard close to Asmara, Eritrea. The only thing that had originally been done in both cases was to fence off the areas for keeping out livestock (mainly goats) and people collecting firewood. The result in both cases was a much more abundant vegetation than in all the surrounding, otherwise totally identical areas.

One thing has to be pointed out clearly: regrowth of vegetation, and with it the approach to choose and the level of effort required for site rehabilitation, depends very much on the local climate. While in an area with abundant rainfall the spontaneous growth will generally be fast and almost unstoppable, it can be very difficult under more arid conditions. This is illustrated in Figs. 10.3 and 10.4.

Figure 10.3 shows a spoil disposal site in the central part of Cameroon; in this case, the spoil had not been put in place according to best practice; it had basically just been dumped on the slope, without any compacting, with a too steep slope exposed to erosion, without any berms, and without being covered with topsoil. In spite of that, vegetation developed very quickly, providing good erosion protection. On the other hand, the case of Ilisu HPP in South-East Anatolia, Turkey, illustrates much more difficult conditions: due to the lack of rain, vegetation development is very slow, and plantations need to be watered for a considerable time until they become firmly established (Fig. 10.4). Leaving a site under such conditions just to itself,

Fig. 10.3 Site rehabilitation under favourable conditions. Spoil disposal site, Lom Pangar HPP, Cameroon: situation in July 2014 (left) and in November 2015 (right): spontaneous development of vegetation

Fig. 10.4 Site rehabilitation, Ilisu HPP, South-East Anatolia, Turkey

for vegetation to develop spontaneously, might lead to bare soils being exposed to erosion for quite a long time.

Update Plans When Required

As mentioned above, rehabilitation plans have to be made in time, in order to have them when they are needed. However, and this is not only true for rehabilitation plans, they need to be checked and, if required, adapted to the prevailing situation. The example of the quarry for the Lom Pangar HPP in Cameroon (Fig. 10.5) illustrates this requirement.

Fig. 10.5 Quarry of the Lom Pangar HPP, Cameroon, at the end of the construction period

The rock face created by exploiting the rock, visible in the background and on the right hand side of the picture, is 10 m high or more. When the rehabilitation plan was prepared, it had been decided to fence the quarry off after construction, with the aim of preventing accidents with people or animals falling into the pit. When exploitation ended, the situation was re-checked. The following main points were observed: the quarry was on a newly created peninsula, surrounded on three sides by water of the reservoir; the access road was partly drowned, no direct access possible; the nearest village was over 20 km away from the site; although the site was located partly inside or very close to a protected area, there were only few animals there, no livestock among them; and finally, during the construction a very dense strip of vegetation had developed along the upper border of the quarry and the edge of the forest surrounding it, which made access to the cliff almost impossible. The conclusion was that for all these reasons the risk of accidents was very small. In addition, to build the fence one would have to remove the protecting vegetation, and it seemed incongruous to create access with the only purpose of preventing access. Therefore, the idea of fencing the quarry was abandoned. One alternative idea had been to replace this dense strip of vegetation with thorn bushes. However, there were none growing in this area, and the local vegetation would very soon take over again even if it had been replaced by thorn bushes; the latter, not adapted to the local situation (they would have had to be brought from the savannas in the Northern part of the country) would not be able to withstand. So here again: better left alone!

10.2 Terrestrial Fauna

10.2.1 Baseline

The statements made in the section on vegetation and its importance for impact assessment in dam projects are largely valid for the terrestrial fauna as well, since animals depend on their habitats, and vegetation is an important part of it, if not even the most important part.

In the framework of an ESIA, it is never possible to identify all the animal groups living in the study area (see Sect. 10.2). The efforts have to concentrate on groups of animals which can be identified more or less readily and whose habitat requirements are known well enough, and which can therefore serve as indicator organisms for the state of their habitats. As for plant species, the emphasis lies on either economically important species (food or other uses, pests) and on rare, endangered and/or endemic species. The most important groups in this sense are the terrestrial vertebrates (mammals, birds, reptiles and amphibians). For the entire aquatic fauna, fish serve as indicator organisms. Depending on the site and on the situation on site, representatives of other groups—either systematic groups as e.g. butterflies or ecotypes like benthic organisms—can be of relevance.

One approach that is often used is the "flagship" or "umbrella species" concept. This a an approach also used in conservation, and it is based on the idea that if a certain species, usually a rare and endangered one that needs protection, gets adequate conditions for maintaining its local population, then all the other species, which may be less in the focus of conservationists or the public, will also benefit (Barlow et al. 2020). This concept can also be used in ESIAs for HPPs, and then hopefully be applied in the case of implementation of this project, and it is an procedure which will—and does—often produce good results. In the case of Lom Pangar HPP in Cameroon, e.g., the focus of habitat and biodiversity protection was on the gorilla, the lowland species *Gorilla gorilla*, since an important population of this primate had been identified as living in the area nearby the dam site (ISL 2005); during implementation of the project, an important program was carried out for providing mitigation for the impacts on this and some other species, which aimed mainly at improving habitat conditions and protection. In the case of Ayago-Nile HPP, located in the largest national park of Uganda, one focus was on the elephant (*Loxodonta africana*) population, which had come under pressure due to the lack of protection, which in turn was a consequence of the complete breakdown of tourism, and with it funding for conservation, during the reign of Idi Amin (EWI 1981). In conservation, the results are not always clear and not necessarily positive. For the case of habitat management of the greater sage grouse (*Centrocercus urophasianus*), Pillod et al. (2020) note that a considerable number of reptile species living in the same habitat benefited directly; this was also the case for a small bird restricted to sage habitat, the Brewer's sparrow (*Spizella breweri*); for this latter species, however, it had been demonstrated that the management efforts for the umbrella species did not cover all the requirements of this other species, and that additional specific measures have to be taken for its conservation. On the other hand, it has been shown that protection measures for the great panda (*Ailuropoda melanoleuca*), one of the most prominent flagship and umbrella species, did not offer any benefit to the other large carnivores (leopard, *Panthera pardus*; snow leopard, *P. uncia*; wolf, *Canis lupus*, and dhole, *Cuon alpinus*) whose distribution range overlaps with the one of the panda; the populations of all these latter species had declined over their entire range, showing that they require different conservation measures (Li et al. 2020).

It has to be stated clearly that an ESIA for a specific project in one specific site cannot possibly go into the details of habitat requirements of a number of locally present species. For this reason, and even if the umbrella species concept does not work in every case, it is still a good approach for habitat evaluation and impact assessment for HPPs that affect specific natural habitats. If it is possible to ensure that a locally important, endangered and/or specialised "flagship" species will survive, then the measures taken will certainly also be beneficial to a large number of other species living in the same habitat. Going into more ecological details will in any case have to be left to science.

10.2.1.1 Mammals

Mammals are one of the different groups of terrestrial animals that are usually covered in an ESIA, in any case if habitats are affected which are natural or show relatively little influence by human activities. Mammals, in terms of objectives of an investigation for an ESIA, have a number of advantages: there is a limited number of species, their ecology, in most cases, is rather well known (although this is not necessarily the case for all groups), there is a good stock of literature available, and they are often rather easy to identify, either directly or by signs of their presence. While these characteristics are true for many, and at least for the larger, species of animals, it gets immediately more difficult if one wants or has to include smaller and more elusive species, like e.g. insectivores (shrews), rodents or bats. One difficulty, in addition to their small size, is the fact that most of them are nocturnal. But even large mammals may be secretive, and this is especially true for many rare and endangered species, and even more so if they are hunted and therefore actively avoid to be seen by humans.

The following are the techniques usually applied for identifying mammals living in an area which will be affected by an HPP:

- Literature search: this would be in most cased the first step. At first it is important to get an idea of what animals will have to be expected to live in a certain area, and there a field guide, or even more so a brochure or poster in the case of a project located in a protected area, may provide valuable information. On this basis, it is then possible to look for more specific information, like e.g. on local populations of specific—and specifically relevant—species.
- Field work: this can be very important and useful, depending on the situation. In the case of the Ayago-Nile HPP, for instance, located within the large Murchison Falls National Park, Uganda, moving around the project site gave a very good overview of the species which were not only present in the park, but were actually using the project site and its surroundings. However, this was an especially favourable condition: a savanna environment, with numerous species of medium sized and large mammals, most of which were not afraid of people (or of cars). It is much more difficult e.g. in dense forest, where even large animals can hide very well, and in all situations where the main focus e.g. is on rare and/or endangered animals. Already the fact that they are rare means that they may be seen only occasionally, just by luck, they are usually efficient in hiding themselves, especially in dense forest, and some of them, as e.g. the larger carnivores, use very large territories and are very mobile. In such cases, fieldwork with the aim of gathering observation records of such animals would be much too time consuming and too unreliable. In some specific cases, it can be possible to use signs of the presence of certain species, which are easier to find as the animals themselves; for large apes, gorillas

and chimpanzees, it is often easier to find the nests where they spend the nights (every night in a different place). For small mammals, the only way finding and identifying them is usually to set up traps.

- Interviews with local people: this can be a very relevant source of information. Especially hunters (or poachers, for that matter) usually have a very detailed knowledge of the local fauna. Still, such information has to be checked carefully. One potential problem, in any case where less prominent and less well-known species are concerned, is simple identification of species. Vernacular names can lead to confusion, and especially smaller species are very often not distinguished from each other. But also "educated" people can provide obviously wrong information: in Sarawak, somebody once told me that he had been attacked by a black mamba, one person insisted that there were zebras in Mongolia, and a local zoologist in Cameroon provided a list of animals living in the river in the area of a future reservoir, which was located upstream of a series of rather impressive waterfalls, which included the African manatee (*Trichechus senegalensis*). An additional issue can be that poachers, aware of the fact that their activity is illegal, may be very reluctant to provide information on species they hunt.

There can be basically two main objectives a section on mammals in an ESIA tries to cover. For projects located in an area which is neither protected, considered as intact (critical) natural habitat, nor the known habitat of any relevant (i.e. rare, endangered, protected, endemic, …) species, the aim of the baseline investigation is usually just to provide an overview of the species living or supposed to be living there, focusing on species and their habitats which in any way might be affected by the project. If however the project area is actually the home of such a "relevant" species, then the focus is different: collection of data (applying all methods listed above) will concentrate on this or these species, and more detailed information may be required about population size, daily or seasonal movements, actual use of the habitat, etc. This is then more the "umbrella species approach" shortly mentioned above, whereby it is assumed that if such species can survive in a given place, then this will apply to the entire set of other species living in the same habitat as well.

One important point for these baseline investigation is certainly always to have a good expert on the team who is familiar with the local situation. Such a person will already know a lot about the local fauna (in terms of what species to expect in a certain habitat), and will also be more efficient in finding signs of their presence.

Much of what is said here about mammals also applies to other groups of terrestrial fauna, the most important of which are shortly mentioned hereunder.

10.2.1.2 Birds

Birds are a group of animals which is covered in most if not all green-field projects. They have a number of advantages:

- Most are diurnal and can be more or less readily observed.

- Next to visual identification, many species can be heard at least during some times of the year, mainly during the reproductive period, which allows identification even in dense vegetation.
- Ecology, like habitat requirements, of most bird species is well known.
- There are many indicator species, whose presence or absence provides relevant information about the status of a specific area.
- There is ample literature on birds (very good field guides for most areas as well as more specific scientific literature focusing on single species or important habitats).
- There are local bird specialists almost everywhere who can provide important information.
- Status (in terms of IUCN Red List categories) of most birds is known, making many of the species suitable as umbrella species.

One specific issue that often needs to be considered, and this especially in the case of birds, is the potential impact of a project on migrating species. This is insofar of importance that a specific site—as is the case for many wetlands—can be of high importance for the survival of animal populations and species even if they do not live there permanently, but use this area only at well-defined times of the year, e.g. as resting and feeding places during migrations, as overwintering areas or as breeding habitats. The most important of such wetland areas are included in the List of Wetlands of International Importance, or the Ramsar convention (https://www.ramsar.org).

10.2.1.3 Reptiles and Amphibians

Reptiles and amphibians are often included in ESIAs as well, especially in cases where areas of known high diversity of these two groups, or habitats of specific species, are affected. This is, among other reasons, justified by the high proportion of rare and endangered species within these two groups. They are often mentioned together under the term of "herpetofauna", which is not really a term of biological classification; as a matter of fact, from an evolutionary point of view birds and reptiles are closer to each other than reptiles and amphibians. The approach is probably more based on the fact that reptiles and amphibians are often kept as pets in terrariums; illegal trade of such animals, and to a large extent of rare and legally protected species for the pet market is one of the major threats many of these species are facing; in this context see also the story of the spray toad (Sect. 18.3). Amphibians may be of special importance in the context of a hydropower project due to the fact that they inhabit moist areas and, at least temporarily, mainly for reproduction, water bodies of different types, and that therefore they are very likely to be affected by changes in the aquatic habitat.

10.2.1.4 Other Groups of Terrestrial Animals

The terrestrial vertebrates discussed shortly above are usually included in ESIAs, and they are seen as indicator or umbrella species assuming that if and where these species—and especially the rarer and therefore presumably more sensitive ones—find suitable conditions, all other species inhabiting the same habitat will as well. This may not necessarily be true in any case, but as mentioned as well it is not possible to cover all species in such a study. Under specific conditions, however, it can be indicated to put the focus on some other groups. In the EIA for Curciusa (EWI 1987), e.g., a reservoir that was planned in an alpine environment in Switzerland, butterflies were included, for the simple reason that under such circumstances it is possible that a few species live in very specific and spatially restricted habitats, which can be dictated for instance by the presence of the fodder plant of the caterpillar, and sometimes in small isolated populations; the investigation confirmed that this was not the case there. In another, recent ESIA for an HPP in Africa butterflies were included as well; the study identified 98 species in the project area, without however giving a reason why this group had been studied, and without coming to any conclusions related to this work, leaving you to wonder why butterflies and not, e.g. dragonflies, who as a group were much more likely to have their habitat negatively affected by the project. Again, within the framework of an ESIA including a specific group of animals makes sense only in places where the taxonomy and the ecology, including spatial distribution, of potentially affected species is well known; see to this the example of mosses (Sect. 3.12). In the case of the latter project, it gave the impression that a butterfly specialist just happened to be available, but not one for dragonflies.

10.2.2 Impacts

Impacts of an HPP on animals—and on any other group of living organisms—can be of different nature, the most important ones being:

- Loss of habitat, mainly destruction of terrestrial habitats by submersion; in addition to the reservoir, the construction site for dam and powerhouse, including appurtenant structures, can be of relevance. The importance of such an impact will be assessed according to the types of habitat affected, their size and state, and of course depending on the species living there. In a specific case, the entire focus may be on one single species, like e.g. the newly discovered third orangutan species (*Pongo tapanuliensis*, see Sect. 15.3), which is already endangered by a hydropower project.
- Disturbance: this can be relevant mainly during the construction period, when traffic and general activities on the construction site, but mainly the presence of a large number of people, can present a serious source of disturbance for animals living in the surroundings of the site. This effect will obviously be much more marked if members of the work force engage in hunting.

- Indirect impacts: there are mainly three, namely, (i) impacts on downstream habitats, mainly floodplain or delta areas, that can affect biodiversity (see Sect. 18.2); (ii) choice of resettlement areas, which can be quite considerable depending on the size of the population to be relocated; and (iii) construction of access roads; this latter is a potential impact of special importance and will be discussed in more detail in Chap. 15.

10.2.3 Mitigation

As always, mitigation depends on the importance of the impact—or at least that is what it should. And as always we have to keep in mind that "importance" is not an absolute value which can be determined in a mathematically correct and unquestionable way, it depends on two very relevant other parameters, which are personal value scales of the responsible stakeholders and the weight which is given one argument as opposed to another one.

Here, one potential outcome can clearly be to avoid the impact by not building a dam. This, if the project otherwise looks promising, would certainly be done only if the impact would really be considered as very serious, and the loss of a species which would be directly attributable to such a project could be such a case; one species which actually was extinct in the wild, which was then bred in captivity and released back to its original habitat, supported by a massive effort for artificially recreating its habitat, is the so-called spray toad (see Sect. 18.3). One story which got rather famous was that of the snail darter (*Percina tanasi*), a small fish of the perch family, formerly endemic to the lower part of the Little Tennessee River which was discovered during the construction of the Tellico Dam on this river. Since completion of the project would have led to the extinction of this restricted area species, which would have been in violation of the newly enacted US Endangered Species Act (ESA), it was decided that the project was not allowed to go on; this decision led to comments of a type which can be heard frequently in discussions about the effects of the ESIA, reaching from "provides specific limits to environmental damage which publicly allowed to cause" to this being part of an inflexible law that "would cost jobs, stop public work projects and deter progress" (IUCN 1978). In the end, the Tennessee Valley Authority (TVA), the owner of this project, got permission to complete the project based on an explicit exemption of the snail darter from the ESA; the species apparently was transferred successfully to another river, where a population became permanently established (IUCN 1979; https://en.wikipedia.org/wiki/Snail_darter); it is classified as vulnerable by IUCN.

As mentioned, while it is certainly not the aim of the ESIA to "destroy" projects, the entire process must be able to identify situations where a project should not be implemented, even if such a decision should usually be taken before an entire ESIA was carried out (see Sect. 3.9). However, I do not know of any hydropower project which was actually abandoned on the reason that it might have wiped out a species. In both cases described above, the project was implemented.

Minimisation might occasionally be an option, if e.g. modification of a project would allow to spare a habitat of specific importance. Two such cases were Kukule Ganga and Rusumo Falls HPPs (Sect. 14.1.2), where the lowest dam height was chosen in order to minimise the area to be impounded; in these examples, this was decided with the aim of minimising resettlement, but of course it also reduced the impact on terrestrial habitats. Doing this obviously means to accept that electricity generation will not be what might have been possible with a higher dam.

One measure which is very relevant, especially in the case of reservoirs submerging tropical forests, is the pre-impoundment clearing of the reservoir area. The main idea behind this, and the main reason for doing it, is the removal of biomass to prevent this from being submerged, which can (and in the absence of such measures usually does) lead to massive GHG emissions (see Sect. 17.5). One additional reason can be to make use of the timber, which is not the part of biomass causing GHG emissions, but which is basically a valuable resource which can and should be used instead of just being lost. But if this clearing is done well, meaning starting at the lowest parts of the reservoir, gradually working uphill and not leaving behind any larger islands with denser vegetation, it can also serve the purpose of driving animals out of the area of the future reservoir before the water rises. In order to achieve this, it is important that not only the lower vegetation layers (the soft biomass presented by herbaceous and shrub vegetation) is being cleared, but also a considerable number of trees are felled and removed. Doing this in good time before impoundment will give a chance to animals, whose first reaction may be to seek refuge on high trees, to leave the area in search of better conditions. Such a clearing was unfortunately not done in the case of Lom Pangar in Cameroon, and shortly after impoundment a chimpanzee was found dead in a tree sticking out of the water of the reservoir; it had obviously tried to save itself from the raising waters by climbing this tree and then found itself trapped; not being able to swim and like all apes afraid of water, it apparently starved there. The alternative to pre-impoundment clearing could be to bring teams into the reservoir during impoundment, who catch animals which got trapped on islands or are found swimming in the water, and bring them to higher ground. However, this may be a spectacular activity, but it is of doubtful success, as e.g. the examples form Kariba (known as Operation Noah) and Tucurui have shown (WCD 2000: 75). At a much smaller scale, such a rescuing operation was undertaken during the filling phase of the Bujagali reservoir in Uganda, which I had thought to be not required: it is a very small reservoir, and filling it basically only consisted in raising the water level along approximately 10 km of the river Nile downstream of Owen Falls, Lake Victoria, by between 0 and maybe 20 m at the dam. A narrow stretch of river bank land was submerged, from where tree and bush vegetation had been cleared previously. A small number of mainly small animals were rescued, most of which would probably have been able to regain land on their own. I took the fact that even some Nile monitors (*Varanus niloticus*) were "rescued" as an indicator of how desperate the rescue teams were for finding anything worth saving; I do not know where these animals were released, but I'm sure that the first thing the monitors did was to find a water body where they could hide. Still, even if in that case the measure might not have been required, it is still preferable to do something of that kind rather not to do anything in a case where it would actually have been necessary. In this same project, it would also not have been necessary to remove the vegetation from the land to be submerged, since it was a small surface and water turnover in the reservoir (Bujagali is a ROR scheme) is so high that any formation of anaerobic water layers is not possible. Still, this action may actually have saved a few animals, and, in this case more importantly, the cut wood was piled up in suitable places and made available to the local population as fuel wood, a service that was well accepted.

There may be mainly two reasons why rescuing animals from a reservoir and releasing them above FSL might not have the expected success. On the one hand it is possible that the habitat conditions at higher elevations are not the same as the ones in the area to be submerged, making this habitat unsuitable for the rescued species. On the other hand, it could be that the land surrounding the reservoir was already occupied up to carrying capacity with animals of the same species, simply leaving no space for the newcomers (or only for those who might have been capable of driving out a resident from its territory). These two reasons can certainly also apply to the approach of driving animals out by removing the vegetation (where of course also the presence of people working there plays a role). Still, this approach probably is still better, since it leaves more time for the inhabitants of the reservoir area to move out and look for suitable alternatives, while rescuing animals when the waters rise leads to something more like a panic reaction, in animals as well as humans involved. In any case, as for other impacts, the action needs to be planned on beforehand, and the best approach for the given situation has to be identified and implemented. In situations where most or all species in the reservoir area and its surroundings are depleted to some degree, there is a high probability that many of the individuals that were driven out could find a suitable alternative.

Finally, whenever terrestrial biodiversity is an important issue, there is the possibility of compensating the impact. Basically, this means to identify another area, not impacted by the project and offering similar conditions, and to upgrade this area, mainly by granting it actual protection; legal protection is good, but without protective activities on the ground, mainly by access control and game wardens patrolling the area, usually not effective for protecting a habitat. Such mitigation measures are now being formalised and implemented, in the case of hydropower projects, by the concept of biodiversity offset (see Sect. 6.3).

10.3 Fish

> *Dams and reservoirs and their up- and downstream propagation of fragmentation and flow regulation are the leading contributors to the loss of river connectivity.*
>
> Deinet et al. (2020: 7)

The fish fauna of a river is in every case affected in a major way by a dam project. The two main impacts are the interruption of the river by a solid structure which is an obstacle—or actually a complete barrier—to active (migration) and passive (drift) movements of aquatic organisms. For this reason there is no doubt that the fish fauna has to be one major focal point of most ESIAs for HPPs, the possible exception to this rule only being the rare cases where a dam is going to be built on a stream without fish, which can basically only be the case in small high mountain streams where obstacles in the downstream area (waterfalls) prevent fish from colonising the area, and/or streams which are extremely seasonal, with intermittent flow of water.

10.3.1 Baseline

The baseline for dealing with the fish fauna mainly has to cover the following three points:

1. Identification of species present in the area to be affected, taking into account aspects of diversity (focus on rare, endangered or endemic species), ecological importance (predators, important parts of the food chain, species fulfilling specific ecological functions), and economic importance (which can be importance as food for the local human population, marketable goods, or of importance for leisure like sport fishing; economic importance obviously is an aspect to be considered in the socio-economic impact assessment, and will not be dealt with in any detail here).
2. Habitat conditions, considering that fish species are usually not distributed evenly in a river, but in many cases preferably inhabit zones with specific physical conditions.
3. Existing threats, if any, which are not related to the project; this can be other measures changing the river (like other dams, existing or as projects, river training measures, dikes separating the river from floodplains, etc.), problems with water quality, and, to an increasing extent, overfishing.

One first issue for establishing a good baseline, even today, is often the fact that there is little information on the fish fauna of a specific place, be this a section of a larger river or a tributary. Fish species in the rivers of Lao PDR are a good illustration of this issue. The investigations made for Nam Theun 2 had recorded, from only a short period of fieldwork, a total of 185 species, among which 25 species recorded for the first time in Laos and 11 new species not known from anywhere else. In earlier publications, a total number of 205 species of fish had been recorded from the territory of Lao PDR (SEATEC 1997). From this, it could have been concluded that the rivers investigated for this project (Nam Theun and Xe Bangfai) were highly diverse fish habitats with an extraordinarily high proportion of endemic species. But this is not necessarily so. It is absolutely possible that the new species found there were actually quite widely distributed in the Mekong basin, or that, at the very least, they occur in other tributaries of the Mekong as well. However, because the fish fauna of the water courses of Laos was not yet well investigated, there was no way of knowing it at that time. This also meant that it is almost impossible, with this

Table 10.1 Number of fish species recorded in the 4 Mekong mainstream dam areas

HPP	XHPP	XHPP	PBHPP	PLHPP	XHPP	SHPP	SHPP	LPHPP
Year	2010	2013a	2013b	2017	2018a	2018b	2018c	2019
Total No. of species recorded	54	119	43	27	57	41	30	80
New species in this monitoring	54	85	14	5	7	8	3	26

See Table A.1 in Annex 2 for details

knowledge, to establish a reliable concept for maintaining the fish diversity of the country. Again, an EIA for a specific project cannot produce this kind of information.

The situation has improved considerably over the last 20 years, one very important source of information being the fish base (http://www.fishbase.org/search.php). But even so, there remain a lot of uncertainties, given the size of the area, its diversity in physical conditions and its richness in fish species. This is illustrated here with the examples of the ESIAs for four dams on the Mekong mainstream in Laos, namely Xayaburi (XHPP) (in operation since 2019) Pak Beng (PBHPP; NCG 2015), Pak Lay (PLHPP; NCG 2018) and Luang Prabang (LPHPP; Pöyry 2020) (in various stages of project preparation); see Table 10.1 for details and the map in Fig. 24.4 showing the locations of these projects. Luang Prabang HPP might be the second dam to be built on the main Mekong in Laos; Don Sahong HPP, which is sometimes mentioned as a "mainstream Mekong dam", is a special case which cannot be compared with these four "true mainstream projects"; see Sect. 22.4).

These 6 sampling or reporting periods, one each for the four ESIA plus a construction period sampling (2013a) and a first fish pass monitoring (2018) for Xayaburi HPP, listed a total of 191 species, which makes it clear that none of them came close to the total number of species present. It should be considered, however, that the entire stretch under scrutiny here comprises of several hundred kilometres, and that not all the species listed occur necessarily over this entire area.

This was an example for the Mekong, which presently is probably the one of the major rivers of the world at the forefront of hydropower development (see e.g. Mather et al. 2009; Poulsen et al. 2002; Stone 2016; and Chap. 22). The situation might be somewhat similar for the Amazon basin, and is probably much more difficult in the case of the Congo, which might be the next major focal point for hydropower development once the political situation allows it.

One other issue is the fact that it is often not possible, in the short period available for the preparation of an ESIA, to really get a full picture of the local fish fauna (a fact which, of course, is just as applicable to other groups of organism like e.g. birds or mammals). The main reasons for this are that some species may be present only during a specific season, that some are, for whatever reason, more difficult to catch than others, and that rare species, by definition, are rare and therefore found less often. Actually, it is usually not terribly difficult to show that a certain species is present at a certain location, but it is as good as impossible to prove that it does

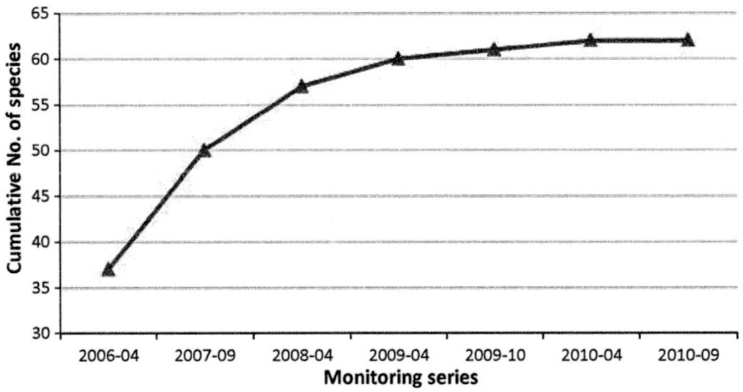

Fig. 10.6 Bujagali HPP: cumulative number of fish species captured. *Source* NaFIRRI 2010

not occur there. Usually, an ESIA report has one year for preparation (and there are many examples where time was much shorter!), and it should sample all relevant seasons or periods of the year. For the reasons mentioned above, however, this may not always be sufficient for reaching a complete picture of the situation. Figure 10.6 shows the results of fish monitoring for Bujagali HPP, Uganda, carried out during 5 years of the construction period. It shows that more than half of the species were recorded during the first monitoring event, but only 4 years later no new species was added to the list any longer, indicating that at that time probably all species present had been detected.

The stretch of the Mekong in Laos covered by the investigations for the 4 HPPs mentioned above, although it spreads over several hundred kilometres, can be considered as one river habitat. This is justified since the main aspect of the river does not really change over this stretch; the river flows in a rather narrow and well defined bed between often rather steep hills, without adjacent floodplains or meandering zones, and there is no natural obstacle like a waterfall which could be a hindrance to fish migrations. Therefore, we could apply this type of cumulative curve to the data shown in Table 10.1. The result is given in Fig. 10.7.

From this curve it could be concluded that the point where all species living in this habitat are on record has not yet been reached. Only extensive additional investigations could clarify this point. As was mentioned in other Sections, this is work that cannot be completed within the framework of an ESIA, the basis for a comparison would have to be provided by scientific work not related to a specific project.

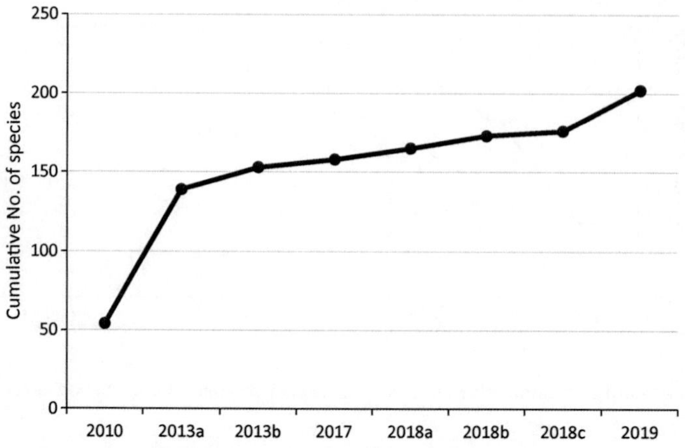

Fig. 10.7 Cumulative number of fish species in four Mekong HPP studies

10.3.2 Main Impacts

The main impacts of dam projects on fish are the following:

1. Disruption of migrations, which affects both active upstream migrations and downstream migrations, whereby the latter can be active or passive, i.e. in the form of drift.
2. Change of a part of the river from river (lotic) to lake-like (lentic) conditions. While some species can adapt easily to this type of habitat, and some might be able to rapidly colonise this new habitat, others cannot cope with this change in situation and will therefore diminish in numbers or will disappear altogether from this area.
3. Breaking up of populations of non-migrating fish species into a number of more or less isolated sub-populations.
4. Change in river flow pattern downstream of the dam, e.g. in cases where seasonally flooded areas serve as breeding grounds, and when this flooding no longer takes place due to the regulating effect of dams.
5. Indiscriminate introduction of exotic species (which is sometimes done as mitigation measure for a dam project, but very often independently of that).

Su et al. (2021) examined human impacts on fish biodiversity in river basins worldwide and noted an increasing threat mainly due to fragmentation of rivers (i.e. damming) and the introduction of exotic species. As for fish, habitat conditions change also for all other aquatic organisms. This is described in Chap. 16.

10.3.3 *Interruption of Rivers*

10.3.3.1 Impacts

One main effect of a dam and reservoir project on fish populations—and arguably the most serious impact of many dams on the biological environment—is the fact that the dam will constitute an obstacle to migration (Deinet et al. 2020). Many species of fish carry out migrations, and many of them depend on these migrations for reproduction. Concerning migrations, we have to distinguish between upstream and downstream migration. Upstream migrations are blocked completely by any type of dam, and even by rather small weirs, unless there is a fish ladder included in the structure. Downstream migration is not necessarily blocked completely, but will have to go either over or through the spillways (which is restricted to the period when there is actually spilling) or through the turbines, which presents a considerable risk of injury or death.

Migrating fish have to do both, although in different stages of their lives. The most prominent upstream (anadromous) migrations are certainly those of the salmons in the temperate and cold parts of the northern hemisphere; adult, sexually mature animals migrate upstream, often over long distances, to reach small streams in the upper part of the catchments of main rivers, where they reproduce. One main reason for this migration pattern is the fact that eggs and larvae require high concentrations of oxygen, and this can be found mainly in small, rather fast flowing cold mountain streams. The young fish then move downstream, in many species to finally reach the sea, where they live until they reach maturity and start their own upstream migration. The best known example of migrations in the opposite direction, where adult specimens migrate downstream (catadromous migration) is the eel (*Anguilla anguilla*). In the case of the European eel, reproduction takes place in the Sargasso Sea, a part of the Western Atlantic, the larvae then use the Gulf Stream to reach the European coast, where they start migrating into suitable rivers. In both cases, dams block such migrations, which led to the disappearance of salmons and of the eel form many rivers where they originally were abundant. The Atlantic salmon (*Salmo salar*) is widespread in the northern hemisphere around the Atlantic ocean. Its population in the Rhine River basin reached many rivers in Switzerland, in the upper parts of the basin; however, the Rhine Falls near the city of Schaffhausen were a natural insurmountable obstacle. During its migrations, it was very abundant in the Rhine near Basel and an economically important species. However, the story that the city of Basel had decreed that inhabitants of the city were not allowed to feed salmon to their servants more than twice a week is just that: a story, of unknown origin, but apparently very persistent (Weiss 2018). This species has disappeared completely from the Rhine basin; the last salmon in Basel was caught in 1958, and since then it is extinct in Switzerland, mainly due to the cascade of dams erected on the river, which completely blocked access to its spawning grounds (Weiss 2018; Pedroli et al. 1991). The situation is improving now, since these dams reach an age where they have to be rehabilitated (see the example of Rheinfelden in Sect. 13.2; Boller and Dönni 2011), and new concessions include the requirements for fish pass facilities; however, the "Rhine salmon", the population that lived in this river basin, is lost, and efforts to re-establish salmon populations are done with animals stemming from other rivers (https://www.regionatur.ch/Themen/Einzelne-Tiere/Lachs). The European eel is also getting increasingly rare, and this also mainly due to dams. A recent investigation (Clavero and Hermoso 2015) has shown that in Spain until the nineteenth century the eel occurred in almost all the rivers up to their sources, while presently it is restricted to about 20% of its former range, and it is restricted to parts of rivers in the coastal areas and completely absent from inland rivers, and this mainly due to dams.

In the case of the eel, as mentioned, the downstream movement is an active migration, adult and nearly mature animals actively swimming downstream to reach the sea again. However, a major part of downstream movements, especially of fish eggs and larvae, is not an active movement, but merely drift, i.e. being swept downstream by the movement of the water. If river flow is considerably slowed or even eliminated by the transformation of a part of the river to a lake, flow speed may be too low to maintain drift. Fish larvae and especially eggs can sink to the bottom under such circumstances, where they often perish, especially in cases where the water quality in the deeper parts of the reservoir is bad due to the lack of oxygen (see reservoir stratification, Chap. 17). But even for fish species which do not migrate and can survive in the remaining parts of the river, there is still the risk that the population of such a species in a river is being separated into several sub-populations, which will become almost or totally genetically isolated from each other, an effect that can, in the middle an long term, eliminate some species (see the case of the endemic fish *Diplomystes nahuelbutaensis* in Chile, Sect. 10.3.5).

One additional impact on fish, related to powerhouse operation, has to be mentioned briefly. Very frequently, water restitution from the turbines, or spilling from dams, results in a very high turbulence. This can have the positive effect of reoxygenation of the water from the reservoir, but it can also lead to an oversaturation of the water with gases (mainly O_2 and N_2). This, in turn, can adversely affect fish by causing the gas bubble disease (too high partial pressure of these gases in the blood, leading to the formation of air bubbles in the circulatory system). This can be lethal for fish (Zhong and Power 1996; Reichenbach-Klinke 1980, p. 358).

10.3.3.2 Mitigation

The obvious solution for this impact are fish passes and fish ladders, installations that allow migrating fish to pass this artificial obstacle and to reach, in spite of it, their spawning grounds in the headwaters of the rivers (such fish ladders are mainly for anadromous fish species, i.e., species that migrate upstream for reproduction like salmon, and only of limited usefulness for catadromous species such as the eel, which migrate downstream for reproduction). Of course, fish ladders are not only of use for those species that migrate from the sea to rivers, but also potentially for a great number of species that cover much more limited distances within the rivers themselves (the so-called potamodromous species). For the design of a fish pass, it is important to know what species are targeted by the measure, because not all species are able to use the same type of structure. There is an extensive literature on fish passes (real fish ladders being only one possibility among several), namely, a vast Russian experience with such structures, but increasingly also other papers not restricted to non-salmonid fish species (Pavlov 1989; DWA 2014; Schmutz and Mielach 2015; Morier-Genoud et al. 2016). This is not the place to deal with details of fish pass design, which, however, is essential for the success of the measure.

In spite of their usefulness in many places, there are limitations to fish passes. First, most designs are still applicable only for passing obstacles of not much more than 10 m in height (Thorncraft et al. 2005). A large dam, i.e. one with a height of 15 m or more according to the ICOLD definition, therefore, is increasingly difficult to cross along with increase in height, or cannot be crossed by means of such a structure. Here, other ways have to be found, such as fish lifts, or the capture of fish and their transport (by road) around the obstacle, to release them upstream from the dam. Even today, notwithstanding the growing body of information on other types or fish, most experience gained so far with fish ladders and similar structures still stems from temperate (Northern) climates. Little is known so far about the effectiveness and limitations of such structures in tropical waters; very often, knowledge of relevant aspects of fish biology is still lacking as well for these rivers; this issue was pointed out, e.g., for of fish ladders in the Brazilian Amazon (Fraser 2018).

As mentioned, fish ladders are useful mainly for upstream migrations, not for downstream movements (neither for juvenile nor adult fish). The main problem involved is that, in the reservoir, the water movement caused by the fish ladder in comparison to that caused by the water intake for turbines is too weak, which makes it impossible for the fish to find the way. Therefore, the downstream migration will mostly go through the turbines, which implies a very high risk of injuries and a high mortality. There are mainly two reasons for death or injuries among fish passing through the turbines. One is the physical contact with the turbine blades; here, incidence of injuries and deaths depends largely on the turbine type and the size of the fish (see Table 10.2). The second problem consists in the very high pressures, and mainly in the differences of pressure, the fish undergo when passing through the turbines (Montén 1985; Berner and Geiger 1988). Modern low-head fish friendly

Table 10.2 Probability for fish to be injured or killed when passing through turbines

Turbine type	Fish size (cm)	Injured or killed (%)
Francis	10–15	> 50
	30–50	100
Kaplan	10–15	5–10
	30–50	15–30
Improved turbine[a]	10–15	< 5
	30–50	5–10

[a]Newly developed turbine type, to be installed in the upgraded Rheinfelden hydropower plant (Rhine River, Swiss-German border). From Montén (1985) and Berner and Geiger (1988)

turbines reach a survival rate of 98–100% for fish with a considerable range of sizes (https://wa21.ch/wp-content/uploads/2017/10/Biel-2014-14-Amaral.pdf). The only further mitigation of this impact consists in providing structures and measures (fish screens) that will prevent fish from entering the waterways leading to the turbines, and offering alternative ways for downstream movements.

Considerable efforts were undertaken recently to come up with better concepts triggered by the Xayaburi HPP, the first dam on the Lower Mekong, i.e. outside of China (see Chap. 22 for more details on these projects). MRC, the Mekong River Commission, collected information on experiences with fish pass structures in large rivers and discussing applicability on the situation in the Mekong, pointing out the specific situation, and therefore the potential problems such structures have to consider, which are mainly the large number of species, their very different sizes (from a few centimetres to 3 m), their widely varying swimming abilities and the differences in seasonality and pattern of migration (Schmutz and Mielach 2015). A fish monitoring program had started earlier already (Baumann 2013). The final dam project then included a fish pass structure which included several possibilities for upstream migration, taking into account the different sizes and abilities of fish supposed to potentially use this structure (Morier-Genoud et al. 2016), but also for downstream movements. MRC (2019) reviewed the design changes which had been made since the first time the project had been launched, and IRN reviewed specifically the fish pass structure (Hirsch and Hensengerth 2019). The one major issue pointed out by these reviews is the fact that the structure has not yet proven its suitability for the task; this will have to be done by the monitoring program, results of which will hopefully be published in due course. Results on performance of this structure are important since other dams are in the project pipeline, to be built up- and downstream of Xayaburi, and these will have to be equipped with a similar—or improved, as the case may be—device for fish migrations.

Measures of a different kind altogether consist in artificially breeding fish species for restocking purposes. This measure has proved to be quite efficient—albeit rather expensive—in cases where natural reproduction was no longer granted, at least not to an extent that would have been sufficient for maintaining the fish stock at the desired high level. This has been, for instance, the case in Switzerland since the 1950s, mainly due to water pollution (see e.g. Pedroli et al. (1991: 17), and it was done only for "interesting" species (i.e., for those commercially attractive or valued for sports fishing). At least in the long run, it is not a real substitute for intact aquatic habitats that allow the natural reproduction of fish.

10.3.4 Change from River to Lake

10.3.4.1 Impacts

As for other groups of organisms, reservoir impoundment has dramatic consequences on the local fish fauna (see Zhong and Power 1996, for an overview). Of the species

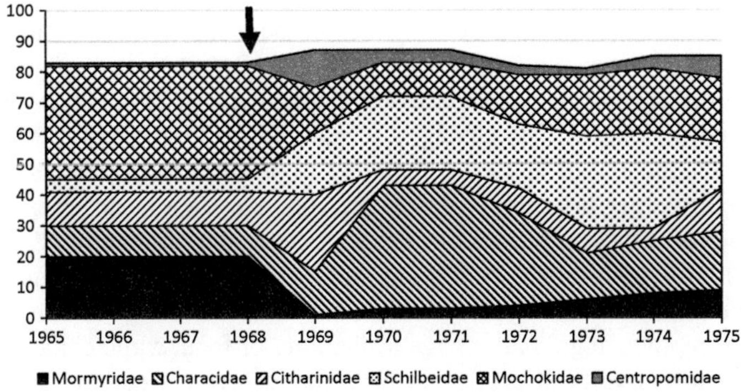

Fig. 10.8 Change of fish fauna at impoundment of Kainji reservoir. Redrawn after Zwahlen (2003), based on data from Visser (1973; % specimen numbers) for the river before impoundment, and from Lewis (1974) and Blake (1976, 1977) (% of weight of catches) for evolution in the reservoir. The arrow indicates impoundment

living in the river and therefore adapted to this habitat, not all will be able to live in the reservoir. Although a few species might actually benefit from this change and become much more numerous than they were before, others will decrease in numbers or even disappear altogether. This is not necessarily the case for migrating species only, and to make species-specific predictions in this respect is impossible if their biology—and especially habitat requirements and the reaction to such changes in habitat quality—is not known. This knowledge can only be gained from investigations of specific species. Figure 10.8 shows the changes in composition of the fish fauna in one example, the Kainji Reservoir after impoundment in comparison with the Niger River before impoundment.

The information provided in this figure should not be over-interpreted, since the data used have some serious shortcomings. First, the two sets cannot be compared directly, since the pre-impoundment data represent % numbers, while the post-impoundment data are % weight of catches; it is not possible, based on these data, to come to any conclusion on actual abundance. Secondly, it is assumed in the graph that the 1965 values represent a stable situation that was maintained until the impoundment in 1968, which may or may not have been the case; no values are available for the period 1966–1968. Finally, the data represent families, not species, and it cannot be assumed that all species within a family will necessarily react to changes in the same way. Still, in lack of a better example, this at least gives an idea on what such changes could be, with some groups diminishing, others increasing in importance. Antwi and Ofori-Dansin (1993) investigated the change from river to reservoir for the case of the Kpong reservoir in Ghana; however, this is a small, shallow reservoir (2000 ha), with a water turnover time of only 5 days, which cannot be taken as a model case for large reservoirs. Still, even there a number of species that lived in the river did not survive in the reservoir. Development of fish populations was investigated for Cabora Bassa (Jackson and Rogers 1976), but this was limited to the initial reservoir filling phase.

Balon (1974) reported on the development of the fish fauna in the Lake Kariba, a very large reservoir on the Zambezi River. He also pointed out the fact that after impoundment, fisheries yields increase very markedly, only to break down sharply after this first period, then stabilising at a much lower level. This presumably is due to the high amount of nutrients stemming from submerged biomass and soils, available shortly after first filling of the reservoir. This first of all leads to a very large amount of decaying biomass in the water, meaning a large food supply for detritivorous fish; since such species can often cope with bad water quality, and namely with low levels of oxygen in the water, they can find good conditions for growing and reproducing rapidly: large availability of food, while competitors and mainly predators have been eradicated from the reservoir. Once the biomass has broken down, this eutrophication can still lead to a high primary productivity in the new lake and therefore to a high supply of food for fish. So in spite of the water-quality problems (anoxic conditions, and sometimes even the presence of highly toxic hydrogen sulphide) in the hypolimnion of the reservoir, there are good conditions for fish near the surface, although for only a very restricted set of species. After a number of years, however, once the nutrients have been washed out, conditions become more oligotrophic. This usually means an improvement of water quality in the lake, but also, because of a shortage of food, a diminishing fish stock. At that time, at least some of the species that have been driven from the reservoir by bad water quality may become reestablished; however, this is usually not the entire set of species which was there before impoundment.

One example that shows such a marked change in fish fauna at the transition from river to lake is the case of Brokopondo reservoir in Suriname. Mol et al. (2007) document long-term changes in the fish fauna of this reservoir, the first large reservoir (1560 km^2) created in tropical rainforest. Before closure of the dam in 1964, the fish fauna of Suriname River consisted of 172 mostly small-sized species, but no single species was dominant in numbers. Large catfishes were dominant in biomass. Before impoundment, there were different habitat types in the reservoir area, like rapids, tributaries of various sizes and the main channel, all of which had their specific fish fauna. Four years after closure of the dam, only 62 fish species were collected from Brokopondo reservoir, and the composition of the fish fauna was still changing. The reservoir fauna in 2005 was very similar to the one in 1978, indicating that a stable equilibrium had been reached about 14 years after closure of the dam. At that time, only 41 species were left in the reservoir. Most species of Suriname River and its tributaries with strict habitat requirements did not survive in Brokopondo reservoir.

Eletronorte published a very nice book on the fishes of Rio Tocantins (Dos Santos et al. 2004), describing the situation 20 years after the construction of Tucuruí dam, which had created a reservoir with a surface of 2875 km^2 and a maximum depth of 75 m. 217 species of fish had been identified in the lower Tocantins River, which includes the reservoir, while in a study in 1984 270 species had been registered. Unfortunately, the book does not directly compare these two studies, there is no indication of what species have no longer been identified, and there is also no information on what species actually live in the reservoir, and which were found in the remaining parts of the river. Such a lack of information on the transition from river to reservoir fish faunas has also been reported for Indonesia (Whitten et al. 1996: 458 ff.), where basically the same was observed: disappearance of many specialised river inhabiting species, proliferation, at least for a limited time, of some species, and in addition to that the tendency to stock reservoirs with exotic species.

All these examples show, more or less clearly, that the creation of a reservoir will affect the fish fauna, and will usually result in a loss of biodiversity; this is not least due to the fact that a reservoir is a much more uniform type of habitat than the river system in the pre-impoundment reservoir area was, consisting of river sections of various types, faster and slower flowing ones, rapids, deep ponds, and tributaries of various size, all with their specific fish fauna. Unfortunately, all of them also show that the transition from river to lake, covering the phase of rapid changes after impoundment till the time where a new equilibrium has been established, is not really documented. I personally do not know of any study that covers the entire process, from the situation in the river before impoundment (for a sufficiently long period to be aware of natural fluctuations), during the construction, and into the operation period, until such a time when stability is finally reached. Such studies would be urgently needed, and if possible from many different places. However, these would have to be scientific investigations, an ESIA is not a suitable instrument for filling this gap. A good monitoring program could provide relevant information in this respect, provided it would start before construction starts, be carried out during the construction period and cover the operation period until such a time where stable conditions in the reservoir will have been reached, a process that can take several years. One obstacle here is sometimes the fact that the responsibility for a project changes. One such case was Bujagali HPP, Uganda. One entity was responsible for the construction. During this period, a serious monitoring of the fish fauna in the area of the future reservoir as well as in the downstream area was carried out. However, in spite of the fact that the POE for this project had repeatedly pointed out that monitoring should be continued into the operation period, and that the results should then be made publicly available, preferably in the form of a scientific publication, the exercise stopped when responsibility passed into other hands for the operation of the project.

10.3.4.2 Mitigation

The fact that living conditions for fish will be changing fundamentally in each project which includes the creation of a reservoir cannot be properly mitigated. However, efforts should be taken to make conditions in the reservoir as good as possible, and this as fast as possible. This is required for enabling a fish population, whether quite similar to the original fish stock in the river or quite different from it, to become established there. This should be done in the interest of biodiversity (which is obviously not limited to fish), but often also in the interest of developing fisheries in the reservoir, which can also be a relevant component of the compensation to be made for negatively affected persons (see Scudder 2005: 80, 156, 189, for the potential, but also for problems related to this issue). One of the very important measures in this respect, which has to be planned and carried out carefully before impoundment, is biomass removal from the area of the future reservoir with the aim of reaching suitable water quality conditions as soon as possible (see Chaps. 17 and 18).

It is a well-known—but apparently not a very well-documented—fact that fisheries yields in reservoirs, after a peak period following impoundment, drop sharply to stabilize at a far lower level, as suggested by the graph provided by Balon (1974: 137) illustrating the case of Lake Kariba. The best known reservoirs in this respect are the large African ones (see, e.g., Lelek 1973; Lelek and El-Zarka 1973; Lewis 1974; Petr 1974; Blake 1976; Kaptesky and Petr 1984). However, considering the problems in actually monitoring fish stocks in view of all the intervening factors—such as catch effort and reliability of reporting—the situation is not easy to analyse. Fishing in reservoirs, of course, is not limited to large reservoirs; small ones can also be very important, mainly for supplementing the diet of the local population or for supplying local markets (Baijot et al. 1994). One important and not always unproblematic way to increase fisheries yield in reservoirs consists of introducing exotic species. The aim here is to have species that will thrive in a lake, especially in cases where there were no species present in the original river system that easily adapt to this situation; Balon (1974: 136) describes the history—which, in this case, was a success—of the introduction of *Limnothrissa miodon*, a pelagic sardine from Lake Tanganyika, to Lake Kariba, whereas the introduction of *Tilapia macrochir* failed. However, there is always a risk involved with the introduction of exotic species, mainly because this can have very adverse effects on indigenous species, either because the new species actively preys on them or competes for food, or because diseases are introduced with the exotic species (Dumont 1986; Worthington 1989; Vila Pinto 1989; Worthington and Lowe-McConnell 1994). In addition, aquaculture by means of floating fish cages is another way of increasing productivity, as practiced e.g. in the Cirata and Saguling reservoirs in Indonesia or in the Nam Ngum 1 reservoir in Laos. As especially the case of the two Indonesian reservoirs has illustrated, if the number of fish cages gets too high input into the reservoir mainly stemming from feeding these fish can represent a substantial contribution of eutrophication of the reservoirs. All these activities, of course, interfere with the lake ecosystem and can have negative impacts.

10.3.5 Additional Potential Impacts on Fish

Creating obstacles or actually completely blocking fish migrations and the habitat change brought about by the formation of a reservoir are in many cases the main problems caused for fish populations associated directly with the construction of a dam. However, a number of other issues can also arise, and the potential for such impacts have to be checked and, if considered of relevance, to be dealt with in the ESIA.

10.3.5.1 Isolation of Populations

While a large number of fish species depend on more or less long migrations mainly for reproduction, or maybe more precisely because suitable habitats for reproduction and for finding sufficient food, quality- as well as quantity-wise, for growing up to the reproductive stage are often widely separated from each other, other species do not migrate at all, but stay more or less their whole lives within close range. In such cases, it is still important to point out that the members of one species, possibly distributed over a long stretch of river, genetically form one population in spite of their apparent immobility, with exchanges among the different sub-populations. Dams and reservoirs can now lead to the formation of isolated small populations between which gene flow is no longer possible. One such case was the siluriform (catfish) *Diplomystes nahuelbutaensis*, which is endemic to the Biobío River and a few other rivers in south-central Chile, whose population was cut into several isolated sub-populations by the Ralco and Pangue dams on Biobío River. Since it had already come under pressure by the introduction of exotic salmonids (Vila et al. 1978) and was considered as threatened, a monitoring and artificial breeding program was started as a mitigation measure with the aim of maintaining this species (Zwahlen and Durán 1998). Impacts from habitat fragmentation are of course not limited to fish species. A number of publications emphasise the problem for fish, other species and the river habitat in general, as e.g. Best and da Silva (1993) for the Amazon dolphin (*Inia geoffrensis*), Gouskov et al. (2016) for the Rhine River; Datry et al. (2017) describing the special case of ephemeral and intermittent streams; Fraser (2018) for the Amazon, Belletti et al. (2020) for European rivers; Diaz et al. (2019) for Chilean rivers; Bino et al. (2019) for the special case of the platypus (*Ornithorhynchus anatinus*) in Australia; and Su et al. (2021) for global freshwater fish diversity.

10.3.5.2 Downstream Impacts

Without going into details here, the changes discussed shortly below can have negative consequences for fish populations in the downstream area of a dam and power house; as always, it should not be forgotten that fish in this discussion serve as a kind of "umbrella group", whereby it is assumed that if conditions in a river are good for the natural fish fauna, this will also be the case for the other inhabitants of the same habitat, plants as well as animals.

- Change in seasonal discharge pattern: as illustrated with a number of cases (e.g. Fig. 9.3), a reservoir can, and often does, result in a reduction of high flow discharge and an increase in low flow discharge, and in extreme cases can even reverse the pattern. This can have a direct impact on fish, namely on species in which migrations for reproduction are triggered by the initial pulse of increasing flows at the onset of the high flow season. If this signal is lacking, or is not clear enough, the entire migration pattern and with it reproduction of such species can be negatively affected.

- Reduction of seasonal discharge differences and especially reduction of average discharge in the high flow season: here, the impact is an indirect one, as described in Sect. 18.2 on floodplains. If flow during the high flow season is reduced, the portion of the floodplain that would normally be flooded will be smaller, and this in turn means that fish which normally use this area, for reproduction and/or for food, will be faced with a shrinking habitat.
- Short-term variations in discharge: can affect fish in different ways. The harshest conditions in this respect are caused by peaking plants (see example in Fig. 9.10, Atatürk). The surge wave will flush out the river bed in short intervals, and this will affect fish directly, by flushing them out as well or by leaving them stranded at the fast reduction of flows, but mainly indirectly by eliminating food organisms and destroying breeding grounds. If in an extreme case discharge will go down to zero in the off hours of the power plant, life in the affected river stretch will be almost nil. Datry et al. (2017) clearly emphasise the importance of intermittent rivers and ephemeral streams for a large set of organism, plants as well as animals, which are adapted to this type of habitat, or even depend on it (some of them mainly due to the fact that fish, which would otherwise be the main predators on them, are absent from such habitats), but in these cases this is usually a system that follows, more or less reliably, a seasonal pattern, not in any way comparable to intermittence caused by a peaking plant.
- Water quality issues: water taken from deeper layers of a reservoir can be charged with a number of pollutants which can negatively affect fish downstream of the dam. In the first years after impoundment, or in the case of eutrophicated reservoirs, lack of oxygen can be a problem (see Chap. 17). In rare cases, the opposite may happen: an oversaturation with O_2 and N2, usually due to extreme turbulence in spillways, can cause the gas bubble disease in fish (Zhong and Power 1996).
- Temperature: lower temperature of the water taken from a reservoir can cause problems for fish, especially in the case of alpine high pressure plants where the reservoir can be located at an elevation of around 2000 m asl, much of the water stemming from snow and glacier melt, to then be released into a river at an elevation of around 400 m asl, as is e.g. the case for Mauvoisin HPP in Switzerland (see Sect. 25.1). A different type of issue caused by water temperature, although not related to hydropower, was observed at the outlet of the cooling water of a nuclear power plant in Switzerland. Since there water temperature was higher than in the remaining part of the river, productivity, first of all of sessile algae, but then also of benthic organisms feeding on them, was higher, which led to a concentration of fish feeding on these benthic organisms, and this in turn led to a higher concentration of fishermen at that site; finally, fishing too close to the cooling water outlet had to be banned (Büttiker and Meier 1983).

10.3.5.3 Exotic Species

As one mitigation measure, with the aim of compensating negative impacts on local fish populations and/or increasing benefits from the reservoir, exotic fish species have often been introduced. The best known examples of such introductions, although not necessarily related to hydropower development, are certainly the introduction of salmonid fish species, foremost the rainbow trout (*Oncorhynchus mykiss*), originating from North America, in a large number of river systems in Europe, Asia and South America. Such introductions were often to the detriment of local species (see Sect. 10.5.4 on invasive species). While this was a success in some instances, as the introduction of the freshwater sardine *Limnothrissa miodon* from Lake Tanganyika into Lake Kariba (the former being a natural lake, the latter a reservoir), others have resulted in very negative impacts on locally present species. If an introduction of an exotic species should be envisaged anywhere, at the very least a very careful analysis of possible risks should be carried out before such a measure would be implemented. Examples of such introductions have often shown that some species, which will then be termed as "invasive" as they turn into pests, cannot be eliminated again once they have become established in a new habitat. In no case should a recommendation for such an introduction be made as an apparently easy compensation measure.

10.4 Disease Vectors and Water Related Diseases

10.4.1 Main Vectors and Diseases

A considerable number of diseases of different types are related, in one way or another, to water, and this more so in tropical than in temperate regions. Since every dam project interferes with water, changing the situation usually to a very considerable extent (see e.g. Hunter et al. 1993), the issue of impacts of medical importance has to be taken into consideration in ESIAs for such projects, and here again especially the case of projects located in tropical or subtropical regions.

The WHO (1992: 115ff) characterises water related diseases as follows:

1. Waterborne diseases: infections arising from water contaminated with human or animal urine or faeces containing pathogenic viruses or bacteria; *Escherichia coli*, a common type of intestinal bacteria, can usually be found in waste water and in any surface water body receiving such effluents. Cholera and typhoid are other examples of such infections. These infections are spread by the use of unsafe drinking water; note that this is not necessarily the only way of infection, e.g. for cholera.
2. Water-washed diseases: these are diseases that are favoured in situations of scarcity or inaccessibility of water. Among others, waterborne diseases can also be considered as belonging to this category, since they are most serious when

access to good quality drinking water is limited or non-existent. Water washed diseases diminish whenever an adequate supply of water is available and used.

3. Water-based diseases: parasitic diseases caused by helminths whose intermediate hosts live in water. The most important type here is schistosomiasis or bilharzia, caused by different species of parasitic flukes whose intermediate hosts are aquatic snails; people risk an infection by getting in contact with such water.
4. Water-related diseases: diseases transmitted by mosquitoes, which need water for their development in the larval stage. The most important disease in this category is malaria, caused by a protozoan; there are also some mosquito-borne viral infections like dengue, chikungunya or Zika.
5. Water-dispersed infections: this is a rather "modern" type of infection, caused by agents which can proliferate in fresh water and enter the human body through the respiratory tract. One main way of propagation of such diseases is through air-conditioning systems.

The main types of such diseases in relation with dam projects are those of the categories 3 (water-based) and 4 (water-related). The most important diseases in this respect are malaria and schistosomiasis. Hydropower projects do play a role in this, but irrigation schemes are often even more important (see e.g. Tyagi and Chaudhari 1997; Ghebreyesus et al. 1999; Lautze et al. 2007; Kibret et al. 2012; Zhang et al. 2014). A few examples are discussed below.

10.4.2 Risks During the Construction Period

One of the risks presented by a large construction site, as for a dam project, is the proliferation of mosquitos, and with it the occurrence of diseases transmitted by them. One main reason for this is the fact that such a site potentially provides a multitude of suitable breeding places for mosquitos. These consist mainly in lack of drainage (which is especially problematic if it concerns the surroundings of the living quarters of the work force), and in a multitude of smaller and larger recipients (like empty cans etc., but also used tyres which are left lying around in the open). Approximately 10 days are usually sufficient for eggs laid into such containers by female mosquitos to develop into larvae and a new generation of mosquitos (Fig. 10.9).

Malaria is a widespread disease, and still a major health issue and a main cause of mortality in tropical areas around the world. Although progress was made in the fight against malaria, like a reduction in cases in Africa by 40% between 2000 and 2015, the threat has by no means been eliminated. Two different anti-malaria strategies are being followed, namely (i) the fight against the vector (mosquito species of the genus *Anopheles*) and (ii) the treatment of the disease, i.e. the fight against the causing agent, the parasitic protozoans of the genus *Plasmodium*. Both approaches have produced good results, but both basically suffer from the main problem: the vector as well as the parasite develop resistance against insecticides and drugs, which presents the

Fig. 10.9 Breeding places for mosquitos on construction sites. Insufficient drainage in the workers' camp. Old barrel partly filled with rain water. Spent tyres not kept under a roof

constant possibility of a resurgence of the disease (Peters and Gilles 1981; Nájera et al. 1993; Bhatt et al. 2015; Maxmen 2018; Bigirinama 2020). The most widely distributed and most aggressive from, the malignant (cerebral) malaria caused by *Plasmodium falciparum*, can be a major health issue on such a site, as demonstrated in Fig. 10.10.

This graph demonstrates not only the relevance of malaria, but also the importance of a sufficient monitoring and a suitable way of presentation of the results. In the beginning, the monthly report just mentioned the cases registered in the respective month, which as a general rule did not tell anybody very much. Then, the suggestion of the POE was followed to show the data in a graph as provided above, with the aim of not only showing the most recent data, but the evolution of the situation over time, which is actually an extremely important piece of information. The figure shows clearly that cases do not occur randomly or equally distributed over the year, but

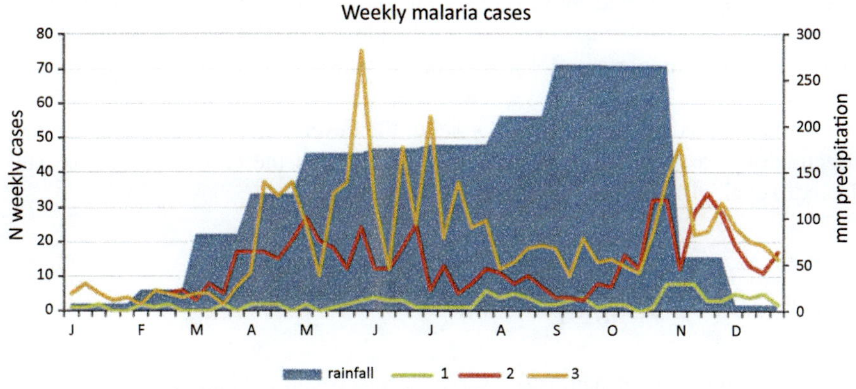

Fig. 10.10 Malaria on a dam construction site: monitoring results of one year. Lom Pangar HPP construction site, Cameroon, data for 2014. The weekly incidence is shown against average rainfall data. 1 = Owner's and Owner's Engineer staff on site (~70). 2 = Contractor's expatriate staff (~360). 3 = Local work force on site (~650).

follow a seasonal pattern. If compared with the data on average monthly rainfall, it is also clear that incidence raises first when the rains start and raise, and then a second time at the end of the rainy season. These data suggest that the conditions for the malaria vector, *Anopheles* mosquitos, is best at the onset and again at the end of the rainy seasons, and in this way they help optimising the timing for mosquito control measures.

It is evident that such data are an indication for a serious need of control measures. While the possibility to offer treatment on site, requiring laboratory facilities for diagnosis, a sufficient stock of drugs and trained personnel, is very important, it is equally important to carry out measures to reduce propagation of the vector. Here the most important measures are:

- Good drainage: the drainage system, especially in the camps where people live, must be adequately planned, designed and implemented, and must be maintained (e.g. by removal of clogs in the drainage channels) on a continuous basis.
- General good housekeeping, removal of all waste, and especially all material (cans and other recipients, used tyres, etc.) where water can accumulate and serve as mosquito breeding places.
- All containers that, for whatever reason, must contain water, have to be emptied at least once a week for eliminating mosquito eggs, larvae and pupae.
- Spraying of living quarters during the peak times of transmission.
- Providing bednets, preferably treated ones, to all person living on site and instructions on how to use them. How important control can be has been shown e.g. by Van Bortel et al. (1996) reporting on a case where bednets had been sold at a very low price, whereby many of them were then resold instead of being used.

It is important that all of these measures are being applied on a continuous basis.

Anti-mosquito measures are effective not only against malaria, but also against other diseases transmitted by mosquitos (Jamison et al. 1993; Curtis et al. 1998; Guyatt et al. 2002; Yohannes et al. 2005). In addition to the malaria vector, *Aedes aegypti* is another important mosquito species, transmitting viral infections such as Zika and dengue. This species has spread widely and is now also widely distributed in Latin America; as mentioned for *Anopheles*, the development of resistance against insecticides is a problem of increasing importance in the case of *Aedes* as well (Leandro et al. 2020).

10.4.3 Risks During Operation

The following effects of dam and reservoir projects can fundamentally change the situation for intermediate hosts or vectors of diseases, and therefore for the risk of the local population of becoming infected:

- Change of a river into a lake-like situation (to this topic see also Chap. 16).

- Having stagnant water year-round in places where before this was the case only seasonally, if at all.
- Seasonal drawdown of a reservoir creating small temporary puddles in the drawdown area.
- Submerging places with high turbulence (like cataracts or rapids), or creating such places (at water outlets, spillways etc.).

The creation of reservoirs as well as the hydrological changes in the downstream areas of dams influence a number of habitats, and therefore have effects on vectors, intermediate hosts or parasites themselves and lead to changes in the prevalence of certain diseases, but, as e.g. the study by Zhang et al. 2014, for the case of the Three Gorges HPP in China has shown, it can be difficult to actually identify the exact contribution of the dam project to such changes. As mentioned above for malaria, hydropower projects can certainly have impacts in this respect, but irrigation schemes even more so (Stürchler 1981: 149ff; Peters and Gilles 1981: 162ff; Oomen 1983; Piekarski 1987: 162; WHO 1993). One reason for this is certainly the fact that irrigation schemes create a larger number of often small permanent or temporary water bodies (reservoirs, irrigation channels, temporarily irrigated fields, drainage ditches) which all can serve as breeding points for mosquitos and the permanent ones as habitat for aquatic snails, the intermediate hosts of schistosomiasis (also called bilharzia). Another important reason is that in drainage projects people are more likely to live close to and/or get in touch with small water bodies more often.

One such case was a project in Burkina Faso, which aimed at rehabilitating about 30 existing small irrigation schemes and creating 3 new ones (EWI 1989; Zwahlen 1992). Figure 10.11 shows a typical scene, and some of the relevant observations made in this case are summarised in Fig. 10.12.

As can be seen from the figure, expectations and outcome were widely different. The main expectation had been that the project would achieve a massive improvement in the nutrition of the local population, by increasing harvests (3 yearly instead of only one before) and livestock, both effects a direct function of water availability year-round. Both, however, had improved only marginally, due to a number of reasons. One was the requirement of paying for the water in order to pay back the project loan; a second one was the fact that before, many men had gone to seek seasonal work during the dry season outside of their villages, which now they could not do any longer, or if they did, there was not enough labour available for actually cultivating and harvesting three times a year. Finally, many of these schemes had broken down after a rather short time of operation; these were mainly the regulating structures for drawing the water from the reservoirs and distribution to the fields, and it turned out that even in the case of rather simple problems the means and the skills for repairing the damage were not available, and the schemes simply stopped working.

Concerning health issues, while some problems had been expected, these turned out to be much more serious than anticipated. Availability of drinking water had been increased, since before only a few wells had been available for that purpose, many of which dried up towards the end of the dry season. Now the reservoirs provided a permanent supply of water, but water quality was bad, especially due to the faeces

Fig. 10.11 Drainage channel as source of water. Towards the end of the dry season, people often have no choice other than using such a place for getting water

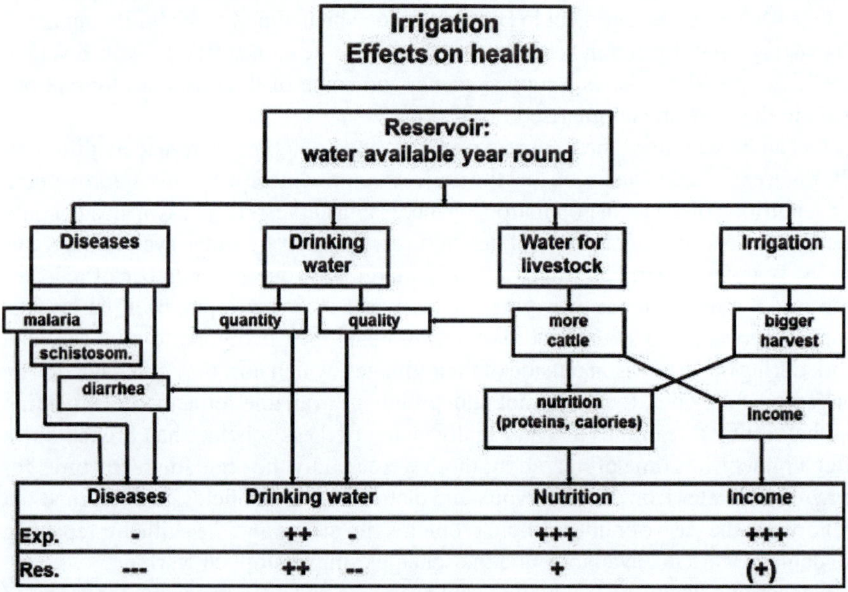

Fig. 10.12 Expectations and results in small irrigation schemes. Exp. = expectations during the planning phase; Res. = results observed at the evaluation for rehabilitation; both expressed in a scale from very negative (−−−) to very positive (+++). Further explanations in the text

of cattle that concentrated close to the reservoirs; diarrhea was one of the problems resulting from that. In addition, there was an increase in malaria cases, since now breeding places for mosquitos were available the whole year, while before it had been a seasonal problem; the use of clay from the drawdown area for making mud bricks added to the problem, since this created small puddles around the reservoirs which were suitable breeding places for mosquitos, especially since there were no fish there that could have predated on the larvae. One major issue, however, was schistosomiasis; before, this serious disease had not existed in this area, since the long dry season had prevented the development of aquatic snail populations which are the intermediate host of this parasite. In the reservoirs, where different types of water plants developed very rapidly, they found very suitable conditions, and in fact populations of *Bulinus* sp., the intermediate host of *Schistosoma haematobium*, the parasite causing the urinary schistosomiasis, could easily be found in all these reservoirs. The main groups at risk were fishermen, women (by getting water and washing clothes) and children (swimming in the reservoirs).

Schistosomiasis is a disease caused by parasitic flukes who all need different types of aquatic snails as intermediate hosts. The two most important species are *Schistosoma haematobium* (causing the urinary schistosomiasis, widely distributed in Africa, from where it has spread to some South and South East Asian countries) and *S. mansoni*, causing the intestinal schistosomiasis, another species of African origin which spread to tropical South America, where it apparently found suitable intermediate hosts. There are three more species of more restricted distribution and somewhat less public health concern, namely *S. intercalatum* (Central Africa), *S. japonicum* (some South East Asian and Western Pacific countries) and *S. mekongi* in the lower parts of the Mekong river (Alonso-Lormand et al. 1995; Biays et al. 1999; Stich et al. 1999).

Measures against these diseases mainly follow three major lines, namely:

- Environmental management: cleaning of irrigation channels from water plants, prevention of forming stagnant pools; drainage of temporary puddles, etc.
- Control of snail populations; this is done either by applying moluscicides, i.e. pesticides specifically aiming at eradicating snails, or by introducing snail-eating fish into permanent water bodies like reservoirs. Fish have been more or less suitable for this purpose in a number of places, although they may reduce snail populations without however eliminating them (McMahon et al. 1977; Jordan et al. 1980; McCullough 1981).
- Improving hygiene, mainly providing suitable toilets and propagating their use; this, in addition to other hygienic benefits, interrupts the infection cycle by preventing eggs in urine (*S. haematobium*) or faeces (*S. mansoni*) from getting into water, where snails can be infested.
- Reducing or preventing contact with (potentially) infested water, from where people could get the parasite; one stage of larvae of these helminths swims freely in water and penetrates the skin of a suitable host, in order to then develop into sexually mature specimens.
- Treatment of the disease; luckily now suitable drugs are available for this purpose.

A combination of different or all of these measures may be possible, and have to be continued especially in areas where one of these parasites is highly endemic (El Gaddal 1985; WHO 1993).

One additional special case of a vector-borne disease worth mentioning is onchocerciasis; this is a disease also caused by a parasite, *Onchocerca volvulus*, which is a nematode (round worm) transmitted by the blackfly, *Simulium damnosum*. It lives under the skin, where adult worms form clearly visible nodules; there they reproduce, producing numerous minute larvae, the so-called mircrofilariae, which move in the subcutaneous tissue, from where they can be taken up by biting blackflies, thus being transmitted to other persons. In cases of heavy infestations, damage caused by microfilariae to the eyes can lead to blindness, which gave the disease the name river blindness. It occurs in tropical Africa and in some parts of Central and Northern South America (Stürchler 1981: 182). Unlike the mosquitos or snails mentioned so far, the blackfly requires fast moving, turbulent water with high oxygen content for its larvae to develop. These usually sit on rocks in cataracts and rapids. According to WHO (2019), between 1974 and 2002, the disease was brought under control in West Africa through the work of the Onchocerciasis Control Programme (OCP), using mainly the spraying of insecticides against blackfly larvae (vector control) by helicopters and airplanes. This has been supplemented by large-scale distribution of the drug ivermectin since 1989. The application of insecticides on a large scale had also negative effects (see e.g. Asibey 1977). Presently, the main strategy for fighting onchocerciasis is the administration of ivermectin, supported by the application of

Fig. 10.13 Re-aeration structure downstream of the Lom Pangar dam, Cameroon. The usefulness of the structure for reaeration is questionable. However, it might turn into a breeding place for the blackfly *Simulium damnosum*, the vector of river blindness. The project is located within the endemic area of onchocerciasis (Stürchler 1981: 182)

environmentally safe methods for vector control where required (Dye and Baker 1986; WHO 1987; Guderian et al. 1997; Thylefors 2004; Makenga et al. 2015).

Dam projects do not seem to play a very important role in relation with onchocerciasis. Submerging fast-flowing river sections which provide suitable habitat for the blackfly, i.e. transforming sections of rivers with rapids into lake-like habitats, can actually eliminate *Simulium* breeding places. On the other hand, some structures, like spillways, surface water intakes, tailrace channels or possibly re-aeration structures (e.g. in the case of Lom Pangar, see Fig. 10.13; Savatier et al. 2016) can potentially create again suitable habitats for *Simulium* larvae (Ebomoyi and Omonisi 1991).

10.5 Biodiversity

Il y a des milliards de fourmis qui voudraient pas qu'on les oublie
There are billions of ants who would not want to be forgotten
From the song "J'avais rendez-vous",
Carrousel (Sophie Burande and Léonar Goignat) 2012

10.5.1 What Is a Species?

Biodiversity can be defined as follows: "The total of variation in organisms, in past times and present, in locations up to and including the entire planet, and organised at three levels: ecosystems, species comprising the ecosystems, and genes prescribing the traits of the species" (Wilson 2016: 227). Biodiversity is described and discussed in detail, but not defined, in Wilson (1988). Overall, biodiversity is declining, as has once more been emphasised by the latest Living Planet Report (WWF 2020), which followed the populations and numbers of 4000 vertebrate species (mammals, birds, reptiles, amphibians and fishes) and found an average decline in numbers of 60% between 1970 and 2016. The biological diversity of an area is made up by the number of species of plants and animals that exist there, and by the number and extent of habitat types which they inhabit and help to form. Biodiversity is often addressed on a regional, country or even global level, mostly with the focus on its protection (e.g. Holm-Nielsen et al. 1989; McNeely et al. 1990; Baatz 1992; Ramamoorthy et al. 1992; Yap 1992; Chape 1996; Huntley et al. 2019).

The most important category for assessing impacts on biodiversity, besides habitat or ecosystem, is the species. This is the category which is used e.g. in lists of animals and plants found (or supposed to be) living in a given habitat, and an important way of addressing impacts of a project is to find out whether any rare, threatened, endangered or endemic species are or will be affected by a project, and if so, to what extent.

At this place it may be useful to make a short (and necessarily incomplete) excursion into what we understand under the topic of "species". The usual, most utilised definition of a species is that this is a population of animals (or plants, as the label is applicable to them as well) which will breed freely and produce fertile offspring under natural conditions. This means that different species are isolated from each other, either genetically, geographically or through behavioural mechanisms (e.g. by not recognising individuals of a different, although rather similar species as potential sex partners, although interbreeding might be possible and in some cases would even lead to fully fertile offspring). This means that species which will exceptionally interbreed in captivity are still considered as "good" species. The best known example for such mixed breeding are probably horse and donkey, but it is also possible between lion and tiger, although in both of these cases with usually infertile offspring. In other cases however, as e.g. crossing domestic dogs (and possibly wolves) with coyotes, offspring are fertile (Silver and Silver 1969), but interbreeding does not happen under natural conditions. A good example for further illustrating the topic are the two camels, the one-humped dromedary (*Camelus dromedarius*) and the two-humped Bactrian camel (*Camelus bactrianus*), which are usually considered as two species, although it is not clear whether they originally stem from two different ancestral wild species or from subspecies of one originally widespread species (Zeuner 1963; Walker 1968: 1374; Benecke 1994), and which, at least under conditions of animal husbandry, hybridise and produce fertile offspring (Dioli 2020). Still, there are many cases of hybridisation between "good" species under natural conditions; see, as recent examples, Herren and Reist (2020) on interbreeding of two shrike (*Lanius*) species, or Chen and Pfennig (2020) on two toad (*Spea*) species or Skovrind et al. (2019) who confirmed a hybrid between a narwhal and a beluga, two species that are even considered to belong to two different genera (beluga or white whale, *Delphinapterus leucas*, and narwhal, *Monodon monoceros*).

While this definition of species is undoubtedly useful and applicable in practice, it is, as just shown, by no means unequivocal. This is quite understandable if we consider two things. First, in evolutionary terms, a species is not something that is well defined for all eternity, but something that changes over time, especially when exterior conditions change (to recognise the basic way in which this happens was the great achievement of Charles Darwin, described for the first time in his book entitled "On the origin of species by means of natural selection", originally published in 1859). Second, "species" is ultimately not a concept of nature as such, but an attempt by humans to categorise and label living organisms. Since Linné, the great Swedish naturalist, species are identified by means of binominal names, the first identifying the genus, the second the species (as examples: *Canis lupus* for the wolf, *Canis latrans* for the coyote, two species belonging to the same genus, i.e. being closely related, just like *Homo sapiens*, Modern Man, i.e. us, and *H. neanderthalensis*, the Neanderthal, our unfortunate cousin whom we might have helped on his way to extinction, and this, according to Shipman (2015), possibly with the help of dogs, domesticated *Canis lupus*). However, it soon became apparent that populations of many species, especially ones with a large geographic range, show some clearly distinctive traits which allow to identify a single specimen not only to the species level, but also to the level of (geographically more or less isolated) populations. This has led to the introduction of a third name, for identifying subspecies or races; for those interested in a more detailed, but still easy to understand account on what species or races are, "The grasshopper's tale" (Dawkins and Wong 2016: 452 ff.) is a good read. Darwin, in his work cited above, was well aware of this problem, and the fact that species are not really clearly defined was an important pillar for his theory.

Let's make an example: the one large animal which unequivocally represents Africa, since it is almost impossible to mistake it for anything else, is the giraffe, *Giraffa camelopardalis*. Populations of this species are shrinking at an alarming extent (Lee et al. 2013; Brown et al. 2019), but it is still widespread over large parts of Sub-Saharan Africa, a typical inhabitant of savannas. Populations from different parts of its range are clearly different, in colour and in the form of their characteristic dark patches. This led to the distinction of a number of subspecies, of which usually nine are generally accepted, among them the "nominal" subspecies *Giraffa camelopardalis camelopardalis*, the Nubian giraffe from South Sudan and Ethiopia, *G. c. reticulata*, the reticulated giraffe from Kenya, Ethiopia and Somalia, *G. c. rothschildi*, the Rothschild giraffe from Uganda and Kenya, *G. c. tippelskirchi*, the Masai giraffe from Kenya and Tanzania, *G. c. angolensis* from Botswana and Namibia, and *G. c. giraffa*, the South African giraffe, from Botswana, Zambia, Zimbabwe and South Africa.

So far, so good, but now it gets more complicated. A recent study of the genome of all extant giraffe subspecies (Fennessy et al. 2016) came to the conclusion that genetic differences between these populations or subspecies of giraffes are so marked as to justify splitting them into four separate species. The proposed names for these four species are *Giraffa camelopardalis*, *G. reticulata*, *G. tippelskirchi* and *G. giraffa*. This means that instead of one widespread species with several subspecies now suddenly we have to deal with four species, some of which still with a number of subspecies. Such an opinion often does not remain uncontested (Bercovitch et al. 2017), other specialists defending the one species position, leading to a typical discussion between splitters (scientists with a tendency to separate subspecies into several species) and lumpers (scientists doing the opposite; see Zachos et al. 2013). This, by the way, is the case for our own species: the discussion of whether the Neanderthal is a species of its own or a subspecies of modern man, i.e. whether it should be called *Homo neanderthalensis* or *Homo sapiens neanderthalensis* (which would make us *Homo sapiens sapiens*; see Pääbo 2014: 237; Reich 2018: 56) is still ongoing. The fact that Neanderthals and modern humans mixed, leaving traces of Neanderthal DNA in all of today's non-African humans, might provide support for the subspecies approach, following the definition of species given above. Another recent example of this type is a study which comes to the conclusion that the Chinese giant salamander, the largest amphibian species in the world, is actually a group of three species (*Andrias davidianus*, *A. sligoi* and a third, so far unnamed species; Turvey et al. 2019); the question might be of relevance for the conservation of these by now critically endangered animals. There is also a discussion among conservationist on whether conservation should take place at the subspecies level, meaning that cross-breeding of individuals of different subspecies should be prevented, and what should be done about hybrids, discussed for the case of two species of great apes, chimpanzees and orangutans (Palmer et al. 2021). As a side-line to this discussion: it is interesting that there is a discussion between splitters and lumpers also in linguistics, when it comes e.g. to define what is a dialect of a language as opposed to a separate language, or what are language families (Grimes 1996: vii; Deutscher 2006: 55; Reich 2018: 173).

Now, this was just an example for illustrating some of the problems associated with the definition of species. When it comes to ESIAs for HPPs, this is usually of little or no practical concern; should a population of giraffes be heavily impacted by the construction of a dam, there would be little or no difference in the outcome whether the report would deal with that issue as affecting a subspecies or a population of a separate species.

However, the question of species or subspecies can become very decisive indeed. Very recently, a population of the Sumatran orangutan was identified as a separate species, named *Pongo tapanuliensis*, which genetically seems to be more isolated than the two currently recognised species, the one from Borneo (*Pongo pygmaeus*) and the one from Northern Sumatra (*Pongo abelii*), of which it had before been considered a subspecies (Nater et al. 2017; Stokstad 2017). This newly described species has a very limited distribution range and a population of only about 800 individuals, which makes it very vulnerable. Soon after its discovery, concerns were raised that it might actually be in danger of becoming extinct due to human interference with its habitat, prominently among which ranges a hydropower project; as so often, the main risk identified was not the dam and power house as such, but the access road which would have to be built (Brainard 2018; Laurance 2018; see also Chap. 15 on access roads). Endangering a population of one of the great apes (gorillas, chimpanzees and orangutans) is a very serious impact for any project, but threatening an entire species with extinction is certainly much more so. If a single species will have such a high importance for assessing project impacts and defining mitigation measures in an ESIA, it will obviously serve as "umbrella species" as discussed shortly in Sect. 10.2.1, insofar as can be assumed that if a highly endangered species can survive, then so will all or at least most other species in the same habitat. Ultimately, biodiversity conservation is not about single species, but about intact communities of living organisms capable of sustaining them.

10.5.2 Why Conserve Biodiversity?

As just mentioned, protection and conservation of biodiversity is not limited to single species (and certainly not to individual animals or plants), but about maintaining habitats or ecosystems, i.e. communities of different species which also include all the relationships and interactions between them. According to Sánchez-Bayo and Wyckhuys (2019), the main drivers of biodiversity loss are: (i) habitat loss and conversion to intensive agriculture and urbanisation; (ii) pollution, mainly that by synthetic pesticides and fertilisers; (iii) biological factors, including pathogens and introduced species; and (iv) climate change. How fast biodiversity can change, and what role people play in this has been shown for a number of islands, where rapidity of change in vegetation cover increased by a factor of 11 after their colonisation (Nogué et al. 2021). Still, reasons for biodiversity conservation (as opposed to the conservation of individual species) may not necessarily be obvious in any case, or, to put it otherwise, it may be rather easy to suggest that biodiversity, in many respects, might not be so terribly important for us. Just to make a very simple example: you may or you may not have heard about the dodo (*Raphus cucullatus*) a flightless bird about the size of a turkey, which lived in large numbers on Mauritius and went extinct approximately 200 years after the first colonisation of the island by Europeans (Brehm 1911, Vol. 7: 431). Have you ever missed it? Probably not, just as the many other extinct species, and especially those of which one has never heard. Still, a number of reasons can be listed for protecting biodiversity, to be covered here only very shortly, and not with the intention of giving a final, single, good for all reason.

- Scientific: in discussions of losses due to projects, dam projects prominently among them, one can often here the argument that especially in highly diverse and not yet very well explored environments only a fraction of species present is actually known (e.g. Kipping et al. (2019), who come to the conclusion that less than 80% of the dragonflies living in Angola are known), and species disappear before we even know that they existed. From a scientific point of view, this is certainly a loss, as long as we consider "basic science", i.e. the science we do not with a specific (economic) goal in mind, but just because we want to know, as having an intrinsic value.
- Stability of ecosystems: elimination of species can lead to destabilisation of ecosystems, and long-term stability often depends on the presence of an entire set different species (Belskky 1986; Remmert 1987; Pomeroy and Alberts 1988: 321). When studying forest regeneration on la Réunion, Albert et al. (2020) came to the conclusion that the extirpation of frugivorous animals led to the disappearance of large-fruited trees, since seed dispersal no longer took place.
- Bio-medical: a large number of plants and other types of organisms were crucial for the development of drugs, and one argument for protecting biological diversity is the possibility that "out there" there are still numerous species which might become important for our future health, provided we ever find them. Even if modern synthesising methods have become more efficient, there are certainly still such discoveries to make, as e.g. some bacteria from the guts of vultures capable of breaking down toxins, which might indicate ways for treating food poisoning (Levin et al. 2021).
- Economic: trying to prove the importance of species on economic grounds, e.g.in the form of a cost–benefit analysis, is very likely to fail in most cases, whereby "fail" here means that the result will not be in favour of the specie's conservation, as shown by Clark (1973) in his model calculation showing that it might be preferable, on strictly economic reasons, to extinguish the blue whale (which however he does not recommend!) rather than to let the stocks recover for being used sustainably. This should not be an encouragement towards exterminating

species for economic reasons, but rather a warning against trying to calculate the "value" of a species in monetary terms.

- Tourism: this is ultimately a merely economic reason, but it can provide a rather powerful argument for biodiversity protection in regions where wildlife tourism is an important, and often the most important, source of income. There the reasoning is simple and straightforward: tourists come and spend money for seeing wildlife, but they only come as long as there is wildlife to see, and therefore wildlife has to be protected. A good example are the mountain gorilla sanctuaries in Uganda and Rwanda: seeing gorillas there is quite expensive, but keeping them—and their entire habitat—intact and making them easily accessible and visible for visitors requires a tremendous effort. COVID-19 has caused great problems in these areas by making it impossible for tourists to go there, curtailing the income of almost all inhabitants there.
- Religious: most people know, at least in meaning, the sentence from Genesis 1.28: "Be fruitful and multiply, and fill the earth and subdue it; and have dominion over the fish of the sea and over the birds of the air and over every living thing that moves upon the earth." Much less known is the phrase in Genesis 2.15: "… took the man and put him in the Garden of Eden to till it and keep it". Our species was—and continues to be—very good where multiplying and subduing is concerned, while for keeping we do a much worse job.
- Ethical: it could be argued that ethical arguments are the same as those related to religion, however, ethical questions can be addressed without referring to religion. In the case of biodiversity, one can argue that since we know what we are doing, and we have the capacity to recognise the adverse consequences of our actions, we also have the responsibility to take actions for avoiding or at least minimising these adverse consequences (and, finally, this is exactly what we strive to do when preparing an ESIA!). As Ehrenfeld (1988: 215) puts it, we "must come to understand the inherent wrongness of the destruction of biological diversity". And once we have understood that, we will have to do something about it.
- Human well-being: a recent study (Methorst et al. 2021) demonstrated a correlation between human well-being (as expressed by subjective assessment of a large number of persons asked to characterise their own personal well-being) and the biodiversity of the area where they live; the best positive correlation was found between well-being and bird diversity, whereby it was noted that bird diversity could also be a proxy for landscape diversity and attractiveness.

In the context of ESIAs it could, obviously, be argued that all these considerations are not really required, since there is one much simpler, down-to-earth reason why biodiversity conservation is a subject in such a study: not for any more or less sophisticated philosophical reasons, but simply because the applicable standards and guidelines say so, and because the TOR ask one to do it, and because many species, and not necessarily only rare or threatened ones, have a status of legal protection, according to legislation which has to be respected in the framework of an ESIA (Sect. 7.2). Still, no harm done in thinking a bit further than just the pure necessities.

10.5.3 Endemic Species

When dealing with project impacts on biodiversity, few words are as important and as meaningful as "endemic" in all its forms (endemic species, endemism), few are potentially of such consequences for a project (keeping in mind that if a project would actually threaten to cause extinction of an endemic species this could well—and probably should—mean that this project would have to be abandoned), and few are used as carelessly as this one.

The first issue is that "endemic" as such does not tell us anything, and the word should never be used like this: it is always "endemic to something", and the really important part is in the "something". Endemic means "only occurring in ...", and therefore this must be defined each time we use the term. So, it is e.g. possible to define endemism of a biogeographic unit, i.e. the percentage of species which live exclusively in a specific area defined by a common set of geographical and ecological characteristics (see e.g. Huntley 1988: 250, saying that 80% of the flowering plants and 45% of the mammals occurring in the Guineo-Congolian biogeographic unit of Africa are endemic to this area). Another way of using this expression is to say that a species is endemic to a country, e.g. when Jalili and Jamzad (1999: 5) state that "in total there are 1727 endemic taxa" in Iran. This definition of endemism by country is sometimes used when discussing conservation of rare species, as a way to point out the importance of a country and its responsibilities for specific conservation measures. Obviously, such descriptions do not really tell us very much either, and it is certainly a very big difference whether we speak of a species endemic to China, or one endemic to Bhutan. To make just one example: a google search for "endemic to Australia" provides 26.9 million results, among them the Wikipedia text saying "The fauna of Australia consists of a huge variety of animals; some 83% of mammals, 89% of reptiles, 90% of fish and insects and 93% of amphibians that inhabit the continent are endemic to Australia" (https://en.wikipedia.org/wiki/Fauna_of_Australia). A well-known representative of this fauna is the platypus (*Ornithorhynchus anatinus*), which is endemic to Eastern Australia and Tasmania (Bino et al. 2019). Likewise, it does not tell us much when we say that the Komodo monitor (*Varanus komodoensis*) is endemic to Indonesia, but it does tell something very relevant about this species when we hear that it is endemic to Komodo and a few other small Indonesian islands. We could consider the enigmatic Rhinogradentia, an order of mammals now unfortunately extinct, which lived on an isolated archipelago in the southern Pacific, as an extreme case of endemism, since no member of this order lived anywhere else (Stümpke 1961).

In the context of environmental impact assessments for dam projects, endemicity of a species would really be of concern if a species would be endemic to the area of influence of the project, or be even more restricted, living only in the area of the future reservoir, the proximity of the dam, or the river stretch heavily impacted by a change in discharge dynamics by dam and powerhouse operation. Species with such a restricted area of distribution are the exception rather than the rule, but they do exist, whereby at times it is not clear, especially where rare, small or otherwise not very “prominent” species are concerned, whether they are really that rare and restricted in their distribution, or whether there is just a lack of information on them (see e.g. the example of mosses in a rock fall area mentioned in Sect. 3.12). A few examples:

- The newly discovered third species of orangutan, the *Pongo tapanuliensis*, of which only a small population living in a very limited area in North-central Sumatra is left, and which is threatened by a hydropower project, or more precisely probably by the access road that has to be built for this project (Sect. 15.3).
- The spray toad (*Nectophrynoides asperginis*), which was extinct in the wild by changes caused in its habitat due to a hydropower project (Sect. 18.3).
- In the ESIA for a project in Cameroon, Memve’ele, a plant growing directly on rocks within the waterfalls was found and identified as *Ledermanniella annithomae*; the botanist involved in the field work who had found and determined it pointed out its very limited range and proposed to carry out a search for this species in other water falls and rapids along this river (Pöyry and Alpha Technology 2010). The species belongs to a genus which is rather widely distributed in Africa, but strictly limited to rocks in water falls which are covered by water most of the year. A short literature search leads to the impression that *Ledermanniella* can be found in every suitable water fall, and the differences between the populations seem to justify the description of a new species rather frequently when investigating water falls which have not been checked before (Kita et al. 2008), describing *L. ntemensis* from Ntem water fall, southern Cameroon, and Cheek et al. (2017), describing *L. yiben* from a rapid in the Seli (Rokel) River in Sierra Leone, upstream from the Bumbuna reservoir, which is threatened by the planned Yiben dam).

While endemism and the protection of endemic species is of real concern when dealing with project impacts, it is just as important that such findings are carefully evaluated. Unfortunately, the term can also be, and sometimes is, misused, as the following sentence from an ESIA report shows, which said: “The endemics will suffer the loss of habitats, and local endemics would face pressures of extinction. The value of the ecosystem as a cauldron of endemism and its evolutionary potentials will dissipate.” In this case, there was actually not one single species threatened by the project (and none was actually mentioned in the report), it was just an attempt to prevent implementation of the project. Even if the author of the study was of the opinion, for whatever reason, that no dam should be built at this site, unjustified use of such an argument can definitely not be tolerated.

10.5.4 Exotic and Invasive Species

En ces temps-là, la ville était envahie

par tous les rats venus du fond du pays

In those days, the town was invaded by all the rats come from all over the country

Hugues Aufray, French singer-songwriter, 1966

Potential problems with introduced species were mentioned shortly in Sect. 10.3.5, considering the introduction of fish as "mitigation measure". Another issue with exotic, invasive species is described in Sect. 16.3: floating plants causing problems on reservoirs. Here, the topic shall be extended to some extent.

Shortly, any animal or plant species will be considered as an exotic or alien species if populations are found somewhere outside of its natural range of distribution, in places where it did not occur naturally, and which it could not have reached without human interference. This definition might not be easily applicable in every case, but for most of the species to be considered under this heading it will do. The definition clearly excludes vagrants, i.e. individuals found far off their normal range, but who got there entirely on their own. This is known mostly from birds, where not too rarely individuals appear far outside of their normal range; a reason for that may often be specific weather conditions like e.g. wind, which might force individuals, mainly of flying species, to cover long distances in directions they would not usually go. Still, there are cases where a species extended its range without obvious "help" form humans, as e.g. the Eurasian collared dove, *Streptopelia decaocto*, an originally Asian species which started to expand its range into Europe in the 1930 and is now very frequent in most of Western Europe (Knaus et al. 2018: 160); the fact that it lives mainly in human settlement, however, might be an indication that it did get some encouragement, after all.

Many of these exotic species are just that: exotic. They may be rather remarkable because they are something new or unfamiliar, but they remain limited in numbers, and many of them disappear again. Some, however, prove to be vigorous, the population is increasing, and they are able to spread in their new environment. Such spreading can be at the cost of local species, and in that case this is a threat to biodiversity. These species are then called "invasive", and they have come to stay (IUCN 1998). Invasiveness of an alien species is by no means a rare event, as the following few figures illustrate. Wittenberg (2006) lists 800 alien species found in Switzerland, 107 of which are considered as invasive. In a survey of alien species in Norway, Sandvik et al. (2020) found that from a total of 1519 exotic species recorded 1183 were reproducing in the country, and 9% of these were assessed as having a severe impact, 7% a high and another 7% a potentially high impact. Wilgen et al. (2020) list 1400 alien species in South Africa, and Poland et al. (2021: 435) list 1799 species of plants and 221 species of animals, the latter ranging from earthworms to mammals, introduced to the Caribbean islands alone.

There are many ways in which species reach new places outside of their normal range. The most important, with just a few examples for each category (note that the categories can overlap to some extent), are the following:

- **Accidental introduction:** a number of species have travelled as "blind passengers", e.g. hiding in the cargo of ships, and were in this way introduced. The most successful in this respect were probably a few species of rats, which travelled in boats and got in this way almost to all places that were ever visited by seafaring people. Especially on islands rats are one of the worst threats to biodiversity, e.g. by predation on breeding birds and their eggs and nestlings, as in Hawaii and other pacific islands, or on eggs and hatchlings of sea turtles or the Galapagos giant tortoises (Van der Geer 2018; Raine et al. 2020). Another example of accidental introduction is the zebra mussel (*Dreissena polymorpha*), which travelled attached to ships and expanded in this way its range from southern Russia and Ukraine to Western Europe and North America. Due to its habit of forming thick clusters of shells well attached to solid substrates, it can cause problems in all types of structures (like water pipes or trash racks of power stations, see Fig. 10.14). In a number of lakes and reservoirs, it provided good food for different water birds and may have been instrumental, e.g., for the expansion of the tufted duck (*Ahytia fuligula*) from Siberia and Northern Europe southwards to Central and Western Europe. Another species, closely related and also stemming from the Black Sea region, the quagga mussel (*Dreissena rostriformis bugensis*) has reached Switzerland just a few years ago and is expanding rapidly. On a completely different level, disease agents like e.g. smallpox were introduced by Europeans to newly colonised lands, where such diseases killed a large part of

Fig. 10.14 Zebra mussels (*Dreissena polymorpha*) on the trashrack of an HPP. Zigulievskaya HPP on Volga River, central Russia

the native population which did not have any resistance against them (Crosby 1986; Crawford 1992: 87). One rather recent event of an accidental introduction is that of the brown tree snake (*Boiga irregularis*), a species from Australia and New Guinea, which was inadvertently brought to Guam island in 1950, where in absence of any natural enemies it proliferated enormously, to the point where it probably has already brought to extinction several species of birds endemic to Guam, and is threatening other ones (Mehrtens 1987: 213; Rodda et al. 1992). None of the various attempts to get rid of it again has been successful so far.

- **Intentional introduction:** here as well, there is a large list of species that could be mentioned. Intentional introductions have been done mostly for one of three reasons: as game animals, as countermeasure against pests, or just because somebody liked to see a species, which was possibly familiar from the original home of the person, in the new home. The first category comprises, among many other species, bringing the European rabbit (*Oryctolagus cuniculus*) to Australia, the red deer (*Cervus elaphus*), along with some other attractive game animals, to New Zealand, the nilgai antelope (*Boselaphus tragocamelus*) to Texas, or different game birds to Hawaii (Myers 1970; Lewin 1971; Sheffield et al. 1971). In the same category are a number of fish species, which have been introduced either for sport fishing (here the probably most prominent species, i.e. the most widely introduced, is the North American rainbow trout (*Oncorhynchus mykiss*), which was brought to Europe, South America, South Africa and other places. It is known to be a threat to a number of autochthonous species, e.g. in Chile, where local species like the Chilean trout or pejerrey (*Percichthys trucha*) cannot compete with it, the problem being predation by the exotic species as well as competition for food (Vila et al 1978; Soto and Campos 1996). One of the worst cases of introduction of a predator, and, when considering just the distance, not even of a really alien species, was the introduction of the Nile perch (*Lates niloticus*) into Lake Victoria. The Nile perch has a rather wide natural distribution (Nile, Niger, Volta, Lake Chad and Congo basins; Durand et Lévêque 1980: 745), but in the Nile it had never been able to reach the part of the river upstream of Lake Albert; Murchison Falls (see Sect. 26.1.1) and Owen Falls, the outlet of Lake Victoria, had been unsurmountable obstacles, until the time where people lent a hand. Lake Victoria is, geologically speaking, of relatively recent origin, and had been isolated from the downstream part of the Nile ever since. This isolation had let to rapid evolution of the limited fish fauna there; the overwhelming part of fish in Lake Victoria, as in some other East African lakes, are species belonging to the Cichlid family, which have evolved very rapidly to colonise all the different types of habitat available in this large lake. The result was a very elevated number of closely related species, many of which have already become extinct due to the large predator, the Nile perch, which was released in the lake and developed there very well (IUCN 1986; Worthington 1989; Worthington and Lowe-McConnell 1994). Likewise, a small pelagic fish from Lake Tanganyika, the Tanganyikan anchoveta (*Limnothrissa miodon*), was introduced to Lake Kivu and Lake Kariba, as a measure for increasing fisheries in the lake as well as in the large reservoir (Balon 1974: 85; Dumont 1986). In Lake Kivu, this species is suspected to contribute to the accumulation of CO_2 and methane in the lake since it diminishes zooplankton, which leads to more planktonic algae to sink to the bottom of the lake, where they are being broken down under anaerobic conditions (Schmid 2006). Species introduced in the hope that they would fight pests have been introduced rather often, not always with satisfying results. Milne (1965: 43ff.) tells the story of fighting rats in the sugar cane plantations of Jamaica (of which at least 3 of the 4 species living there had been accidentally introduced by Europeans), by introducing a number of other species, with more or less success, but always to the detriment of the local fauna. The same has happened e.g. with the European fox (*Vulpes vulpes*) brought to Australia for fighting the rabbits; as it turned out, local marsupials were a much easier prey. Another such species is the cane toad (*Rhinella marina*), a large South American toad which had been brought—unsuccessfully—to Jamaica to fight rats, and later on to Australia in the hope that it would reduce other vermin; here as well, it turned on locally available prey, and in addition to that it is a threat for the (mostly rather small) Australian predators who try to feed on it, since the toxin produced in the toads' skin can be lethal for them; the toad is one of the main threats for the increasingly rare quoll (*Dasyurus maculatus gracilis*), the largest surviving Australian predator (Uzqueda et al. 2020). More successful was the introduction of a beetle, *Cyrtobagus salviniae*, with the aim of fighting the exotic aquatic weed *Salvinia molesta* (see Sect. 16.3), which did not eradicate the weed, but at least resulted in a reduction of its frequency in a number of places (Room and Thomas 1985; Thomas and Room 1986; Julin et al. 1987; Room 1988). To close this short enumeration of intentional introductions: animals brought just because they are nice and were familiar in the old country include quite a number of song birds which were brought to the United States, among them the European starling (*Sturnus vulgaris*).

- **Feral domestic animals**: in many parts of the world, escaped domestic animals, mostly brought from Europe, have founded feral populations, in most cases again to the detriment of the local fauna and flora. The three probably worst species in this respect are pigs, cats and goats. Again, the most damage by these species was done on islands, e.g. the Galapagos, where numerous local species were threatened by them, e.g. rats as well as pigs preying on eggs of the Galapagos giant tortoise, or the feral cats in Madagascar, which might have been brought there by Arab sailors as long ago as 1000 years, and had developed into the largest predator on this island (Johnson 1975; Fowler and Johnson 1985; Gade 1985; Baber and Coblentz 1986; Coblentz and Baber 1987; Stone and Loope 1987; Sokol 2020). Robinson Crusoe just gladly accepted the fact than on his island there were goats, without wondering about how they might have gotten there, but they were undoubtedly proof that he was not the first European ever to have reached this place. Other examples of feral domestic animals are the camels in Australia and horses in the USA and Australia.
- **Fur farms**: a number of fur farms have contributed to the list of exotic and partly invasive animals, especially in Europe, among them the muskrat (*Ondatra zibethicus*), originally from Canada, and the South American coypu (*Myocastor coypus*); both of these largely aquatic rodents cause damage in agriculture and,

through their digging, on flood protection dykes (Baumann 1949: 233; Wendelspiess 1986; Graf and Fischer 2021: 382ff.). The raccoon (*Procyon lotor*) from North America and the East Asian raccoon dog (*Nyctereutes procyonides*) are other introductions from the same source.

- **Pets**: quite a few invasive species were originally brought in as pets, and they either escaped or were released, especially when they got too big. This is certainly the way how the Burmese python (*Python bivittatus*) got to the Everglades in Florida, where it found very good conditions and is now a major threat to a large variety of indigenous species. The rose-ringed parakeet (*Psittacula krameri*), stemming from Africa and India, has firmly established populations in cities in Western Europe and elsewhere. Young specimens of the red-eared slider (*Trachemys scripta elegans*), a semi-aquatic turtle from the South-West of the United Sates, are very attractive as babies, with a carapace diameter maybe between 5 and 10 cm, the attractive green and black pattern and the bright red patch at the sides of their heads, but they grow rapidly to more than twice that size, which then often makes them too big for the cage where they were kept, and as adults they get more to an unattractive dirty grey; this may have motivated many of their keepers to release them, and they can be found now in many places in Europe. In China, where they may have been introduced for food since they are easy to breed, they now threaten a number of local species (Jun et al. 2020). Numerous fish species have escaped—or were released—from aquariums as well and have survived where they found suitable conditions (Kushlan 1986). A very special kind of "pets" are the hippos (*Hippopotamus amphibius*) which were kept in the private zoo of drug lord Escobar in Colombia, and which after his death finally escaped and obviously found very good conditions in nearby rivers where they are multiplying. Interestingly, there is a discussion now about their role in nature. Why some see them as a threat to people and to biodiversity, others are of the opinion that they are taking the place in the ecosystem which had been empty since the time people, after immigrating to south America about 10,000–14,000 years ago, wiped out all large herbivores (Lundgren et al. 2020). In spite of their size, hippos in their native Africa are rather good at surviving even under hunting pressure, and once well established in Colombia it might be rather difficult to get rid of them again. There is no information as of yet that they have colonised any reservoir, but it could well be that some reservoirs might turn out to be suitable habitats for them.

For those who would like to get some additional information on invasive species and on the problems related to attempts of getting rid of them while at the same time reading an entertaining novel, Boyle (2012) can be recommended.

In the context of hydropower plants and ESIAs for such projects, there are certainly two groups of invasive species to be considered mostly. One is the exotic floating plants which have spread all over the topics and subtropics and can cause major problems on reservoirs (see Sect. 16.3). The other is fish, which have often been introduced into reservoirs in an attempt—successful at least in a number of cases—to increase fisheries for the benefit of the local population. Even if the outcome here is not necessarily bad in any case, there is enough experience that there is always a risk, and as good as always such introductions went to the detriment of local species. Before recommending the introduction of an exotic species as a mitigation measure, the possible consequences should be examined very carefully, and in most cases it will be preferable not to recommend such an action.

10.6 Protected Areas

Impacts of a project on protected areas have to be evaluated very carefully, mainly for the following reasons:

- Protected areas have a legal protection status, and interference with them in any way that is not in compliance with this legal protection therefore has to be considered as an illegal activity.
- Protected areas have been declared as such because they contain some exceptional values (most often in terms of biodiversity, but occasionally also because of scenic, cultural or historical value) which need to be protected. Impacts of a project might contradict the main protection aims of such an area, e.g. by interfering with habitat conditions on which the biodiversity of this area depends, or by affecting archaeological objects.

For both reasons, potential project impacts need to be identified, and mitigation measures need to be taken in case of adverse impacts.

Project impacts on protected areas can be manifold, as the following examples try to show:

- Occupation of land: e.g. for a construction site
- Disturbance of habitats and wildlife by the presence of people: e.g. poaching and other illegal activities within a PA by the work force.
- Habitat destruction by submersion when creating a reservoir:
- Change in hydraulic conditions of wetlands (floodplains, swamps, etc.) by change in river discharge pattern downstream of a dam. Examples: tugai in Tajikistan (Rogun; see Sect. 18.2); Kafue flats (Itehzi-Tehzi; see Sect. 9.1.1.2); Sanaga (Lom Pangar), Rufiji, etc.
- Opening the area by means of access roads: often the most serious impact of an HPP (Chap. 15).

As a general rule it can be said that any occupation of land within a protected area is certainly to be considered as a negative impact which requires mitigation. However, not every area is of the same value or importance. It is of high relevance to carefully check where, in relation to the entire protected area, an occupied surface would be located, and what part of the reserve would actually be affected. Usually, an occupation of land close to the border of the protected area would be less problematic than a project site closer to the centre (see to this the case studies in Sect. 26.1).

While identification of any protected areas in the vicinity of a project, and the assessment of potential impacts on such areas is very important, mainly for addressing issues related to biodiversity conservation, this must by no means lead to the conclusion that such an assessment could be limited to protected areas. While we can assume that areas designed for protecting nature are of importance for biodiversity conservation, we must not assume that land outside protected areas is of no value for biodiversity and therefore does not have to be included in the investigation. This error was made, e.g. in the preparatory stages of an investigation of different highway options in Switzerland (Kommission Biel 1981), where in the initial list of criteria to be applied impact on nature was proposed to be evaluated and quantified by the surface of protected areas the project would occupy. While it is certainly true that in landscapes very extensively used for agriculture biodiversity value of land outside protected areas is often really very limited, this does not necessarily apply to all unprotected land. The work to be done for the topics outlined in Sects. 10.1–10.3 has by no means to be limited to protected areas, and the "critical habitat" approach e.g. in the IFC safeguards covers habitats based on their condition, and not on any status of legal protection.

References

Albert, S., O. Flores, and D. Strasberg. 2020. Collapse of dispersal trait diversity across long-term chronosequence reveals a strong negative impact of frugivore extinctions on forest resilinece. *Ecology* 108 (4): 1386–1397. https://doi.org/10.1111/1365-2745.13359.

Alonso-Lormand, M.J., et al. 1995. Rehabilitation of health services and the problem of Schistosoma mekongi infection in Kratié region, Northern Cambodia. Von der Impfung zur Immunologie. Schweiz. Ges. für Tropenmedizin und Parasitologie, Sarnen, 14, 9.-11.11.95 (Abstract).

Antwi, L.A.K., and P.K. Ofori-Dansin. 1993. Limnology of a tropical reservoir (the Kpong Reservoir in Ghana). *Tropical Ecology* 34 (1): 75–87.

Asibey, E.O.A. 1977. The blackfly dilemma. *Environmental Conservation* 4 (4): 291–295.

Baber, D.W., and B.E. Coblentz. 1986. Density, home range, habitat use and reproduction in feral pigs on Santa Catalina Island. *Journal of Mammalogy* 67 (3): 512–525.

Baijot, E., J. Moreau, and S. Bouda. 1994. *Aspects hydrobiologiques et piscicoles des retenues d'eau en zone Soudano-Sahélienne*. CTA: Wageningen.

Balon, E.K. 1974. *Fishes of Lake Kariba*, 144 pp. Neptune City: Africa. TFH Publications.

Barlow, N.L., C.P. Kirol, K.E. Doherty, and B.C. Fedy. 2020. Evaluation of the umbrella species concept at fine spatial scales. *Journal of Wildlife Management* 84 (2): 237–248.

Baumann, F. 1949. *Die freilebenden Säugetiere der Schweiz*., 492 S. Bern: Huber.

Baumann, P. 2013. *Fish biomass and migration study 2012–2013. Xayaburi hydroelectric power project*. Xayaburi Power Company Limited. Vientiane, Laos PDR.

Belletti, B., C. Garcia de Leaniz, J. Jones, et al. 2020. More than one million barriers fragment Eurpe's rivers. *Nature* 588 (7838): 436–441. https://doi.org/10.1038/s41686-020-3005-2.

Belskky, A.J. 1986. Population and community processes in a mosaic grassland in the Serengeti Tanzania. *Journal of Ecology* 74 (3): 841–856.

Benecke, N. 1994. *Der Mensch und seine Haustiere: Die Geschichte einer jahrtausendealten Beziehung*, 470 pp. Stuttgart: Theiss.

Bercovitch, F.B., P.S. Berry, A. Dagg, et al. 2017. How many species of giraffe are there? *Current Biology* 27: R137–R138.

Berner, P., and W. Geiger. 1988. *Fischereigutachten Neubau Kraftwerk Rheinfelden*. KWR, EWI, 106 pp.

Best, R.C., and V.M.F. da Silva. 1993. Inia Geoffrensis. *Mammalian Species* 426: 1–8.

Bhatt, S., et al. 2015. The effect of malaria control in *Plasmodium falciparum* in Africa between 2000 and 2015. *Nature* 526 (7572): 207–211.

Biays, S., A.H.R. Stich, P. Odermatt, L. Chan, C. Yersin, M. Chan, S. Caem, and J.D. Lormand. 1999. Foyer de bilharziose à Schistosoma mekongi redécouvert au nord du Cambodge: I. perception culturelle de la maladie; description et suivi de 20 cas cliniques graves. *Tropical Medicine and International Health* 4 (10): 662–673.

Bigirinama, R.N., J.A. Ntaongo, N.A. Batumbo, et al. 2020. Environmental and anthropogenic factors associated with increased malaria incidence in South-Kivu province, Democratic Republic of the Congo. *Tropical Medicine and International Health* 25 (5): 600–611. https://doi.org/10.1111/tmi.13379.

Bino, G., R.T. Kingsford, M. Archer, et al. 2019. The platypus: Evolutionary history, biology and an uncertain future. *Journal of Mammalogy* 100 (2): 308–327. https://doi.org/10.1093/jmammal/gyz058.

Blake, B.F. 1976. Lake Kainji, Nigeria summary of the changes within the fish population since the impoundment of the Niger in 1968. *Hydrobiologia* 53 (2): 131–137.

Blake, B.F. 1977. The effect of the impoundment of Lake Kainji, Nigeria, on the indigenous species of mormyrid fishes. *Freshwater Biology* 7: 37–42.

Boller, L., and W. Dönni. 2011. *Potentialabschätzungen und Massnahmen für die Rückkehr des Lachses in die Schweiz.* Studie im Auftrag des WWF Schweiz, 33 pp.

Boyle, T.C. 2012. *When the killing's done*, 384 pp. New York: Random House.

Braatz, S. 1992. *Conserving biological diversity. A strategy for protected areas in the Asia-Pacific Region.* World Bank Technical Paper No. 193: 66 pp. Washington: The World Bank.

Brainard, J. 2018. Dams may threaten endangered apes. *Science* 361 (6398): 112.

Brehm, A. 1911. *Brehms Tierleben, siebenter Band. Die Vögel, zweiter Band. 4. Auflage, bearbeitet durch O. von Strassen*, 492 S. Leipzig und Wien: Bibliographisches Institut.

Brown, M.B., D.T. Bolger, and J. Fennessy. 2019. All the eggs in one basket: A countrywide assessment of current and historical giraffe population distribution in Uganda. *Global Ecology and Conservation.* https://doi.org/10.1016/j.gecco.2019.e00612.

Brownell, R.L., Jr., R.R. Reeves, P.O. Thomas, B.D. Smith, and G.E. Ryan. 2017. Dams threaten rare Mekong dolphins. *Science* 355 (6327): 805 (letter).

Büttiker, B., and W. Meier. 1983. *Mästung von Regenbognforellen, Karpfen und Aalen mit erwärmtem Aarewasser des Kernkraftwerks Beznau.* Bern: Bundesamt für Umweltschutz, Bundesamt für Veterinärwesen, 21 pp.

Chape, S. 1996. *Biodiversity conservation, protected areas and the development imperative in Lao PDR: Forging the links.* IUCN, Vientiane, Lao PDR. Discussion Paper 1, 61 pp.

Cheek, M., X. van der Burgt, J. Momoh, and A. Lebbie. 2017. *Ledermanniella yiben* sp. nov. (Podostemaceae), Critically Endangered at the proposed Yiben Reservoir, Sierra Leone. *Kew Bulletin* 72 (31): 7. https://doi.org/10.1007/S12225-017-9699-0

Chen, C., and K.S. Pfennig. 2020. Female toads engaging in adaptive hybridization prefer high-quality heterospecifics as mates. *Science* 367 (6484): 1377–1379. https://doi.org/10.1126/science.aaz5109.

Clark, C.W. 1973. Profit maximisation and the extinction of animal species. *Journal of Political Economy* 81 (4): 950–961. http://www.zo.utexas.edu/courses/THOC/Readings/Clark_Extinction1973.pdf.

Clavero, M., and V. Hermoso. 2015. Historical data to plan the recovery of the European eel. *Journal of Applied Ecology* 52 (4): 960–968.

Coblentz, B.E., and D.W. Baber. 1987. Biology and control of feral pigs on Isla Santiago, Galapagos Ecuador. *Journal of Applied Ecology* 24 (2): 403–418.

Crawford, M.H. 1992. *Antropología biológica de los Indios Americanos*, 314 pp. Madrid: Colección Indios de América. Editorial MAPFRE S.A.

Crosby, A.W. 1986. *Ecological imperialism. The biological expansion of Europe, 900–1900.* Cambridge: Cambridge University Press (Studies in environment and history), 368 pp.

Curtis, C.F., C.A. Maxwell, R.J. Finch, and K.J. Njuwna. 1998. A comparison of use of a pyrethroid either for house spraying of for bednet treatment against malaria vectors. *Tropical Medicine and International Helath* 3 (8): 619–631.

Darwin, C. 1859. *On the origin of species by means of natural selection*. Reprint 2006, Dover Thrift Editions, 321 pp.

Datry, T., N. Bonada, and A. Boulton. 2017. *Intermittent rivers and ephemeral streams. Ecology and management*, 597 pp. London: Academic Press, Elsevier.

Dawkins, R., and Y. Wong. 2016. *The Ancestor's Tale. A pilgrimage to the dawn of life*, updated ed. London: Weidenfeld and Nicolson, 771 pp.

Deinet, S., K. Scott-Gatty, H. Rotton, et al. 2020. *The living planet index (LPI) for mirgratory freshwater fish*. Technical Report. The Netherlands: World fish Migration Foundation. https://worldfishmigrationfoundation.com/wp-content/uploads/2020/07/LPI_report_2020.pdf.

Deutscher, G. 2006. *The unfolding of language. The evolution of mankind's greatest invention*, 360 pp. UK: Arrow Books.

Díaz, G., P. Arriagada, K. Górski, O. Link, B. Karelovic, J. Gonzalez, and E. Habit. 2019. Fragmentation of Chilean Andean rivers: Expected effects of hydropower development. *Revista Chilena de Historia Natural* 92 (1). https://doi.org/10.1186/s40693-019-0081-5.

Dioli, M. 2020. Dromedary (*Camelus dromedarius*) and Bactrian camel (*Camelus bactrianus*) crossbreeding husbandry practices in Turkey and Kazakhstan: An in-depth review. *Pastoralism: Research, Policy and Practice* 10 (6), 20. https://doi.org/10.1186/s13570-020-0159-3.

Dos Santos, G.M., B. de Mérona, A.A. Juras, and M. Jégu. 2004. *Peixes de baixo río Tocantins: 20 anos depois da Usina Hidrelética Tucuruí. Eletronorte*, 215 pp, Brasilia.

Dumont, H.J. 1986. The Tanganyika sardine in Lake Kivu: Another ecodisaster for Africa? *Environm. Conserv.* 13 (2): 143–148.

Durand, J.R., and C. Lévèque. (eds.). 1980. *Flore er faune aquatiques de l'Afrique sahélo-soudanienne*, vol. 1: 389 pp., vol 2: 873 pp. Paris: ORSTOM.

DWA. 2014. *Fischaufstiegsanlagen und fischpassierbare Bauwerke - Gestaltung, Bemessung, Qualitätssicherung. Merkblatt DWA-M 509*. Deutschland: Deutsche Vereinigung für Wasserwirtschaft, Abwasser und Abfall e. V., 334 pp. ISBN 978-3-942964-91-3.

Dye, C., and R.H.A. Baker. 1986. Measuring the capacity of blackflies as vectors of Onchocerciasis in southwest Sudan. *Journal of Applied Ecology* 23 (3): 883–893.

Ebomoyi, E.W., and M.K. Omonisi. 1991. Human ecology and behavior in Onchocerciasis control in isolated villages of Kwara State, Nigeria. *Ambio* 20 (1): 43–47.

Ehrenfeld, D. 1988. Why put a value on biodiversity? In *1988: Biodiversity*, ed. Wilson, E.O., 212–216. Washington D.C.: National Academy Press.

El Gaddal, A.A. 1985. The Blue Nile Health Projeca comprehensive approach to the prevention and control of water-associated diseases in irrigated schemes of the Sudan. *Journal of Tropical Medicine and Hygiene* 88: 47–56.

EWI. 1981. *Ayago-Nile Hydroelectric Project*. Uganda: Ecology. Electrowatt Engineering, 75 pp.

EWI. 1987. *Saisonspeicherwerk Curciusa: Bericht zur Umweltverträglichkeitsprüfung (UVP)*. Zürich, für Misoxer Kraftwerke: Elektrowatt Ingeinieurunternehmung AG.

EWI. 1989. Aménagement hydroagricole de 35 barrages: Santé. Report prepared by Electrowatt Engineering for Ministère de l'Eau, Burkina Faso, 45 pp.

Fennessy, J., T. Bidon, F. Reuss, V. Kumar, P. Elkan, M. Vamberger, U. Fritz, and A. Janke. 2016. Multi-locus analyses reveal four giraffe species instead of one. *Current Biology* 26: 2543–2549. https://doi.org/10.1016/j.cub.2016.07.036.

Fowler de Neira, L.E., and M.K. Johnson. 1985. Diets of giant tortoises and feral burros on Volcan Alcedo, Galapagos. *Journal of Wildlife Management* 49 (1): 165–169.

Fraser, B. 2018. Dams nudge Amazon's ecosystems off-kilter. *Science* 359 (6375): 508–509.

Gade, D.W. 1985. Man and nature on Rodriguez: Tragedy of an island common. *Environmental Conservation* 12 (3): 207–216.

Ghebreyesus, T.A., M. Haile, K. Witten, et al. 1999. Incidence of malaria among children living near dams in northern Ethiopia: Community based incidence survey. *British Medical Journal* 319: 663–666.

Gouskov, A., M. Reyes, L. Wirthner-Bitterlin, and C. Vorburger. 2016. Fish population genetic structure shaped by hydroelectric power plants in the upper Rhine catchment. *Evolutionary Applications* 9 (2): 394–408. https://doi.org/10.1111/eva.12339.

Graf, R.F., and C. Fischer (Hrsg.) 2021. *Atlas der Säugetiere Schweiz und Liechtenstein. Schweiz. Ges. für Wildtierbiologie (SGW),* 488 pp. Bern: Haupt Verlag.

Grimes, B.F. (ed.). 1996. *Ethnologue. Languages of the world*, 13th ed. Dallas: Summer Institute of Linguistics, Inc, 966 pp.

Guderian, R.H., et al. 1997. Successfull control of onchocerciasis with community-based ivermectin distribution in the Rio Santiago focus in Ecuador. *Tropical Medicine and International Health* 2 (10): 982–988.

Guyatt, H.L., S.K. Corlet, T.P. Robinson, S.A. Ochola, and R.W. Snow. 2002. Malaria prevention in highland Kenya: indoor residual house-spraying vs. insecticide-treated bednets. *Tropical Medicine and International Health* 7 (4): 298–303.

Guyatt, H.L., S.A. Ochola, and R.W. Snow. 2002. Too poor to pay: Charging for insecticide-treated bednets in highland Kenya. *Tropical Medicine and International Health* 7 (10): 846–850.

Herren, B., and E. Reist. 2020. Brut eines Neuntöters *Lanius collurio* mit einem Rotkopfwürger *L. senator* im Kanton Bern. (Brood of a red-backed shrike *Lanius collurio* with a woodchat shrike *L. senator* in the canton of Berne.) *Ornithologischer Beobachter* 117 (1): 66–70.

Hirsch, P., and O. Hensengerth. 2019. *Review of design changes madefor the Xayaburi hydropower project*. Oakland, USA: International Rivers Network, 27 pp. https://www.internationalrivers.org/wp-content/uploads/sites/86/2020/08/xayaburi-redesign-commentary-for-website.pdf.

Holm-Nielsen, L.B., I.C. Nielsen, and H. Balslev (eds.). 1989. *Tropical forests. Botanical dynamics, speciation and diversity*. London: Academic Press. *Proceedings of Symposium on Tropical Forests: Botanical dynamics, speciation and diversity*, Aarhus University, 8–10 August 1988, 380 pp.

Hunter, J.M., L. Rey, K.Y. Chu, E.O. Adekolu-John, and K.E. Mott. 1993. *Parasitic diseases in water resources development. The need for intersectoral negotiation.* Geneva: World Health Organization, 152 pp.

Huntley, B.J. 1988. Conserving and monitoring biotic diversity. In *Biodiversity*, ed. E.O. Wilson, 248–260. Washington D.C.: National Academy Press.

Huntley, B.J., V. Russo, F. Lages, and N. Ferrand (eds.). 2019. *Biodiversity of Angola. Science & conservation: a modern synthesis*. SpringerOpen, 549 pp. https://doi.org/10.1007/978-3-030-030 83-4.

ISL, Oréade-Brèche, SOGREAH. 2005. *Etude environnementale du barrage de Lom Pangar*. Rapport de Sythèse. République du Cameroun, Ministère de l'Energie et de l'Eau.

IUCN. 1978. In the wake of the snail darter. *IUCN Bulletin, N.S.*, 9 (7/8): 33–34.

IUCN. 1979. Tellico goes ahead—snail darter loses last battle. *IUCN Bulletin, N.S.* 10 (9): 82

IUCN. 1986. Big fish eat out Lake Victoria. *IUCN Bulletin* 17 (1–3): 8.

IUCN. 1998. Invaders from planet earth. Doble issue on invasive species. *World Conservation* 28 (4), and 29 (1): 1–63

Jackson, P.B.N., and K.H. Rogers. 1976. Cabora Bassa fish populations before and during the first filling phase. *Zoologica Africana* 11 (2): 373–397.

Jalili, A., and Z. Jamzad (eds.). 1999. Red Data Book of Iran. Research Inst. of Forests and Rangelands, Tehran, Iran. Publ. No. 1999–215, 748 pp.

Jamison, D.T., W.H. Mosley, A.R. Measham, and J.L. Bobadilla (eds.). 1993. *Disease control priorities in developing countries*. New York: Oxford University Press, 746 pp.

Johnson, D.L. 1975. New evidence on the origin of the fox, Urocyon littoralis clementae, and feral goats on San Clemente Island, California. *Jorunal of Mammalogy* 56 (4): 925–928.

Jordan, P., J.D. Christie, and G.O. Unrau. 1980. Schistosomiasis transmission with particular reference to possible ecological and biological methods of control. *A Review. Acta Tropica* 37: 95–135.

Julin, H.M., A.S. Bourne, and R.R. Chan. 1987. Effects of adult and larval Cyrtobagus salviniae on the floating weed Salvinia molesta. *Journal of Applied Ecology* 24 (3): 935–944.

Jun, Wu, et al. 2020. China's wild turtles at risk of extinction. *Letter to Science* 368 (6493): 838.

Kapetsky, J.M., and T. Petr, eds. 1984. *Status of African reservoir fisheries. CIFA Technical Paper No. 10.* Rome: FAO.

Kibret, S., J. Lautze, E. Boelee, and M. McCartney. 2012. How does an Ethiopian dam increase malaria? Entomological determinants around the Koka reservoir. *Tropical Medicine and International Health* 17 (11): 1320–1328.

Kipping, J., V. Clausnitzer, E. Fernandes, and K.D.B. Dijkstra. 2019. The dragonflies and damselflies of Angola: An updates synthesis. In *Biodiversity of Angola. Science & conservation: A modern synthesis*, ed. by Huntley, B.J., V. Russo, F. Lages, and N. Ferrand. SpringerOpen, 549 pp. https://doi.org/10.1007/978-3-030-03083-4

Kita, Y., S. Koi, R. Rutishauser, and M. Kato, 2008. A new sepcies of Ledermanniella (Podostemaceae) from Cameroon. *Acta Phytotaxonomica et Geobotanica* 59 (3): 223–227. https://www.researchgate.net/publication/280811086

Knaus, P., S. Antoninazza, S. Wechlser, J. Guélat, M. Kéry, N. Strebel, and T. Sattler. 2018. *Schweizer Brutvogelatlas 2013–2016*. Vogelwarte Sempach, 648 p

Kommission Biel. 1981. *Schlussbericht der Kommission zur Überprüfung von Nationalstrassenstrecken (NUP).* Switzerland: EDMZ Bern, 288p.

Kushlan, J.A. 1986. Exotic fishes of the Everglades: A reconsideration of proven impact. *Environmental Conservation* 13 (1): 67–69.

Laurance, W. 2018. If you can't build well, then build nothing at all. *Nature* 563 (7731): 295.

Lautze, J., M. McCartney, P. Kirshen, D. Olana, G. Jayasinghe, and A. Spielmann. 2007. Effect of a large dam on malaria risk: The Koka reservoir in Ethiopia. *Tropical Medicine & International Health* 12 (8): 982–989.

Leandro, A.S., J.A. Rios, A.S. Britto, et al. 2020. Malathion resistance in Aedes aegypti: Laboratory conditions and in situ experimental approach through adult entomological surveillance. *Tropical Medicine and International Health* 25 (10): 1271–1282. https://doi.org/10.1111/tmi.13474.

Lee, D.E., J.T. Fennessy, and M.L. Bond. 2013. Standig tall for giraffes. Research and conservation of an overlooked African icon. *The Wildlife Professional* 7(2), 35–39.

Lelek, A. 1973. Sequence of changes in fish populations of the new tropical man-made lake, Kainji, Nigeria, West Africa. *Archives of Hydrobiology* 71 (3): 381–420.

Lelek, A., and S. El-Zarka. 1973. Ecological comparison of the preimpoundment and postimpoundment fish faunas of the River Niger and Kainji. In *Man-made lakes: Their problems and environmental effects*, vol. 17, ed. by Ackermann, W.C.G., F. White, E.B. Worthington, and J.L. Ivens, 655–660. Geophysical Monographs.

Levin, D., N. Raab, Y. Pinto, et al. 2021. Diversity and functinal landscapes in the microbiota of animals in the wild. *Science* 372 (6539). https://doi.org/10.1126/science.abb5352.

Lewin, V. 1971. Exotic game birds of the Puu Waawaa Ranch, Hawaii. *Journal of Wildlife Management* 35 (1): 141–155.

Lewis, D.S.C. 1974. The effect of the formation of Lake Kainji (Nigeria) upon the indigenous fish population. *Hydrobiologia* 45 (2–3): 281–301.

Lewis, W.M. Jr. 1987. Tropical limnology. *Annual Review of Ecology, Evolution, and Systematics* 18: 159–184. https://www.annualreviews.org/doi/pdf/10.1146/annurev.es.18.110187.001111.

Li, S., W.J. McShea, D. Wang, X. Gu, X. Zhang, et al. 2020. Retreat of large carnivores across the giant panda distribution range. *Nature Ecology & Evolution*. https://doi.org/10.1038/s41559-020-1260-0.

Lundgren, E.J., D. Ramp, J. Rowan, et al. 2020. Introduced herbivores restore Late Pleistocene ecological functions. *PNAS*. https://doi.org/10.1073/pnas.1915769117.

Makenga Bof, J.C., V. Maketa, D.K. Bakajika, et al. 2015. Onchocerciasis control in the democratic Republic of Congo (DRC): Challenges in a post-war environment. *Tropical Medicine and International Health* 20 (1): 48–62.

R. Mather, S. Kritsavanarin, G. Pangare, and D. Weerapong. 2009. *Siphandone: The Mekong under threat.* Mekong Region Water Dialogues Series Publication No. 1, IUCN, Gland, 180 pp.

Maxmen, A. 2018. The enemy in waiting. *Nature* 559 (7715): 458–465.

McCullough, F.S. 1981. Biological control of the snail intermediate hosts of human Schistosoma spp.: a review of its present status and future prospects. *Acta Tropica* 38 (1): 5–13.
McMahon, J.P., R.B. Highton, and T.F. Marshall. 1977. Studies on biological control of intermediate hosts of schistosomiasis in western Kenya. *Environmental Conservation* 4 (4): 285–289.
McNeely, J.A., K.R. Miller, W.V. Reid, R.A. Mittermeier, and T.B. Werner. 1990. *Conserving the World's Biological Diversity*, 191 pp. Gland, Switzerland: IUCN.
Mehrtens, J.M. 1987. *Living snakes of the world*, 480 pp. New York: Sterling.
Methorst, J., K. Rehdanz, T. Mueller, B. Hansjürgens, A. Bonn, and K. Böhning-Gaese. 2021. The importance of species diversity for human well-being in Europe. *Ecological Economics*. https://doi.org/10.1016/j.ecolecon.2020.106917.
Milne, L.J. and M. Bopp. 1965. *Das Gleichgewicht in der Natur*. Parey, Hamburg udn Berlin, 288 p. (Original: the balance of nature. Knopf, New York, 1960).
Mol, J.H., B. de Mérona, P.E. Ouboter, and S. Sahdew. 2007. The fish fauna of Brokopondo reservoir, Suriname, during 40 years of impoundment. *Neotropical Ichthyology* 5 (4). https://doi.org/10.1590/S1679-62252007000300015
Montén, E. 1985. *Fish and turbines: Fish injuries during passage through power station turbines*. Vattenfall.
Morier-Genoud, G., M. Raeder, T.S. Coe, and T. Poomchaivej. 2016. Upstream fish passage facilities at Xayaburi HPP. An exlanation of the design concept. In *Paper presented at the Lower Mekong Fish Passage Conference, Vientiane*, November 2016.
MRC, 2019. *Review of the design changes made for Xayaburi hydropower project*. Technical Reference Paper No. 65, 56p. Vientiane: MRC, Lao PDR. https://www.mrcmekong.org/assets/Publications/Review-of-the-design-change-made-for-Xayaburi-hydropower-project_technical-ref-paper_2019.pdf.
Myers, K. 1970. The rabbit in Australia. Proceedings of Advanced Study Institute on Dynamics Numbers in Populations, pp. 478–506.
NaFIRRI. 2010. *Aquatic Ecology Monitoring: Water quality, fish, fish catch, sanitation and disease vectors. Monitoring Report No. 7*. National Fisheries Resources Research Institute (NaFIRRI). Report prepared for Bujagali Energy Limited (BEL), Jinja, Uganda.
Nájera, J.A., B.H. Liese, and J. Hammer. 1993. Malaria. In *Disease control priorities in developing countries*, ed. D.T. Jamison, W.H. Mosley, A.R. Measham, and J.L. Bobadilla, 281–302. New York: Oxford University Press.
Nater, A., M.P. Mattle-Greminger, A. Nurchyo, et al. 2017. Morphometric behavioral, and genomic evidence for a new Orangutan species. *Current Biology* 27: 3487–3498. https://doi.org/10.1016/j.cub.2017.09.047.
NCG. 2015. *Environment Impact Assessment for Pak Beng Hydropower Project*. Vientiane: National Consulting Group.
NCG. 2018. *Paklay Hydropower Project, Environmental Impact Asssessment (EIA)*. Report prepared by National Consulting Group, Vientiane: Powerchina Resources Ltd.
Nogué, S., A.M.C. Santos, H.J.B. Birks, et al. 2021. The human dimension of biodiversity changes on islands. *Science* 372 (6541): 488–491. https://doi.org/10.1126/science.abd6706.
Oomen, J.M.V. 1983. Rapport de la mission de courte durée "Aspects sanitaires du projet 40 petits barrages en Haute Volta". Rapport polycopié, 82 pp. Euroconsult, Arnhem, NL.
Pääbo, S. 2014. Neanderthal man. In *Search of lost genomes*. New York: Basic Books, 275 pp.
Palmer, A., V. Sommer, and J.N. Msindai. 2021. Hybrid apes in the Anthropocene: Burden or asset for conservation? *People and Nature* 3 (2): 573–586. https://doi.org/10.1002/pan3.10214.
Pavlov, D.S. 1989. *Structures assisting the migration of non-salmonid fish: USSR*. FAO Fisheries Technical Paper No. 308, 97pp.
Pedroli, J.-C., B. Zaugg, and A. Kirchhofer. 1991. *Verbreitungsatlas der Fische und Rundmäuler der Schweiz. Atlas de distribution des poissons et des cyclostomes de Suisse*. Documenta Faunistica Helvetiae 11, Neuchâtel: CSCF, 206 pp.
Peters, W., and H.M. Gilles. 1981. *A colour atlas of tropical medicine and parasitology*, 2nd ed. London: Wolfe Medical Atlases, 17:400 pp.

Petr, T. 1974. Distribution, abundance and food of commercial fish in the Black Volta and the Volta man-made lake in Ghana. *Hydrobiologia* 45 (2–3): 303–337.

Piekarski, G. 1987. Medizinische Parasitologie in Tafeln. 3. Aufl. Springer, Berlin, 364 S.

Pilliod, D.S., M.I. Jeffries, R.S. Arkle, and D.H. Olson. 2020. Reptiles under the conservation umbrella of the greater sage grouse. *Journal of Wildlife Management* 84 (3): 478–491. https://doi.org/10.1002/jwmg.21821.

Poland, T.M., T. Patel-Weynand, D.M. Finch, C.F. Miniat, D.C. Hayes, and V.M. Lopez (eds.). 2021. Invasive species in forests and rangelands of the United States. Cham: e-book Springer Nature Switzerland, 455 p. Open Access. https://doi.org/10.1007/978-3-030-45367-1.

Pomeroy, L.R., and J.J. Alberts (eds.). 1988. Concepts of ecosystem ecology. In *Ecological Studies*, vol. 67. New York: Springer, 384 pp.

Poulsen A.F., Ouch Poeu, Sintavong Viravong, Ubolratana Suntornratana, and Nguyen Thanh Tung. 2002. *Fish migrations of the Lower Mekong River Basin: implications for development, planning and environmental management*. MRC Technical Paper No. 8, Mekong River Commission, Phnom Penh. 62 pp. ISSN: 1683-1489

Powerchina, 2017. *Paklay Hydropower Project*. Feasibility Study Report. Report prepared for POWERCHINA.

Pöyry, 2020: Luang Prabang HPP: Environmental impact assessment. Vol. 4.1 of the feasibility study. Report prepared for Luang Prabang Power Company Limited.

Pöyry et Alpha Technology. 2010. Projet Hydroélectrique de Memve'ele. Etude d'Impact Environnemental et Social. Rapport Final EIES. Cellule d'Appui à la Maîtrise d'Ouvrage du Projet d'Aménagement Hydroélectrique de Memve'ele, Cameroun.

Radford-Smith, J., L.P. Shoo, and J.M. Dwyer. 2021. Better left alone: trying to control pasture grasses in untended rainforest plantings incurs multiple costs and delivers few benefits. *Ecological Solutions and Evidence* 2 (2). https://doi.org/10.1002/2688-8319.12062.

Raine, A.F., S. Driskill, M. Vynne, D. Harvey, and K. Pias. 2020. Managing the effects of introduced predators on Hawaiian endangered seabirds. *Journal of Wildlife Management* 84 (3): 425–435. https://doi.org/10.1002/jwmg.21824.

Ramamoorthy, T.P., R. Bye, A. Lot, and J. Fa (eds.). 1992. *Biological diversity of Mexico. Origins and distribution*. Oxford University Press, 812 pp.

Reich, D. 2018. *Who we are and how we got here. Ancient DNA and the new science of the human past*. Oxford, UK: Oxford University Press, 335 p.

Reichenbach-Klinke, H.H. 1980. *Krankheiten und Schädigungen der Fische*, 472 pp. Stuttgart: Gustav Fischer.

Remmert, H. 1987. Sukzessionen im Klimax-System. *Verh. Ges. für Oekologie Bd.* XVI: 27–34.

Rodda, G.H., T.H. Fritts, and P.J. Conry. 1992. Origin and population growth of the Brown Tree Snake, Boiga irregularis, on Guam. *Pacific Science* 46 (1): 46–57. https://scholarspace.manoa.hawaii.edu/bitstream/10125/1672/v46n1-46-57.pdf.

Room, P.M. 1988. Effects of temperature, nutrients and a beetle on branch architecture of the floating weed Salvinia molesta and simulations of biological control. *Journal of Ecology* 76 (3): 826–848.

Room, P.M., and P.A. Thomas. 1985. Nitrogen and establishment of a beetle for biological control of the floating weed Salvinia in Papua New Guinea. *Journal of Applied Ecology* 22 (1): 139–156.

Sánchez-Bayo, F., and K.A.G. Wyckhuys. 2019. Worldwide decline of the entomofauna: A review of its drivers. *Biological Conservation* 223: 8–27. https://doi.org/10.1016/j.biocon.2019.01.020.

Sandvik, H., O. Hilmo, S. Henriksen, et al. 2020. 2020: Alen species in Norway: Results form quantitative ecological impact assessment. *Ecological Solutions and Evidence* 1: e12006. https://doi.org/10.1002/2688-8319.12006.

Savatier, J., M. Lino, S. Erpicum, M. Pirotton, S. Descloux, C. Daux, G. Gwet, and A. Emadak. 2016. Aeration weir at Lom Pangar dam. "Congrès SHF "HydroES 2016"". Grenoble. https://www.researchgate.net/publication/306403250.

Savill, P.S., and J.E.D. Fox. 1967. *Trees of Sierra Leone*. Omagh, Co. Tyrone, 316 pp. (published by authors).

Schmid, M. 2006. Gefährliche Gasmassen in der Tiefe des Kivu-Sees. *Digital: EAWAG News 62d/Dezember* 2006: 24–26

Schmutz, S., and C. Mielach. 2015. *Review of existing research on fish passage through large dams and its applicability to Mekong mainstream dams.* MRC Technical Paper No., 48: Mekong River Commission, Pnom Penh, Cambodia. 149 pp. https://www.mrcmekong.org/assets/Publications/technical/tech-No48-Review-of-Existing-Research-on-Fish-Passage.pdf.

Scudder, T. 2005. *The future of large dams. Dealing with social, environmental, institutional and political costs*, 389 pp. London: Earthscan.

SEATEC. 1997. *Nam Theun 2 Hydroelectric Project. Environmental Assessment and Management Plan (EAMP).* Australia: Draft Final Report. SEATEC International in Association with Sinclair Knight Merz, ECI, Inc., and EDAW.

Sheffield, W.J., E.D. Ables, and B.A. Fall. 1971. Geographical and ecologic distribution of Nilgai antelope in Texas. *Journal of Wildlife Management* 35 (2): 250–257.

Shipman, P. 2015. *The invaders. How humans and their dogs drove Neanderthals to extinction.* Cambridge, Massachusetts, London, England: The Belknap Press of Harvard University Press, 266 p.

Silver, H., and W.T. Silver. 1969. Growth and behavior of the coyote-like canid of Northern New England with observations on canid hybrids. *Wildlife Monographs* 17: 41. https://www.jstor.org/stable/pdf/3830473.

Skovrind, M., J.A. Samaniego Castruita, J. Haile, et al. 2019. Hybridization between two high Arctic cetaceans confirmed by genomic analysis. *Scientific Reports* https://www.nature.com/articles/s41598-019-44038-0.pdf.

Sokol, J. 2020. Madacascar's misterious, murderous cats identified. *Science* 367 (6483): 1178.

Soto, D, and H. Campos. 1996. Los lagos oligotróficos del bosque templado húmedo del sur de Chile. In *1996: Ecología de los bosques nativos de Chile*, ed. J.J. Armesto, C. Villagrán, M.K. Arroyo. Santiago, Chile: Editorial Universitaria, pp. 317–334.

Stich, A.H.R., S. Biays, P. Odermatt, M. Chan, S. Cheam, S. Kiev, S.L. Chuong, P. Legros, et al. 1999. Foci of Schistosoma Mekongi, Northern Cambodia: II. Distribution of infection and morbidity. *Tropical Medicine and International Health* 4 (10): 674–685.

Stokstad, E. 2016. Power play on the Nile. *Science* 351 (6276): 904–907.

Stokstad, E., 2017: New ape found, sparking fears for its survival. *Science* 358 (6363): 572–573.

Stone, C.P., and L.L. Loope. 1987. Reducing the negative effects of introduced animals on native biotas in Hawaii. *Environmental Conservation* 14 (3): 245–258.

Stone, R. 2016. Dam-building threatens Mekong fisheries. *Science* 354 (6316): 1084–1085.

Stümpke, H. 1961. *Bau und Leben der Rhinogradentia*, ed. by G. Fischer, Stuttgart, 90 p. Nachdruck 2011: Spektrum Akademischer Verlag Heidelberg.

Stürchler, D. 1981. *Endemiegebiete tropischer Infektionskrankheiten. Karten und Texte für die Praxis.* Bern: Huber, 246 S.

Su, G., M. Logez, S. Tao, S. Villéger, and S. Brosse. 2021. Human impacts on global freshwater fish biodiversity. *Science* 371 (6531): 835–838. https://doi.org/10.1126/science.abd3369.

TA. 2010. Es wächst ein Wald in der Wüste. Tagesanzeiger 2010-10-06: 14, Zurich.

TEAM. 2010. *Environmental impact assessment for Xayaburi hydroelectric power project.* Lao PDR: TEAM Consulting Engineering and Management Co., Ltd.

Thomas, P.A., and P.M. Room. 1986. Successful control of the floating weed Salvinia molesta in Papua New Guinea: Useful biological invasion neutralizes a disastrous one. *Environmental Conservation* 13 (3): 242–248.

Thorncraft, G., L. Baumgartner, and T. Marsden. 2005. Fish passage and fishways in the Mekong basin: getting past the barriers. In *Proceedings of 7th Technical Symposium on Mekong Fisheries Ubon Ratchathani*, Thailand, 15th–17th November 2005, pp. 237–251. http://archive.iwlearn.net/mrcmekong.org/download/free_download/proceedings_7TechSymp/paper19-Fish-passage-n-fishways.pdf.

Thylefors, B. 2004. Eliminating onchocerciasis as a public health problem. *Tropical Medicine and International Health* 9 (4): A1–A3.

Turvey, S.T., M.M. Marr, I. Barnes, et al. 2019. Historical museum collections clarify the evolutionary history of cryptic species radiation in the world's largest amphibians. *Ecology and Evolution* 00: 1–15. https://doi.org/10.1002/ece3.5257.

Tyagi, B.K., and R.C. Chaudhary. 1997. Outbreak of falciparum malaria in the Thar Desert (India), with particular emphasis on physiographic changes brought about by estensive canalization and their impact on vector density and dissemination. *Journal of Arid Environments* 36 (3): 541–555.

Uzqueda, A., S. Burnett, L.V. Bertola, and C.J. Hoskin. 2020. Qiantifying range decline and remaining population of the large marsupial carnivore of Australia's tropical rainforest. *Journal of Mammalogy* 101 (4): 1021–1034. https://doi.org/10.1093/jmammal/gyaa077.

Van Bortel, W., M. Barutwanayo, C. Delacollette, and M. Coosemans. 1996. Motivation à l'acquisition et à l'utilisation des moustiquaires imprégnées dans une zone à paludisme stable au Burundi. *Tropical Medicine and International Health* 1 (1): 71–80.

Van der Geer, A.A.E. 2018. Changing invaders: Trends of gigantism in insular introduced rats. *Environmental Conservation* 45 (3): 203–211. https://doi.org/10.1017/S03768929000085.

Vila P., Zeiss E. Irma, and H. Gibson. 1978. Prospecciones de sistemas hidrograficos para la introducción del "salmón" en Chile. *Biol. Pesqu. Chile* 10: 61–73

Vila Pinto, I. 1989. *Diagnóstico y evaluación de la fauna íctica del Rio Maule*.Pehuenche S.A.: Universidad de Chile. Informe Policopiado.

Visser, S.A. 1973. Preimpoundment features of the Kainji area and their possible influence on the ecology of the newly formed lake. In *Man-made lakes: Their problems and environmental effects*, vol. 17, ed. by W.C. Ackermann, G.F. White, E.B. Worthington, and J.L. Ivens, 590–595. Geophysical Monographs.

Walker, E.P. 1968. *Mammals of the world*, 2nd ed. Baltimore: The Johns Hopkins Press, vol. I: 1–646, Vol. II: 647–1500 pp.

WCD. 2000. *Dams and development. A new framework for decision-making*. The Report of the World Commission on Dams. London: Earthscan, 404 pp.

Weiss, H. 2018. *Basel und der Lachs. Basler Stadtbuch, Dossier 2018*. Christoph Merian Stiftung. www.baslerstadtbuch.ch.

Wendelspiess, M. 1986. Die Bisamratte (Ondatra zibeticus) in der Schweiz. *Wildtiere* 4 (4): 9–14.

Whitten, T., R.E. Soeriaatmadja, and S.A. Afiff. 1996. *The Ecology of Java and Bali*. The Ecology of Indonesia Series Vol. II. Singapore: Periplus Editions (HK), Ltd., 969 pp.

WHO. 1987. *Comité OMS d'experts de l'onchocercose*. Tech.Rep.Ser. 752, 187 pp, Genève: WHO/OMS.

WHO. 1992. *Our planet, our health*. Geneva: WHO Commission on Health and Environment, World Health Organization, 282 pp.

WHO. 1993. The control of schistosomiasis. Second Report of the WHO Expert Committee. WHO Technical Report Series No. 830: 86 pp.

WHO. 2019. Onchocerciasis. https://www.who.int/news-room/fact-sheets/detail/onchocerciasis.

Wilgen, B.W., J. Measey, D.M. Ricahrdson, J.R. Wilson, and T.A. Zengeya (eds.). 2020. *Biological invasions in South Africa*, vol. 14. Springer Series in Invasion, 975 pp. Springer Open. https://doi.org/10.1007/978-3-030-32394-3.

Wilson, E.O. (ed.). 1988. *Biodiversity*. Washington D.C.: National Academy Press, 521 pp.

Wilson, E.O. 2016. *Half Earth. Our planet's fight for life*. New York: Liveright, 259 pp.

Wittenberg, R. (Hrsg.). 2006. Gebietsfremde Arten in der Schweiz. Eine Übersicht über gebietsfremde Arten und ihre Bedrohung für die biologische Vielfalt und die Wirtschaft in der Schweiz. Bundesamt für Umwelt, Bern. Umwelt-Wissen Nr. 0629: 154 S.

Worthington, E.B. 1989. The lake Victoria Lates saga. *Environmental Conservation* 16 (3): 266–267.

Worthington, E.B., and R. Lowe-McConnell. 1994. African lakes reviewed: Creation and destruction of biodiversity. *Environmental Conservation* 21 (3): 199–213.

WWF. 2020. *Living Planet Report 2020—Bending the curve of biodiversity loss*. Gland, Switzerland: WWF, 159 pp.

Yap, S.K., and S.W. Lee (eds.). 1992. In harmony with nature. In *Proceedings of the International Conference on Conservation of Tropical Biodiversity*, 12–16 June, 1990, 656 pp. Malayan Nature Society, Kuala Lumpur.

Yohannes, M., M. Hailel, T.A. Ghebreyesus, K.H. Witten, A. Getachew, P. Byass, and S.W. Lindsay. 2005. Can source reduction of mosquito larval habitat reduce malaria transmission in Tigray, Ethiopia? *Tropical Medicine and International Health* 10 (12): 1274–1285.

Zachos, F.E., M. Apollonio, E.V. Bärmann, et al. 2013. Species inflation and taxonomic artefacts—A critical comment on recent trends in mammalian classification. *Mammalian Biology* 78 (1): 1–6. https://doi.org/10.1016/j.mambio.2012.07.083.

Zeuner, F.E. 1963. *Geschichte der Haustiere. German edition 1967*, 448 pp. München: BLV Bayerischer Landwirtschaftsverlag.

Zhang, X., L. Peng, W. Liu, et al. 2014. Response of primary vectors and related diseases to impoundment by the Three Gorges Dam. *Tropical Medicine and International Health* 19 (4): 440–449.

Zhong, Y., and G. Power. 1996. Some environmental impacts of hydroelectric projects on fish in Canada. *Impact Assessment* 14 (3): 285–308.

Zwahlen, R. 1992. Failure of irrigation projects and consequences for a different approach: A case study. *Ecological Economics* 5: 163–178.

Zwahlen, R. 2003. Identification, assessment and mitigation of environmental impacts of dam projects. In *Modern trends in applied aquatic ecology*, ed. R.S. Ambasht, 281–370. New York: Kluwer Academic Publishers.

Zwahlen, R., and P. Durán. 1998. “Greening” Ralco. *International Water Power and Dam Construction* 50 (4): 22–23.

Chapter 11
Human Environment

Involuntary resettlement is the most serious issue of hydropower.
Thayer Scudder (1997)

As mentioned in the introduction, the focus of this book is on environmental impacts of dam projects and their assessment. The issues related to social impacts and resettlement have found broad coverage in the literature (McDowell 1996; Cernea 1999; Cernea and McDowell 2000; Scudder 2005). However, the interlinkages of social with environmental issues are too many and too strong as to be ignored when discussing the latter; for this reason, a number of issues concerning social impacts are brought up briefly throughout the text. Here, only a few relevant points shall be mentioned and discussed shortly, without however going into details on any of them. The literature cited can provide a deeper insight into these issues. Specifically for resettlement problems related to large dam projects, and the often unsatisfactory outcome of such programs Scudder (2005) is essential reading.

11.1 Conflicts Over Land Use

Several times I started an argument with one or the other of my engineering colleagues by telling them the following: people tend to concentrate in broad valleys along rivers mainly because this is where the good, fertile alluvial soils are. Now you come and insist on building dams in these same valleys, submerging these alluvial soils with the result that people have to move out, often to higher elevations within the same valleys, where the soils are by far not as good, making agriculture more difficult and less profitable. Why don't you make an effort for building your dams up there, leaving the valley floors undisturbed? More than one then tried to explain me why, for simple hydraulic reasons, this is not possible.

The background for this "argument", however, is real. In many places of the world alluvial soils are the ones most suitable for agriculture, because they are fertile, usually flat, and water is not far away. The conflict caused by projects which occupy these very soils is frequent, and resettlement required for a project is often one, or

R. Zwahlen, *Assessing the Environmental Impacts of Hydropower Projects*,
Environmental Earth Sciences, https://doi.org/10.1007/978-3-030-91185-0_11

even the most important reason for public resistance against a project. One probably extreme case illustrating this problem is the Three Gorges HPP, where approximately 1.2 million people hat to be relocated, half of whom were farmers; the aim was not to dislocate them to some faraway place, but just to "move them uphill", an plan which proved to be very difficult to implement since most or all the good soils were lost (Chau 1995). One reason for the problem was the fact that replacement land had been identified by remote sensing and apparently without any ground truthing, and when at the moment of implementation it was discovered that this land was not really suitable, "quota had already been fixed, and they had to be met by the local officials in charge" (Duan and Steil 2003). Tamondong-Helin (1996) describes a much less well-known case of a project in the Philippines, where people were moved from fertile alluvial lands to a much less fertile high plateau.

Sometimes, the choice of a ROR scheme can be a way out, by minimising land loss and with it resettlement; this obviously comes at a cost seen from the energy supply point of view and will certainly not be a feasible solution in every case, but is an alternative worth thinking about (see the two cases of Kukule Ganga and Rusumo Falls in Sect. 4.4.2). In some other places, it might be possible to opt for alternative power sources, wind and solar, to be put at higher elevations not suitable for agriculture.

11.2 The Farmer's View

Many resettlement programs bring about a fundamental change in the life—and especially in the livelihood—of affected farmers, and many people involved in resettlement implementation programs are frustrated about the attitude of these farmers, who are reluctant to embrace the new development which was conceived specifically with the aim of improving their economic situation, e.g. by turning subsistence farmers into coffee or bamboo shoots producers, in this way making much more money than they ever would have been able to do remaining in their old site and continuing their old habits. The problem, or at least one of the problems here is that people from an urban setting do not understand the situation and the concerns of the farmers, let alone their strategies.

I think that I learned this, at least to some extent, at an early age. The family of my godmother were farmers, and as a boy I used to spend my autumn holidays there. They cultivated land which was partly located in the lowland plain, partly on the slope of a gentle hill. Each year, they planted a portion of land in both sites with potatoes; potatoes, at that time, might have been even more important than they are today for Swiss farmers: it was the staple food, the main fodder component for the pigs, and the surplus could be sold. Having these two potato fields was a simple, but very effective risk minimisation strategy. In a normal year, they would get a reasonably good harvest on both plots. In a wet year, the lowland plot, which was located in a former swamp drained long ago for agricultural purposes, was too wet, many potatoes there were rotting, resulting in a small harvest, while the harvest on the hillside would be very good. In a dry year however, the hillside would be too dry, producing only a small harvest of small potatoes, and then the harvest on the lowland plot would be very good. Like this, they usually had a sufficient supply of potatoes under any circumstances.

This type of risk-minimising strategy is crucial for farmers, and especially for subsistence farmers, and according to Shermer (2018: 171f), risk averse behaviour is part of our evolutionary history. Every farmer in such a situation is constantly aware of two main issues, namely (i) having a bad harvest is a very real threat to his family, and (ii) with what he is doing he and his family survived so far. This inevitably leads to a perception of a change in habits as a real risk, which is all the more difficult to assess as he is not familiar with the new system that he is supposed to embrace. Reluctance against change, in such a situation, is very understandable.

This means, in the very least, that in case such a change in livelihood is inevitable, early involvement of the affected population in the development of the new livelihood system, and a thorough training and supervision, is crucial for the success of such a change. It would certainly also be helpful if the change could be made in steps, e.g. using the several years of the construction period for a large dam project for a stepwise introduction of the new methods and approaches, while fields could still be used in the traditional way.

11.3 Caught in the Limbo: Project Delays

Hydropower projects tend not to have simple, straightforward development histories, many of them can be compared with bumpy roads with a lot of twists and turns. This can cause considerable problems for people affected—or potentially affected—by them, and often creates hardships and sometimes long lasting uncertainties. This is also an issue which is not well covered in the technical literature. Three cases are shortly described here to illustrate this point.

11.3.1 Bumbuna

Bumbuna dam was built in the early 1990s. In 1992, the company intending to build it went bankrupt; at that time, the dam was largely built, but impoundment had not yet taken place and the construction of the power house had not yet started. That was the moment when the World Bank was asked for financing. One of the first things it did, after an evaluation of the project, was to commission an ESIA, something that had not been done yet. One main emphasis of this study was on the socio-economic impact of the project (EWI 1996). During that period, a civil war was raging in the country, which was mainly about control of the diamond fields in the Eastern part. However, before the study was finalised, rebel bands also reached the project site, and in a raid the town of Bumbuna, located just below the Bumbuna falls a short stretch downstream of the dam site, was destroyed to a great extent, many of the houses being looted and burned down. The last of the work force still remaining on site had to leave, and not much of the construction site installations remained intact. During almost 10 years, nothing happened there; then it was decided to finally complete construction, and in 2004 the RAP was updated (or, rather, a new RAP was commissioned). It was quite clear that after such a long period, nothing concerning the information on the affected population could still be considered as valid (Electrowatt-Ekono and Techsult 2005).

The reservoir to be formed by the dam was long, straight and very narrow, located in a narrow valley between hills, with rather steep flanks on either side (see Figs. 16.2a and 17.3). The villages were all at somewhat higher elevations on the more or less flat hill tops, mostly due to the fact that there was no suitable place in the valley for a village and no good soils for cultivation. Only in the topmost part of the future reservoir there was a village very close to the river which obviously would have to be relocated before impoundment, and there was another one which was supposed to be submerged at least in part. One major problem there was the absence of a reliable, exact topography, which would have allowed to verify exactly up to what point the water would eventually reach. Just before the 2005 study was finalised, such data were finally available, and the inhabitants of the village could be informed that after all, the village would not be affected by the reservoir which was soon going to be formed. For over 10 years they had lived with the information that most probably at least a part of the village would be under water and would therefore have to be relocated (Fig. 11.1).

11.3.2 Bujagali

For the construction of Bujagali HPP in Uganda, a short stretch downstream of Owen Falls, the place (and since 1954 a hydropower plant) where the Victoria Nile leaves Lake Victoria, two villages had to be relocated. Relocation started in 2002, but in 2003, before construction of the dam began, the developer pulled out. At that moment, the new village had largely been built, the people had moved there, but a number of essential things were still missing, land allocation for farming was not finalised and many people were left without land title and clear information about the land they were entitled to use, and all programs concerning livelihood restoration had been put on halt. One seemingly secondary problem concerned firewood. The site of the new village had been cleared of all woody vegetation before construction started, there were no trees or bushes left. This meant that people—and, as is usually the case, mostly women—had to walk long distances for getting firewood, and it was not clear if and where in the nearest forest, which was basically a forest reserve, they were entitled to collect firewood, a situation which cost a lot of time and caused considerable problems for the inhabitants of this new village. Only in 2006, when a new developer took over, things started to move again.

Fig. 11.1 Women in Kamayimbo village. When, after 10 or more years of uncertainty, the inhabitants of the village were informed that their village would not be submerged after all, some of the women spontaneously started to dance

11.3.3 Rogun

Construction of Rogun dam in Tajikistan (see Chap. 23) had started in the early 1980, and during this period resettlement had also started. A considerable number of people were moved to newly formed collective farms in the lowland part of the country. When Tajikistan got its independence at the end of the Soviet Union, construction work stopped. Civil war started shortly afterwards, and during that time a considerable part of the infrastructure of the country was damaged or destroyed. No work was carried out on the construction site, and many of the people who had already been relocated moved back to their original villages. In 2010, World Bank decided to finance a study for reassessing the feasibility of the project in the new geopolitical situation. Some work was resumed at that time, but it was agreed that this should not be construction work as such, but should be directed only at maintaining the structures which had already been built in order to prevent further damage to them.

The resettlement program, which had been resumed by the government of Tajikistan a few years earlier, was also stopped, to be resumed only if and when a decision would have been reached on whether or not to build the dam and power station.

During all this time, however, one thing had not stopped: when the original decision had been taken to build Rogun HPP, all the villages in the future reservoir area were informed that they were no longer entitled to build new houses or to make substantial improvements on existing houses, and no more investments were made for maintaining, let alone improving, the infrastructure in this area. This situation had been maintained, and when this new evaluation was done starting in 2011, everything was in a state of advanced decay. Such an announcement of a cutoff date is a measure usually taken in a resettlement program, first of all because it would not make sense to build new structures only for destroying them again after a short while, but also for preventing unjustified compensation costs. Only in this case, this situation had finally lasted for over 20 years, which had made life for the inhabitants very difficult. Luckily, this was then taken into account once Tajikistan resumed the resettlement program for this project: unlike what is done in most cases, the focus here was not on households as basic units, but on families. Whenever there was a situation where more than one family found themselves crammed in a house, because grown-up children, when they got married and had children of their own did not have any option than to continue to live with their parents, each family was entitled to compensation for a house, the rule "house for house" was implemented in a different way, in favour of the beneficiaries (Fig. 11.2).

Fig. 11.2 Girl fetching water. After 20 years with a complete lack of investment, infrastructure in the villages within the Rogun HPP reservoir area had completely broken down

11.4 Victims and Profiteers: Dealing with People

At first sight, or seen from outside, i.e. before getting into touch with a population affected by a project (usually called "project affected persons" or PAP), they may seem to be a quite homogenous group: living, e.g. in one or a few villages looking similar to each other, in houses of a similar style, and with similar livelihood stiles (e.g. farmers, mostly for subsistence but partly also creating a cash income). Once one gets a chance to take a closer look, by carrying out a socio-economic survey or later on during direct negotiations for compensation, this impression of homogeneity soon disappears. Taking the risk of oversimplification, but with the main aim of illustrating a few important points, three categories are shortly presented and discussed here.

The first, and usually largest of these categories are those who usually live, under the circumstances considered and not taken as a score on any absolute category, in a situation which could be termed as satisfactory; these are the ones who are reasonably well situated, far from being rich, but comparatively just or almost as far from being poor. They may or may not show opposition against the idea of being relocated, and overall they make efforts to defend their position and try to negotiate a fair compensation package, and once they have accepted the fact of relocation they tend to be more or less cooperative. That this is by no means a homogenous group either is shown e.g. by Lassally-Jacob (1996) reporting on a project in Africa where originally all resettlers had been given the same amount of land, but where in spite of that the outcome was not the same for everyone: "...differences in living standards among tenants having increased between those who, capable of profiting from new opportunities, now form an elite supervising huge domains, and the others who, excluded, only dispose of the labour their families can provide".

The second, usually much smaller group are those who have obvious problems, they are (again seen in comparison with the other people in the same population, but usually also in comparison to a defined threshold value, the so-called poverty line (Sect. 11.5) poor, and they often see themselves burdened with a number of additional constraints like old age, physical or mental handicap, sickness, being a single woman with children (and, much rarer and sometimes not really taken into account, but potentially just as problematic: a single man with children).

Finally, the third and usually smallest group are persons who are in most cases relatively well off (although not necessarily always the comparatively richest in the society), who got along quite well with the situation so far, and who are likely to do so in the future, when the situation changes. This is fine, but there are always a few who will try, often quite successfully, to get an additional advantage for themselves in every situation, one that goes beyond fair compensation, and mainly one which is not for all in the community; in a resettlement program, trying to get a good deal, which then if possible should define the standard for the entire community, is absolutely understandable and acceptable, and so is trying to get the best out of it for oneself, as long as it is not unfair and mainly not to the detriment of others who are just as affected by the project. There is, however, no clear limit between "a good deal" and "an unfair advantage".

All internationally accepted and applied guidelines and standards for resettlement make specific provisions for the second group, by pointing out the special attention which has to be given the so-called "vulnerable persons". Two definitions or descriptions of vulnerable persons are: "This disadvantaged or vulnerable status may stem from an individual's or group's race, color, sex, language, religion, political or other opinion, national or social origin, property, birth, or other status. The client should also consider factors such as gender, age, ethnicity, culture, literacy, sickness, physical or mental disability, poverty or economic disadvantage, and dependence on unique natural resources." (IFC 2012, PS1, p. 4, footnote 18), and "... pay particular attention to the need of disadvantaged or vulnerable groups, especially those below the poverty line, the landless, the elderly, female headed households, women and children, Indigenous Peoples, and those without legal title to land." (ADB 2009: p. 49). Important key words in these descriptions are: may stem from, factors such as, especially those; this means that neither of the description is thought to be a complete, exhaustive and universally applicable catalogue of vulnerable people, defining exactly who is an who is not. Whoever is carrying out such an investigation has to take this as a guideline, being aware that it is necessarily incomplete in two ways: a person or a household (the household being the category which is usually considered as the basic unit to be dealt with) who does not fall into one of the specific categories listed can still be vulnerable (as the single father with children mentioned above), and a person or a household definitely falling into one of these broad categories does not necessarily have to be vulnerable. One example to shortly illustrate this point: in one case, where a local consultant was charged with carrying out the socio-economic survey there was an incredibly high proportion of vulnerable households. When checking the basic data it turned out that in this area there was a rather large number of households which consisted of one or two elderly persons without any people of working age to support them. However, many of those had adult children who had moved to the city, but continued to support their parents, or a substantial number who were well to do for any other reason; all of these had automatically been included in the vulnerable group, a procedure which undoubtedly does not really reflect reality.

While, as said, all guidelines and standards make reference to vulnerable people, none of them makes any direct reference to those PAP who try to get an unjustified benefit; here, it is obviously the responsibility of the project proponent to pay attention to such cases (which, on the other hand, must not lead to attempts of curtailing justified requests). One provision, however, which guidelines do make and which goes in that direction is the cutoff date: in a normal procedure (which, in this respect, Rogun mentioned above was not), once the PAP have been informed on the project and its impacts concerning them and the socio-economic survey carried out, it has to be officially announced that this survey will be the basis for compensation, and that structures etc. which were not there at the time of the survey and which therefore have not been included in the register will not be entitled for compensation. If the project proponent does either not announce such a cutoff date, or does it in a way as to leave uncertainties, there will certainly be people who will try to take advantage of this situation—and one cannot really blame them for that. Two examples: in a case in Viet Nam, the water level of the future reservoir had been marked in order to make it clear to everybody what would be lost—and therefore entitled for compensation—and what would not, but a cutoff date had not officially been announced. Over the following few months a surprisingly lager number of "houses" appeared within this demarcation, usually a concrete slab (to give it a "serious" character, and on top of that a kind of a house just nailed together from planks; still, structures deserving compensation. In another project in Turkey, where there had been doubts about the cutoff date, one farmer whose land was located directly in the dam axis planted a large field with hundreds of pistachio saplings; he was well aware of the fact that his land was among the first which would have to be acquired for the project, but he was also aware that it might well take a few more years to that moment, and that although these plants would certainly never reach the productive age, they would still glean a nice benefit. Well, such things should probably be considered as attempts to make the best of an—overall not very pleasant—situation rather than as an abuse (Fig. 11.3).

One characteristic of what is labelled above as the third group is their penchant to complaining loudly, which is usually done with the more or less direct intention to get a benefit. While in a resettlement situation complaints are, probably to the largest part, absolutely justified (and efforts have to be made not to ignore those who would have reasons to complain, but who, for whatever reason, do not), there is also a proportion of complaints which are not justified.

This is an impression one sometimes gets when reading a report covering a resettlement audit. One striking example was such a report prepared for the Rogun project in Tajikistan. A local consultant had been mandated with a survey of the ongoing resettlement process, mainly the question how the company in charge of the project handled socio-economic impacts. One of the central findings of this study was that the company did not offer jobs to the local population, or at least not to a sufficient

Fig. 11.3 Typical "house" built in the reservoir area

extent. When checking on that conclusion it turned out that apparently one young man who had been interviewed during this survey had complained about that, and this had been turned into a general statement. The socio-economic survey which had been carried out as part of the ESIA for this project had revealed that 28% of the men living in the surrounding villages worked on the construction site, and this at a time when construction activities were far from being at their height; in addition, with the exception of a small number of civil servants these were the only paid jobs in the area (Pöyry 2014). Obviously, when people who are interviewed in such an audit notice that the interviewers specifically look for complaints and week points in the program, this is exactly what they will get, especially if there is a possibility that there might be a reward for complaining. Undoubtedly, in every resettlement program there are good reasons for at least some of the affected persons to complain, and it is very important that such concerns are discovered and are being taken seriously. Still, the validity of complaints also has to be checked, they cannot always be taken at face value, precisely due to the fact that while there are people who tend to suffer without saying anything, there are also always people around who will leave no situation without trying to obtain an additional benefit.

11.5 The Most Vulnerable of All: Poor People

Después de un desatre las sociedades del mundo pobre no retornan al estado anterior, sino que parecieron no alcanzar el nivel más alto previo a la calamidad.

Rosenzvaig (1996)

As mentioned above, vulnerable groups or vulnerable persons are an important topic of all the international standards and guidelines on mitigation of social impacts of development projects of all types. Rosenzvaig (1996: 495), in the statement provided above, gives a good reason for this saying that after a disaster of any kind (whether natural or man-made) the poor very often are not able to return to their previous situation, but will find themselves and remain poorer than they were before. One of the problems, in addition to all the difficulties they already face, which especially poor vulnerable people have is that it is very difficult for them, for one reason or another, to voice their problems or to be heard if they do.

The first of the 17 Sustainable Development Goals adopted by the United Nations in 2015 (https://www.un-page.org/page-and-sustainable-development-goals) is entitled "end poverty in all its forms everywhere". This is a goal that will certainly never be reached in this complete and absolute form, but it will have to be pursued constantly. In this respect, all guidelines for resettlement and compensation stress the two following points: people who need to be relocated due to a project should in the end at least not be worse off than they were without the project, and if possible they should benefit, meaning that compensation should lead to an improvement of their situation. The basic meaning of this may seem simple enough, but in reality it often turns out that it is by no means as easy as that. Hong (1987: 169 ff.) describes problems in this respect with projects in Sarawak (Malaysian Borneo), and Scudder (2005) discusses in detail a number of cases. The reasons for failure in reaching this goal are manifold. One issue can be that during project development and concession negotiations a budget is defined, which may even seem to be rather generous, but certainly sufficient. However, as is often the case in large projects of any type, during implementation a number of problems become manifest which all tend to increase costs, and then the developer tries to limit expenditures wherever possible. In such a situation the PAPs (Project Affected Persons) suddenly appear only as a cost factor, and efforts are made to reduce costs. This inevitably goes at the expense of the resettlers, and if there is no strong institution on the monitoring side, which finally can also enforce decisions, the outcome will no longer be what the original plan said it would. The importance of sufficient funding as a *sine qua non* condition for the outcome of a resettlement program is obvious (see e.g. Cernea 1999; Sherbinin et al. 2011; Sparkes 2012). There are other reasons as well, as e.g. the fact that developers, officials and affected persons do not necessarily agree as to what "improvement" means. Without going further into the details here, in every hydropower project requiring resettlement every effort will have to be made for reaching the objective of the guidelines mentioned above. One approach that definitely does not work is the "accountant's approach": in a discussion about the resettlement caused by the Tarbela dam in Pakistan one representative of a European company allegedly said that "70,000 people have lost, but 120 million people have benefitted, therefore it was a good project".

One final point concerning the issues around poverty. Poverty is often measured or defined according to the international or national poverty line, i.e. people living on less than 2 USD per day, or a nationally defined amount in local currency, are considered as living in extreme poverty. There is however one category of people, the truly "traditional" societies, which are not necessarily adequately covered with this definition. What is meant here are the (increasingly rare) traditional societies which are not monetised, living as hunter-gatherers, as subsistence farmers, or as a combination thereof. Examples for this are the Penan in Sarawak, Borneo as described by Manser (1992), some other tribal groups in Sarawak (Hong 1987), New Guineans and Australian Aborigines (Diamond 1997: 297ff.), !Kung in the Kalahari, Yanomami in Venezuela and other groups (Eibl-Eibesfeldt 1997) or the Pygmies in the rain forests of central Africa (Heymer 1995). Such groups are—or, for the most part, were—not necessarily poor, since their way of living and their habitat provided for everything they need. However, they are the ones who are most at risk for being pushed into poverty in case they lose their habitat and are forced into an alternative way of living. In addition to their traditional livelihood they often have to face other issues like being an ethnic minority, often considered as backward by people belonging to the mainstream society, speaking a different language, having no formal education and not being familiar with handling money; all that will work to their disadvantage, and will often be deliberately used against them.

11.6 Gender Issues

In many societies women are especially at risk, since decisions are taken by men, and men are the ones participating in discussions. This was made very clear to me once in a project in Togo, where a village official, when discussing exactly this issue, told me "jamais la poule n'a chanté dans la basse-cour" (never the hen has been singing in the farm yard): women have to remain silent. Examples of that type are plentiful (Fig. 11.4).

It cannot possibly be the aim of an ESIA, or of the social part of it, to change traditional patterns of a society affected by a project. The planners will have to find ways for doing their job arranging themselves with the given structures, including social ones. This means that if official meetings are held exclusively by men, then so it is. It is still important, however, to include the women in a suitable way, in order to hear and understand their views and their problems (or, for that matter, their suggestions on how to solve a problem). One possibility is certainly to engage in informal discussions as the one mentioned in Fig. 11.4 (which, by the way, went along rather well up to the moment when a man joined the group), but it is definitely not sufficient. A good way to do it is to organise focus group discussions, forming groups of different categories, like e.g. young people (yet another group which often has problems in getting heard), specific occupations like fishermen, and women. Two other conditions have to be met: it takes a woman for talking to women, and the discussion has to be led in the local language; this last point is important in any case,

Fig. 11.4 Men have the say. Official meetings related to resettlement issues: Tajikistan (Rogun, top left); Bumbuna (Sierra Leone, top right); Pakistan (Spat Gah, bottom left): only men attend. Bottom right: informal encounter with a group of women led to a good discussion on their views and concerns (Bumbuna, Sierra Leone)

but for discussions with women even more so, to take the sometimes comparatively limited education of women into account.

11.7 Ethnic Minorities

Ethnic minorities or indigenous peoples are usually recognised as being vulnerable, or as having a high risk of belonging in this category. Different terms are being used to identify such groups, as indigenous peoples, ethnic minorities, indigenous cultural communities, aboriginals, hill tribes, minority nationalities, first nations, autochthonous peoples, scheduled tribes, or tribal groups. According to ADB (2009: 58), such groups are usually defined by showing, to various degrees, the following characteristics:

i. self-identification as members of a distinct indigenous cultural group and recognition of this identity by others;
ii. collective attachment to geographically distinct habitats or ancestral territories and to the natural resources in these habitats and territories;
iii. customary cultural, economic, social, or political institutions that are separate from those of the dominant society and culture; and
iv. a distinct language, often different from the official language of the country or region.

Ethnic minorities—this term is used here with the aim of including all of the labels used for such groups listed above—are vulnerable in more than one sense. They are often few in numbers, which is one reason why they have little or no political power, they speak a different language which makes communication more difficult, they are considered as backward by the dominant group, and very often their level of education is low. All this puts them in a weak position in case they are affected by a project. In many cases, such groups were victimised by large development projects which went ahead in their traditional land, often forested areas, often destroying the traditional livelihood of these peoples without offering an acceptable alternative, as has happened e.g. in India (Fernandes 1997). This type of "development" has often led to resistance by the affected groups (Gray 1996; Leslie 2005). All these reasons have led to the requirement of formulating a special Indigenous Peoples Development Plan in case such a group will be impacted by a project. Consultation with such people is essential, because their view may not necessarily be the view of those in charge of planning and implementing resettlement. In the case of Ralco HPP, Chile, where a small group of Pehuenche (a subgroup of the Mapuche, the original inhabitants of South-Central Chile and the adjacent parts of Argentina; Bengoa 1992; Marielo et al. 1995; Orellana 1994; Villalobos 1989) had to be relocated, they insisted on having a traditional settlement structure: houses in the traditional style, widely spaced, each house surrounded by the family's land, the dwellings not concentrated in a village (Zwahlen and Durán 1998). Contrasting with that, Downing (1996) mentions a case from Mexico where the resettlers did not agree with the design of the new houses offered to them: as they saw it, for their original houses they had done as best they could, but as a compensation for the forced relocation they expected something better. Another point of discussion or misunderstanding can be what is often called "integration into mainstream society" in the meaning of offering a real step forward in development, but has also been called "forced integration", or, as Oliver-Smith (1996: 91) describes it: "In some cases, the control and integration of ethnic minorities is a secondary goal, although somewhat covert, of development-project resettlement". For Murum HPP, a small group of Penan (already no longer the forest nomads they used to be, now living in "mainstream" longhouses, with a livelihood based on agriculture) would have had to be relocated. They lived near to the upstream end of the future reservoir (see Fig. 21.4). The aim of the officials in charge was to move them downstream in order to better integrate them into the national society, also by providing them with health services and schools. They themselves, however, preferred to move even further upstream, since they feared that otherwise they would find themselves at the very bottom of the social hierachy, having problems to compete with the members of the other ethnic groups in this area, and not wanting to lose their identity (Chang et al. 1995). Providing development with resettlement is a widely accepted goal, still it has to be done carefully, and especially not by forcing people into a "better" situation that they themselves do not want, and do not experience as an improvement. In the case of traditional societies, development invariably also means changes in the traditional lifestyle.

11.8 Cash Compensation

> *Payment of compensation in cash to tribal people, whose economy is largely nonmonetized, has invariably left most of the money in the pockets of brokers, bootleggers, gamblers, or confidence tricksters.*
>
> Mahapatra (1999: 202)

It can be considered a well-established fact that compensation in cash, especially if the affected persons are a rural community based on subsistence farming, is not a good approach, because it finally often leaves the people in question with nothing at all. Hong (1987: 169ff) describes the fate of a number of people who had been resettled due to Batang Ai HPP in Sarawak, Malaysia, and who had received generous cash compensations for their houses and land, which gave them the impression that now they could afford all kind of luxury and did no longer need to work. Once the money spent, they found themselves in a very difficult situation, without a livelihood to sustain them. One condition that goes in the same direction is the fact that rich people have a lot of friends who know exactly what do to with so much money. The preferred aim in such cases is always the in-kind compensation, i.e. providing a house for a house and, most importantly—but often also the most difficult to achieve—land for land wherever agriculture is the most important and often only livelihood basis.

Still, cash compensation is often implemented in spite of known problems. One obvious reason can be that the legislation of the country clearly stipulates that every person whose assets have to be expropriated due to a project has the right to choose between cash compensation and the option of being integrated in a resettlement program; this is the case, e.g., in Turkey (Altinbilek et al. 1999), Laos and Zambia, and certainly in a number of other countries. In such a situation, it is not possible for the agency in charge of resettlement to refuse the choice of cash compensation. The developer also has limited interest at best to try to convince affected households to opt for the resettlement package: it is much simpler to define a sum of money, to sign a contract with the affected person and to pay him or her the money, and in most cases this is ultimately the less costly alternative, which can very well lead to an attitude where the choice to take the money is encouraged. From the side of the resettlers, there are different reasons why they might chose cash rather than in-kind compensation. One of these reasons, and certainly the worst of all, is the fact that they never had any "real" money, and the amount offered to them seems to be just enormous; this is exactly what then leads to a situation as described above, where after a short while of good life in luxury the money is gone, and everything else with it, pushing the people into poverty. Another, not much less dangerous situation can be the fact that such a choice may have to be made early in the development of the resettlement program, where people see how much money they can get, but where it is not clear at all what the outcome will be if they opt for the resettlement package. In such a situation many may want to stay on the safe side and go for the money. Of course there are situations—or people—where cash compensation is not a problem: if the affected person is in a business of any kind, and has experience and the required knowledge for investing the money. This condition, however, is very often not fulfilled.

11.9 Impacts on the Spiritual World

If in the footprint of a project a cemetery is affected, it will have to be relocated in most cases, and very often specific procedures and ceremonies have to be observed in order to do so. The involvement of the affected community in this process is always essential. However, sometimes there are impacts on the spiritual world which may be somewhat less obvious and easy to understand, especially from a "Western" or "Northern" perspective, but dealing with these may be just as important.

In the case of Bumbuna HPP, Sierra Leone, it became clear rather quickly that each village had a secret society, and that every one of these societies had its own sacred meeting place. Obviously, it being a secret society, outsiders were not allowed to know the locations of these places, let alone to visit them. But it was very clear that all these sites needed to be relocated, and while this was not really a problem, it still required a ceremony. For these ceremonies, a number of ingredients were needed, and here every village presented a very precise list of what would be required for that purpose. These lists, for three of these villages, looked as follows (EWI 1996):

Gbangbani: 2 cows, 4 goats, 5 containers of palm oil, 10 bags of salt, 5 packets of Maggi cubes, 20 bags of rice.

Korlon: 5 bags of rice, 2 goats, 2 packets of Maggi cubes, 4 bags of salt, 2 containers of palm oil.

Saigereh: 3 cows, 2 goats, 5 bags of rice, 5 bags of salt, 3 containers of palm oil, 4 chickens and 2 packets of Maggi cubes.

I was quite astonished to learn that apparently, in these traditional African rituals, Maggi cubes seemed to play an important role. Obviously, since the exact position of these sites was not known, we had no way of knowing whether a specific site would actually be affected or not, but in such a case it is certainly better to go along with the request.

Another impact, of quite different nature, came up in the case of Bujagali HPP in Uganda. Here, the project made use of an island in the river: the plan was to close one arm of the river for building there the main structures (the part of the dam including power house and flood gates), using the other arm as a river deviation, and then to close this arm for building the remaining part of the dam. However, this island was the haunt of a spirit, who had to be relocated before work could start there. One obstacle had to be overcome before this relocation could be done: on either side of the river there lived a sorcerer, and each of them claimed that he was the only one entitled to do this. Once this question cleared, the ceremony was held, and work could start.

It is easy to laugh about such beliefs or practices, but there is not much justification for this. As good as no plane has a seat row No. 13, and not many hotels and other buildings have a 13th floor. Quite recently, there was a problem on a dam site, where the tunnel specialist of the planning engineering company was a woman: many miners and tunnelers all over the world seem to firmly believe that giving a woman access to a tunnel will cause any type of bad things to happen; interestingly, they all rely on Saint Barbara, definitely a woman, to protect them exactly from such mishap.

References

ADB. 2009. *Safeguard policy statement. Policy Paper*. Asian Development Bank, 92 pp.

Altinbilek, D., M. Bayram, and T. Hazar. 1999. The new approach to development project-induced resettlement in Turkey. *Water Resources Development* 15 (3): 291–300.

Bengoa, C.J. 1992. *Quinquén: 100 anos de historia Pehuenche*. Santiago de Chile: ChileAméica CESOC, 136 pp.

Cernea, M.M. (ed.). 1995. *Putting people first. Sociological variables in rural development*, 2nd edn., reviesed and expanded. New York: The World Bank, Oxford University Press, 575 pp.

Cernea, M.M. 1999. Why economic analysis is essential to resettlement: a sociologist's view. In: *The economics of involuntary resettlement*, ed. by Cernea, M.M. Washington D.C.: World Bank.

Cernea, M.M., and C. McDowell. (eds.). 2000. *Risks and reconstruction. Experience of resettlers and refugees*. Washingotn D.C.: World Bank, 487 pp.

Chang, Y.T., K.J. Khoo, A.C. Loo, G.A. Piggott, and R. Zwahlen. 1995. Human resettlement in dam projects—An overview of issues and sensitivities. In *Proceedings of the MWA International Conference on Dam Engineering*, 549–553, 1–2 Aug. 1995, Kuala Lumpur, Malaysia.

Kwai-Cheong, C. 1995. The Three-Gorges project of China: Resettlement prospects and problems. *Ambio* 24 (2): 98–102.

Diamond, J. 1997. *Guns, germs, and steel. The fates of human societies*. New York: Norton, 480 pp.

Downing, T.E. 1996. Mitigating social impoverishment when people are involuntarily displaced. In *Understanding impoverishment. The consequences of development-induced displacement. Refugee and Forced Migration Studies*, vol. 2, ed. by McDowell, C., 33–48. Oxford: Berghahn Books.

Duan, Y.F., and S. Steil. 2003. China Three Gorges project resettlement: Policy, planning and implementation. *Journal of Refugee Studies* 16 (4): 422–443.

Eibl-Eibesfeldt, I. 1997. *Die Biologie des menschlichen Verhaltens. Grundriss der Humanethologie. 3*, 1118 pp. Weyarn: Auflage Seehammer.

Electrowatt-Ekono and Techsult. 2005. *Bumbuna hydropower project: Resettlement action plan for the Reservoir and dam area*. Report prepared for Government of Sierra Leone and World Bank.

EWI. 1996. *Environmental impact assessment for the Bumbuna hydroelectric project*. Final Report February 1996. National Power Authority, Republic of Sierra Leone. Electrowatt Engineering Services Ltd. and Techsult and Company Ltd.

Fernandes, W. 1997. Post-independence policies, planned development and tribal deprivation. In *Tribes and Government Policies*, ed. by Bahndari, J.S., and S.M. Channa, 157–206. New Delhi.

Gray, A. 1996. Indigenous resistance to involuntary relocation. In *Understanding impoverishment. The consequences of development-induced displacement*, ed. by C. McDowell, 99–122. Refugee and Forced Migration Studies Vol. 2. Oxford: Berghahn Books.

Heymer, A. 1995. *Die Pygmäen. Menschenforschung im afrikanischen Regenwald*. München: List, 539 pp.

Hong, E. 1987. *Natives of Sarawak. Survival in Borneoś Vanishing Forests*, 265 pp. Malaysia: Institut Masyarakat.

IFC. 2012. *Performance standards on environmental and social sustainability*. Performance Stadards PS1-8, Guidance Notes GN1-8. http://www.ifc.org/performancestandards.

Lassally-Jacob, V. 1996. Land-based strategies in dam-related resettlement programmes in Africa. In *Understanding impoverishment. The consequences of development-induced displacement*, ed. by C. McDowell, 187–199. Refugee and Forced Migration Studies Vol. 2. Oxford: Berghahn Books.

Leslie, J. 2005. *Deep Water. The epic struggle over dams, displaced people, and the environment*, 352 pp. New York: Picador.

Mahapatra, L.K. 1999. Testing the risk and reconstruction model on India's resettlement experiences. In *The economics of involuntary resettlement*, ed. by M.M. Cernea, 189–230. World Bank.

Manser, B. 1992. *Stimmen aus dem Regenwald. Zeugnisse eines bedrohten Volkes*. Bern: WWF/Zytglogge, 299 S.

Marielo, A., A.M. Bacigualupo, R. Salas, R. Curivil, C. Prker, and A. Saavedra. 1995. *Modernización o sabiduría en tierra Mapuche? Colección Cultura y Religón, No. 2*, 198 pp. Santaigo de Chile: San Pablo.

McDowell, C. (ed.).1996. *Understanding impoverishment. The consequences of development-induced displacement*. Refugee and Forced Migration Studies, Vol. 2. Providence: Berghahn Books, 257 pp.

Oliver-Smith, A. 1996. Fighting for a place. The policy implications of resistance to development-induced resettlement. In *Understanding impoverishment. The consequences of development-induced displacement*, ed. by C. McDowell, 77–97. Refugee and Forced Migration Studies Vol. 2. Oxford: Berghahn Books.

Orellana, R.M. 1994. *Prehistoria y etnología de Chile. Colección de Ciencias Sociales, Universidad de Chile*, 240 pp. Santiago de Chile: Bravo y Allende.

Pöyry. 2014. *Environmental and social impact assessment for Rogun hydropower plant*. ESIA Vol. I: Text. Vol. II: Annexes. Report prepared for OSHPC "Barki Tojik", Republic of Tajikistan.

Rosenzvaig, E. 1996. *Etnias y árboles. Historia del universo ecológico Gran Chaco*, 618 pp. La Habana, Cuba: Casa de las Américas.

Scudder, T. 1997. Social impacts of large dam projects. In *1997: Large Dams: Learning from the past, looking at the future*, ed. by Dorcey, et al., 41–67. Gland and Washington: IUCN/The World Bank.

Scudder, T. 2005. *The future of large dams. Dealing with social, environmental, institutional and political costs*, 389 pp. London: Earthscan.

de Sherbinin, A., M. Castro, F. Gemenne, M. Cernea, et al. 2011. Preparing for resettlement associated with climate change. *Science* 334 (6055): 456–457.

Shermer, M. 2018. *Heavens on earth. The scientific search for the afterlife, immortality and utopia*, 305 pp. London: Robinson.

Sparkes, S. 2012. Lessons learnt from the Theun-Hinboun Expanison resettlement in Laos. *Hydropower and Dams* 19 (2): 70–78.

Tamondong-Helin, S.D. 1996. State power as a medium of impoverishment. The case of the Pantabangan dam resettlement in the Philippines. In *Understanding impoverishment. The consequences of development-induced displacement*, ed. by C. McDowell, 169–183. Refugee and Forced Migration Studies Vol. 2. Oxford: Berghahn Books.

Villalobos, R.S. 1989. *Los Pehuenches en la vida fronteriza*, 269 pp. Santiago: Ediciones Universidad Católica de Chile.

Zwahlen, R., and P. Durán. 1998. "Greening" Ralco. *International Water Power and Dam Construction* 50 (4): 22–23.

Chapter 12
The ESMMP

It is concluded that there is a need for more study of impacts downstream of Tucuruí, also for some assessment of the effects of increasing the number of turbines operation there, and for an examination of why some predicted difficulties have not in practice been avoided.
Barrow (1988)

As shortly described in Sect. 4.2.4, the construction period is the time where most of the impacts become manifest, either in the form of impacts directly caused by construction activities—many of which will then be over and no longer of any relevance—or as a preparation of the permanent changes which will then characterise the new situation once the project goes into operation. The ESIA has to do two main things as a preparation of the construction phase: it has to identify the impacts that are going to be of relevance, and it has to define the measures which will be required for mitigating these impacts. This is being done in the form of the ESMMP (Environmental and Social Management and Monitoring Plan). This is usually a chapter of an ESIA Report which outlines the activities that will be required in order to deal with all these issues at hand during the construction period, activities which aim at reducing the various negative consequences to an acceptable level or, where there are legally defined threshold values, to a level which is in compliance with these norms. It also identifies the responsibilities of the various actors, of which there are mainly three, namely, (i) the project proponent or owner, who is the ultimately responsible entity in any case, (ii) the contractor (which can be one company, or a consortium or joint venture of several companies, including subcontractors, if any), and finally (iii) the authorities in charge of monitoring compliance; for environmental matters, this latter is usually the Environmental Protection Agency (which can have different names depending on the country), acting at the national, provincial or local level. Other entities may play a role in this, especially where the socio-economic aspects of project implementation are concerned.

In the following Sections, important parts of this ESMMP are outlined, and the responsibilities of the parties are described.

R. Zwahlen, *Assessing the Environmental Impacts of Hydropower Projects*, Environmental Earth Sciences, https://doi.org/10.1007/978-3-030-91185-0_12

12.1 EHS Management: The Contractor's Responsibility

Table 12.1 provides a list of plans which are usually required for a proper environmental, health and safety (EHS) management of a construction site. The sequence in which these plans are listed does not necessarily indicate an order of importance, nor is the list necessarily complete. Depending on the situation additional specific plans may be required, and, as indicated at some places in this list, it may be advisable to prepare separate plans for points included in one of the items listed. However, most of these plans will be required on any large construction site, and the responsibility for preparing and implementing them will be the contractor's. In order to ascertain that all these plans will be prepared, implemented and respected, it is important that the requirements will be clearly included in the tender documents.

As mentioned above, the responsibility for developing these plans in detail and for implementing them is the contractor's. The project proponent, however, has the responsibility to examine the drafts of these plans, to comment on them, to approve them once they are satisfactory, and to closely monitor their implementation. Ultimately, and also towards the authorities and the lenders and funding agencies, the project proponent will always be the responsible party. Keeping this in mind, and also the fact that all the relevant conditions should be included in the tender documents, it can well be that such conditions need to be clarified in more detail than just what is provided in Table 12.1. It is not possible here to list all possible pitfalls in this endeavour, however a few items, some of great importance, others of maybe rather secondary relevance, but all based on situations that have actually happened, shall be listed and discussed briefly:

- Detailed tender documents to allow bidders to include all relevant conditions: most if not all of these conditions come at a cost (although many of them may cause some costs upfront, but will prevent additional costs, and potentially higher ones, from becoming manifest at a later stage). A bidder who follows these requirements should not be punished for doing so, but unfortunately this may happen. It is now common practice, and is certainly a policy followed by the World Bank, to "choose the lowest bidder". This is certainly justified, but actually this should clearly be understood as "the lowest bidder who has presented a complete offer", i.e. one which includes, in addition to all technical requirements, also all EHS requirements. Unfortunately, this is not necessarily always the case, and missing items in the originally proposed budget are sometimes some of the details presented under the following points.
- Binding conditions for subcontractors: all these contractually defined plans are binding conditions for the contractor. However, it has to be clearly stated that the contractor is also liable for guaranteeing that all subcontractors, some of whom may be hired at a later stage for specific parts of the work, have to act in compliance with these conditions as well.
- Timely preparation of EHS plans: ideally—and this should be a contractual obligation as well—all these plans should be prepared, approved and applicable at the time when work on site starts. This, probably, is the least respected of all these

Table 12.1 List of plans for EHS management of a construction site

No.	Topic	Objectives	Main measures
1	Waste Management	Prevention, reduction and proper disposal of any kind of waste	• Put in place a suitable waste management system • Separate waste, recycle to the extent possible • For non-recyclable domestic waste, use a suitable waste landfill for disposal; make the necessary arrangements with the administration in case a communal waste disposal site should be used (which in general is not recommended due to problems of access control) • Special care to be given to hazardous waste. Recycle to the extent possible, dispose of according to best practice and legal requirements. A separate plan may be required for hazardous waste
2	Hazardous Material Management	Prevention of accidents (spills etc.) caused by hazardous materials	• Put in place a suitable waste management system • Store carburant and lubricants in an appropriate location (away from water, under a roof, and in a sealed place capable of retention in case of leaks) • Use such material in places equipped for the purpose (e.g. maintenance of vehicles and machines, changing of oil etc., in a workshop with sealed floor draining into an oil separator) • Store and use explosives, in case any will be used, according to relevant safety rules

(continued)

Table 12.1 (continued)

No.	Topic	Objectives	Main measures
3	Spoil Disposal	Disposal of spoil in a way as not to create additional impacts	• Remove topsoil and store separately for later use (see No. 9) • Disposal of spoil in a way as to reduce erosion risk; no spoil disposal in or in immediate proximity to water courses; compact spoil • Sites for spoil disposal will have to be identified and agreed upon with authorities and land owners. Compensation may have to be paid • Dispose of spoil in a way as to facilitate site rehabilitation at the end of works
4	Water Resources Management	Prevent pollution of water	• Provide safe drinking water to work force (either by providing bottled water or by applying suitable treatment) • Collect and treat waste water before letting it flow into a water body. Use infiltration pits for domestic waste water (e.g. from toilets) as long as there is no risk of surface or ground water pollution. A specific waste water treatment plan may be required • Use sedimentation ponds etc. where required for clearing water with heavy load of suspended solids • Take measures to protect water when working in or near a water body • Prevent contamination of ground water

(continued)

Table 12.1 (continued)

No.	Topic	Objectives	Main measures
5	Erosion and Sediment Control	To minimise erosion and sediment input into water bodies	• Minimise, to the extent possible, open surfaces exposed to rain, especially on slopes • No disposal of spoil etc. in the vicinity of water courses (see No. 3) • Site rehabilitation for temporarily used sites, to be carried out as soon as a site is no longer used
6	Air Pollution and Dust Control	Prevent air pollution and reduce dust, especially in or near inhabited areas	• Maintain vehicles and machines in good working order • During the dry season, sprinkle roads for reducing dust • Cover lorries loaded with fine dry material (if any) when transporting it through inhabited areas, or moisten loads so as to prevent dust
7	Noise and Vibration	Reduce noise and vibration to acceptable limits, especially in work places and in or close to inhabited areas	• Maintain vehicles and machines in good working order • No noisy work to be carried out close to inhabited areas during the night • Take special measures, if required, in proximity of hospitals and similar infrastructure • Provide suitable protection (PPE), especially hearing protection, to workers who work in noisy places • A special noise and vibration plan may be required when and where explosives are used

(continued)

Table 12.1 (continued)

No.	Topic	Objectives	Main measures
8	Vegetation Clearing	Minimisation of impact on vegetation	• Clearly identify areas from which vegetation will have to be removed. Limit such areas to the surfaces actually required. Do not remove vegetation outside of these areas • Clearly mark and fence off areas on which vegetation must be kept intact (erosion protection, protected vegetation types) • Coordinate tree felling with local forestry authorities in order to use any commercially valuable timber • Where feasible, make wood from cut trees available to the local population • Dry and burn all other vegetation that will have been cut • Specific measures may be required e.g. for TL corridors (removal of tall vegetation, land use restrictions) or areas used temporarily outside of the construction site as such (e.g. access roads)

(continued)

Table 12.1 (continued)

No.	Topic	Objectives	Main measures
9	Site rehabilitation	Rehabilitation of all areas which may have been used temporarily during the construction period as soon as they are no longer being used	• When preparing the sites, remove topsoil and store separately for later use (see No. 3) • When no longer used, restore the aspect of temporarily used sites (removal of all material, landscaping if required) and cover with topsoil for facilitating revegetation • Depending mainly on climatic and soil conditions, spontaneous development of vegetation may be sufficient; however, in specific situations (visually exposed, erosion prone, etc.) planting of suitable plants may be required
10	Biodiversity Management	Reduction / prevention of negative impacts on natural habitats, on flora and fauna, especially on protected and/or endangered species	• Such a plan may be required if the construction site is located inside or in proximity of a protected area (as was the case e.g. for NT1, Sect. 22.3, or Lom Pangar, Sect. 17.2) • In such a case, two main measures are required, namely, (i) limitation of any project related activities strictly to the required area, which needs to be fenced in, and (ii) strictly forbidding any hunting and other activities by project staff within the protected area • Seek close cooperation with the management of the protected area (e.g. for anti-poaching measures)

(continued)

Table 12.1 (continued)

No.	Topic	Objectives	Main measures
11	Traffic and Access	Reduce negative impacts and accident risk	• Contractor will develop a traffic Management Plan, which will have to cover traffic within the construction site as well as on access roads • Maintain vehicles and machines in good working order • Instruct drivers on traffic rules, enforce compliance; especially: no excessive speed within inhabited areas • Minimise damage to land in case temporary access routes have to be opened
12	Emergency Preparedness	Prepare an emergency plan	• Carry out planning for risks (fire, accidents, oil spills, etc.) • Have required material ready on site (e.g. for the case of an oil spill; fire extinguishers, firefighting equipment and personnel) • Prepare a referral plan (for emergency evacuation in case of a serious accident) • Instruct personnel accordingly
13	Construction and Management of Workers' Camps	Providing suitable living conditions for the work force	• The contractor will be fully responsible for providing suitable living conditions for the work force, whether in a camp prepared specifically for this purpose or in rented quarters • Main requirements: hygienic conditions, good housekeeping, waste management, water supply and waste water treatment, suitable accommodation including toilets and showers • Specific structures (living quarters, toilets and showers) for women

(continued)

Table 12.1 (continued)

No.	Topic	Objectives	Main measures
14	Training and Awareness	Ensure compliance with relevant rules, minimise risks	• Instruct all personnel involved in the works on the relevant ESMMP regulations and rules • Training on Code of Conduct for relation with surrounding communities • Provide instruction on the use of PPE and on first aid to workers (see No. 14)
15	Human resources management plan	Recruit and manage human resources according to relevant standards	• The contractor will comply with national labour and employment regulations, including social security conditions • Define clear recruitment policy (which will include, e.g., priority to be given to locally hired labour, provided the required qualifications can be found) • Recommendation to have a recruitment office in a nearby town and not on site • Clear regulations and procedures concerning gender-related issues and prevention of child labour

(continued)

Table 12.1 (continued)

No.	Topic	Objectives	Main measures
16	Project Personnel Health Plan	Minimise risk of accidents, provide health care in order to protect the health of the work force	• Instruct all workers on safety rules and risk awareness • Provide suitable PPE depending on the specific work of a person, supervise and enforce use of PPE • Provide instruction on first aid, provide suitable first aid material • Provide appropriate health services (clinic, ambulance, dispensary, health staff) on site • Make arrangements for referral of serious cases (accidents or diseases) to a health post, hospital etc. suitably equipped • Provide information on risks of HIV/AIDS and other STDs as well as on COVID-19 protection if required
17	Cultural Heritage	Protect and prevent damage to objects or sites of cultural, historical or archaeological importance	• Give adequate protection to known such sites • Prepare a chance find procedure, which must include instruction of personnel (e.g. foremen responsible of opening new sites or for excavations), agreements with respective services, and clear instructions on how to proceed in case a site is found or suspected • Carry out community consultations before working near cultural sites as e.g. graveyards • Make arrangements with the communities in case known sites of cultural importance will be affected

rules; in fact, situations where two years after the start of construction works there is still no valid site-specific waste management plan, and therefore no state of the art waste management on site (e.g. using "temporary" landfills, which by no means are up to standards), are too frequent.

- Topsoil storage: as mentioned in the measures to be taken, whenever land is prepared for being used, topsoil, i.e. the first approximately 50 cm of earth to be removed, should be stored separately. In reality, this if very often not done (in part due to the lack of the corresponding plan in the early stages of the work). In addition, once work starts everybody is in a hurry for having sites prepared in a way that allows the start of the "real work". This leads to just dumping the topsoil in the places which have been chosen for spoil disposal, and then all other material (i.e. subsoil) is put on top of that. In the end, when one of the obligations is to rehabilitate all sites that were used temporarily, there is no topsoil to spread on these sites in order to create conditions under which vegetation can develop rapidly. This is not only an aesthetic problem, it can also hinder erosion protection, and it is certainly one of the points that finally might cause higher costs than what would have been required if the procedure had been followed according to plan.
- Food for workers: it is a trivial observation that food, especially for workers on a construction site who carry out hard physical work, needs to be sufficient in quality, diversity and quantity. A number of problems can be avoided if food is provided to workers for free. Most of the workers on such a site want to save as much money as possible for supporting their families. This also means that they try to save money if they have to pay for food. They can do this in different ways, e.g. by asking for larger portions with the aim of saving some of the food for another meal, or by trying to get some additional food, e.g. by setting snares for catching small wildlife, which they will then prepare in places not suitable for this purpose. In one case, this has led to a fire in living quarters, destroying a number of rooms. An additional problem with such practices is lack of hygiene, e.g. by temporarily storing food in the rooms, which can lead to health problems. If such a condition on providing food was not included in the original contract, the obvious response of the contractor to an attempt to introduce this procedure at a later stage will be that there is no budget for this.
- Details of living quarters: like the food, living quarters should be adequate, and here it might be surprising what kind of details, in hindsight, should have been defined at a very early stage. Two examples: workers complained about the quality of their beds (matrasses too thin, or even lack of any), and on the layout of toilets and showers. In one case, no matrasses, the answer of the contractor was that there is an obligation to provide beds, but there is nothing about matrasses. And it should be pointed out that this is not ill will: there are cultural differences, and in some places of the world a plank with a thin bamboo mat is considered as an adequate bed, in other places it is not. Likewise for the toilets and showers: the African workers had complained that the toilets as well as the showers were separated only by breast high walls and lacked a door or at least a curtain, a situation which was not considered as adequate and which was then changed. However, in the quarters used by the contractor's own personnel, this was exactly how the toilets

looked. The contractor was a Chinese company, and for them this was apparently an acceptable situation. Trivial as this may seem, in one case these problems lead to a strike by the workers, which paralysed the entire construction site for two or three days, an event which causes quite some considerable costs.

12.2 Off-Site Measures: The Owner's Responsibility

The list provided in the previous Section covers activities and measures to be taken on the construction site as such, directly related to activities on the site, and therefore in the responsibility of the contractor. However, there are a number of issues, which are related to or caused by the project as well, but not with construction or construction related activities. These issues are clearly not within the responsibility of the contractor charged with construction, here the project owner is directly responsible, although of course for some of these issues other parties may be involved, either officially or as contractors again.

The most important of these off-site issues are all activities related to resettlement, whereby here resettlement is understood in the broad sense of the term, including physical relocation (i.e. the case where people actually need to be moved to a different site) as well as economic relocation (i.e. the case where people have to be compensated for assets, but do not need to relocate to a different place). All these activities are covered by the RAP (the Resettlement Action Plan), which is usually part of the ESIA. Depending on the situation, other more specific plans may be required, as e.g. a Stakeholder Engagement Plan, a Livelihood Restoration Plan, or an Indigenous Peoples Development Plan. These are very important aspects, but not the core subject of this book, and therefore not discussed in any more detail. For more information on these issues see e.g. Cernea (1996, 1997, 1999), Cernea and McDowell (2000), McDowell (1996), Scudder (2005), or the respective guidelines and handbooks of the major development banks (ADB 1998; IFC 2002; WB 2017).

Here, a number of issues are highlighted which usually arise in the context of a major hydropower project and which require the attention of the developer, usually to be defined in a specific plan.

Camp Followers Plan: a large construction site for a typical dam and HPP project will last for a number of years, and at the height of activities several hundred or even a few thousand people may live there. This can attract people who do not work on the site, but who want to benefit from this situation by offering services to the work force. In this way, a considerable number of buildings, mainly shops, restaurants and bars, may spring up. There may be no way to prevent this from happening, but it is definitely in the interest of the good functioning of the construction site to try to have at least a minimum of organisational and administrative standards observed in this settlement. It is the project proponent's obligation to get in touch as early as possible, ideally before disorganised development has started, with the local authorities in order to take influence on what happens there. It may be good to have some control over who is going to settle there, and obviously this has to be done by local authorities, the project as such has little if any influence. An important measure for limiting influx of people from outside is not to recruit staff on site, but to have a recruitment office e.g. in the nearest town; people who come to the site in the expectation to find work there may remain nearby if they are not successful, still hoping to make a living there. One major risk in an unorganised setup is hygiene, especially food hygiene. Providing food to the workers for free is a good way for minimising the temptation to go for food outside of the camp, which is especially high if there food can be had at a lower price than within the camp. But even so, people will occasionally go out, and having health standards applied in the restaurants and bars outside the camp will help in limiting health problems. Given the fact that a workers' camp is populated mostly by men, prostitution will inevitably be among the services offered, which poses another health problem. In addition to that, especially when camp followers come from outside of the area, there is a risk of conflicts with the local population. All these points have to be addressed in the Camp Followers Plan.

Biomass Clearing Plan: this is another point in which the contractor usually will not be involved, and which is entirely the concern of the project proponent. Whether or not such a plan is required will normally have been decided in the ESIA. As described in Sect. 17.4, basically a look into the future reservoir area will be sufficient for deciding whether or not a pre-impoundment biomass removal will be required, but some more sophisticated analysis may be required for clarifying this question. The plan to be made for this activity will basically be the tender document for getting proposals from companies who can do this kind of work. As for the main contract, this document should also clearly specify EHS related aspects of this work; focal points here may be workers' safety and measures to prevent cutting vegetation or spreading of fires outside of the future reservoir area. Dense vegetation is usually also a good habitat for a large number of different species of animals, and removal of vegetation should also aim at driving animals out of the zone at risk. Leaving islands of vegetation within the area that will be submerged may offer temporary refuge to animals, who will then be at a higher risk of drowning during impoundment. Timing of vegetation removal is another essential but difficult issue: the programme must start early enough for giving sufficient time for carrying out the work. On the other hand, removing vegetation too early can lead to a massive re-growth, e.g. when cleared land remains open for more than one rainy season; such regrowth, which will invariably consist exclusively in soft biomass, may undo much of the effort.

Downstream Area Management Plan: such a plan will be required if the project has a considerable impact on the downstream area. In the case of ROR schemes, such impacts will be usually limited, since flow conditions in the river downstream of the power plant will not change. Storage projects, however, will change this situation, whereby three major impacts can be distinguished, namely (i) change in seasonal flow pattern, (ii) peaking, or (iii) interbasin projects which permanently reduce flow in one river and increase it in another one; all three effects, which can arise in any form of combination, have the potential to change the living conditions for the affected riparian human population in a fundamental way. According to Scudder (1996) this is the group of project affected people which is the most likely for being overlooked in the impact assessment (as was e.g. the case for Lom 1 village mentioned in Sect. 24.1), although the consequences of the project can be quite far-reaching for them.

Watershed Management Plan: such a plan is often asked for in the TOR for an ESIA for a dam project, or is a condition in the Concession Agreement for the project. Here, it has to be seen clearly that activities in the watershed area can influence a hydropower project, and most of all its reservoir, the two most relevant ways being an increase in sediments brought into the reservoir, e.g. due to deforestation, which can reduce the life span of an HPP, or the deterioration of water quality due to the inflow of not or insufficiently treated waste water. However, the project proponent—or, at a later stage the operator of the power plant, which is not necessarily the same entity—has very limited if any possibility for influencing development and activities in the catchment area. This means that a "watershed management plan" prepared for an HPP can actually analyse the prevailing situation as well as the likely developments in the catchment area, identify potential problems arising from there, and make suggestions or recommendations for measures to be taken or types of development to be prevented, but the project owner does not have the legal power for taking measures accordingly. This will have to be done by the respective authorities, who may take such suggestions or recommendations into consideration, but who on the other hand may have priorities of a different kind.

Biodiversity Offset Plan: this point is discussed in Sect. 6.3. Such a plan will be required if the project has considerable impacts on biodiversity which cannot be avoided or minimised to a sufficient extent, so that compensation elsewhere will be required.

12.3 Monitoring: A Shared Responsibility

The importance of monitoring has been mentioned several times already, as e.g. in Sect. 9.3.5 concerning water quality. Here, only the issue of responsibility for monitoring in the process of ESMMP implementation is discussed shortly. Such a monitoring needs to be done on three levels:

1. Primary level: internal monitoring done by the contractor. On a large construction site, the contractor will usually have an EHS unit, adequately staffed and equipped, on site. This unit is in charge of implementing the different plans listed above, or has the obligation to ascertain that these plans are being applied and implemented. This responsibility also includes monitoring, i.e. a constant supervision of how the measures are implemented, and the obligation to take corrective action if the aims are not reached. These corrective actions can again be on two levels; the most frequent will be to intervene wherever measures are not applied (here, e.g., the use of appropriate—and mandatory!—PPE, the personal protection equipment, is one rule often ignored, which requires ongoing control and insistence). However, it is also possible that monitoring shows that

originally planned measures do not bring the expected or required results, which means that the respective plan needs to be adapted; one such case could be that monitoring reveals that in spite of treatment waste water does not reach the legally required standard, meaning that more efficient treatment measures, like a better waste water treatment plant, may be required.

2. Secondary level: as mentioned, the ultimate responsibility for ESMMP implementation lies with the project proponent, who therefore will also need to have a group of specialists on site whose main obligation is to follow up on all activities, identify and report non-compliance, and take actions to remedy the situation. This will usually be done by a combination of own staff of the project proponent with specialists of the OE (Owner's Engineer), a company mandated with the management of the project in the owner's interest.
3. Tertiary level: external monitoring. This is a type of monitoring done by occasional visits of external experts who will check on all relevant issues at hand. This is often being done, e.g. in World Bank financed or co- financed projects, by a so-called POE (Panel of Experts, sometimes called IPOE, Independent Panel of Experts, for emphasising the fact that this is a group of external experts which represent neither project proponent nor contractor), or the LTA (Lender's Technical Advisors) in the case of a financing by private banks. In both cases, these groups will usually comprise technical as well as environmental and social experts, who will concentrate their efforts on their respective area of expertise. In addition to that, monitoring can also be done by regular or occasional visits of specialists from the competent authorities, e.g. the EPAs. In both cases, these experts will not take direct action on the construction site, but will note their observations and communicate them in discussions with the responsible persons as well as in their reports, and will check the outcome in subsequent visits (Fig. 12.1).

Fig. 12.1 Health and safety risks on dam construction sites. From left to right: complete lack of any PPE; substandard scaffolding, no safe access; the originally red colour of the box has disappeared, and so has the fire extinguisher which is supposed to be in it

References

ADB. 1998. *Handbook on resettlement. A guide to good practice*. Manila: Asian Development Bank, 115 pp.

Barrow, C. 1988. The impact of hydroelectric development on the Amazonian environment: With particular reference on the Tucuruí project. *Journal of Biogeography* 15: 67–78.

Cernea, M. 1996. Understanding and preventing impoverishment from displacement. Reflections on the state of knowledge. In *Understanding impoverishment. The consequences of development-induced displacement*, ed. by C. McDowell, 13–32. Refugee and Forced Migration Studies Vol. 2. Oxford: Berghahn Books.

Cernea, M. 1997. The risks and reconstruction model for resettling displaced populations. *World Development* 25 (10): 1569–1588.

Cernea, M.M. 1999. Why economic analysis is essential to resettlement: a sociologist's view. In *The economics of involuntary resettlement*, ed. by M.M. Cernea. Washington D.C.: World Bank.

Cernea, M.M., and C. McDowell. (eds.). 2000. *Risks and reconstruction. Experience of resettlers and refugees*. Washingotn D.C.: World Bank, 487 pp.

IFC. 2002. *Handbook for preparing a resettlement action plan*, 79 pp. Washington D.C.: IFC, International Finance Corporation.

McDowell, C. (ed.).1996. *Understanding impoverishment. The consequences of development-induced displacement*. Refugee and Forced Migration Studies, Vol. 2: 257 pp. Providence: Berghahn Books.

Scudder, T. 1996. Development-induced impoverishment, resistance and river-basin development. In *Understanding impoverishment. The consequences of development-induced displacement*, ed. by C. McDowell, 49–74. Refugee and Forced Migration Studies Vol. 2. Oxford: Berghahn Books.

Scudder, T. 2005. *The future of large dams. Dealing with social, environmental, institutional and political costs*, 389 pp. London: Earthscan.

WB. 2017. *The World Bank environmental and social framework*. Washington, D.C.: International Bank for Reconstruction and Development/The World Bank, 106 pp. http://documents.worldbank.org/curated/en/383011492423734099/pdf/114278-WP-REVISED-PUBLIC-Environmental-and-Social-Framework.pdf.

Chapter 13
Decommissioning

> *...there was a long tumultuous shouting sound like the voice of a thousand waters - and the deep and dank tarn at my feet closed sullenly and silently over the fragments of the house of Usher.*
> Edgar Allan Poe, The Fall of the House of Usher (1839)

Dams, reservoirs and hydropower plants have a finite lifetime, and for each project the moment will come when the question of decommissioning will have to be addressed. In fact, an increasing number of dams now reach this moment, meaning that the issue will become more prominent, an important field of action. In an attempt to anticipate this development, ESIA guidelines increasingly ask for a section in the report specifically dedicated to decommissioning. Basically, there are three potential solutions, namely (i) replacing an aged structure with a new one, (ii) improving and changing it in a way as to extend its useful life, or (iii) removing the dam; a potential fourth possibility, namely, just abandon a structure and leave it as it is, is most probably not really a solution, since sooner or later such a course of action might lead to a serious safety issue. These possibilities related to decommissioning are discussed in this chapter, along with the question of how much sense it makes to include this topic in an ESIA.

13.1 ESIA and Dam Decommissioning

As mentioned above, many standard TORs for ESIAs for dam projects now demand a chapter on decommissioning. This addition was probably based on the idea of a lifecycle analysis, which is required, and rightly so, for an increasing number of products, where rather often the most lasting and severe impacts are caused at the end of the life cycle, when the product must—or should be—disposed of in a proper way. This is certainly a very good approach for a large variety of products, and thinking of end-of-life disposal has led in many cases to an improvement of such products, like e.g. multi-component packages, which are suitable for production, transport and sale, but difficult if not almost impossible to recycle, and which often proved as

R. Zwahlen, *Assessing the Environmental Impacts of Hydropower Projects*, Environmental Earth Sciences, https://doi.org/10.1007/978-3-030-91185-0_13

long-lived and environmentally unsafe. The other point that may have contributed to including this requirement in standard TORs is probably the fact that a number of dams have reached an age where they are no longer safe and need to be removed. In fact, this has been done in a number of cases already (see Sect. 13.4).

However, although basically a life-cycle approach is a sound way of thinking about any project, I am not convinced that the requirement for a chapter on decommissioning in every ESIA for an HPP is necessarily a good idea. In my opinion, this cannot go beyond some very basic statements, reflecting the different options that might be available, but it cannot possibly be case-specific, as an ESIA should be: identifying project-specific impacts and proposing feasible, project specific solutions. This cannot really be done for a larger, complex dam and hydropower project at the feasibility or detailed design stage, and for this reason I consider including a detailed account of decommissioning of a dam and related structures in the ESIA, i.e. in the planning stage, as mostly useless. This opinion is based on the following facts:

- A hydropower dam, and especially a large one, is usually designed and meant to last for about one hundred years.
- Development of the structure during the lifetime and state at the end of it will depend on a number of factors, starting with planning, then construction, and maintenance during operation.
- Some impacts on these structures, like e.g. siltation of the reservoir or potential life span of certain components, can certainly be anticipated, and options for dealing with such a situation in the future can be listed. However, nobody can foresee what else is going to happen during this long time span, since basically everything can change, from the political and economic situation (power market, economy in general), the environmental and social situation in the project area, the legal situation, to the development of new technologies etc.
- In any case, the removal of a dam, and especially a larger one, will require an in-depth analysis of options and potential outcomes at the time when it will actually have to be done; such a study will have to include technical aspects, and it will certainly have to include an ESIA; this latter, then, will be done taking into account the prevailing situation, and it will not be based on at best very vague, at worst completely wrong assumptions on the situation anticipated a hundred years ago. I'd be very surprised if any of the decommissioning studies done in the planning phase would still be of any use at decommissioning a century later, save maybe for giving the future planners a good laugh.

This situation will most probably lead to a chapter on decommissioning which is pretty much a standard text, at best adapted a little bit to the specific project, but without any really useful information. It will add to the (already often exaggerated) bulk of ESIA reports, without any real value added.

This is illustrated with an example: the case of the Rheinfelden Hydropower Plant.

13.2 The Replacement of Rheinfelden HPP

The Kraftwerk Rheinfelden (Rheinfelden HPP) is located on the Rhine River between the Lake of Constance and the city of Basel, where the river forms the border between Switzerland and Germany. It consisted in a dam across the river and a headrace channel on the German side leading to the power house about 500 m downstream of the dam. It started operation in 1898, and with its 10 MW capacity it was the worldwide first, and for a long time largest, run-of river hydropower plant. Later on, installed capacity was raised to 25.7 MW. Conceived and built as a run-of-river plant, as all the other HPPs built later on this stretch of the Rhine River, the dam just provided some additional head, but very little if any regulation; regulation is done naturally by the large Lake of Constance upstream of the cascade.

In the 1980s, auditing of the structures showed that the foundations of the dam had started to weaken, and that therefore the end of the lifetime of this structure was approaching; it was, however, not in any way an emergency: thanks to good and regular maintenance and monitoring, the owners of the project got an early warning, with ample time for deciding on how to proceed. Still, a first attempt for replacing the power plant was abandoned in 1984: the opening of the European power market made economic success of this project questionable. It remained clear, however, that the existing structure could not be maintained much longer.

A study was commissioned for developing feasible technical alternatives, and this was accompanied by an EIA (EWI 1987; EWI and Simultech 1988). Since mere maintenance work turned out not to be sufficient for guaranteeing safety of the structure it became clear rather quickly that there were two main alternatives, namely, (i) removing the dam and returning to the "natural state", or (ii) replacing it with a new dam; the latter still offered the alternatives of keeping the present water level in the upstream area, or raising it to some extent.

Technical feasibility for the project of a replacement dam, to be built a very short stretch downstream of the existing one, was established. The analysis also showed the project of building a new dam and power house at this place was marginally interesting from an economic point of view. One important difference of the new structure, from an environmental point of view, was that it would have to include a fish pass; due to a number of dams across the river, fish migrations in the river Rhine had no longer been possible for close to 100 years, which had led to the almost complete disappearing of the European salmon (*Salmo salar*) from this river system, and it was now clear that every power plant of the cascade, on renewing the concession and/or replacing the structure, would have to include a fish pass with the aim of ultimately restoring fish migrations.

Environmental investigations revealed some important information. One was that, obviously, nature had adapted to the new situation over these 100 years, with the new water level at a maximum of about 10 m above the former natural water level in the river. One consequence of this was the groundwater level in the area surrounding the reservoir, which was higher than it had been originally. Of more importance, however, was the situation of the man-made, built environment. A large number of buildings of all types had been constructed along the shorelines and in the surroundings of the reservoir. This, first of all, ruled out any plan of increasing the water level, since this would not have been possible without massive interference with many of these structures, causing enormous compensation payments. On the other hand, there was also a risk in removing the dam, since lowering the water level in the river would have led to a lower groundwater level in the surroundings, which in turn might have led to damage to some of the buildings adapted to this artificial situation.

Finally, one additional fact was discovered: In the 1920s, among other industries, a chemical factory had been built on the German side of the reservoir, which produced mainly goods for the German army (see also Cioc 2002: 132). During a long period, waste water from the factory, presumably all of it untreated, drained directly into the reservoir; protection of the environment, if any, was certainly a very low priority during this entire period. This situation led to contaminated sediment in the reservoir, which then was covered again by about 40 years of uncontaminated sediment. Nobody knew exactly what kind of substances might lay buried in the reservoir, and probably nobody really wanted to know. However, one thing was clear: as long as the sediment remained undisturbed where it was, there was no major risk of contamination. Removing of the dam and restoring original water level and flow velocities would necessarily lead to mobilising this sediment, and whatever there was would have been swept downstream. This would not have been acceptable. This meant that before removal of the dam it would have been necessary to investigate in detail the nature of these sediments, and to remove, treat and dispose of them in an environmentally safe manner. Costs for dealing with such contaminated sites can be quite enormous. Here, the additional question would have been who would have to pay. The company which had caused the damage no longer existed. So it would have been either the Kraftwerke Rheinfelden AG, or one or both of the bordering states. Conclusion: better leave it alone.

There was also a discussion about the old headrace channel and power house: it had been listed as a historically valuable monument documenting industrial development, and there had been requests to maintain it. However, among other reasons because the only available space for inserting a fish pass into the new structure was the place where this channel had been, it was finally decided to remove it.

In 2003, construction of the new structure, dam and integrated power house, started. This included a fish pass in the form of a natural bypass stream, which represents a new type of structure for this purpose, and also a beaver pass: during the planning and construction stage of the new scheme, which had lasted for close to 20 years, the beaver had become re-established along a number of Swiss rivers, and also along this part of the Rhine. The old dam and power house were removed, and the new 100 MW power plant went into operation in 2010. Power generation had thereby risen from 185 GWh from the 25.7 MW structure to 600 GWh per year.

This should not be understood as saying that the new dam was built only to keep the contaminated sediment in place, but it is quite certain that the fact of this contamination and the potential consequences of mobilising these sediments helped tipping the balance that way. The relevant point, however, is that nobody, even if there had been an ESIA, including a chapter on decommissioning, for the original scheme in the 1890s, would have been able to foresee the development over the coming 100 years, and would then have come to conclusions which would have been of any use in the discussions about how to decommission this plant in the early twenty-first century. It cannot be assumed that what is being done now in this respect for new projects is of any more relevance.

13.3 Rehabilitation

Rehabilitation of an existing HPP will not usually be part of the TOR for a new project. However, if an HPP reaches the end of the concession period, which is not necessarily the same as reaching the end of its life span, rehabilitation may be required. In this sense it is important that already in the original project aspects like maintenance were defined. This is certainly of importance in the cases of so-called BOT (Build–Operate–Transfer) projects. The idea here is that an entity, usually a private developer, builds an HPP and gets a concession for operating it, e.g., during 30 years in order to get the investment paid back, and after this period the project will be transferred to the country in which it is located, for being operated for the remaining part of its useful life span. In such a case the original contracts, largely covering technical aspects, have to define the maintenance requirements and the condition in which the plant will have to be handed over, to avoid a condition where the concession holder would do just the necessary to keep the installation working till the end of the concession period, and then transferring it in a condition which would require major investments to bring it up to a good working condition again.

However, one other point needs to be considered as well, and this will also apply to a dam which, possibly after a much longer operation period, is considered to be structurally still in a condition that allows to continue to use it. During the operation period, legal conditions may have changed. Concerning hydropower, the two main (environmental) conditions that may have changed, and that actually have changed in some places, are the legal requirements concerning residual flow and fish pass facilities. Many of the older power plants were built at a time when such requirements were not yet defined legally, or only at a level which is no longer considered as acceptable today. Legally it is often, maybe usually, not possible to retroactively impose such new regulations on existing structures, at least not if this would mean an economic loss; increasing the residual flow, e.g., will do exactly that, if a part of the water then would need to be restored to the river instead of being used for generating electricity. However, when a new concession is required, this would be the moment for adapting the structure and its operation pattern to the new standards. This is an important part, e.g. in the structures within the Rhine River basin, most of which are rather old and need to be rehabilitated, and which now have to integrate fish passes, which are a prerequisite for re-enabling the migration of the salmon in this river and its main tributaries; other fish species, which may carry out less long and less spectacular migrations, but which nevertheless depend to some extend on possibilities to move within the river, will benefit from such changes as well.

This means that for a rehabilitation project, even if a full-fledged EIA might not be required, the studies required for answering to new legal requirements will have to be done.

Other situations where a rehabilitation may be required are cases where structures, due e.g. to political unrest, transition periods or war, were either neglected, e.g. due to a lack of spare parts, or where they have suffered more or less extensive damage, which affected their proper functioning and/or their safety. In such a case, a careful assessment of the situation, technically as well as environmentally and socially, will have to be carried out in order to assess the situation: technical and economic feasibility of rehabilitation and measures required to bring it up to standard. A few examples are listed below, whereby the main aim is not to go into the details of the different cases, but to use each one of them to illustrate a few items which were of importance.

Examples:

- **Enguri**: this 1300 MW power plant in Georgia, which with a height of 271.5 m includes the highest double curved arch dam, was in a rather bad shape when the EBRD commissioned a rehabilitation study (EWE 1997). The dam is located on the Enguri River, and from there the water is transferred by a 15 km long headrace tunnel to the power house, the tailrace ending in the Okumi River; both rivers drain into the Black Sea. Only one of the five units was still working, and even this was in a precarious condition. Damage had been caused by the civil war after the breakdown of the Soviet Union, and afterwards the acquisition of spare parts had turned out to be as good as impossible. The situation was further aggravated by the fact that the dam was in Georgia ("as such"), but the power house was in Abchasia, the Western province which had some kind of de facto independence. During this period, it was not possible to drive the approximately 15 km between power house and dam in one car, arrangements had to be made for a car from the other side to come and pick one up. All structures, with the exception of the dam body as such, were in a bad condition, and access to all structures—underground cavern power house, dam, surge shaft—was very dangerous. There had been a fire fighting system, but all parts of it were out of order; as one example: the small reservoir which was supposed to provide water in case of a fire was empty, probably due to excessive leaking. The plant seemed to run mainly on vodka. The large underground power house was completely without any light, everything was covered with a slippery water and oil film, and all types of debris were lying everywhere. Rehabilitation, in addition to the technical measures which must have included replacement of a number of the larger structures in order to make it fully functioning again, was also about reducing and eliminating environmental, health and safety risks, some of which are illustrated in Fig. 13.1. The first picture in this Figure also illustrates an additional shortcoming: at the end of the construction phase, nothing had been done for rehabilitating temporarily used areas; everything still looked very much like a construction site, structures in various stages of decay still littering the landscape.

Fig. 13.1 Problems in Enguri HPP in 1997. Top, from left: view of the dam, with some residues of structures as e.g. the conveyor belt (top centre) used during construction still in place; dangerous access at the foot of the dam; damaged catwalks, a danger to those walking below. Bottom, from left: opening of the surge shaft freely accessible, mesh wire grid covering it partly missing; unmarked manhole in a dark corner, handrails partly rusted away; insulation of pipes, probably asbestos, badly damaged

- **Mount Coffee, Liberia**: this project, a run-of-river scheme, was constructed between 1963 and 1966 and went in operation in 1967. The initial generating capacity was 30 MW provided by two turbines, which was increased by two additional turbines to 64 MW in 1973. It was the most important power plant in Liberia until the civil war started in 1990. On June 28, 1990, the National Patriotic Front of Liberia (NPFL) took control of Mount Coffee HPP, and the plant operator was ordered to stop supplying electricity; on June 30th, 1990, the last unit was taken out of operation. With no units on-line the spillway gates could not be raised and the forebay dam was overtopped and breached August 12, 1990, and the reservoir emptied completely. Since then powerhouse, substation and spillway structures have suffered considerable looting and damage. The power house was completely stripped of all metal structures, including turbines and generators, literally down to the last handrail and cable, only the concrete structure was left (Figs. 13.2 and 13.3). Refurbishment meant to repair the dam and to completely rebuild and re-equip the power house. In the over 20 years without water, some parts of the reservoir had been taken over again for cultivation, which included some rubber tree plantations, but luckily nobody had moved in yet to live there, which made compensation not too difficult, since it did not require any physical relocation (Pöyry 2012).

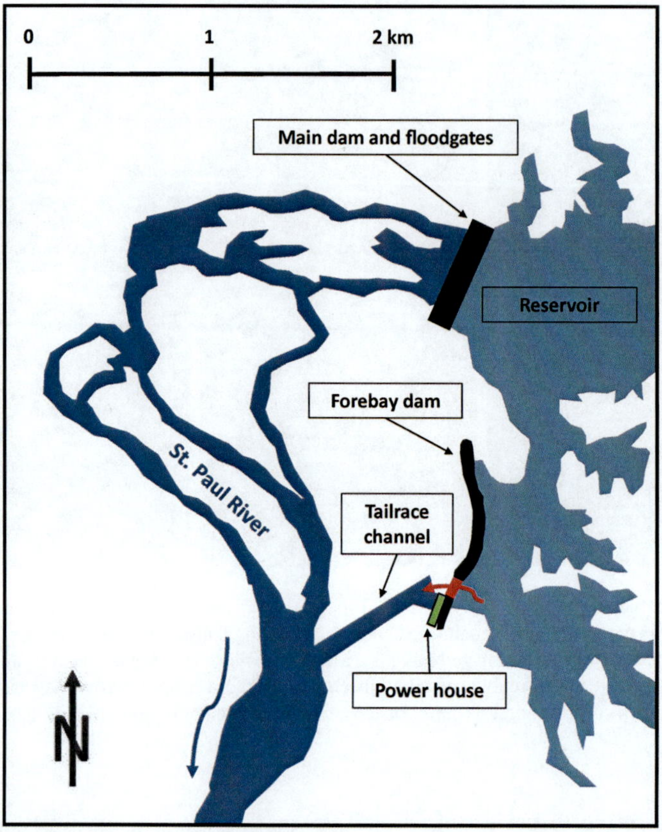

Fig. 13.2 Layout of Mount Coffee HPP, Liberia. The red block and arrow indicate the place where the forebay dam was breached

Fig. 13.3 Mount Coffee HPP, Liberia. Left: site of the dam breach; the forebay dam was an earth dam, the concrete structure on the right is a part of the intake; view from downstream. Right: remaining shell of the power house, breach to the left, part of the forebay dam visible on the right behind the power house

- **Nurek**: after the breaking up of the Soviet Union, Tajikistan found itself in a difficult situation in more than one sense. The energy sector had not been devised as an independent system, but rather as a part of a large Central Asian grid. The largest hydropower plant, Nurek, which with 300 m had the then highest dam in the world, had been conceived mainly for providing electricity to a large aluminium smelter and peak power to the entire grid, while the thermal power plants in Uzbekistan were designed for providing baseload power. After independence, this exchange stopped, and the country had to try to get along on its own resources. This was further aggravated by the civil war which soon broke out. In this situation, domestic energy production was by far not sufficient, and the country went through a very difficult period. In spite of that, Nurek was quite well maintained, e.g. in comparison with Enguri mentioned above, although here as well it was very difficult or actually impossible to get any spare parts. When approximately 10 years after independence first assessments of the situation were made, it turned out that the issue was actually much more than just to repair any damage that Nurek HPP might have suffered. In fact, large parts of the distribution grid had literally vanished, since people hat made use of anything that was available, e.g. by turning wooden poles of community grids into fuelwood (EWE and SMEC 2000).
- **Pamir**: in the Pamir area, Gorno Badachshan, the South-Eastern province of Tajikistan, the situation was even worse than in the rest of the country. There was only one major power plant, Pamir 1, built during the Soviet period; it had been planned for 28 MW, of which 14 MW had been installed, the rest to be added later. Additional electricity needs had been covered by diesel generators. However, after independence diesel was no longer available for free, and nobody could afford to buy any. The result was an immediate energy shortage especially in winter, which led to load shedding in Khorog, the provincial capital, and in cutting off all the villages in the area from electricity supply completely. In 1991, IFC and AKFED, the Aga Khan Foundation for Economic Development, financed a feasibility study for investigating the conditions for upgrading the situation (EWE 2001). The aim was to bring Pamir 1 to full capacity and to rehabilitate and improve a few other, minor HPPs in the region. Part of the studies undertaken was a so-called Willingness to Pay study, in order to check whether the project would have a chance of being economically feasible, i.e. whether the loans could be paid back via payments from consumers. In this study, it was made clear to the population that the project would bring electricity year-round, but at a cost for all consumers (a notion which after 70 years of a communist regime was difficult to grasp), and the people were asked whether they would agree with the project under these conditions. Not surprisingly, nobody ever said that they would rather not have the project, although from the results of the socio-economic survey, as well as from the facial expressions of the people in the meetings, it was absolutely clear, given the extremely low level of income, if any at all, that without an improvement of the economic situation as good as nobody would be able to pay the bill. As in the study carried out for Nurek, and as in so many other similar studies, the denomination of the study was wrong: the issue was not about Willingness to Pay,

but about Ability to Pay. Luckily, the project was implemented anyway, and in the meantime ability to pay has been restored at least to some considerable degree. See Sect. 5.7.3 for a discussion of the WTP approach and Sect. 25.2 for some additional, mainly environmental, issues around this project.

13.4 Removal of Dams

While, as just explained, a detailed assessment of decommissioning in the framework of an ESIA for a dam project may be of little use, the decommissioning as such is certainly an issue of growing importance. Dams have been built and operated since a very long time (see Sect. 3.1), but construction of a very large number of increasingly large dams, mainly for hydropower, has started only a little bit more than 100 years ago, and has increased very much over the past decades. This also means that a number of dams, as mentioned in the example of Rheinfelden HPP, is now reaching the end of their life span, and we will increasingly be faced with the question of decommissioning. Since the first dams built were comparatively rather small ones, not surprisingly we also first have to deal with decommissioning of small dams. But over the next decades, some of the large dams will start to get into this phase of the lifetime end of such structures, and solutions will have to be found how on to deal with this situation. There can be no doubt that, as also just mentioned above, decommissioning of dams will require a detailed assessment, of technical, financial, economic as well as environmental and social issues at hand.

A considerable number of dams have been removed already, according to Vahedifard et al. (2020) about 1700 alone in the USA. All of them, so far, were small to medium sized dams, and very often the main aim was to remove the dams in order to restore rivers as salmon habitats, i.e. to re-establish the migration routes of these fish. There is a growing number of cases where this has been done successfully already. The main question in these cases was usually abut mobilisation and removal of sediments, and this includes concerns about removal of the sediments accumulated in the former reservoir as well as the impact of a sudden massive increase in sediment loads in the downstream area of the former dam. In many cases reported in the literature, the return to a natural or near-natural state came about with surprising speed (Learn, 2016; Foley et al. 2017a, b). However, Vahedifard et al. (2020) state that "[r]ecords show that local river and ecosystem do not fully revert to their pre-dam conditions"; two things have to be mentioned here: first, with dams having been built > 90 years ago it is questionable that information on pre-dam conditions is really available, and second, a return to real pre-dam conditions, after 90 years of impoundment, cannot reasonably be expected to be reached within just a few years.

As mentioned, the cases where this has been done so far were mostly small to medium sized dams, on rather small rivers. To the best of my knowledge, so far no really large dam, let's say concrete dams with a height of over 100 m which formed a reservoir with a storage volume of several km^3, and which after a useful life time of a century were finally filled with sediments, has been removed so far. This will be a challenge of a completely different order of magnitude, and there will not be that many options. Just building a new dam a short stretch downstream of the existing one, and then removing the old structure and continue to operate it, as was the case in Rheinfelden, most likely will not be an option in most cases. This will only leave two alternatives: either leave the dam in place (and, in the best of cases, using the structure as a ROR power plant since the river will still be there), or removing it. Leaving it may turn into a safety issue for the downstream area, since the stability of the dam might no longer be guaranteed. Removing it will make mobilisation of the accumulated sediment unavoidable, and this could be a risk in itself, but, in the case of a cascade, would also jeopardise structures in the downstream area, which might be younger and could still be operated for a long time, but may not be able to cope with such a massive sediment input. A lot of work is waiting here.

At least an indication of effects of a dam removal on the downstream area is provided by Foley et al. (2017) for the case of the Elwha River in Washington State, USA. There, starting in 2011, the 32 m high Elwha dam, commissioned in 1912, and the 64 m high Glines Canyon dam, commissioned in 1927, were removed over a period of almost 4 years. The removal of these dams mobilised the 30 million Mg of sediment which had accumulated in the reservoir. The situation in this case was rather special since the lower of these two dams, the Elwha dam, is located only about 12 km away from the sea. The very large amount of sediment which was transported to the estuary over a very short time had a profound effect on the delta, increasing it considerably, which in turn changed habitat conditions which had become established over the period of almost 90 years of reduced sediment influx in a rather fundamental way, shifting community composition from brackish to freshwater-dominated species. The authors suspect that in such a case reverting to a situation close to pre-dam conditions, or at least to a new equilibrium, might not take just a few years, but could well take several decades.

References

Cioc, M. 2002. *The Rhine. An eco-biography 1815–2000.* University of Washington Press, 255 pp.

EWE and SMEC. 2000. Power sector development project. ADB 3207 TAJ. Vol. 1: Energy sector review and policy formulation. Vol. 2: Diagnostic analysis of Barki Tajik. Vol. 3: Rehabilitation and investment program. Vol. 4: Environmental and socio-economic analysis. Vol. 5: Feasibility report. Reports prepared by Electrowatt Engineering and SMEC International for Asian Development Bank and Government of Tajikistan.

EWE. 1997. *Enguri Dam and hydroelectric power station, Georgia.* Environmental, Health and Safety Audit. EBRD.

EWE. 2001. Evaluation of the GBAO electric networks and assistance in project preparation. In *Environmental and social measures*, vol. 3. International Finance Corporation and Government of Tajikistan.

EWI und Simultech. 1988. *Neubau Kraftwerk Rheinfelden. Ergänzungsbericht zum UVP-Bericht 1987.* Kraftübertragungswerke Rheinfelden AG.

EWI. 1987. *Neubau Kraftwerk Rheinfelden.* UVP-Bericht. Kraftübertragungswerke Rheinfelden AG.

Foley, M.M., J.R. Bellmore, J.E. O'Connor, et al. 2017. Dam removal: Listening in. *Water Resoources Research* 53: 5229–5246. https://doi.org/10.1002/2017WR020457.

Foley, M.M., J.A. Warrick, A. Ritchie, et al. 2017b. Coastal habitat and biological community response to dam removal on the Elwha River. *Ecological Monographs* 87 (4): 552–577. https://www.jstor.org/stable/26358501.

Learn, J.R. 2016. *Hydroelectric Dam Demolition Helps American Dippers*. The Wildlife Society, Posted on January 28, 2016 http://wildlife.org/hydroelectric-dam-demolition-helps-american-dippers/.

Poe, E.A. 1839. *The fall of the house of Usher*. https://www.poemuseum.org/the-fall-of-the-house-of-usher.

Pöyry. 2012. *Mount Coffee HPP ESIA and RAP*. ESIA Report. Rehabilitation of the 64 MW Mount Coffee Hydropower Plant, Liberia. West African Power Pool (WAPP).

Vahedifard, F., K. Madani, A.A. Kouchack, and S.K. Thota. 2020. Preparing for proactive dam removal decisions. *Science* 369 (6500): 150. https://doi.org/10.1126/science.abc9953.

Chapter 14
Comparison with Alternatives

Quand on n'a pas de pétrol, il faut avoir de l'eau.
(When there is no oil, one needs to have water.)
Cans (1997: 166)

The comparison of alternatives is a requirement of all ESIA guidelines. While this is certainly justified, it is not always an easy thing to do, since very often there is one project to be studied, with no real alternatives, especially in the later stages of project development like feasibility study or detailed design. Some of the possibilities that may exist are presented below, and discussed on the basis of suitable examples. While project-specific alternatives like different dam sites or project dimensions may be under discussion only exceptionally, two potential alternatives can—and should be—discussed in any case, namely (i) the possibility of generating the same amount of electricity with a different energy source, and (ii) the no-project alternative, meaning the solution of not doing anything.

14.1 Direct Project Alternatives

14.1.1 Different Dam Sites

The choice of the dam site, for a number of reasons, can be very decisive in terms of environmental and social impacts. Because of this, in cases where there is such a choice, the alternatives have to be evaluated carefully. Very often, however, while in the project identification and prefeasibility stage a number of dam sites may be considered, this is most often no longer the case in later project development stages. While hydrology (i.e. availability of water) and topography (i.e. availability of landscape structures favourable for the placement of a dam, like e.g. a narrow gorge) are decisive criteria in the project identification phase, the specific dam site is then usually defined according to geological criteria. If a potential dam site has been excluded due to such an evaluation, there is no more point in trying to take it into

R. Zwahlen, *Assessing the Environmental Impacts of Hydropower Projects*,
Environmental Earth Sciences, https://doi.org/10.1007/978-3-030-91185-0_14

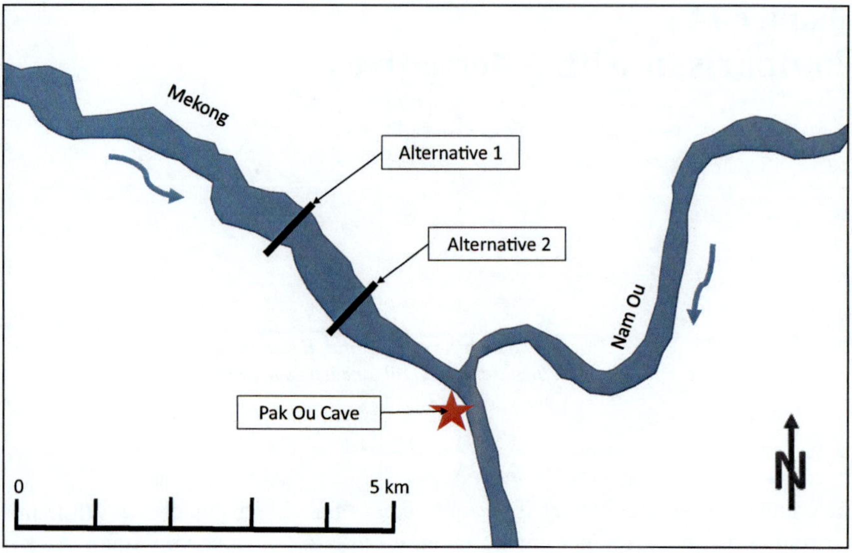

Fig. 14.1 Alternative dam sites for Luang Prabang HPP, Laos

account for environmental or social reasons. Dam safety is a major concern, and this depends to a large extent on the geological conditions.

Still, in some cases more than one dam site are being evaluated, and if they offer similar conditions from a geological or general engineering point of view, environmental and social criteria may actually be decisive. Such a case was Luang Prabang HPP in Laos (Fig. 14.1).

This project is located about 25 km North (upstream) of the historical city of Luang Prabang, a UNESCO World Heritage site. Two possible locations for the dam had been identified, which are about 1 km away from each other. Alternative 1 was finally chosen, mainly for engineering reasons (site allowing for a simpler and therefore faster construction procedure), but also because this site was "just around the bend", no longer visible from the Pak Ou caves, a cultural site and a considerable tourist attraction, accessible only by boat (Pöyry 2020).

14.1.2 Different Project Layout

One option that is rather frequently encountered is a choice of different dam heights at one given site. From an engineering point of view, a higher dam will often be preferred since it allows for more storage and will provide more head. From an environmental and social point of view, however, a lower dam will usually have a smaller impact, since the area to be occupied by the reservoir will be smaller. The relevant figures for three such examples are given in Table 14.1.

Table 14.1 Three cases with alternative dam heights

Rusumo[a]			
Alternative	Full	Intermediate	ROR
FSL m asl	1325	1323.5	1320
Reservoir area ha	28,053	17,118	977
Agricultural area lost ha	9152	2228	187
Villages affected	263	167	8
Persons to be resettled	87,250	36,650	4500
Installed capacity MW	90	85	80
GWh/y	507	497	446
Kukule Ganga[b]			
Alternative	High dam	Low dam	ROR
FSL m asl	242	230	206
Dam height m	56	40	16
Reservoir area ha	2100	714	27
Installed capacity MW	150	100	70
HH to be resettled	2991	714	27
Persons to be resettled	11,455	3570	135
Rogun[c]			
Alternative	High dam	Middle dam	Low dam
FSL	1290	1255	1220
Dam height m	335	300	265
Reservoir area km^2	170	114	68
Installed capacity MW	3600	3200	2800
Yearly energy output TWh/y	14.4	12.4	10.1
Persons to be resettled	42,000	18,000	13,000
Minimal life span years	115	75	45

Source [a]ARTELIA (2013); [b]EWE (1992); [c]Pöyry (2014)

In the case of Kukule Ganga, three alternative dam heights had been studied during the feasibility phase (see Fig. 14.2 and Table 14.1). As can be seen, increasing the dam height would have led to a massive increase of the reservoir area and also in terms of people to be relocated. An additional criterion here was the fact that the land to be submerged would have been mostly agricultural land, and here again the largest part of it the most valuable type, namely irrigated paddy land, suitable for rice cultivation. Replacement of this amount of land would have been as good as

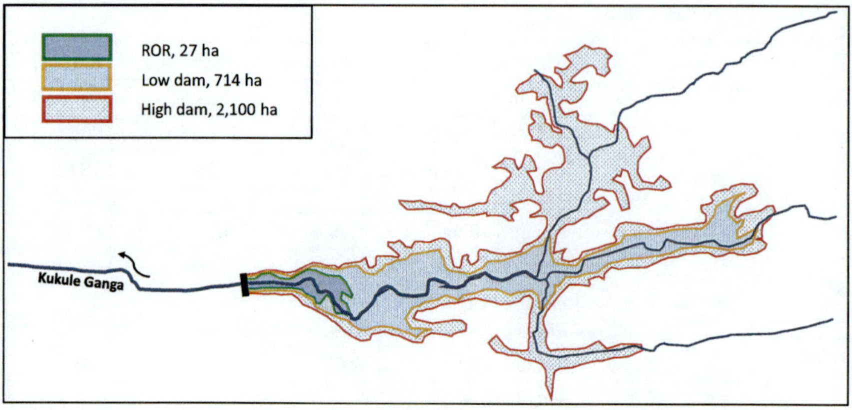

Fig. 14.2 Kukule Ganga HPP, Sri Lanka: reservoir alternatives. See further data in Table 14.1

impossible, since all the good land, i.e. all the alluvial soils along the rivers, were occupied, and the remaining area was steep and rather infertile land covered with forest, and once cleared would not only have had a low productivity, but would also have been erosion-prone. It was therefore decided to choose the ROR option. Not all of our colleagues from the engineering side were happy with this decision. Still, in the official announcement of the commissioning of the power plant it was stated that "Kukule Ganga will also be used as a model in future hydro power developments owing to the stringent controlling measures adopted in the project development to minimise negative impacts on the environment".

More recently, Rusumo Falls HPP, a project shared between Rwanda, Burundi and Tanzania, led to a similar decision, mainly also due to the disproportionate increase of the social impact the higher dam alternatives would have had. As already stated in Sect. 3.11.2, in the case of Rogun, in Tajikistan, the decision was in favour of the high dam; in that case, the expected life span of the three alternatives was considered as being the most relevant argument.

14.2 The No-Project Alternative

One alternative that is always available, at least in theory, is the no-project alternative, which simply assumes that the project in question would not be implemented, and which usually, and very often probably unjustified, assumes that in this case "nothing would happen". Such a comparison can be useful, especially in the case of an HPP in a remote location, when it actually can be assumed that if the idea of this project would be abandoned, the site envisaged for its implementation would be left untouched. This can make a very considerable difference, especially if a long access road would have to be built in the first place for reaching this site, with all the secondary effects such a road can have (see Chap. 15).

Still, just assuming that really nothing would happen by dropping the idea of an HPP will, in most cases, be too simple. Two things may at least have to be considered, namely (i) other types of development, whether new or ongoing and independent of hydropower, would still continue, but probably not in exactly the same way as with the project, and (ii) the need for energy might have to be satisfied in a different way. While the first of these two issues cannot usually be addressed in an ESIA for a specific project, the second may at least be evaluated to a certain degree, of course without going as far as actually developing alternative projects. This is addressed shortly in the following section.

14.3 Different Sources of Energy

14.3.1 Comparison

Different forms of primary energy can be used for producing electricity, and they are associated with different types and intensities of impacts. One generalised comparison between these is shown in Table 14.2.

As the authors of this comparison (Borsch and Wagner 1992: 156) clearly state, this is a simplified way of comparing the alternatives, and one can be of different opinion regarding specific scores; I personally would consider as adequate a score of ++ (if not even +++!) for waste disposal for nuclear energy, and a ++ for the accident risk for nuclear energy and possibly hydropower, given that the probability of a major accident might not be very high, but the potential damage can well be. Likewise, as mentioned in Sect. 8.2.3, under specific conditions hydropower projects can be important sources of greenhouse gases, mainly methane. Still, such general comparisons are certainly useful, as long as the uncertainties involved are made clear. In any case, the table leads to two important conclusions, namely, (i) that there are considerable differences in the environmental impacts caused by the different alternatives, and (ii) that no alternative goes without any impacts.

Hydropower is include here under the heading "renewables"; I agree with this view, being aware that the question has led to intense discussions. This point is raised in Sect. 21.2.

14.3.2 Fossil Fuel

One important alternative to hydropower for generating electricity is the use of fossil fuel, coal, oil and gas. One question that can be asked in every case of hydropower is: how much GHGs can be avoided by using hydro, or, putting it the other way round, how much additional GHGs would be released to the atmosphere if the amount of energy to be produced by the HPP in question were generated by using fossil

Table 14.2 Impact caused by different forms of primary energy

Primary energy	Fossil			Nuclear	Renewable				
	Coal	Oil	Nat. gas	Uranium	Hydropower	Photovoltaic	Wind	Biomass	Geothermal
Fuel production[a]	+	+	–	+				+	
Transport[a]	+	+	–	(+)				–	
Required space	++ +	(+)	(+)	+	++ +	++	++	++	+
Transformation to secondary energy[a]	++	++ (+)	(+)	(+)	–	–	–	++ (+)	++
Material for structures[a]	+	(+)	(+)	+	++	++	++	+	+
Waste disposal[a]	++	++	(+)	(+)	–	–	–	++	–
Greenhouse effect	++	++	+	–	–	–	–	–	–
Accident risk[b]	+	+	+	+	+	–	–	–	–
Comparative evaluation[c]	24 22	19 16	9	11	9 7	8	8	18 15	10

Source Adapted from Borsch and Wagner (1992: 156)

Impact: – none; (+) very small; + present; ++ considerable and/or major efforts required for mitigation. Two qualifications for the same item indicate different alternatives (as e.g. dam vs. ROR in hydropower)

[a]Impact due to emissions

[b]Risk of accident with >100 deaths

[c]Not included in original source; scale used: – = 0, (+) = 1, + = 2, ++ = 4

fuel instead. Given the present situation, where the effects of global warming can definitely be felt, where there can no longer be any doubt that the effect is entirely man-made, and where efforts are being made (or, sad to say, should be made) to curb this development, this is a very relevant question.

Although the values in Table 14.3 show some differences, one important point is very clear: all renewable sources (including biogas, geothermal and hydro) rank much better in terms of overall GHG emissions than the fossil fuel alternatives; while this may not give a very strong argument in favour of one of the renewable sources as compared to the others, it definitely gives a strong argument against any of the fossil fuels. That the different sources give rather clearly differing values for hydro may be caused by the fact that the axpo values specifically refer to the situation in Switzerland, where due to climatic conditions hydro reservoirs as a rule are no significant emitters of GHGs, while on a global level this is not necessarily the case (see also Sect. 8.2.3).

Using such figures, it can easily be calculated how much GHG emission can be avoided by generating the electricity with hydropower instead of using any of the fossil fuels. For the case of Rogun, with an expected yearly output of 14.4 TWh/y, the avoided GHG emission amounts to 6.7 Mt $CO_{2\ equiv.}$/y (Pöyry 2014: 358).

An interesting point is raised by Golden et al. (2006), who point out that an often neglected side effect of thermal power is the fact that it is an important consumptive user of water, meaning that it uses water which is then not given back to the water

Table 14.3 GHG emission from different primary energy sources

Primary energy	g $CO_{2\ equiv.}$/kWh			
	axpo[a]	IPPC 2012[b]	Wik.[c]	Moomaw[d]
Coal	903	1001	820	861.5
Oil		840		
BTTP* with natural gas	628			
Gas combined cycle power plant	423	469	490	456.5
BTTP with biogas	75			
Solar photovoltaic	60	46	48	54
Geothermal	45	45	38	41.5
Wind	16	12	12	9
Nuclear	6	16		
Storage HPP	4	4 (2–170)	24	14**
ROR HPP	3	8 (4–12)		

Source [a]axpo (2009)
[b]IPCC (2012): Table A.II.4, p. 928
[c]Wikipedia: https://en.wikipedia.org/wiki/Life-cycle_greenhouse-gas_emissions_of_energy_sources
[d]Moomaw et al. (2011)
*Block-type thermal power station
**No differentiation between hydro types

body from where it was taken. This is water that is lost to evaporation in the cooling systems. In an investigation made in Phoenix, Arizona, they find that in this way each household consumes a yearly total of 6550 gallons (approximately 24,800 l) of water only due to energy consumption for cooling, i.e. air conditioning, which is likely to aggravate the risk of water shortage in this arid region.

14.3.3 Wind Farms

Wind is a promising—and increasingly used—source of renewable energy. However, as recent experiences from Switzerland show, it is facing a problem similar to the one nuclear energy was facing maybe some 50 years ago. When in Switzerland the new energy legislation was accepted and based on it the energy strategy 2050 was formulated, this latter contained two very important aims: stepwise elimination of nuclear power generation and strong increase in capacity of renewables, wind as one of them. In the meantime, wind has met with increasing opposition: while everybody basically thinks that wind is a good alternative, nobody wants to have a wind farm nearby, and there is much resistance against wind farms from a point of view of landscape protection. One main argument of prospective neighbours of wind farms is noise, the other is aesthetics. Whether a wind farm is considered as a technical endeavour which also has aesthetic qualities, or whether it is just plain ugly and a nuisance from a visual point of view is obviously a matter of personal perception and preferences, and there is no generally accepted standard or scale for the beauty of such structures, as Kreuzer (2011) has discussed for dams. But one thing is certainly clear. Wind farms, possibly even more than dams, are very conspicuous structures and clearly visible in the landscape. With this, development of wind generation capacity is well behind schedule.

In addition to aesthetics and noise, there is one other problem with wind farms: the negative effects of these structures on birds and bats. Although it has been shown that at least some birds can adapt their flight or migration paths in a way as to minimise the risk of collision (Plonczkier and Simms 2012; Cabrera-Cruz and Villegas-Patraca 2016), and that under certain conditions hunting bats fly below the height of the rotor blades (Reynolds 2006), mortality of birds and bats caused by wind turbines is an issue. Zimmerling and Francis (2016) have shown that migrating bats seem to be more vulnerable to wind turbines, Graff et al. (2016) show that bird and bat mortality is higher in the case of wind farms in grassland than in agricultural areas. The importance of avoiding good bird habitats when choosing sites for wind farms is also pointed out by Loring et al. (2014). Not all species react in the same way; according to Pearce-Higgins et al. (2012) three species monitored (*Lagopus lagopus, Galllinago gallinago* and *Numenius arquata*) were negatively affected during the constriction of wind farms, the first one recovered afterwards, while the two others did not. Schmuecker et al. (2020) show that the use of upland habitats, where wind farms are preferably placed, by bald eagles (*Haliaetus leucocephalus*) depends on weather and migration pattern, making the species more vulnerable during some parts of the year and under certain weather conditions. Some papers also point out the difficulty of adequately assessing this impact in the pre-construction phase (Ferrer et al. 2012; Green et al. 2016), or the problems with monitoring the effect in existing structures, so e.g. Smallwood et al. (2020), who demonstrated that dogs are much more efficient at finding carcasses of small birds and bats than people and in this way allow for a more accurate monitoring.

Impacts on non-flying animals seem to be generally less marked. Milligan et al. (2020) found no consistent negative response of habitat selection or avoidance due to wind farms by pronghorn (*Antilocapra americana*), no negative impact on populations of a desert lizard (*Uta stansburiana*) by wind farms was detected (Kheen et al. 2019), who also showed that roads—whether associated with construction and operation of wind farms or not, did have a significant negative effect. In the case of an offshore wind farm it was detected that harbour seals (*Phoca vitulina*) were displaced by construction activities, but returned to the site again once this disturbance was over and the turbines were operating (Russell et al. 2016). But at least some bird species seem not to suffer much from wind turbines: a study of breeding populations of ducks (*Anas platyrhynchos*, *Anas discors*) showed raptors as the main cause of mortality of breeding females irrespective of the presence or absence of wind turbines (Gue et al. 2013).

14.3.4 Solar

Solar, whether used for photovoltaic electricity generation or directly as heat, is certainly a renewable source of energy, and it is available in large quantities, although spread rather thinly over wide areas. It is also comparatively environmentally friendly, as long as can be assured that solar panels can be produced, and later on decommissioned, in an environmentally friendly way. Still, "[a] photovoltaic energy revolution which would make hydroelectricity less necessary, and sea-water desalinisation cheaper (in economic and environmental terms), would resolve both the conflict and the problem" (Martinez-Alier 2002: 128) certainly has some potential, but it might not be quite as easy as that.

Solar as well as wind are not without environmental impacts. Impacts of such power plants on desert vegetation are reported by Grodsky and Hernandez (2020), on desert inhabiting lizards by Kheen et al. (2019). Li et al. (2018) show that massive solar and wind development in the world's deserts could even lead to regional climate change; here, however, I have problems in seeing the real risk: in their modelling study, they assume an installed capacity of 79 TW spread over the entire Sahara desert and directly covering 20% of its surface. Comparing this to the 1.3 TW installed hydro capacity in 2019 (IHA 2020), I think we have a long way to go to reach this aim, which, however, would be able to cover today's entire worldwide energy need. Still, the point to be made here: one cannot assume that solar would be environmentally friendly to the point as not to cause any adverse environmental impacts, and it would be wrong to assume that deserts are just bare land good for nothing, where anything

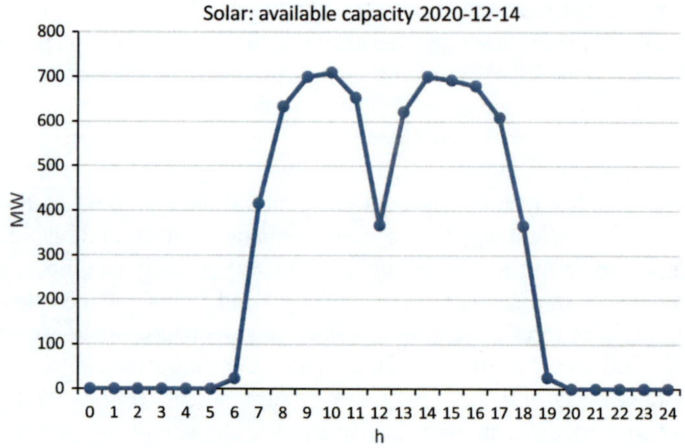

Fig. 14.3 Effect of a solar eclipse, Argentina, 2020-12-14. Based on data from https://despachorenovables.cammesa.com/renovables/

could be done without causing harm. Even this environmentally friendly energy alternative will require a careful environmental and social impact assessment. One way for reducing such impacts, i.e. the competition for space with desert inhabiting organisms, is certainly the increasing use of roofs and other surfaces already occupied by man-made structures for placing solar panels.

Solar, again like wind, has an additional problem: it has to be used when it is available, which is not necessarily when there is demand for the energy. It could be said that unlike wind it has some predictability, insofar as it is clear that during the night there will be no production. Other natural phenomena can affect solar as well.

On December 14, 2020, there was a solar eclipse which in South-central Argentina and Chile was visible as a 100% eclipse. Figure 14.3 shows the available capacity of solar energy production in Argentina on that day. During the duration of the eclipse, this capacity dropped by 32%, different solar power plants being affected by between 20 and 60% depending on their position relative to the path of the total eclipse. This shows that solar production depends not only on day and night, but also on cloud cover and on, admittedly rare, events like solar eclipses. For a short time compensation of such a drop in productivity, storage hydro is probably the most reliable and flexible alternative.

14.3.5 Combination

As just mentioned, both of the two main alternative renewables, solar as well as wind, have one major drawback: they produce while the sun is shining or the wind is blowing, and not when electricity is needed. This means that a major part of production can arise in moments when demand is low, and that very often there will be no or insufficient production when demand is high. Storage hydro is an ideal source for such a situation, since it can be turned on an off at short notice, and it is much more flexible than thermal generation of any kind. Neither solar nor wind offer this possibility. This means that solar and wind are useful only in combination with capacity for storage of electricity. What kind of storage is required largely depends on the scale of the schemes. For very small ones, exclusively for local use, a car battery may be sufficient (see Fig. 14.4). But already for slightly larger units, as e.g. solar panels on roofs, something bigger is required. There are larger batteries which may be sufficient for storage during the day and use during the night for single-family houses, but to this day no such accumulators for large amounts of electricity are available. If Schmidt et al. (2019: 331), state that "a few square kilometres of solar panels can match a dam's power", this may be true, but the problem of matching production and demand is not solved. The best option in such a case, and almost *a conditio sine qua non* for larger solar and wind parks, is a combination with pump storage, whereby surplus electricity is being used for pumping water from a lower to an upper reservoir, for being used again for electricity generation during times of peak demand. One idea seems very promising; the combination of solar and hydro in the way that floating solar panels could be installed on reservoirs. This would have a number of advantages: no additional space would be required for the solar panels,

Fig. 14.4 Use of small-scale solar, Gobi desert, Mongolia. A solar panel on the yurt, connected to a car battery, will provide enough power for some light in the evening and, more importantly, in combination with a satellite dish, for a few hours of TV

electricity generated by them could directly be used, as long as there is demand, saving hydro for times with lack of solar irradiation, surplus energy could be used for pumping water up into the reservoir, and the cooling effect of the reservoir water on the solar panels would increase their efficiency (Reithe and Grossmann 2019). A recent study has shown that two types of benefits can be gained by covering channels with solar panels: increase of solar panel efficiency by the cooling effects from the water flowing in the channel, and reduction of evaporation loss (McKuin et al. 2021).

In any case, looking for the best option, which often could be a combination of different energy sources, as e.g. postulated by Oppermann et al. (2019) and Schmidt et al. (2019) is certainly a good strategy. This, however, will usually have to be done in the framework of a sector planning, e.g. identifying and assessing energy options for a region or for an entire country, an ESIA for a specific project will normally not be able to do more than just look into such possibilities, without carrying out a detailed analysis.

14.3.6 Nuclear

When in Switzerland the discussion on nuclear power development started, late 1950s and early 1960, this was considered as an inexhaustible, clean source of energy, and nature protection organisations actually requested an immediate ban on further hydro development in favour of nuclear, which would require much less space and, most importantly, have much less impact on rivers. As soon as the first nuclear plants had been built, this attitude started to change. This followed exactly the pattern which McCully (1996: 285) describes in the following way: "…, the conservationists took a position of alternatives which many later strongly regretted—they argued that the power to be generated could be produced more cheaply by nuclear or coal plants". One issue was the fact that rivers were being used, or were intended to be used, for cooling purposes, with a risk of increasing water temperature in the receiving rivers above a value tolerated by fish. The main issues, however, were the discussions about the safety risks, and increasingly so the problem of safe disposal of nuclear waste. In a publication dated 1971, this problem was considered as resolved ("the waste problem has found solutions", SVA 1971). However, resistance against nuclear power increased. The last Swiss reactor was commissioned in 1984. Just when the first ideas for additional nuclear power plants started to be developed, in 2011 the accident at Fukushima happened, and Switzerland decided to phase out nuclear power.

Recently, there were voices again in favour of new nuclear power plants, mostly pointing out the fact that they do not emit any GHGs, and would therefore be the inevitable solution for halting climate change, e.g. in an interview with Steven Pinker (Schöpfer 2018), one with the MIT-economist Andrew McAfee (Grassegger 2020), or an article by a former director of the Swiss Energy Agency (Kiener 2019). They all emphasise the role nuclear energy could—and according to them has to—play for climate protection. This argument seems to be so important that according to another article there are even "scientists who ask for reducing safety standards"

(Eichhorn 2019). A new thrust of this movement seems to be the development of small mobile nuclear units, e.g. 60 MW units that can be transported and of which several could be combined for possibly replacing a typical reactor (Gates 2021; Jäggi 2020). According to these authors, these units require much less space than "conventional" nuclear plants, can be produced more rapidly and in series, and are said to be safe. Here, one just wonders what "safe" means. They must contain nuclear material, and the fact that they are small and can be carried around might make them more vulnerable, if not to accidents, then possibly to terrorist attacks, and although the effect may be smaller as it would be in the case of a conventional nuclear plant, there is certainly still a risk.

The biggest issue of nuclear power is still the nuclear waste, and here especially the highly radioactive waste. None of these newspaper articles only so much as mentions this problem. In Switzerland, in the mid-1980s, it was announced that within a few years there would be a final storage site for such highly radioactive waste, which was going to be stacked safely away somewhere deep underground, in a stable geological formation; in 2020, such a site still does not exist. And even if such a site will eventually be found and used, the waste would not just disappear, it would only be hidden. Unfortunately, this type of waste needs to be kept away from the biosphere for several tens of thousands of years. So far in human history, nothing has been remembered even only for a fraction of this time span, and we have to assume that at some point in the future site and purpose of such storage plants would be forgotten. Just imagine what would happen, 10,000 years from now, if somebody, digging at such a site for whatever purpose, would stumble upon some mysterious, large underground structure dating from a time long past and forgotten. A case in point is the recent discovery of over 100 wooden coffins, stacked in underground shafts, discovered in 2020 in Saqqara, in Egypt: they are about 2500 years old, and nobody remembered them (Science 370, https://science.sciencemag.org/content/370/6522/1252.full). No way that after such a discovery a nuclear waste site would be left alone. Both my brothers are archaeologists, and I think I have a fairly good idea about what the outcome would be.

References

ARTELIA. 2013. Rusumo falls hydroelectric project—dam & powerplant component. In *Environmental and Social Impact Assessment (ESIA). NELSAP (Nile Basin Initiative, Nile Equatorial Lakes Subsidiary Action Program)*, Kigali, Rwanda.

axpo. 2009. Treibhausgas-Emissionen europäischer und Schweizer Stromsysteme. *Sonntagszeitung*, 19 Apr 2009.

Borsch, P., and H.J. Wagner. 1992. *Energie und Umweltbelastung*, 174 pp. Berlin: Springer-Verlag.

Cabrera-Cruz, S.A., and R. Villegas-Patraca. 2016. Response of migrating raptors to an increasing number of wind farms. *Journal of Applied Ecology* 53 (6): 1667–1675.

Cans, R. 1997. *La bataille de l'eau*, 232 pp. Paris: Le Monde Editions.

Eichhorn, C.V. 2019. 1000 AKW auf den heissen Stein. *Tagesanzeiger*, 05 Feb 2019, 42.

EWE. 1992. Kukule Ganga hydropower project. Environmental impact assessment. Joint venture Kukule Ganga: Nippon Koei, Electrowatt Eng., Lahmeyer Int.: Feasibility study, vol. 3. Report prepared for CEB, Sri Lanka.

Ferrer, M., M. de Lucas, G.F.E. Janss, et al. 2012. Weak relationship between risk assessment studies and recorded mortality in wind farms. *Journal of Applied Ecology* 49 (1): 38–46.

Gates, B. 2021. *How to avoid a climate disaster*, 257 pp. New York and Toronto: Knopf.

Golden, J.S., A. Brazel, J. Salmond, and D. Laws. 2006. Energy and water sustainability: The role of urban climate change from metropoitan infrastructure. *Journal of Green Building* 1 (3): 125–138. https://www.scienceopen.com/document_file/32d72a35-6bf8-4e06-9f60-721dfb319254/API/i1943-4618-1-3-124.pdf.

Graff, B.J., J.A. Jenks, J.D. Stafford, K.C. Jensen, and T.W. Grovenburg. 2016. Assessing spring direct mortality to avifauna from wind energy facilities in the Dakotas. *Journal of Wildlife Management* 80 (4): 736–745.

Grassegger, H. 2020. Interveiw mit einem Optimisten. Der MIT-Ökonom Andrew McAfee glaubt, dass die Marktwirtschaft de Welt aus der Klimakatastrophe führen kann. *Das Magazin* 7: 14–18. Tamedia, Zurich.

Green, R.E., R.H.W. Langston, A. McCluskie, R. Sutherland, and J.D. Wilson. 2016. Lack of sound science in assessing wind farm impacts on seabirds. *Journal of Applied Ecology* 53 (6): 1635–1641.

Grodsky, S.M., and R.R. Hernandez. 2020. Reduced ecosystem services of desert plants from ground-mounted solar energy development. *Nature Sustainability* 3: 1036–1043. https://doi.org/10.1038/s41893-020-0574-x.

Gue, C.T., et al. 2013. The effects of a large-scale windfarm on breeding season survival of female mallards and blue-winged teal in the prairie pothole area. *The Journal of Wildlife Management* 77 (7): 1360–1371.

IHA. 2020. *2020 hydropower status report*, 46 pp. London: International Hydropower Association. https://hydropower-assets.s3.eu-west-2.amazonaws.com/publications-docs/2020_hydropower_status_report.pdf.

IPCC. 2012. *Renewable energy sources and climate change mitigation*, 1076 pp. Special Report of the IPCC. Cambridge UP.

Jäggi, W. 2020. Eine neue Generation Kraftwerke Die Schweiz steigt aus der Kernkraft aus, die USA wieder ein. Republikaner wie Demokraten unterstützen die Entwicklung einer neuen AKW-Generation. Was ist davon zu halten? *Tagesanzeiger*, 16 Dec 2020, 36.

Kheen, J.E., K.T. Shoemaker, and C.R. Feldman. 2019. Population-level effects of wind farms on a desert lizard. *Journal of Wildlife Management* 83 (1): 145–157. https://doi.org/10.1002/jwmg.21565.

Kiener, E. 2019. Warum wir leider Atomkraftwerke brauchen. Eine nüchterne Einschätzung. *Das Magazin* 20, 18 May 2019, 23–27.

Kreuzer, H. 2011. Thoughts on the aesthetics of dams. In *Hydropower & dams (reprint), ICOLD 79th annual meeting*, 10 pp, Lucerne.

Li, Y., E. Kalnay, S. Motesharrei, et al. 2018. Climate model shows large-scale wind and solar farms in the Sahara increase rain and vegetation. *Science* 361 (6406): 1019–1022 + suppl. mat. https://doi.org/10.1126/science.aar5629

Loring, P.H., P.W.C. Paton, J.E. Osenkowski, et al. 2014. Habitat use and selection of black scoters in southern New England and siting of offshore wind energy facilities. *Journal of Wildlife Management* 78 (4): 645–656.

Martinez-Alier, J. 2002. *The environmentalism of the poor. A study of ecological conflicts and valuation*, 312 pp. Cheltenham, UK: Edward Elgar.

McCully, P. 1996. *Silenced rivers. The ecology and politics of large dams*, 350 pp. London: Zed Books.

McKuin, B., A. Zumkehr, J. Ta, R. Bales, J.H. Viers, T. Pathak, and J.E. Campbell. 2021. Energy and water co-benefits from covering canals with solar panels. *Nature Sustainability*. https://doi.org/10.1038/s41893-021-00693-8.

Milligan, M.C., A.N. Johnston, K.T. Beck, et al. 2020. Effects of wind energy development on Pronghorn habitat selection. Paper presented at the *Wildlife society's 2020 annual conference*, 28 Sept–2 Oct. https://twsconference.org/sessions/wind-energy-impacts-on-wildlife/.

Moomaw, W., P. Burgherr, G. Heath, M. Lenzen, J. Nyboer, and A. Verbruggen. 2011. Annex II: Methodology. In *IPCC special report on renewable energy sources and climate change mitigation*, ed. O. Edenhofer, R. Pichs-Madruga, Y. Sokona, K. Seyboth, P. Matschoss, S. Kadner, T. Zwickel, P. Eickemeier, G. Hansen, S. Schlömer, and C. von Stechow. Cambridge, United Kingdom and New York, NY, USA: Cambridge University Press.

Opperman, J., J. Hartmann, M. Lambrides, J.P. Carvallo, E. Chapin, S. Baruch-Mordo, B. Eyler, M. Goichot, J. Harou, J. Hepp, D. Kammen, J. Kiesecker, A. Newsock, R. Schmitt, M. Thieme, A. Wang, and C. Weber. 2019. *Connected and flowing: A renewable future for rivers, climate and people*, 104 pp. Washington, DC: WWF and The Nature Conservancy. https://www.nature.org/content/dam/tnc/nature/en/documents/TNC_ConnectedFlowing_Report_WebSpreads.pdf.

Pearce-Higgins, J.W., L. Stephen, A. Douse, and R.H.W. Langston. 2012. Greater impact of wind farms on bird populations during construction than subsequent operation: Results of a multi-sites and multi-species analysis. *Journal of Applied Ecology* 49 (2): 386–394.

Plonczkier, P., and I.C. Simms. 2012. Radar monitoring of migrating pink-footed geese: Behavioural responses to offshore wind farm development. *Journal of Applied Ecology* 49 (5): 1187–1194.

Pöyry. 2014. Environmental and social impact assessment for Rogun hydropower plant. In *ESIA*, Vol. I: Text, Vol. II: Annexes. Report prepared for OSHPC "Barki Tojik", Republic of Tajikistan.

Pöyry. 2020. *Luang Prabang HPP: Environmental impact assessment*, vol. 4.1 of the feasibility study. Report prepared for Luang Prabang Power Company Limited.

Reithe, G., and J. Grossmann. 2019. Floating PV on reservoirs: A threshold analysis by means of a cost-benefit appraisal referencing various potential sites across Africa. Paper 22.06 presented at the *Africa 2019 hydropower conference*, Windhoek, Namibia, Apr 2019.

Reynolds, D.S. 2006. Monitoring the potential impact of a wind development site on bats in the Northeast. *Journal of Wildlife Mangement* 70 (5): 1219–1227.

Russell, D.J.F., G.D. Hastie, D. Thompson, et al. 2016. Avoidance of wind farms by harbour seals is limited to pile driving activities. *Journal of Applied Ecology* 53 (6): 1642–1652.

Schmidt, R.J., N. Kittner, G.M. Kondolf, and D.M. Kammen. 2019. Deploy diverse renewables to save tropical rivers. *Nature* 569 (7756): 330–332.

Schmuecker, S.J., D.A. Becker, et al. 2020. Use of upland and riparian areas by wintering bald eagles and implications for wind energy. *Journal of Wildlife Management* 84 (8): 1578–1589. https://doi.org/10.1002/jwmg.21927.

Schöpfer, L. 2018. Das Leben in der Schweiz kann noch besser werden. *Tagesanzeiger*, 31 Dec 2018, 2–3.

Smallwood, K.H., D.A. Bell, and S. Standish. 2020. Dogs detect larger wind energy effects on bats and birds. *Journal of Wildlife Management* 84 (5): 852–864. https://doi.org/10.1002/jwmg.21863.

SVA. 1971. *Kernenergie: sauber, unentbehrlich, unerschöpflich*, 31 pp. Bern: Broschüre herausgegeben von der SVA (Schweizerische Vereinigung für Atomenergie).

Zimmerling, J.R., and C.M. Francis. 2016. Bat mortality due to wind turbines in Canada. *Journal of Wildlife Management* 80 (8): 1360–1369.

Part III
Specific Issues

Part II lists all topics that usually are standard chapters in any type of ESIA for water resources development projects. There are, however, a number of topics which are or can be of high importance, but do not necessarily fit into one or the other of these chapters, since they can be based on or affect several of the main topics or environmental subjects treated there. In the following chapters, a number of these issues is highlighted and discussed, the ones considered as especially important. Whether, for a given project, any of these will be of relevance will have to be decided on a case to case basis.

Chapter 15
Access Roads

... et passera la route!
(... and the road will pass!)
Hugues Aufray, French singer/songwriter: Debout les gars, 1964

Construction of an HPP requires access, and often this is necessary already in the preparatory stage, e.g. for bringing in equipment like drilling rigs for geological investigation of the selected site. In an inhabited area, access as such is usually not a problem, since very often existing roads can be used; extension of the road network may be required to some degree, or existing roads may have to be upgraded, but the incremental impact of the project under such conditions is usually rather small. There will be an impact during the construction period due to the heavy traffic required for transports to and from the site, mainly noise and dust, but also the increased risk of accidents, especially where such roads are passing through settlements. These are impacts that need to be identified, and, more importantly, managed for the duration of the construction works. This will be done by developing and implementing a traffic management plan as part of the ESMMP (see Chap. 12). On the other hand, improvement of the road network can also be a benefit for the local population.

The situation is different when a potential project site is located outside of populated areas, and where therefore a new access road, often of considerable length, needs to be built. Environmental impacts, direct as well as indirect ones, can be quite severe in such cases, and they can even be more relevant than the impact of the hydropower plant as such. This topic is discussed here, while a few projects where access roads are a major concern are described in Chap. 26, Ayago-Nile in Uganda, Ivindo in Gabon, Sutieska in Bosnia-Herzegovina, Palas Valley in Pakistan, and Amayla in Guyana (Sect. 15.2).

R. Zwahlen, *Assessing the Environmental Impacts of Hydropower Projects*,
Environmental Earth Sciences, https://doi.org/10.1007/978-3-030-91185-0_15

15.1 Impacts of Roads on Natural Habitats

The different effects of roads on natural habitats, groups of animals or individual species are rather well known from a large number of studies. One first direct effect of a road network is the fragmentation of the landscape or the habitats within it, as e.g. described by Knauer and Wolter (1980), who found for a region in Germany that the average size of habitat islands, i.e. of areas not cut by roads, was less than 10 km^2. This can lead to a splitting up of animal populations into more or less isolated sub-populations, which makes them more vulnerable, especially if the original populations are already small. It has been demonstrated for a number of animals that roads act as barriers which are seldom if ever crossed, as e.g. for two species of rodents (*Sigmodon hispidus* and *Microtus ochrogaster*), where in a 9 year study from 462 and 741 individuals trapped more than once only 47 resp. 23 had ever crossed a merely 3.5 m wide gravel road (Swihart and Slade 1984). Betts et al. (2019) have demonstrated that animal species with a long history of exposure to different causes of habitat fragmentation, e.g. by fires, glaciation, hurricanes or deforestation, are less fragmentation-sensitive. A number of studies have shown that certain species are not only limited in their movements by the fact that they do not easily cross roads, but that they also tend to avoid the proximity of roads, as, e.g. caribou (*Rangifer tarandus*) in Alaska, and here especially females during the calving period (Prichard et al. 2020), wild camel (*Camelus ferus*) in China (Yang et al. 2019), grizzly bear (Ursus arctos) and black bear (*Ursus americanus*) (McLellan and Shackleton 1988; Brody and Pelton 1989), ocelot (*Leopardus pardalis*) in Texas (Lombardi et al. 2020) or amphibians and reptiles in the Amazon of Ecuador (Maynard et al. 2016); this avoidance behaviour leads to a road-induced habitat loss which is by far greater than the mere surface directly occupied by the road. Another well-known road impact is mortality due to traffic, which can exert a heavy toll e.g. on hedgehogs (*Erinaceus europaeus*; Reichholf and Esser 1981) or amphibians and other animals (Walter 1977: 39; Healey et al. 2020). There can also be some more surprising indirect effects of roads, like e.g. when a predator is favoured by roadside structures or vegetation and in this way can colonise sites from where it has been absent before. Road construction into formerly closed rainforest within the buffer zone of a national park in Brunei, Borneo, led to the immigration of a large frog species (*Limnonectes ingeri*) which had not been present there before, and which was favoured by open roadside habitat and especially small pools along the road. Among other prey, it was also feeding on a number of smaller frog species, and since it finally lived as far as 500 m away from the road, here again the impact was much larger than just the road corridor (Konopik et al. 2014). In some areas of the USA the common raven (*Corvus corax*) was shown to benefit even more from human activities, since it could increase its population in the presence of increased road density, transmission lines, agricultural activities etc. This led to a reduction in sage grouse (*Centrocercus urophasianus*) populations, since the raven is a main predator of this species (O'Neil et al. 2018). Habitat fragmentation by roads also presents a threat to the spotted-tailed quoll (*Dasyurus maculatus gracilis*), one of the larger carnivores indigenous to Australia, and this species is further threated trough poisoning from the exotic and invasive cane toad (*Rhinella marina*), whose expansion in range seems to be favoured by road construction (Uzqueda et al. 2020). One very different effect of a road corridor can presently be observed in Central America. The raise of the isthmus about 3–4 million years ago has led to an exchange of fauna between the two subcontinents. Opening roads now provides corridors which allow non-forest dwelling species to expand their ranges. In this way, the South American crab-eating fox (*Cerdocyon thous*) and the North American coyote (*Canis latrans*) have both reached Panama, and the last barrier for them is the Darién forest, which is not yet traversed by a road. Opening the Panamerican Highway might form such a corridor especially for the coyote to expand into South America (Hody et al. 2019).

These were just a few points to indicate the range of direct or indirect impacts roads can have on natural habitats and with that on biodiversity. For a broader discussion of issues related to road construction see e.g. Alamgir et al. (2017).

15.2 Main Impacts of Access Roads

Figure 15.1 illustrates a number of problems associated with access roads. For Ivindo HPP, the road was built without the required permits. The developer had the permit to investigate a certain site, which had been identified previously as a potential site for hydropower development. Access was needed for carrying out the required geological investigations, i.e. mainly for bringing in a drilling rig. The developer mandated a logging company with the construction of this access road, without however defining any criteria and without seeking the necessary permits. The result was an access road as shown in the picture on the left, with a total length of about 40 km, approximately half of which in a national park. Such a road has a number of immediate consequences. The first is the destruction of the vegetation that was growing there, and the exposure of the soils to erosion. Then there is the gap in the vegetation cover that is created. In the case of a closed forest this is most limiting for canopy dwelling animals, which rarely if ever come to the ground. For them, such a gap is an insurmountable obstacle. But also many of the animals living on the ground, including bird species living in

Fig. 15.1 Access roads. Left: access road built illegally in a National Park (Ivindo, Gabon). Right: temporary road for pre-impoundment salvage logging (NT1, Laos)

dense vegetation, will not or only exceptionally leave the closed forest in order to cross this now open terrain. In this way, roads lead to breaking up populations into more or less isolated sub-populations; the negative effects such an isolation can have, especially with respect to local extinction of species, has been demonstrated by the Biological Dynamics of Forest Fragments Project (BDFFP) near Manaus, Brazil, described e.g. by Kolbert (2014: 173 ff.). In addition to that, in this case the felled trees were not removed, but just shoved out of the way with bulldozers, tree trunks and major branches forming a barrier on both sides of the road which can be an obstacle even for larger animals, e.g. elephants, keeping them from moving across the road. The fact that the resulting gap was about three times as broad as would have been required at this moment added to the severity of the impact. The picture on the right shows that it is possible to prepare such roads in a way that will limit the impact; it is only just broad enough to allow the vehicles to pass, and the crowns of larger trees can even touch above the road. Once no longer in use, it will be easy to block it to prevent vehicles from entering, and the vegetation will reclaim it very quickly.

Depending on the situation of a site, road impacts can be very different. This is illustrated in Fig. 15.2 with two extreme examples. One is the gap cut into the forest canopy in a tropical rainforest, with the effects described above. The other stems from a completely different setting. Although the situation is changing gradually, Mongolia can be described as a land without roads. Over much of its central part, which is characterised by a steppe vegetation, it is also rather flat; this means that even in the absence of roads driving around the countryside is quite easy, and roads are opened simply by circulating vehicles. However, as soon as an established track gets a bit bumpy, which can happen mainly after occasional rains or snowfall, somebody will prefer to drive just in parallel to the previous track, and others will soon follow. This results in situations as shown in the picture; there can be 10 or more such tracks running next to each other. Vegetation is destroyed quickly by the tires, and given the harsh climate its regeneration is very slow. Such situations can be seen not only in steppes, but maybe even more so in different types of semi-deserts or in high

Fig. 15.2 Roads: different issues in different situations. Left: newly built access road cutting a gap into primary forest (Amayla, Guyana). Right: parallel vehicle tracks in the absence of a defined road, Mongolia

mountain areas, where the passage of a single vehicle can leave a track that may remain visible for years. It is therefore important that access to sites under such conditions is clearly defined, and that drivers are instructed not to leave the tracks.

However, the most decisive and lasting indirect impact of opening a new access road is the fact that it gives access not only for the work to be carried out in relation with the project, but access that can be—and eventually will be—used for other purposes as well. Especially in forested areas people not or not directly related to the project will sooner or later move in, and activities of different types—logging and agriculture as the most important ones—will start spreading on either side of the road, as e.g. has been demonstrated for regions in the Brazilian Amazon (Michalski et al. 2008; Bebbington et al. 2018; Santos et al. 2018; Andrade et al. 2021). Figure 15.3 shows such a development using an example from Laos; it should be noted that this figure only serves the purpose of illustrating indirect road effects, this road was not related to a hydropower project. As in this example, such impacts can be caused by roads built for whatever purpose, as e.g. logging roads mentioned by MacKinnon et al. (1996: 405 ff.) in Kalimantan.

15.3 Mitigation

The severity of the potential impact will have to be evaluated very carefully in each specific case, as discussed in the examples mentioned in the opening paragraphs of this chapter. One recent example is the case of a planned HPP in Sumatra, which would be built in the habitat of the just recently described new orangutan species (*Pongo tapanuliensis*; see Sect. 10.5.1). The access road required for this HPP, and the habitat destruction that might result from it, were identified as the main threats for this isolated endemic species (Stokstad 2017; Rochmyanigsih 2019; Laurance 2018).

The only really efficient way for avoiding the impact when really valuable habitats are at stake may often be the decision not to go ahead with a specific project. Of course, if as in the case of Ivindo, the access road has already been opened before such a decision could be reached, a considerable part of the damage may already have been caused. One precautionary measure, if access is required before a final decision, is certainly to prepare such a road with the minimal standard absolutely required, i.e. as narrow as possible, as shown in Fig. 15.1. In case such a road would then no longer be required, it would have to be effectively blocked, and the surface may have to be ripped open at least in part, since the development of vegetation is much slower on soils compacted by heavy vehicles. Whether tree plantation or any other active revegetation is required will again largely depend on the situation, considering prospects for a spontaneous plant growth.

If a road into a sensitive area is required, access has to be closely controlled. This was done successfully in the case of Lom Pangar in Cameroon, where the access to the dam site led through a part of a national park. The entrance to this park was equipped with a guard post, where every person wanting to use this road was controlled, and

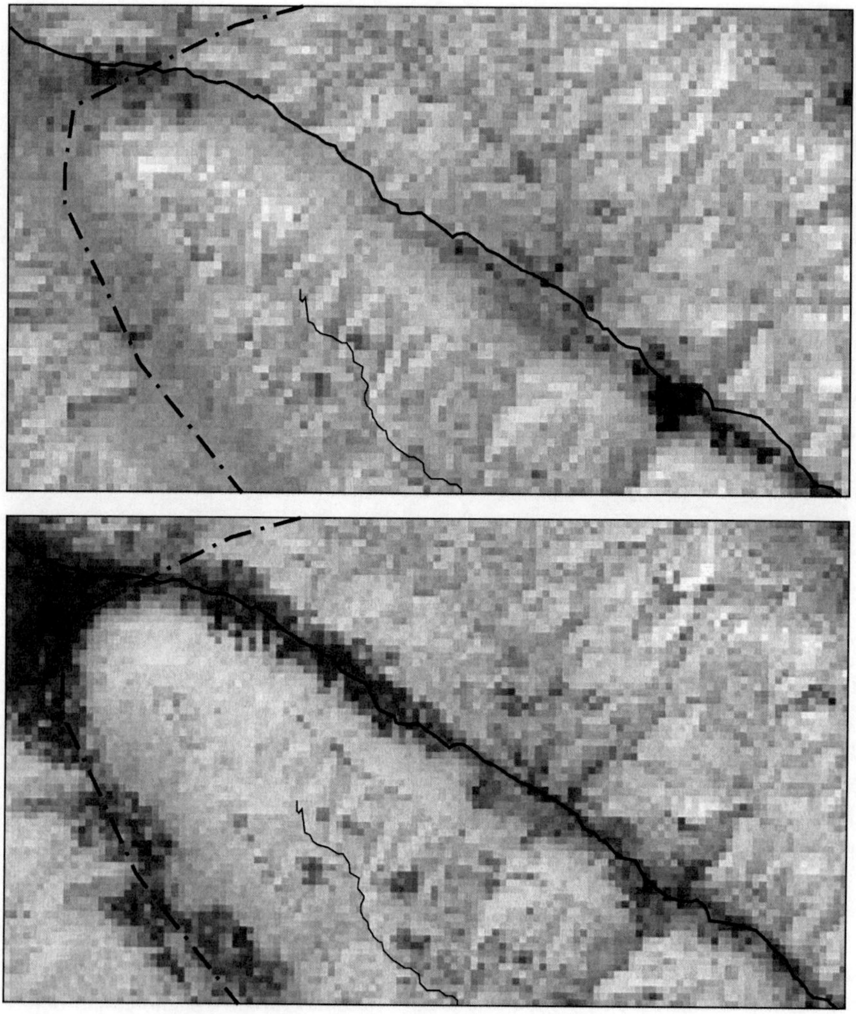

Fig. 15.3 Effect of a road leading into a protected area. Thick solid line: road through protected area, providing access from both sides. Thin black line: track into core part of protected area. Dashed line: border of protected area. White or light gravy: vegetation cover (forest) intact; dark grey: cultivated areas; black: bare soil. Situation in 1990 (top) and 2016 (bottom). Part of Nam Kading Nationally Protected Area, Laos

where access was granted only to those who really were entitled to go there. Still, the promise to control access is not yet a guarantee that this would effectively be done, as e.g. the case of Amayla (Guyana) has shown, where the prospective developer pulled out after roughly half of the access road had been built, leaving it without the envisaged control (see Sect. 15.2). And it is a well-documented fact that if in some remote place there is no road that would allow to take the timber out, this is the better protection of a forest than a control post on a road (see Sect. 26.2 for a discussion of this point).

References

Alamgir, M., M.J. Campbell, S. Sloam, et al. 2017. Economic, socio-political and environmental risks of road development in the tropics. *Current Biology* 27: R-1130-R1140. https://doi.org/10.1016/j.cub.2017.08.067.

Andrade, M.B.T., L. Ferrante, and P.M. Fearnside. 2021. Brazil's highway BR-319 demonstrates a crucial lack of environmental governance in Amazonia. *Environmental Conservation* 48 (3): 161–164. https://doi.org/10.1017/S0376892921000084.

Bebbington, A., J. Humphreys Bebbington, D. Sauls, et al. 2018. Resource extraction and infrastructure threaten forest cover and community rights. *PNAS* 115 (52): 13164–13173. https://doi.org/10.1073/pnas.1812505115.

Betts, M.G., C. Wolf, M. Pfeifer, et al. 2019. Extinction filters mediate the global effects of habitat fragmentation on animals. *Science* 366 (6470): 1236–1239. https://doi.org/10.1126/science.aax9387.

Brody, A.J., and M.R. Pelton. 1989. Effects of roads on black bear movements in western North Carolina. *Wildlife Society Bulletin* 17 (1): 5–10.

Healey, R.M., J.R. Atutubo, M.D. Kusrini, L. Howard, F. Pagen, N. Hallisey, and N.E. Karraker. 2020. Road mortality threatens endemic species in a national park in Sulawesi, Indonesia. *Global Ecology and Conservation* 24. https://doi.org/10.1016/j.gecco.2020.e01281.

Hody, A.W., R. Moreno, N.F.V. Meyer, K. Pacifici, and R. Kays. 2019. Canid collision—expanding populations of coyotes (*Canis latrans*) and crab eating foxes (*Cerdocyon thous*) meet in Panama. *Journal of Mammalogy* 100 (6): 1819–1830. https://doi.org/10.1093/jmammal/gyz158.

Knauer, N., and H. Wolter. 1980. Oekologische Auswirkungen des Strassennetzes dargestellt am Beispiel von Schleswig-Holstein. *Forstarchiv* 51 (12): 250–255.

Kolbert, E. 2014. *The sixth extinction. An unnatural history*, 319 pp. London: Bloomsbury.

Konopik, O., K.-E. Linsenmair, and T.U. Grafe. 2014. Road construction enables establishment of a novel predator category to resident anuran community: A case study from a primary lowland Bornean rain forest. *Journal of Tropical Ecology* 30 (1): 13–22.

Laurance, W. 2018. If you can't build well, then build nothing at all. *Nature* 563 (7731): 295.

Lombardi, J.V., M.E. Tewes, H.L. Perotto-Baldivieso, J.M. Mata, and T.A. Campbell. 2020. Spatial structure of woody cover affects habitat use patterns of ocelots in Texas. *Mammal Research* 65: 555–563. https://doi.org/10.1007/s13364-020-00501-2.

MacKinnon, K., G. Hatta, H. Halim, and A. Mangalik. 1996. The ecology of Kalimantan. Indonesian Borneo. In *The ecology of Indonesia series*, vol. III, 802 pp. Singapore: Periplus Editions (HK).

Maynard, R.J., N.C. Aall, D. Saenz, P.S. Hamilton, and A. Kwiatkowski. 2016. Road-edge effect on herpetofauna in a lowland Amazonian rainforest. *Tropical Conservation Science* 9 (1): 264–290. https://doi.org/10.1177/194008291600900114.

McLellan, B.N., and D.M. Shackleton. 1988. Grizzly bears and resource-extraction industries: Effects of roads on behaviour, habitat use and demography. *Journal of Applied Ecology* 25 (2): 451–460.

Michalski, F., C.A. Peres, and I.R. Lake. 2008. Deforestation dynamics in a fragmented region of southern Amazonia: Evaluation and future scenarios. *Environmental Conservation* 35 (2): 93–103.

O'Neil, S.T., P.S. Coates, B.E. Brussee, et al. 2018. Broad-scale occurrence of a subsidized avian predator: Reducing impacts of ravens on sage-grouse and other sensitive prey. *Journal of Applied Ecology* 55 (6): 2641–2652. https://doi.org/10.1111/1365-2664.13249.

Prichard, A.K., B.E. Lawhead, E.A. Lenart, and J.H. Welch. 2020. Caribou distribution and movements in a Northern Alaska oilfield. *Journal of Wildlife Management* 84 (8): 1483–1499. https://doi.org/10.1002/jwmg.21932.

Reichholf, J., and J. Esser. 1981. Daten zur Mortalität des Igels (Erinaceus europaeus), verursacht durch den Strassenverkehr. *Zeitschrift für Säugetierkunde* 46 (4): 216–222.

Rochmyanigsih, D. 2019. Accusations fly over threat to rarest great ape. *Science* 365 (6458): 1064–1065.

Santos, M.A.D., A.M. Yanai, F.O. Sousa, et al. 2018. *BR-319 como propulsora de desmatamento: simulando o impacto da rodovia Manaus-Porto Velho*, 54 pp. Manaus, Brazil: IDESAM. https://www.researchgate.net/publication/330740789.

Stokstad, E. 2017. New ape found, sparking fears for its survival. *Science* 358 (6363): 572–573.

Swihart, R.K., and N.A. Slade. 1984. Road crossing in *Sigmodon hispidus* and *Microtus ochrogaster*. *Journal of Mammalogy* 65 (2): 357–360.

Uzqueda, A., S. Burnett, L.V. Bertola, and C.J. Hoskin. 2020. Quantifying range decline and remaining population of the large marsupial carnivore of Australia's tropical rainforest. *Journal of Mammalogy* 101 (4): 1021–1034. https://doi.org/10.1093/jmammal/gyaa077.

Walter, J. 1977. Amphibien unserer Heimat. *Neujahrsblatt der Naturforschenden Gesellscahft Schaffhausen* Nr. 28: 78 pp.

Yang, J., Q. Fu, D. Wang, X. Liu, and Q. Xu. 2019. China's highways threaten wild camels. *Science* 364 (6447): 1242. https://doi.org/10.1126/science.aay1264.

Chapter 16
Reservoirs as Habitats

> *[L]akes and to some extent also reservoirs add significantly to biodiversity on Earth and act as important foraging areas for many terrestrial animals and waterfowl.*
> Jeppesen et al. (2010: 376)

When the filling of a reservoir starts, a transition from lotic (river-like) to lentic (lake-like) conditions starts, and by this development the living conditions of aquatic organisms change dramatically. It is important to insist on the fact that a reservoir will be "something like a lake", but not exactly the same. To what extent there will be similarities of a new reservoir with a natural lake, and in what respect there will be differences, depends on a number of conditions, reservoir morphology and HPP operation pattern being the most important ones. Similarities and differences between the two types of water bodies are discussed, as well as the main consequences for organisms which may or may not find suitable conditions for using it as a habitat. The ESIA will have to give some indications on the expected qualities of the reservoir to be formed, for anticipating relevant management activities which will have to be taken in order to optimise its development, or for pointing out some additional potential uses of or potential risks in the future artificial lake.

16.1 Reservoirs Types

The morphology of a future reservoir is determined largely by the shape of the river valley, in which the dam is placed. Three typical examples are shown in Fig. 16.1, and main characteristics of these valley types and dams and reservoirs potentially to be built there are described shortly in Table 16.1.

In reality there are many special cases and very often it will not be possible to describe a reservoir as belonging to one or the other of these three categories; every type of intermediate situation is possible. So e.g. a rather narrow valley (type A) can have wider sections with inhabitable (and often inhabited) river terraces, often only on one of the river banks, and whether a wide valley is "still" of type B or "already"

R. Zwahlen, *Assessing the Environmental Impacts of Hydropower Projects*, Environmental Earth Sciences, https://doi.org/10.1007/978-3-030-91185-0_16

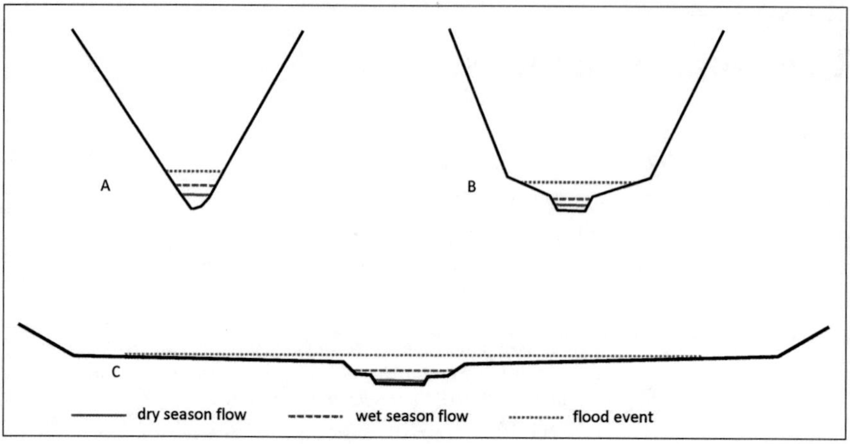

Fig. 16.1 Cross sections of typical river valleys

of type C will often be a matter of definition. Still, especially for a rapid assessment of potential impacts of a dam project in the screening and scoping phase it is useful to consider these differences dictated by the local topographical situation.

As for the cross-sections, more or less typical forms of reservoirs can be described. Again, there are mainly three different types, and here again with all possible intermediate forms. The three basic types are:

1. Long, narrow reservoirs which submerged basically one valley or gorge of type A above. Such reservoirs are often rather deep and straight. Examples are a and g in Fig. 16.2. An extreme example of this type is the Three Gorges reservoir in China.
2. Highly dendritic reservoirs which have several arms, valleys of tributaries that were submerged. Such reservoirs can also be deep and mostly narrow, however, with a complex form (as b, c, d and h in Fig. 16.2). The submerged valleys or parts thereof can be of any combination of the types A to C described above.
3. Wide, more or less irregularly shaped reservoirs, often located in rather flat areas (type C), and therefore shallow. The examples f and h in Fig. 16.2 are a mix of the dendritic and wide, shallow type i is a shallow reservoir which submerged a rather wide part of the river valley. Balbina reservoir in Brazil, located in a very flat region, is certainly an example of this type: a 2920 m long and only 33 m high dam creates a reservoir with a surface of 2360 km^2, for an installed capacity of only 250 MW (Fearnside 1989) (Table 16.2).

A short description of these reservoirs follows.

Bumbuna HPP, Sierra Leone: dam on the Seli River. The reservoir submerged a very long and narrow river valley with numerous but very small lateral tributaries. Only minor resettlement was required since the villages scattered along both banks of the river were not located in the narrow valley with rather steep flanks, but on higher ground. The ESIA for this project had been prepared in 1994/95 (EWI 1996);

Table 16.1 Main characteristics of valley types and dams

Type	A	B	C
Main valley characteristics	Narrow, gorge-like valley with steep flanks	More or less broad valley with rather steep flanks and alluvial banks along the river	Vast floodplain
River characteristics	Narrow, often with steep gradient, fast flowing to torrential	Narrow, gradient from rather steep to gentle	Little gradient, slow flowing, often meandering or braiding
Difference wet to dry season	River still narrow, increased flow speed. River banks exposed during dry season often bare rock	Between A and C, variable, depending on specific valley and river morphology	River often considerably wider during wet season; river banks exposed during dry season often fine alluvial soils, often used for dry season river bank gardening
Flood event	River still narrow, with greatly increased flow speed and erosive power	Depending on river morphology, width of the valley and magnitude of the event. Flooding of river banks, can accumulate sediment (gravel, sand, silt) or erode away river banks	River floods a large part of the floodplain, depending on the magnitude of the event. Slow flowing to almost stagnant, deposits fine silty sediment
Human settlements	Usually not settled, since valley flanks are too steep, no suitable soils for agriculture	Depending on valley morphology. River terraces (alluvial soils) often used for agriculture, settlements on higher, less flood-prone parts of the terraces	Floodplain above normal (yearly flood) high water level often extensively used for agriculture. Settlements on higher, less flood-prone parts of the floodplain, but still vulnerable to higher floods
Biological value	Not possible, just based on these three main types of valley morphology, to make any statements concerning differences in value in terms of importance for biodiversity. Each of these types can harbour habitats of high importance, and this is influenced also by prevailing forms of human activities. Still, there will usually be a gradient in biodiversity (as expressed in number of habitat types and consequently in number of species present), increasing from A to C		

(continued)

Table 16.1 (continued)

Type	A	B	C
Dam and reservoir type	High dams with short crown length; forms a narrow, often long and very deep reservoir	High dams with rather short crown length possible, depending on chosen dam site (often placed in a natural restriction of the valley). Reservoir can be deep and long	Low dam, but very long crest. Reservoir very large and shallow
Main impacts[a]	Submerged area usually rather small Direct social impact (loss of agricultural land, resettlement) nil or very small Limited water losses due to evaporation	Submerged area can be of considerable size Direct social impact (loss of agricultural land, resettlement) often important Intermediate water loss due to evaporation	Submerged area often very large Direct social impact (loss of agricultural land, resettlement) potentially very important High to very high water loss due to evaporation
Examples[b]	Nurek (Tajikistan) Mauvoisin (Switzerland) Jannah (Lebanon) Kafue Gorge Lower (Zambia) Manwan (Lancang, China) Bumbuna (Sierra Leone)	Egiin Gol (Mongolia) Rogun (Tajikistan) Lom Pangar (Cameroon) Xayaburi and Luang Prabang HPPS (Lao PDR) Ilisu (Turkey) Nam Theun 1 (NT1, Lao PDR)	Kulpawn (Ghana) Aswan High Dam (Egypt) Balbina (Brazil) Amaila (Guyana)

[a]The main “normal” dam impacts (interruption of the river with potential negative effect on fish migration, submerging of terrestrial ecosystems, accumulation of sediments, impact on downstream river dynamics) apply to all three types and are not listed here. Impacts on water quality and related issues (reservoir stratification, anoxic conditions, potential for GHG emission) depend more on site climate, water turnover rate in the reservoir and vegetation submerged during impoundment that on the form of the reservoir

[b]Examples of existing or planned dams and reservoirs discussed or at least shortly mentioned in this book

at that point in time, the civil war was already under way, and during the study period Bumbuna, the town near the dam site, was looted and most houses were destroyed. This civil war eventually stopped all work on the project; the dam was already built, but not the power house, and reservoir impoundment had not yet started. In 2004/05 a new RAP was prepared for this project (Electrowatt-Ekono and Techsult 2005), which was then finalised. Towards the end of this study there finally was a reliable topography, and inhabitants of a village located near the upper end of the reservoir could be informed that, unlike the information they had had for over 10 years, their village would not be submerged after all.

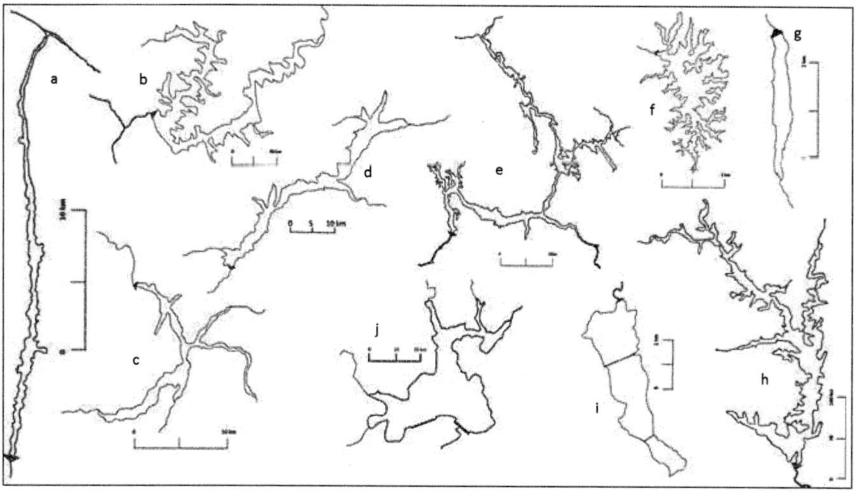

Fig. 16.2 Reservoir shapes. **a** Bumbuna, Sierra Leone; **b** Lom Pangar, Cameroon (ISL et al. 2005); **c** Ralco, Chile; **d** Rogun, Tajikistan; **e** NamTheun 1, Laos; **f** Houay Ho, Laos; **g** Mauvoisin, Switzerland; **h** Volta, Ghana, Akosombo dam; **i** Sihlsee (Lake Sihl), Switzerland; **j** Kulpawn, Ghana. *Note* Different scales!

Table 16.2 Basic data to reservoirs shown in Fig. 16.2

No.	Reservoir	Country	Dam height m	Res. area km^2	Res. volume Mm^3	Inst. capacity MW	Rate MW/km^2
a	Bumbuna	Sierra Leone	88	21	445	50	2.4
b	Lom Pangar	Cameroon	45	540	7500	51	0.09
c	Ralco	Chile	155	34.7	1223	570	16.4
d	Rogun	Tajikistan	335	170	13,300	3600	21.2
e	Nam Theun 1	Laos	177	93.6	3009	300	3.2
f	Houay Ho	Laos	80	38	596	150	2.6
g	Mauvoisin	Switzerland	250.5	2.3	300	330	143.5
h	Volta	Ghana	134	8502	153,000	1038	0.12
i	Sihlsee	Switzerland	33	11.5	96	140	12.2
j	Kulpawn	Ghana	25	1025	7200	36	0.04

Lom Pangar HPP, Cameroon: very large and rather shallow reservoir submerging large parts of the valleys of the Lom and the Pangar Rivers, whose confluence was just a short stretch upstream of the dam. The lower (Southern) half of the reservoir was entirely covered with dense forest, the Northern part was savanna. I was on the Environmental and Social Panel of Experts for this project from 2012 to 2018, during the construction phase. From the confluence of the Lom with the Djerem about 20 km downstream of the dam the river is called Sanaga, and this is the largest river in Cameroon. The main purpose of the dam is to provide seasonal storage, to make more water available in the dry season for a cascade of power plants along the Sanaga (ISL et al. 2005).

Nam Theun 1, Laos: high dam located at the point where the Nam Theun or Nam Kading river leaves the mountainous area of the Annamite region and flows into the Mekong valley to join the main river about 30 km downstream of the dam site. In its Southern part, the reservoir submerges a series of narrow and deep valleys, tributaries of the Nam Theun, leading to a type A reservoir; the Nam Theun itself has already been dammed twice further upstream, by the Theun-Himboun dam and the Nam Theun 2 dam (see Sect. 22.3). The North-Eastern part of the reservoir is comparatively shallow, a typical type B reservoir, and there the river banks were inhabited, the affected population had to be relocated. Over time, a series of studies had been prepared for this project, for different developers, the first in 1994, then in 2003/04 and finally in 2017, and the results of these latter studies were then implemented during construction.

Ralco HPP, Chile: high dam in a narrow valley upstream of the Pangue HPP, on the Bío Bío River. Reservoir in a series of rather narrow valleys in the Chilean Andes. The ESIA for this project was prepared in 1996, and I was involved in a dispute between the project proponent and IFC in 1997. One major issue of the project was the fact that it was located in an area reserved for the Pehuenche, a section of the Mapuche group of indigenous peoples (Villalobos 1989; Orellana 1994) in this region.

Rogun HPP, Tajikistan: very high dam (presently under construction, once finished it will be the highest dam in the world, with 335 m). Impounds a part of the valley of Vakhsh River just upstream of the existing Nurek HPP. Rather large and very deep reservoir, with over-annual storage and a large capacity for flow regulation (see Chap. 23 for more details). The project was originally planned by the Soviet Union, and construction had started in the 1980s. It came to a complete stop at the breaking up of the Soviet Union. In 2011, WB started a comprehensive technical as well as environmental and social study (Pöyry 2014, for the latter), and construction was resumed shortly thereafter.

Houay Ho HPP, Laos: reservoir located on the Boloven Plateau in southern Laos, using the difference in elevation between the plateau and the Xe Kong River valley, where the power house is located. The reservoir is formed by a large number of small streams which drain the plateau to form the Houay Ho River (EWI 1996). The reservoir is mostly rather shallow.

Mauvoisin HPP, Switzerland: very high dam impounding a steep and narrow valley. The dam was built in the early 1950s, a double arch dam with a height of 237 m. In the early 1990s, it was heightened by 13.5 m to reach its actual height (EWI 1987). The aim was to create additional storage for shifting energy production from summer to winter. The power plant is used for peaking, mostly in winter. It uses a small alpine stream and some lateral inflows, but with a total of 1485 m (used in 2 steps of 469 and 1016 m) it has a very high head; FSL is at 1975 m asl. Most water is snow and glacier melt in addition to the summer rains, as good as no inflow to the reservoir in winter.

Volta, Ghana: Lake Volta, impounded by the Akosombo dam and HPP. The reservoir is very large, impounding a number of tributaries of the Volta River, Black Volta, White Volta, and Oti River being the major ones. In spite of the dam of considerable height, the dendritic reservoir is mostly rather shallow, due to the generally flat landscape in which it is located.

Sihlsee (Lake Sihl), Switzerland: power plant commissioned in 1937. It is combined with a second reservoir which acts as pumped storage reservoir. The power plant produces electricity for the Swiss railways. The reservoir is shallow, with a maximum depth of 17 m. It is crossed by two bridges (indicated by the two double lines in the figure).

Kulpawn, Ghana: on the White Volta, upstream of Volta reservoir. The dam would be very low (25 m), but very long (9 km), and in addition a very long saddle dam would be required; the two dams are indicated as solid straight lines in the figure. The reservoir would be very large, but extremely shallow. Electricity output would be 166 GWh/y, but it would lead to a reduction in production at Akosombo, resulting in a net addition to the national grid of only 25 GWh/y (Coyne and Bellier 1993; see observations in Sect. 5.5 explaining why, in my opinion, this project should be definitely abandoned).

16.2 Reservoirs and Lakes

When the reservoir starts to be filled, the conditions in the directly affected stretch of the river change from a lotic (river-like, flowing water) to a lentic (lake-like, stagnant water) condition (Fisher and Sponseller 2010: 88). Often in ESIA reports, this is described as "flooding such and such an area". I do not think that the word "flooding" should be used in this context. According to a Wikipedia definition, "a flood is an overflow of water on normally dry ground. This is most commonly due to an overflowing river, a dam break, snowmelt, or heavy rainfall." This might not be a terribly good description: while dam brakes, unfortunately, do occur (the 2018 dam break of Xe Pian-Xe Namnoy in Laos or the break of a tailings dam in Brazil (Stewart 2019) as recent examples), such an event luckily is much more rare than floods caused by snow melt or heavy rainfall, and should certainly not be qualified as anything "common". But the relevant point is that, although frequently catastrophic, a flood is a short time, reversible event, while the impoundment of a reservoir permanently submerges the area in question.

Impoundment of a reservoir is a major change in biotic conditions, and usually is the start of an very strong shift in the composition to the species living in this area; see Zhong and Power (1996) for an overview. Many of the species adapted to flowing water conditions, and especially those requiring high oxygen concentrations in the water, will disappear; species that are tolerant of a wide range of conditions, as e.g. fish living in deep, slow flowing rivers, may find acceptable conditions in the new reservoir, while others may colonise this new habitat over time. Still, it has to be pointed out that in general reservoirs are "lake-like", but not really an exact replication of natural lakes. What and how marked the difference is depends on a number of conditions.

One important point is the transition period after impounding. Depending on the type of land—and especially on the type and amount of vegetation—submerged, this can have far-reaching consequences on water quality for a number of years, until the submerged biomass has been broken down and a new equilibrium is established. This is often accompanied by a dramatic change in fish population (see Sect. 16.4).

Most natural lakes have a surface outflow, the river that drains the lake; most reservoirs, and as good as all reservoirs specifically created for hydropower, do not have that. A natural outflow not only regulates the water level in the lake, it also links it directly to the downstream area of the river by allowing movements of organisms out of the lake, whether by drift or by active locomotion, and in most cases it allows actively migrating organisms, mainly fish, to move upstream from the river into the lake and possibly beyond. There are certain exceptions to this rule; one well-known example is Lake Victoria in Africa, where the Owen Falls at the outlet of the lake formed a barrier against any fish migrating upstream, and which in this way protected the very specific fish fauna of this great lake up to the moment when the Nile perch (*Lates niloticus*) was introduced there (Worthington 1989); this natural barrier has in the meantime been reinforced by replacing Owen Falls with a hydropower plant, the Owen Falls HPP. There are some natural lakes which do not have any outflow at all, like e.g. the Aral Sea, which by now has almost disappeared (see Sect. 23.5), and there are a few, mostly small lakes, which only have an underground outflow, as e.g. Lake Ira Lalaro in East Timor, which is drained through a karstic system. What is the exception in natural lakes, a barrier against up- and downstream movements of aquatic organisms, is the rule in reservoirs. Even a surface spillway cannot really replace this, although the occasional fish might get out that way, and might even survive the chute.

In reservoirs, the water intake for the power house is usually located at a certain depth, mainly in order to prevent air from being sucked into the turbines, and this also in a situation where the reservoir is being drawn down to its minimum operation level (MOL). This is a very decisive difference between reservoirs and natural lakes not only concerning migration options for aquatic organisms, but also concerning water quality. Thermal stratification, and very often oxygen stratification, as described and discussed in Chap. 17, can occur in natural lakes as well as in reservoirs. However, in the case of a lake the outflowing water will be from the surface, i.e. warm (depending on season) and with a high oxygen content. In the case of a reservoir and HPP, however, water will be taken from deeper layers, usually from below the thermo- and oxycline. This can mean that the water delivered from the power plant to the river downstream of the dam can be considerably colder than surface water would normally be, and can lack in oxygen; this latter effect is most marked in a newly created reservoir, in cases where a large amount of vegetation was submerged, and where therefore oxygen content is very low or even zero. Both effects, low temperature as well as oxygen deficit, can have negative impacts on fish and other aquatic organisms.

One additional very decisive characteristic of reservoirs is defined by the operation of the hydropower plant: the drawdown. There are mainly two types of drawdown with consequences for the suitability of a reservoir as habitat: daily or seasonal. Daily drawdown means that more or less every day, during the hours of peak electricity demand, the reservoir level is lowered, in some cases very considerably, to be filled again during the non-peak hours. This is the case mainly in small reservoirs used for peaking production, as e.g. the Kafue Lower HPP in Zambia, mentioned in Sect. 9.1.1.2. Seasonal drawdown is mainly done in climates which have distinct wet and dry (or, often more to the point: high and low flow) seasons, where seasonal high inflow from rainfall or snow melt is stored to be used in the low-flow season; most large storage reservoirs are built specifically for allowing this type of storage, and numerous examples are mentioned in different parts of this book. In both cases, drawdown means that the so-called drawdown area is covered with water for some part of the time and completely dry during the rest. This is very decisive for the biota of a "lake", since it directly affects the littoral, the zone of relatively shallow water, where light penetrates to the ground and where there is usually oxygen in sufficient concentrations available, which is usually the most productive zone in a natural lake. Seasonal wetlands, which are covered with water during a part of the year and get dry the rest of it, accompanied by a vegetation adapted to this situation, are widespread and well known, like the Kafue Flats in Zambia (Sect. 9.1.1.2) and the Great Lake/Tonle Sap in Cambodia (Sect. 22.5), to name just two; in the case of the latter, one could actually speak of a very massive "drawdown area", which is of great importance not only as a natural habitat, but also for the local human population (see Sect. 22.5). On rare occasions, a situation which is somewhat similar to this can develop around a reservoir, provided the reservoir is large and shallow enough, the water level fluctuations are sufficiently regular, and a suitable vegetation type is available in the area for colonising such zones. This has happened e.g. around Lake Kainji in Nigeria, where mats of a grass (*Echinochloa staginata*) have developed, which float on the surface of the water during the wet season and are used as fodder for cattle during the dry season (Morton and Obot 1984; Obot and Ayeni 1987). Another such example is the Itezhi-Tezhi Lake in Zambia, a large and rather shallow reservoir created in a wide and relatively flat river valley. Its shores have a very gentle slope, and the drawdown is not very marked. The dam is located on the edge of, the reservoir stretches within the Kaufe National Park and has become an integral part of it (Ashley 2012; see Fig. 16.3). However, in most reservoirs no populations of plants and animals can get established under the conditions of dramatic change due to drawdown, especially in those reservoirs with irregular and short time fluctuations and/or steeper shorelines, since there, in addition to the change from being covered by water to being extremely dry, soil is being washed away by the combined effect of the drawdown and wave action (see Fig. 16.4).

Fig. 16.3 Shoreline of the Itezhi-Tezhi reservoir, Zambia

Fig. 16.4 Drawdown area of Nurek reservoir. Nurek reservoir, Tajikistan, in May. In such a reservoir in a mountainous area and with a marked seasonal drawdown the drawdown area is completely bare rock, all the soil having been washed away, and no plants being able to grow there

16.3 Vegetation, Floating Plants

The conditions mentioned above affect the development of aquatic vegetation in a reservoir. The drawdown effect in many cases prevents the development of rooted plants in the littoral zone of a reservoir. While in natural lakes, at least in lakes with not too steep shorelines, in this zone different types of vegetation develop depending on depth—from reeds at the edge of the water to rooted plants with swimming leaves like water lilies or lotus to totally submerged plants—this is not possible in reservoirs with a marked drawdown. Only shallow reservoirs with gently sloping littoral zones, whose water level is kept more or less stable throughout, can develop these vegetation assemblies typical for natural lakes under similar conditions.

One aspect of water vegetation, however, is very marked in tropical and subtropical reservoirs: the floating plants (usually called "weeds", true to the gardeners' view that everything that grows by itself, and especially if I do not like it, is a weed). There are mainly three species which are famous in this respect, since they have become widespread and have indeed caused problems in many reservoirs—and continue to do so: the water hyacinth (*Eichhornia crassipes*), the water lettuce (*Pistia stratiotes*), and an aquatic fern (*Salvinia molesta*), all three originally from South America, and they can cause major problems mainly on reservoirs (Raynal-Roques 1980; Bhatacharjee 1984; Sale et al. 1985; Abbasi and Nipaney 1986; Goeltenboth and Kristyanto 1987; Bechara and Andreaní 1989). At least for the water hyacinth is it well known that it did not just spread by accident, but was introduced to many places because of its decorative value (Janzen 1983: 236 f.). According to Whitten et al. (1996: 184, and other volumes of this series), it was introduced from Brazil to the botanical garden in Bogor, Indonesia, in 1894, and from there by now probably has spread to all stagnant water bodies in Java and beyond, even to remote islands (Monk et al. 1997: 171 ff.) (Fig. 16.5).

Zauke et al. (1992: 113) describe what happened in two cases: in the Brokopondo reservoir in Suriname, a surface of 20 km^2 was covered with a thick mat of the water hyacinth (*Eichhornia crassipes*), and additional 65 km^2 with the water

Fig. 16.5 The most important floating plants in reservoirs. From left to right: Water hyacinth (*Eichhornia crassipes*), flowering. Water hyacinth (*Eichhornia crassipes*), young plant, inflated leaf stems clearly visible. Water lettuce (*Pistia stratiotes*). Water fern, giant salvinia (*Salvinia molesta*)

fern *Ceratopteris*; a similar development with the water hyacinth took place in the Curuna-Una reservoir in the Brazilian Amazon region. In both cases, the floating mats contributed to the development of anoxic conditions in the water of the reservoirs, which led to occasionally massive fish deaths. One more recent example is the Kafue Gorge Upper reservoir in Zambia, which shortly after impoundment was completely covered with water hyacinths. In that case, a concerted action consisting mainly in physical removing of the floating plant mass, succeeded in getting rid of the problem (Pöyry and Zenith 2016).

One reason for the often massive development of these floating plants in reservoirs is one specific characteristic of reservoirs mentioned above: the fact that, as opposed to natural lakes, reservoirs do not have a surface outflow. Behind the Owen Falls dam in Lake Victoria, Uganda, e.g., *Eichhornia* mats are constantly forming and need to be removed. Before construction of the dam such plants would have been swept out of the lake and passed on to the downstream area, but this is no longer possible; it has to be added, in this case, that the water hyacinth, an invasive species stemming from Brazil, might not even have been there before construction of the dam, but this does not change the mechanism. An interesting "case in the making" right now is Rusumo Falls HPP on the Rwanda-Tanzania border, presently (2020) under construction. In the entire swamp area upstream of the falls *Eichhornia* is present in small quantities, and single plants or small clusters are constantly being swept down the river. Most of them get literally smashed up when they go over the falls, and small mats of leaves, bulbs and stalks of the plants can be seen in still water below the falls. Once the dam closed, this constant evacuation will no longer be possible, the plants will accumulate there and presumably find suitable conditions for reproducing. It remains to be seen what measures might be required to prevent them for developing into a problem.

16.4 Fish

As already mentioned (Sect. 10.3), habitat conditions for fish in reservoirs are very different from those in rivers, which, first of all, means that the fish fauna in a reservoir is often very different from that of the original river. While some species are able to live under both conditions, some species may actually do much better in a reservoir, while others, which thrived in the river, may disappear.

Reservoirs can harbour a very considerable fish stock, in terms of number of species as well as in terms of productivity. Therefore, they can be productive fishing grounds. This applies to rather small and shallow reservoirs, which often have a high productivity, as e.g. the ones described by Baijot et al. (1994) for the case of Burkina Faso, where numerous small reservoirs were created with the aim of providing water for the population year round and for irrigation, but which proved also useful in terms of fishing.

One way for estimating the fisheries productivity of lakes and reservoirs is the use of the Morpho-Edaphic Index (MEI), which is determined by the following formula:

$$MEI = \frac{\text{conductivity } \left(\frac{\mu\text{mhos}}{\text{cm}}\right)}{\text{mean depth (m)}}$$

For African lakes of very different sizes, including a number of reservoirs, capture could be approximated by the formula

$$\text{kg/ha} = \text{MEI}^{0.3813}$$

although data of actual catches show rather large variations, whereby data on fish catches depend not only on the situation of the lake in question, but also on the density of fishermen (Welcomme 1972; Henderson and Welcomme 1974; see also http://www.fao.org/3/e6645b/E6645B01.htm). This approach is applicable to stable waterbodies, i.e. lakes and reservoirs which do not show any large fluctuations in water level and water quality. In newly impounded reservoirs, the development of the fish fauna will develop not only in function of fish fauna which was present in the river, but will depend also on the development of the water quality, which can undergo very considerable changes until a new equilibrium, mainly in terms of in- and output of nutrients, has been reached. The early period of this development is often characterised by a considerable increase in productivity and a severe drop in fish diversity, as has been documented for Lake Kariba (Balon 1974, see Table 9.1 and Sect. 10.3) and as recently observed in the Lom Pangar reservoir (Sect. 24.1).

One special issue related with the development of fisheries in reservoirs is the stocking with exotic species; this problem, which counts with a number of success stories, but also with major failures, was discussed shortly in Sect. 10.3.5.3.

On the other hand, some reservoirs are not suitable fish habitat at all. This was shown clearly for the Mauvoisin reservoir, a rather small, but very deep reservoir in the Swiss Alps (EWI 1987). This reservoir has very steep banks which, in combination with the yearly drawdown, do not allow for any aquatic plants to develop. I addition, the water is extremely turbid due to the very fine suspended matter stemming from the glacier which feeds the reservoir, and which is very slow to settle. In consequence, sunlight cannot penetrate beyond the topmost layer, which prevents plankton from developing. Overall, this means very little food for fish, and in spite of stocking the reservoir with fish these did not develop there.

16.5 Plankton

In many ESIAs for hydropower plants there is a section on plankton, describing the efforts that were made for sampling plankton and the few results obtained from this sampling. For me, this is one aspect of futile effort made in such studies, and this even more so since so far I have never seen an example where this effort would actually have contributed anything to the relevant findings, conclusions and proposed measures of an ESIA report. The only explanations I have for this work is that either such an investigation is asked for in the applicable guidelines, or that the institution carrying out the study has a plankton specialist and wants to occupy him; in this respect, see the observations on noise studies in Sect. 8.3. It should be sufficient to make a statement in the screening and scoping report reflecting what Fisher and Sponseller (2010: 83) say: "… longitudinal transport in streams precludes the development of a true plankton community in all but the slowest (often largest) rivers."

As soon as the reservoir is there, the situation obviously changes, since from that moment onward the main requirement for the development of a true plankton community is given: stagnant water. The few plankton organisms that are swept down to reach the reservoir, or may reach it by other means as e.g. water birds moving between different water bodies, may now find good conditions for establishing stable populations, and plankton, especially in large reservoirs, can develop into an important food source for pelagic fish species (which usually were also not present at that place before the dam was built). How profoundly a plankton community can change due to impoundment, in terms of biomass as well as species composition, was demonstrated by Paterson et al. (2019), even though this long-term study was not really about damming a river, but rather increasing the surface of a pre-existing pond located a very short stretch downstream of a lake. In short, development of plankton may be a topic to follow up upon under the heading of reservoir development monitoring, but it is usually not really a relevant subject in an ESIA for a dam project.

16.6 Benthos

Benthos includes all types of invertebrate organisms (larvae of insects, worms, small molluscs, etc.) which live on the ground of water bodies. Since many of them have very specific requirements, e.g. in terms of oxygen content, flow velocity, pollution etc., they can be used as indicators for the condition in their habitat. After impoundment, conditions for benthic organisms change fundamentally. The typical species inhabiting flowing water—which usually also implies a high oxygen content—will disappear, and a different fauna adapted to lentic conditions will become established. Here as well, water quality, and especially the question of presence of organic matter and oxygen, will be decisive. As in a natural lake, concentrations of benthic organisms will usually be higher in the shallower parts of the reservoir, most species concentrating in the top 20 m; this is also an indication that conditions in a reservoir with a very marked seasonal drawdown might be less suitable as a benthos habitat. Low oxygen concentrations in deeper parts of a reservoir can also be a problem for benthic organisms (Strayer 2010). Benthos can also be a source of food for different water birds. Behney (2020), in a comparison of different types of lakes and reservoirs as a habitat for ducks feeding on benthic organisms, has found that lakes are usually more suitable in this respect than reservoirs, and small reservoirs are better than larger ones. Submerged trees can be colonised at least by some benthic organisms (McLachlan 1971), and mats of floating vegetation can also be inhabited in this way (Bechara and Andreani 1989).

16.7 Birds and Mammals

What was said above for other groups of organisms is also true for birds and mammals. Deep reservoirs with steep banks and a marked drawdown usually do not offer good conditions for a colonisation by these animals, while reservoirs that are shallow or have an extensive littoral area and limited fluctuations in water levels can develop into being good habitats. Three smaller reservoirs in Switzerland, all three ROR structures, are Ramsar sites, i.e. areas of international importance for the protection of water birds, and they reached this status, among other reasons, because of their significance for overwintering ducks and other water birds (Fig. 16.6). On and around just one of these reservoirs, in 2019 alone more than 200 bird species were observed, breeding, migrating or overwintering (Niggeler and Keller 2007; Niffenegger 2020; Ruckstuhl 2020). Reservoirs of that type can also be colonised by beavers (*Castor fiber*). In Spain, populations of the endangered Eurasian otter (*Lutra lutra*) have rebounded during the last few decades after a long period of decline and are now expanding and occupying both natural and man-made habitats, i.e. reservoirs (López-Martín and Jiménez-Pérez 2009; Martínez-Abraín et al. 2020), and the Neotropical otter *Lontra longicaudis* is colonising reservoirs as well as rivers (García-Silva et al. 2021).

Fig. 16.6 Different types of water birds on the Klingnau reservoir, Switzerland. Among them two ruddy shelducks (*Tadorna ferruginea*), an invasive species (see Sect. 10.5.4)

References

Abbasi, S.A., and P.C. Nipaney. 1986. Infestation by aquatic weeds of the fern genus Salvinia—its status and control. *Environmental Conservation* 13(3): 235–241.

Ashley, N. 2012. *The Kafue National Park, Zambia*, p 232. UK: CBC Publishing, Bath.

Baijot, E., J. Moreau. and S. Bouda. 1994. *Aspects hydrobiologiques et piscicoles des retenues d'eau en zone Soudano-Sahélienne*. Wageningen: CTA.

Balon, E.K. 1974. *Fishes of lake Kariba, Africa*, 144. Neptune City: TFH Publications.

Bechara, J.A., and N.L. Andreaní. 1989. El macrobentos de una laguna cubierta por Eichhornia crassipes en el valle de inundación del Río Paraná (Argentina). *Tropical Ecology* 30 (l): 142–155.

Behney, A.C. 2020. The influence of water depth on energy availability for ducks. *Journal of Wildlife Management* 83 (3): 436–447. 101002/jwmg.21811.

Bhatacharjee, A. 1984. You can't fight that weed, can you? *Parks* 9 (2): 7.

Coyne et Bellier, 1993. *White Volta development scheme. Prefeasibility Study*. Report prepared for the Volta River Authority, Republic of Ghana.

Electrowatt-Ekono and Techsult, 2005. *Bumbuna hydropower project: Resettlement action plan for the reswervoir and dam area*. Report prepared for Government of Sierra Leone and World Bank.

EWI, 1987. *Projet de surélévation du barrage de Mauvoisin. Analyse des effets sur l'environnement*. Forces Motrices de Mauvoisin.

EWI, 1996. *Houay Ho hydroelectric power project. Environmental impact assessment*. Seoul, Korea: Daewoo Corporation.

Fearnside, P.M. 1989. Brazil's Balbina Dam: Environment versus the legacy of the Pharaohs in Amazonia. *Environmental Management* 13:401–423 https://doi.org/10.1007/BF01867675

Fisher, S.G., and R.A. Sponseller. 2010. Ecology of flowing waters. In: *2010a: River ecosystem ecology. Encyclopedia of inland waters*, ed. G.E. Likens, 81–88. . Oxford: Elsevier.

García-Silva, O., J.P. Gallo-Reynoso, M. Bucio-Pacheco. et al. 2021. Neotropical otter diet varition between a lentic and a lotic systems. *Therya* 12 (1): 95–103. https://doi.org/10.12933/therya-21-781.

Goeltenboth, F., and A. Kristyanto. 1987. Ecological studies of the Rawa Pening lake (central Java): some biotic parameters of the semi-natural lake. *Tropical Ecology* 28 (1): 101–116.

Henderson, H.F., and R.L. Welcomme. 1974. *The relationship of yield to morpho-edaphic index and numbers of fishermen in African inland fisheries*. CIFA Occasional Paper No. 1: 1-19. Rome: FAO.

ISL, Oréade-Brèche, SOGREAH, 2005. *Etude environnementale du barrage de Lom Pangar*. Rapport de Sythèse. République du Cameroun, Ministère de l'Energie et de l'Eau.

Janzen, D.H. 1983. *Costa Rican natural history*, 816. Chicago: University of Chicago Press.

Jeppesen, E., M. Sondergaard, H.S. Jensen, and A.M. Ventäla. 2010. Lake and reservoir management. In Lake ecosystem ecology. Encyclopedia of inland waters, ed. G.E. Likens, 2010b, 376–390. Oxford: Elsevier.

López-Martín, J.y. and J. Jiménez-Pérez. 2009. *La nutria en España. Veinte años de seguimiento de un mamífero amenazado. Sociedad Española para la Conservación y Estudio de los Mamíferos SECEM*, 494 pp. ISBN: 978-84-612-9094-9.

Martínez-Abraín, A., P.S. Tomillo, and J. Veiga. 2020. Otter diet changes in a reservoir during a severe autumn dorught. *Journal of Mammalogy* 101 (1): 211–215. https://doi.org/10.1093/jmammal/gyz185.

McLachlan, A.J. 1971. Submerged trees as a substrate for benthic fauna in the recently created Lake Kariba (Central Africa). *Journal of Applied Ecology* 7: 253–266.

Monk, K.A., Y. de Fretes, and G. Reksodiharjo-Lilley. 1997. *The ecology of Nusa Tenggara and Maluku. The ecology of Indsonesia series*, vol. 5, 966. Singapore: Periplus.

Morton, A.J., and E.A. Obot,. 1984. The control of Echinochloa staginata (Retz.) P.Beauv. by harvesting for dry season livestock fodder in Lake Kainji, Nigeria—a modelling approach. *Journal of Applied Ecology* 21 (2): 687–694.

Niffenegger, C. 2020. Verlandung und Verschilfung des Klingnauer Stausees. *Ornbithologischer Beobachter* 117 (2): 84–85.

Niggeler, E., und V. Keller. 2007. Winterbestände der Wasservögel am Aare-Stausee Niderried 1951/52–2005/06. *Der Ornithologische Beobachter* 104 (4): 279–300.

Orellana, R.M. 1994. *Prehistoria y etnología de Chile. Colección de Ciencias Sociales, Universidad de Chile*, p 240. Santiago de Chile: Bravo y Allende.

Obot, E.A., and J.S.O. Ayeni. 1987. Conservation and utilization of aquatic macrophytes in Lake Kainji, Nigeria. *Environmental conservation* 14 (2): 168–170.

Paterson, M.J., K.G. Beaty, D.L. Findlay. et al. 2019. Long-term changes in nutrient dynamics and plankton communities following the creation of a new reservoir. *Canadian Journal of Fisheries and Aquatic Sciences* 76 (8). https://doi.org/10.1139/cjfas-2018-0197s13s16w9p34n6x

Pöyry. 2014. Environmental and Social Impact Assessment for Rogun Hydropower Plant. In *ESIA* vol. I: Text, vol. II: Annexes. Report prepared for OSHPC "Barki Tojik", Republic of Tajikistan.

Pöyry and Zenith. 2016. *Kafue Gorge lower HPP ESIA: Environmental and social impact assessment*, 287. Zambia: Report prepared for ZESCO and Kafue Gorge Lower Power Development Corporation, Ltd.

Raynal-Roques, A. 1980. Les plantes aquatiques (plantes à fleur et fougères). In: *1980: Flore er faune aquatiques de l'Afrique sahélo-soudanienne*, eds. J.R. Durand, et C. Lévèque, vol. 1, 63–152. Paris: ORSTOM.

Ruckstuhl, M. 2020. Jahresübersicht 2019. *Klingnauer Stausee Bulletin Nr.* 59: 42. http://www.klingnauerstausee.ch/ornithologie/

Sale, P.J.M., P.T. Orr, G.S. Shell, and Erskine. 1985. Photosynthesis and growth rates in Salvinia molesta and Eichhornia crassipes. *Journal of Applied Ecology* 22 (1): 125–137.

Stewart, J. 2019. 40 Dead, hundreds reportedly missing after Brazilian Mining Company's dam collapses. https://www.npr.org/2019/01/25/688803282/200-people-reportedly-missing-after-brazilian-mining-companys-dam-collapses?t=1548585832510.

Strayer, D.L. 2010. Benthic invertebrate fauna, lakes and reservoirs. In: *2010b: Lake ecosystem ecology*, ed. G.E. Likens, 27–40. Oxford: Encyclopedia of inland waters. Elsevier.

Villalobos, R.S. 1989. *Los Pehuenches en la vida fronteriza*, p 269. Santiago: Ediciones Universidad Católica de Chile.

Welcomme, R.L. 1972. The inland waters of Africa. *CIFA Technical Paper* (1): 117. http://www.fao.org/3/n0084e/N0084E00.htm#TOC

Whitten, T., R.E. Soeriaatmadja, and Afiff, S.A., 1996. *The ecology of Java and Bali. The ecology of Indonesia series* vol. II, 969. Singapore: Periplus Editions (HK), Ltd.

Worthington, E.B. 1989. The lake Victoria Lates saga. *Environmental Conservation* 16 (3): 266–267.

Zauke, G.-P., R.-G. Niemeyer, und K.-P. Gilles. 1992. *Limnologie der Tropen und Subtropen. Grundlagen und Prognoseverfahren der limnologischen Entwicklung von Stauseen*, 171 S. Landsberg/Lech: Ecomed.

Zhong, Y., and G. Power. 1996. Some environmental impacts of hydroelectric projects on fish in Canada. *Impact Assessment* 14 (3): 285–308.

Chapter 17
Reservoir Stratification

Hydropower is the world's most important source of renewable energy. It produces neither greenhouse gases nor the pollution associated with burning fossil fuels, and tends to be more acceptable to people than nuclear energy.
Clarke and King (2004: 42)

Most natural lakes will show a thermal stratification, which depending on climatic conditions can be more or less stable (mainly in tropical areas) or will be reversed twice a year (in temperate and boreal conditions). The same will happen in reservoirs, with the exception of small and shallow ones, especially when operated as ROR schemes, and reservoirs with a high water turnover rate, which is often combined with a marked drawdown. While a stable thermal stratification in a reservoir as such is usually not a problem, it can however lead to the development of an additional phenomenon, namely, oxygen stratification. This is due to the fact that in the upper layers of the reservoir, just as in a natural lake, planktonic algae will produce oxygen by means of photosynthesis as far down in the water column as sunlight is reaching, while in the absence of light in the deeper layers this will not be possible; there, most markedly in newly created reservoirs in moist temperate and especially tropical regions, biomass, i.e. plant material, covered with water during impoundment will be decomposed, which can easily lead to an exhaustion of the oxygen content in the water. This in turn leads to water quality issues and to emissions of GHGs from the reservoir, and this is definitely an issue of major environmental concern. Whether such a situation will or may develop needs to be determined in the ESIA, and mitigation measures have to be formulated and implemented prior to impoundment in order to reduce the risk of such a development, or at least to create a situation which will enable the new artificial lake to develop an acceptable condition as quickly as possible. The stratification process is described, methods for assessing the probable risk mainly of adverse oxygen stratification are discussed, and the main potential mitigation measure, pre-impoundment reservoir area clearing, is described.

R. Zwahlen, *Assessing the Environmental Impacts of Hydropower Projects*, Environmental Earth Sciences, https://doi.org/10.1007/978-3-030-91185-0_17

17.1 Thermal Stratification Process

Large and deep reservoirs in temperate climates, like natural lakes, usually show a thermal stratification during summer, when water temperature at the surface is rather high, while water in greater depth is much colder. Thermal stratification and circulation in lakes and reservoirs in temperate regions are governed by the following parameters and processes:

- Density of water depends on temperature; it is highest (water being heaviest) at 4 °C, lower at both higher and lower temperatures, and lowest when it is in the form of ice. This means that water, when it reaches 4 °C, will sink to the bottom, and in a deep reservoir or lake water temperature at the bottom will always be at or close to 4 °C.
- In summer, solar irradiation will raise water temperature at the surface; this warmer water, since it is lighter, will remain at the surface. Wind and wave action will mix this warmer water with the cooler water below it; however, this affects only the upper layers of the water body. This leads to a stable thermal stratification, characterised by a sudden drop in water temperature at a depth of usually around 10 m.
- In autumn the water at the surface will slowly become cooler. Cooler water will start to sink, bringing the warmer water below it to the surface; this is the start of the circulation, which will gradually reach deeper water as the cooling process goes on.
- When the water reaches 4 °C over the entire water column, it will circulate freely, since there are no layers with different densities any more.
- Further cooling in winter will lead to colder (<4 °C) water at the surface, and thus again to a thermal stratification, the colder water remaining at the surface. When the water starts freezing, the ice will remain at the surface; a closed ice cover will slow down the further cooling process.
- In spring, with a warming of the surface the process will start again (water getting heavier as it approaches 4 °C and sinking to deeper layers, spring circulation, and then remaining at the surface as temperature rises.

More details on the process of stratification and turnover in lakes and reservoirs, which also describe different circulation types depending on climate zones, can be found in textbooks on limnology (e.g. de Santo 1978: 84 ff.; Schwoerbel 1980: 26 ff.; Likens 2010b, and there especially Boehrer and Schultze 2010: 96 ff.)

An empirical dependence of reservoir stratification on residence time (τ) and the maximum temperature difference between the surface and hypolimnion (i.e. water in the deeper areas of the reservoir) was found by Straskraba and Mauersberger (1988; cited after EAWAG 2006) for several reservoirs in the Czech Republic, approximated by the equation:

$$\Delta T_{0-30} = 20(1 - \exp(-0.0126 * \tau))$$

whereby ΔT_{0-30} is the difference in water temperature between the surface and a depth of 30 m, and τ is the residence time of the water in the reservoir.

This can be used for lakes or reservoirs located in a temperate climate, with marked differences between summer and winter temperatures. So e.g., applying this formula to a large an deep reservoir, Rogun HPP in Tajikistan (Pöyry 2014: 116), this wold mean that, with a residence time (τ) of about 252 days, the temperature difference between surface and hypolimnion would correspond to a maximum of about 19 °C. This should not be taken as an accurate value, but rather as an estimate of the upper limit. Nevertheless, given the climatic situation in the area of this project, with hot summers (average July temperature 24 °C) and cool winters (average January temperature −5 °C), it sounds plausible. Such an important difference in temperature would indicate a stable thermal stratification, which however would be seasonal, being reversed in autumn, following the general pattern described above.

In tropical lakes and reservoirs such a stratification can develop as well, although the temperature differences are much smaller. However, stratification under such conditions can be more stable, since the circulation triggered by seasonal temperature differences does not take place.

A model for tropical lakes and reservoirs was proposed by Lewis (1987, cited after Zauke et al. 1992: 16 ff.). One decisive difference between this and the approach described above is the fact that this model, which allows to calculate the temperature to be expected at the bottom of a tropical lake or reservoir, includes geographical latitude as well as elevation. The proposed equation is the following:

$$T_b = 28.9 - 0.43B' - 0.0038H$$

whereby

T_b temperature at the bottom of the water body (°C)
B' corrected latitude (the "meteorological equator" being at 3.4° N)
H elevation (m asl)

Zauke et al. (1992: 17) provide a comparison of values obtained in this way for a number of tropical lakes and reservoirs, which compare quite well with actual values measured in the respective water bodies.

While in an ESIA such calculations may be useful for getting an idea on what the thermal stratification in the future reservoir might be, it is usually preferable to make a comparison with real-world examples, i.e. with results from measurements in existing reservoirs. Examples are plenty, and they have the additional advantage that they do not only give one value, as e.g. a comparison of surface and −30 m temperature, but actually show the difference over the water column. One decisive information, which the calculations cannot provide, is the depth of the thermocline, i.e. of the zone where there is a rapid transition from warm surface water to cool water in lower layers of the water column. Measurements carried out in a lake or reservoir with actual thermal stratification show this very clearly, as in the example in Fig. 17.1.

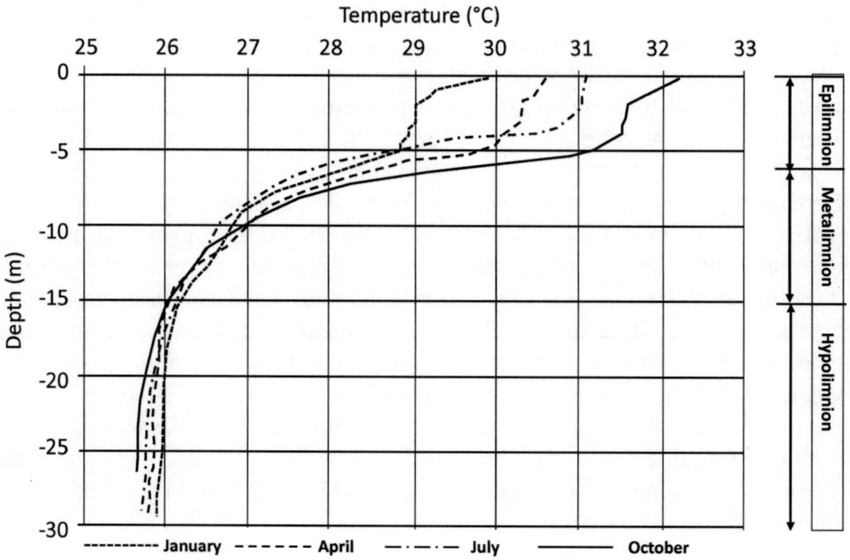

Fig. 17.1 Thermal stratification in a tropical reservoir (Petit Saut, French Guyana). *Source* Based on data from Abril et al. (2008)

Figure 17.1 shows that in this case, temperatures over the year do not vary much. It also shows the three main layers, namely:

- Epilimnion: the top layer with high water temperature, in this case the top 5–6 m.
- Metalimnion: transition zone, here between depths of 5–15 m; in this zone, water temperature drops considerably (thermocline).
- Hypolimnion: the deeper zones of the reservoir, in this case below −15 m; here, temperature is permanently considerably lower than in the epilimnion, and seasonal variations are very small.

Such a situation, with a distinct temperature gradient in every season, would indicate a very stable stratification.

One simple way to determine whether there will be a stratification in the reservoir or not is the calculation of the Froude number (Parker et al. 1975; Orlob 1983; Ford and Johnson 1983; Deas and Lowney 2000). This is done according to the formula

$$F = 320\frac{LQ}{dV}$$

whereby

L length of reservoir (m)
Q mean annual inflow (m^3/s)
d average depth of reservoir (m)
V reservoir volume (m^3)

Table 17.1 Froude number of some reservoirs

HPP	L	Q	d	V	F
	m	m^3/s	m	m^3	
Luang Prabang (Laos)	156,000	3293	38	1,256,000,000	3.4442
Song Mbengué (Cameroon)	7000	1650	15	140,000,000	1.7600
Three Gorges (China)	600,000	30,000	85	39,300,000,000	1.7243
Memve'ele (Cameroon)	12,000	393	10	130,000,000	1.1608
Pangue (Chile)	8900	296	55	175,000,000	0.0876
Kafue Gorge Lower (Zambia)	4500	315	65	83,000,000	0.0841
Ralco (Chile)	23,700	270	70	1,223,000,000	0.0239
Brokopondo (Suriname)	56,000	324	12	22,700,000,000	0.0222
Nam Ngum II (Laos)	36,400	196	80	3,590,000,000	0.0079
Rogun (Tajikistan)	60,000	631	150	13,300,000,000	0.0061
Houay Ho (Laos)	9800	10	35	596,000,000	0.0015

and where $F > 1$ meaning: no stratification
$F < 1$ meaning stratification.

Table 17.1 provides a number of examples of this calculation for different reservoir types.

The table shows that, as would be expected, the large Luang Prabang HPP, to be built on the Mekong mainstream, which is a typical ROR plant, will not show any stratification. Three other reservoirs in this table, although less marked, also show F values greater than 1, indicating no stratification. This seems to be obvious for the two rather small (in relation to inflow) and shallow reservoirs in Cameroon. There are doubts, however, for the case of the Three Gorges reservoir, which, it would seem, is just too large (and mainly: too long!) to be entirely characterised with such a simple figure; it could be imagined that on this stretch of about 600 km, there are parts where thermal stratification may occur. The remaining ones, according to this calculation, would exhibit thermal stratification. This may be correct, but certainly not in one case: the Kafue Gorge Lower (KGL) reservoir, for the following reasons: the reservoir is small, although it is rather deep. Turnover time of the water in the reservoir is approximately 3.5 days, i.e. average daily inflow represents about 30% of the reservoir volume (as a comparison, average daily inflow into the very large Rogun reservoir is only 0.4% of the storage volume, meaning that it takes 250 days on average for a water turnover in the reservoir). With such a high turnover, thermal stratification is not totally excluded, but it is highly unlikely. And in this case there is one more condition which almost certainly precludes stratification: KGL HPP is going to be operated as a peaking plant, meaning that the reservoir will be drawn down to a very large extent every day (at least during the low-flow season) and refilled during the night. With such an operating regime, stable thermal stratification can be excluded. And this is the first concern with this formula: it is too simple, relying only on very few hydrological parameters, and does not take into account any project-specific conditions. One additional difficulty can easily and readily be checked by everyone: the formula is certainly easily applicable to a ROR project like Luang Prabang HPP, whose reservoir basically just consists of the 50 or so km of the river upstream of the dam; however, when one looks at some reservoirs on suitable maps or in google earth, or at the reservoir shapes shown in Fig. 16.2, one will very often wonder how reservoir length should be determined. The highly dendritic Ralco reservoir on the Biobío River in Chile, or the Hoay Ho reservoir on the Boloven plateau in Laos are good examples for this.

Another concern with the Froude number is of a totally different nature. ESIA reports, as a rule, are made public, they have to be accessible for interested parties and, prominently among them, for the directly affected population. For this reason, it is a widely accepted norm that such reports should be prepared in a way that they can be understood by non-specialists. When such a non-specialist now reads that in a certain case the Froud number is 0.7 and therefore thermal stratification has to be expected, this does not tell him or her anything.

This does not mean to say that the number should not be calculated, or that, as the case may be, other simple calculation methods as e.g. the ones mentioned above should not be used; this can be done as a first approach to the question, but it should not be taken as a sufficient explanation or forecast of the future situation. An additional explanation, as the one provided above for KGL, but of course with more detail, is required for really understanding the situation. For this, as always, the project-specific conditions have to be discussed, and in any case, taking into account that one talks about a possible or probable future situation which cannot be confirmed by measurements as yet, a comparison with existing reservoirs similar to the one presently under investigation is certainly helpful.

17.2 Oxygen Stratification

Thermal stratification is important in respect to one very relevant aspect of water quality, namely, oxygen content. This is usually expressed in one of two ways: as dissolved oxygen (DO), measured in mg/l, or in saturation, expressed in %. The capacity of water to dissolve oxygen decreases with increasing temperature. At sea level, 100% saturation is about 14.6 mg O_2/l at 0 °C, and this drops to 8.6 mg/l at a water temperature of 25 °C. In addition to that, solubility is also influenced by atmospheric pressure, meaning that oxygen content will diminish with an increase in altitude.

In a reservoir, oxygen content is influenced mainly by the following processes:

- Oxygen content in inflowing water.
- Resorption of oxygen from contact with the atmosphere; this is greatly influenced by wind, since it is comparatively low as long as the water is very calm, but much higher in conditions of waves agitated by wind.

- Production of oxygen in the water body itself; this, at least in the zones of a reservoir outside of the littoral, the shallow zone where higher rooted plants can grow, is done mainly by algae in the plankton. Photosynthesis of green plants, the process by which oxygen is produced, is limited to the zones that receive sufficient light, which is usually the epilimnion. If the entire reservoir is shallow enough as to receive light down to the bottom, oxygen will be produced in a sufficient amount to prevent depletion, as e.g. mentioned for the Kamburu reservoir in Kenya (Odingo 1979: 67).
- Consumption of oxygen in the water, by some animals, but mainly bacteria breaking down organic material and consuming oxygen in the process.

The two latter processes can be especially important in the case of reservoirs, and here again especially in the first months or years, depending on the situation, after impoundment. In such a situation, these two processes can be very marked, and they can, in a way, mutually reinforce each other. If during reservoir filling a large amount of land vegetation was submerged, dense tropical forest as an extreme case, this plant material will die and will be broken down by bacteria who will consume the oxygen in the water. In the upper layers of the reservoir, in the epilimnion, where light penetrates, green plankton algae will use the light for photosynthesis, producing oxygen in the process. Decomposition of dead vegetation will free different plant nutrients, which will act as fertiliser, in this way increasing photosynthetic productivity. High concentrations of algae in the epilimnion will limit the depth to where light can penetrate, limiting oxygen production to the surface of the water body. Some of the algae will die and sink to the bottom of the reservoir, in this way contributing to the amount of biomass to be degraded, which in turn means that oxygen demand and consumption in the hypolimnion will increase. This can then lead to a stratification of oxygen concentration which is similar to thermal stratification: high oxygen concentrations, up to saturation, in the epilimnion, and low concentrations in the hypolimnion, which in extreme cases can go to a complete depletion of oxygen, i.e. to anoxic conditions. If such a development takes place under conditions of a thermal stratification, the effect will be reinforced, since thermal stratification means little or no exchange of water between the layers with different temperature.

Depending on conditions, the picture of oxygen content in the water of a reservoir can be very different, as shown in Fig. 17.2.

Batang Ai is a reservoir in Sarawak, Malaysia, the graph shows the situation shortly after impoundment. Water to a depth of about 5 m is saturated, then oxygen concentration drops sharply to values near 0, indicating a high rate of oxygen consumption due to a large amount of organic material (biomass). Mauvoisin is a reservoir in the Swiss Alps, with an FSL at almost 2000 m asl. The water flowing into the reservoir is mainly melting water from snow and glaciers, very cold, and can therefore contain higher concentrations of oxygen than the one in Batang Ai. It contains very little if any organic material, therefore oxygen consumption in the water is very low. The higher concentration at a depth of 70 m probably stems from inflowing water that was colder than the water at the surface, and which therefore sank to a deeper zone.

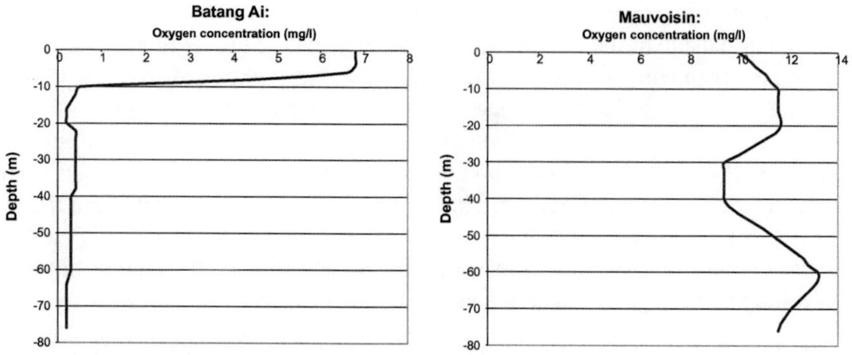

Fig. 17.2 Oxygen concentrations in two different reservoirs. *Data source* Zwahlen (2003: 317) (Batang Ai); EWI (1987) (Mauvoisin); additional explanation in the text

In addition to geographical latitude, elevation, weather conditions (wind) and morphology, factors which mainly determine thermal and oxygen stratification in a lake, in a reservoir this can also be influenced by the configuration of the structures and operation pattern. A deep intake (and water intake for the power plant in a storage reservoir will usually be considerably lower than the normal thermocline) will take water from the hypolimnion, and this can have an influence on stratification. If a reservoir is used for peaking power production, especially in small reservoirs this can lead to high fluctuations in the water level of the reservoir, and this also will have an effect on stratification.

17.3 Impacts

As just discussed, thermal stratification is a natural process which occurs in natural lakes just as well as in reservoirs. Still, while it is usually not considered as a problem in lakes, it can have negative consequences in the case of reservoirs. There are two main issues, and both are related to the fact that, while in natural lakes the effluent is usually at the surface, namely, the river flowing out of a lake, this is as good as never the case for reservoirs, where the main outflow, the water intake for the power house, usually is located at a depth of around 40 m. One reason for this is that it has to be avoided that air is being sucked into the penstock, since this could damage the turbines. Furthermore, drawdown of the reservoir has to be considered. Most large and deep seasonal storage reservoirs are filled in the wet or high flow season (high flow season not necessarily always being the "wet" season, e.g. in situations where much of the precipitation falls as snow, and snow melt is then filling the reservoir even when during this period there is not much precipitation), and the stored water is then being used during the low flow season, which leads to a drawdown of the reservoir which can be several tens of meters. This means that the water intake has to be deep enough for still having the required depth at the time when the reservoir is at its MOL. In reservoirs with thermal stratification the intake will always be below the thermocline.

The first consequence of this is that the water which is passed from the turbines to the river downstream of the power plant can be considerably colder than the surface water in the river would normally be, and this can have negative impacts on fish and other aquatic organisms. This is especially critical in the case of peaking power plants, because each abrupt rise in water quantity at the moment when the plant starts operating will be accompanied by a drop in temperature, increasing the stress which peaking in any case represents for aquatic organisms.

As mentioned above, especially in tropical reservoirs with a stable thermal stratification, the thermocline also defines the oxycline, since the temperature difference prevents water from different depths to be mixed. Therefore, in such a case the water delivered to the downstream area will not only be colder, it will in addition to that have a high BOD (Biological Oxygen Demand) content and a low degree of saturation with oxygen, in an extreme case even being anoxic, i.e. completely free of oxygen. This can kill the fish in the river downstream of the reservoir, and also other aquatic organisms. This effect can potentially affect a long stretch of the river, especially in cases where this river flows without much of a slope, where rapids etc. could lead to a quick reoxygenation. In addition to that, if the concentration of organic material stemming from the reservoir (expressed as BOD, biological oxygen demand) is high, this can mean that whatever oxygen is being brought back to the water is immediately consumed again by the ongoing process of biomass breakdown; this was observed in the case of Lom Pangar, see Sect. 24.1). The impact of a lack of oxygen will very often be much more important than the one caused by a drop in temperature.

The negative effects of this impact are not limited to the downstream are, they can also cause problems in the reservoir as such. Even hydropower reservoirs very often have other (at least potential) uses, fishing often being very important. High concentration of organic material can mean very large amounts of nutrients at least for some fish species (detritivores, i.e. species feeding on decaying plant material).On the other hand, a lack of oxygen will eliminate all the species which are sensitive to low oxygen concentration, and this usually is the largest part of the fish species normally living in a river. If such a reservoir is also used as a source of drinking water, bad water quality will lead to higher costs for water treatment. Finally, lack of oxygen in the water will lead to the production of methane (a breakdown product of anaerobic bacteria which decompose organic material under anoxic conditions, and a very powerful GHG) and possibly other substances, like hydrogen sulphide (H_2S), which is highly toxic as well as very corrosive.

Whether such an impact will become manifest or not depends on different parameters like pre-impoundment vegetation in the reservoir area, morphology and size of the reservoir, hydrology, and operation pattern of the power plant. Two examples are shortly mentioned here to illustrate this, two more are described with some more detail in Chap. 24; see also Fig. 18.3.

Saguling-Cirata-Jatiluhur Cascade, Indonesia: this is a multi-purpose cascade of three reservoirs in Eastern Java: Saguling (commissioned 1985, reservoir surface 56 km^2, installed capacity 700 MW), Cirata (1988, 62 km^2, 500 MW) and Jatiluhur (1967, 83 km^2, 150 MW). In addition to power production, they serve for fisheries (largely through aquaculture in floating cages; see also Whitten et al. 1996: 459 for comments on the development of fisheries) and for drinking water supply. Stratification of these highly eutrophicated reservoirs was very marked, thermo- and oxycline being at a depth of 5–11 m (Lehmusluoto et al. 1999: 133, 145). The food for the fish in the far too numerous floating cages in these reservoirs, in addition to runoff from soils used for agriculture and untreated waste water, contributed to eutrophication. In specific situations however, mainly under conditions of strong winds, the stratification was broken up, meaning that oxygen-free water from the hypolimnion could reach the surface, which often killed a large number of fish in these cages.

Brokopondo, Suriname: construction the dam was completed in 1964, and according to Mol et al. (2007) it was the first large reservoir in a tropical rainforest. The dam is 54 m high, the reservoir is shallow, with a mean depth of 11.5 m, but very large, the surface is 1560 km^2. The entire area was covered by dense tropical rainforest. After impoundment, water quality deteriorated very rapidly (van der Heide 1982; Zauke et al. 1992: 113) and there was a very marked development of H_2S, to the point that for a period the work force at the power plant had to wear gas masks. The number of fish species had been reduced from 172 in the river before impoundment to 62 four years after dam closure; a new stable condition was reached about 14 years after dam closure, with only 41 species of fish firmly established in the reservoir (Mol et al. 2007).

17.4 Assessing the Impact

Since oxygen depletion, as just described, is a potentially serious issue especially in newly impounded reservoirs in tropical regions, this is a point that definitely needs to be addressed in an ESIA for such a project.

The first step is simple and straightforward: look and think. Two different parameters have to be taken into consideration. The first concerns the future reservoir as such, with the main parameters surface area, depths, general morphology (which in this stage does not require a detailed and very exact topography of the area to be covered with water), total reservoir volume and average inflow; the latter two parameters allow to calculate retention time. If the reservoir is very shallow and small in relation to inflow, i.e. with a short retention time, there will in all likelihood be no stratification and no oxygen depletion. This is usually the case in small ROR reservoirs or in in "almost ROR" reservoirs, as the one of Kafue Gorge Lower mentioned above.

The second parameter is vegetation, or, in this context, biomass. The important point is here (unlike for the questions related to biodiversity, see there) not the type of vegetation, but only the quantity, and more precisely the quantity of "soft" biomass, i.e. leaves, twigs, herbs etc.; this is the part of the total biomass which will decompose very quickly and therefore causes anoxic conditions, while the coarser part of the biomass, tree trunks and major branches, will decompose, if at all, very slowly and will therefore not influence the oxygen content of the water in any noticeable way. A few pictures shown in Fig. 17.3 give some examples to show the potential differences in the pre-impoundment period.

The largest amount of terrestrial biomass can be found in moist tropical forests (as in the Lom Pangar example in Fig. 17.3), and in addition to that there we usually have rather high water temperatures, a fact which influences water quality in the future reservoir in two negative ways: warm water has a lower oxygen content than cold water, and at the same time bacterial activity is higher than at lower temperatures, and both these effects tend to increase the problem: faster and more complete removal of oxygen from the water.

Just looking at these two main parameters—amount of biomass and oxygen content—will in most cases already allow to come to a conclusion whether in a specific case reservoir stratification and oxygen depletion will not, might, will probably or will certainly develop. If we have a reservoir of whatever size and retention time which submerges only steppe vegetation or more or less bare soil, as e.g. in Bakhtiari (Fig. 17.3), water quality deterioration due to breakdown of biomass will in all likelihood not be a problem. If, on the other extreme of the possible range we have a case of a reservoir with seasonal storage submerging rain forest, there will very certainly be such a problem, irrespective of all other parameters.

This, as mentioned, is a necessary first step (usually to be done already in the screening and scoping phase of the ESIA). But even if the problem can be excluded with a high probability, and very certainly in all other cases, a closer look at the situation and a more sophisticated analysis will be required. This includes a more precise analysis of the reservoir and hydrology data, as well as a better look at the vegetation. The aim is to get a better estimate of what the development in the reservoir after impoundment will be. This is important to know, since it is the basis for defining mitigation measures.

An early attempt at providing a simple rule for a determination whether or not water quality problems of this type would have to be expected was made by Garzon (1984), which was based on the two parameters of forested area submerged and retention time of the water in the reservoir (Fig. 17.4). His conclusion was that if the retention time is less than a year and if less than approximately 300 km^2 of forest will be submerged, there will be no water quality problem. However, Garzon himself in this publication pointed out that this is a very rough estimate, and that the development of water quality in a reservoir is determined by a lot of additional variables like reservoir morphology, depth, tributaries, etc., to name just a few. In the end, this would give less information than the "just to look at it" approach described above. Limiting closer investigations of this issue to reservoirs which will submerge more than 300 km^2 of forest and will have a retention time of more than 12 months would—incorrectly—have eliminated a large number of reservoirs which definitely had their share of water quality problems due to lack of oxygen.

Fig. 17.3 Examples of reservoir areas showing different types of vegetation. From top left to bottom right: Bakhtiari, Iran: mountainous area, narrow valley with steep slopes, dry and heavily grazed; very little vegetation; Ilisu, Turkey: dam site upstream of (behind) river crossing; dry, pastures and agriculture, little vegetation; Rogun, Tajikistan: wide valley partly filled with massive alluvial (gravel) banks; pastures, some fields; Bumbuna, Sierra Leone: dense bush vegetation, secondary growth (slash and burn fallow); Kafue Gorge Lower, Zambia: Miombo forest; Lom Pangar, Cameroon: water submerging dense primary forest at impoundment

A more sophisticated approach to this issue consists in applying a model. One such model is BioREM, which is mandatory for ESIAs for hydropower projects in Lao PDR, and which is made available on MONRE's (the Lao Ministry of Natural Resources and Environment) webpage (see e.g. https://wis.moepp.gov.mk/BioRem/Default.html). The BioREM model simulates physical, chemical and biological

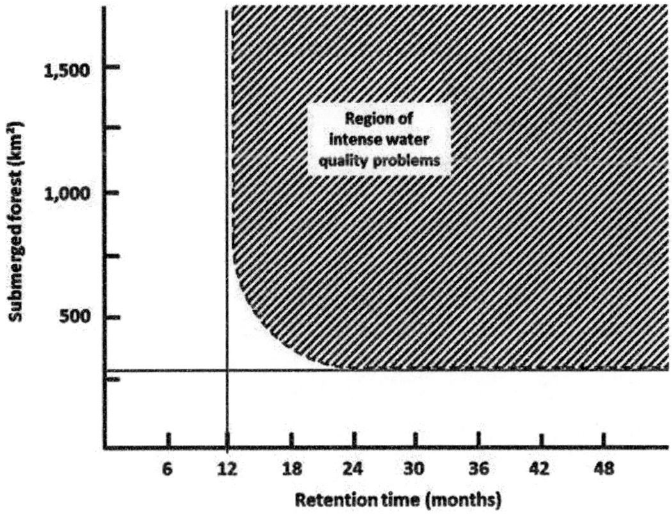

Fig. 17.4 Water quality as a function of reservoir retention time and area. *Source* Modified from Garzon (1984: 2)

processes in reservoirs. It does much more than to determine whether or not oxygen depletion in the reservoir will have to be expected. It simulates the development of water quality (mainly DO) and expected GHG emissions over a longer period, like e.g. a concession period of 30 years, or even longer. The aim is to allow developers to estimate how much biomass must be removed prior to impoundment of a hydropower reservoir in order to obtain a reasonable water quality and moderate greenhouse gas emissions during operation of the plant. One example of a biomass removal plan prepared on the basis of this model is the one for Nam Ngiep 1 HPP which is available on NNP1's website (https://namngiep1.com/wp-content/uploads/2014/10/NNP1_Bio-Mass-Removal-Plan_July2015.pdf). A model which is not country-specific and which allows for determining GHG emissions is the one developed by IHA (https://g-res.hydropower.org/), see comment in Sect. 21.2).

Figure 17.5 illustrates an example for a calculation of the elimination of biomass from a reservoir (theoretical case to match approximately the example shown in the BioREM guidelines) showing the progress of decomposition of soft biomass (leaves, twigs, herbs, litter etc.) and hard biomass (tree trunks and major branches) expressed in terms of BOD, i.e. the amount of oxygen required for breaking down the remaining part of the biomass. In this case, soft biomass would be completely eliminated after about 7 years, while the process for breaking down hard biomass is much slower; in the case shown in the figure this process would take approximately 60 years. The time span indicated in this way for soft biomass seems realistic, while the one for hard biomass is probably on the pessimistic side, the process might take much longer, as would e.g. be suggested by the case of the Nam Ngum 1 reservoir (see Fig. 17.6). This reservoir started operation in 1971, and in 2006 there were still

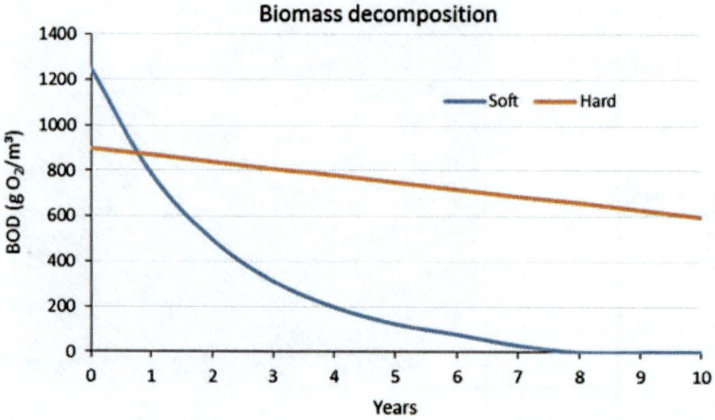

Fig. 17.5 BioREM prediction of biomass breakdown in a reservoir. Theoretical case. *Source* Redrawn after MONRE (2010: 64)

Fig. 17.6 Dead trees in the Nam Ngum 1 reservoir. Picture taken 2006-08-16

dead trees in large numbers standing in the water, and this despite of the fact that in the years before that large quantities of timber, and especially stumps and roots of very large trees, had been extracted from the reservoir for being sold to furniture manufacturers. Especially hard wood will decay so slowly as to be almost without any influence on water quality. On the other hand such submerged trees, and mainly the ones not quite reaching the surface, even 40 or more years after impoundment can still be an obstacle to fisheries, as the nets will tend to be caught and damaged, and a considerable risk for boat traffic.

I see two main potential sources of error in the application of such a model, namely:

1. Reservoir morphology and hydrological conditions. The morphology of the reservoir must certainly play a major role in its future development, which is determined by the form (e.g. the question whether it is a more or less simply oval shaped "lake" with an inflow at the upper end and an outflow at the lower end, or a highly dendritic reservoir with several arms and several inflows of different size (see Fig. 16.2), among other parameters. The hydrological conditions in the reservoir are determined not only by inflow, but also by outflow, and this latter is determined in a decisive way by the layout and operation pattern of the dam and hydropower plant (operation regime as peaking or baseload; frequency of spilling and arrangement of the spill gates, depth of the intake, presence or absence of bottom outlets, etc.). These latter operational parameters can certainly be modelled rather well.
2. Data on biomass in the reservoir area. This is normally done on the basis of vegetation maps of the area to be submerged and on information from the literature on biomass to be expected in different vegetation types. One cannot expect such values to be really accurate, and it is difficult to determine the potential margin of error.

In addition to that, the model also requires data on water quality, like phosphorous concentrations in inflowing or outflowing water. This can be rather easily determined, but values will very often just reflect the moment of the sampling and not necessarily a solid average value, let alone the development of such input over a longer period.

This is not to say that such a model should not be used, but it is important to understand that the results are indications of a potential or probable development, and they should not be taken as an exact representation of the future. As mentioned above, the model probably rather accurately depicts the near future, i.e. the first few years after impoundment, i.e. predicting the breakdown of soft biomass over the first 5–10 years, and a correspondingly high emission of GHGs during this period if nothing is being done, and this is the relevant time span.

Just to mention one more point: water quality 50 years later will in all probability be influenced more by the development of the human population in the catchment area and the question whether the settlements there have waste water treatment plants or not than by the phosphorous concentration in the water during the feasibility study. These long-term forecasts should be completely disregarded; it will be much more important to have a good monitoring system for actually documenting the development and deciding on whether or not additional measures might have to be taken (measures which, by the way, often may not be related to the hydropower plant, but to other types of development in the wider project area). In any case, comparing monitoring data with the forecast obtained from the model will be useful for future projects, since they may improve modelling as well as providing input for improved mitigation measures.

I often used a calculation which is much simpler that the model mentioned above, but provides the answers to the main question with much less effort. This calculation is based on the amount of soft biomass to be submerged in the future reservoir and on some simple hydrological data. The information on biomass is obtained in the

Table 17.2 Soft and hard biomass in various vegetation types

Vegetation type	Biomass					Source
	t/ha			%		
	Soft	Hard	Tot	Soft	Hard	
1. Boreal forests						
Picea forest dry Canada	70	112	182	38.5	61.5	De Angelis et al. (1981)
Picea forest fresh Canada	248	267	515	48.2	51.8	De Angelis et al. (1981)
Picea forest moist Canada	252	481	733	34.4	65.6	De Angelis et al. (1981)
Picea forest wet Canada	153	186	339	45.1	54.9	De Angelis et al. (1981)
Picea boreal Finland	68	68	136	50.0	50.0	Havas and Kubin (1983)
2. Central European forests						
Beech forest, France	228	282	510	44.7	55.3	De Angelis et al. (1981)
Beech forest, Solingen, Germany	77	134	211	36.5	63.5	De Angelis et al. (1981)
Beech forest, Solingen, Germany	72	152	224	32.1	67.9	De Angelis et al. (1981)
Beech forest, Solingen, Germany	69	275	344	20.1	79.9	De Angelis et al. (1981)
3. Mediterranean forests						
Quercus ilex Medit France	50	240	290	17.2	82.8	De Angelis et al. (1981)
Pinus silvestris Spain	68	406	474	14.3	85.7	De Angelis et al. (1981)
4. Tropical dry forests						
Miombo forest, Zaire	84	89	173	48.6	51.4	De Angelis et al. (1981)
Trop. dry deciduous forest, India	69	264	333	58.7	41.3	De Angelis et al. (1981)
Trop. dry deciduous forest, India	23	39	62	37.1	62.9	De Angelis et al. (1981)
5. Tropical moist forests						
Dipterocarp lowland forest, Malaysia	217	327	544	39.9	60.1	De Angelis et al. (1981)
Trop. rainforest, Ivory Coast	13	60	73	17.8	82.2	De Angelis et al. (1981)
Trop. rainforest, Ivory Coast	103	60	163	63.2	36.8	De Angelis et al. (1981)
Trop. rainforest, Brazil	225	299	524	42.9	57.1	De Angelis et al. (1981)
Rainforest Brazil Amazon	155	230	385	40.3	59.7	Fearnside (1995)
Amazon rain forest	141	385	526	26.8	73.2	Walter and Breckle (1984: 58)
Thailand, rain forest	101	262	363	27.8	72.2	Walter and Breckle (1984: 58)
Cambodia, rain forest	103	276	379	27.2	72.8	Walter and Breckle (1984: 58)
Borneo after forest fire 2 years	5	0	5	100.0	0.0	Nykvist (1996)

(continued)

Table 17.2 (continued)

Vegetation type	Biomass					Source
	t/ha			%		
	Soft	Hard	Tot	Soft	Hard	
Borneo after forest fire 5 years	27	0	27	100.0	0.0	Nykvist (1996)
Borneo after forest fire 8 years	34	29	63	54.0	46.0	Nykvist (1996)
Dipterocarp lowland Malaysia	125	490	615	20.3	79.7	Imai et al. (2010)
Malayan lowland rain forest	40	580	620	6.5	93.5	Whitmore (1985: 114)
Malayan lowland rain forest	35	322	357	9.8	90.2	Whitmore (1985: 114)
Malayan semi-evergreen rain forest	57	300	357	15.9	84.1	Whitmore (1985: 114)
6. Grassland/steppe						
Rangeland USA low	0.4	0	0.4	100.0	0.0	Reeves (2002)
Rangeland USA medium	1.2	0	1.2	100.0	0.0	Reeves (2002)
Stipa tenacissima steppe Spain low	3	0	3	100.0	0.0	Gauquelin et al. (1996)
Stipa tenacissima steppe Spain average	8	0	8	100.0	0.0	Gauquelin et al. (1996)
Stipa tenacissima steppe Spain high	17	0	17	100.0	0.0	Gauquelin et al. (1996)
Average for steppes in general	7	0	7	100.0	0.0	Gauquelin et al. (1996)
N. Afr. *Stipa* degraded pasture	1.5	0	1.5	100.0	0.0	Gauquelin et al. (1996)
N. Afr. *Stipa* degraded pasture	0.8	0	0.8	100.0	0.0	Gauquelin et al. (1996)
Turkish rangeland, Erzurum	0.7	0	0.7	100.0	0.0	Koç and Gökkuş (1996)
Turkish rangeland, Erzurum	0.9	0	0.9	100.0	0.0	Koç and Gökkuş (1996)
7. High mountain vegetation						
Pamir alluvial plain 3800 m asl low	0.4	0	0.4	100.0	0.0	
Pamir alluvial plain 3800 m asl high	4.5	0	4.5	100.0	0.0	
Rangeland Himalaya 4000 m asl	0.5	0	0.5	100.0	0.0	Mishra et al. (2004)
Rangeland Himalaya 4000 m asl heavily grazed	0.1	0	0.1	100.0	0.0	Mishra et al. (2004)

Table 17.3 Calculation of oxygen demand and consumption after reservoir impoundment

No.	Parameter	Unit	Input[a]	NT1	Rogun	Lom Pangar			Kafue Gorge Lower		
	Biomass					100%	50%	10%	Low	Medium	High
1	Biomass total (soft only)	T	(2)*(3)	1,563,120	170,000	5,940,000	2,970,000	594,000	7030	15,540	20,905
2	Reservoir area	Ha	N	9360	17,000	54,000	54,000	54,000	185	185	185
3	Biomass average (soft biomass only)	t/ha	N	167	10	110	55	11	38	84	113
4	Mean annual river discharge	m^3/s	N	317	631	250	250	250	276	276	276
5	Water in reservoir, total volume	Mm^3	N	3009	13,300	6200	6200	6200	83	83	83
6	Annual inflow	Mm^3	(4) * 31.5	9986	19,870	7875	7875	7875	8694	8694	8694
7	Oxygen in inflow water	mg/l	N	8	8	7	7	8	6	6	6
8	Oxygen per m^3	g/m^3	(= 7)	8	8	7	7	8	6	6	6
9	Oxygen total at first filling	T	(8) * (5)	24,072	106,400	43,400	43,400	49,600	498	498	498
10	Oxygen total in annual inflow	T	(8) * (6)	79,884	159,012	55,125	55,125	63,000	52,164	52,164	52,164
11	Oxygen demand/t biomass	T	1.07	1.07	1.07	1.07	1.07	1.07	1.07	1.07	1.07
12	Oxygen demand total	T	(11) * (1)	1,672,538	181,900	6,355,800	3,177,900	635,580	7522	16,628	22.368
13	O_2 balance 1: O_2 (1st filling)– O_2 demand	T	(9)–(12)	−1,648,466	−75,500	−6,312,400	−3,134,500	−585,980	−7024	−16,129	−21,870
14	O_2 balance 2: O_2 (annual inflow)– O_2 demand	T	(10)–(12)	−1,592,654	−22,888	−6,300,675	−3,122,775	−572,580	44,642	35,536	29,796
15	O_2 balance 1: % of O_2 required available 1st filling	%	(9) * 100/(12)	1.44	58.49	0.68	1.37	7.80	6.62	2.99	2.23
16	O_2 balance 2: % of O_2 required available yearly	%	(10) * 100/(12)	4.78	87.42	0.87	1.73	9.91	693.48	313.72	233.20

[a]Input: shaded fields indicate project specific values (N) to be obtained from project data and/or field work. The other cells in column “input” indicate the formula to be applied

way mentioned above, namely, a vegetation map of the future reservoir (which is usually prepared based on satellite pictures supplemented with ground truthing, see Sect. 10.1) and information on biomass in different vegetation types obtained from the scientific literature. Some such values are provided in Table 17.2. The calculation as such, with a number of examples, is presented in Table 17.3.

Some comments are required concerning this table. First of all, as a rule they do not exactly correspond to the values provided in the original sources, since there most often they are given split up between a number of different categories and not classified into the two main fractions “soft” and “hard” that are of interest here. While leaves certainly are part of soft and trunks of hard biomass, this may not be so clear for other categories like e.g. “twigs and branches”. The figures are also rounded to the nearest ton, with the exception of the values for the vegetation types with very little total biomass. If we look just at the values for moist tropical forests, which are the most relevant here since reservoirs in such an environment are the most likely to develop situations of stable stratification and anoxic water, we see that the amount of biomass can vary within a wide range; note that the low overall values provided by Nykvist (1996) refer to regrowth after complete destruction of forest by fire. The percentage of soft biomass also varies to a great extent. Some of this variation may be due to the fact that some of the values contain litter and biomass in soils, while others do not. In any case, when making calculations based on such values it is recommended to start with a rather high value for soft biomass, just to be on the safe side.

Explanations to lines in Table 17.3:

1. Total soft biomass in the reservoir area to be submerged at first filling in t
2. Reservoir area in ha
3. Soft biomass in t per ha
4. Mean annual river discharge at dam site in m^3/s
5. Total reservoir volume at FSL in Mm^3
6. Total amount of water flowing into the reservoir per year in Mm^3
7. Mean concentration of oxygen in inflowing water in mg/l
8. Total oxygen per m^3 of inflowing water in g
9. Total amount of oxygen in amount of water required for filling the reservoir in t
10. Total amount of oxygen flowing into the reservoir per year in t
11. Amount of oxygen required to decompose 1 t of soft biomass; this is based on the (simplified) assumption that soft biomass is mostly glucose, meaning that decomposition follows the respiration formula $C_6H_{12}O_6 + 6\ O_2 \rightarrow 6\ CO_2 + 6\ H_2O$, according to which 1.07 t of oxygen would be required to break down one t of soft biomass
12. Total amount of oxygen required to decompose the submerged soft biomass in t
13. O_2 balance at first filling, indicating the amount of oxygen still required for breaking down all soft biomass after first filling in t
14. O_2 balance after one year, indicating the amount of oxygen that would still be required for decomposing all the soft biomass within one year in t
15. After first filling, available percentage of oxygen of amount required for breaking down all the soft biomass submerged
16. Percentage of required oxygen for breaking down all the soft biomass in yearly inflow into the reservoir.

Some short explanations on the cases listed in the table:

Nam Theun 1 (NT1), Laos: this project is under construction as this text is written, planned to be completed in late 2022 (see Pöyry 2015 and Sect. 22.3 for some further details). The vegetation in the reservoir area is tropical moist forest, with a large amount of biomass present. The figures provided show a serious O_2 deficit if nothing is being done. This project was also analysed by means of the BioREM model, which indicated that pre-impoundment biomass removal is required; the plan is being implemented. But even if this is done very thoroughly, anoxic conditions still have to be expected.

Rogun, Tajikistan: very high dam (with 335 m the highest dam in the world), forming a large and very deep reservoir. Retention time in the reservoir is about 250 days. Vegetation in the reservoir area is mainly steppe-like pasture and subalpine grassland, some of it overgrazed; the original forest has been cleared, only single trees left occasionally, therefore there is very little biomass. As the figures show, O_2 in yearly inflow is sufficient for breaking down 87% of the biomass. A very limited zone with anoxic conditions in the deepest part of the reservoir may be possible in the first years after impoundment. However, the water flowing into the reservoir is usually rather cold, since it is snow and glacier melt; its oxygen content may therefore at times be higher than the 8 mg/l assumed here, and the cold water may reach the deeper layers of the reservoir. In addition to that, the reservoir will be filled gradually, over a total period of about 16 years (Pöyry 2014; see Chap. 23 for more details). These conditions will contribute to a good water quality in the reservoir, development of anoxic conditions is highly unlikely.

Lom Pangar, Cameroon: the reservoir is very large (540 km^2) and more than half of the area was covered with dense rainforest. The table shows three cases, namely, 100% biomass submerged, or soft biomass reduced to 50 and 10%, respectively, before impoundment. Even in the case of a biomass reduction by 90%, there would still be a serious oxygen deficit. Finally, as good as no biomass was removed before impoundment. The effect was somewhat mitigated by a re-aeration sill downstream of the dam and by the fact that the reservoir was partly filled and then emptied again for flushing out biomass (which caused some problems in the downstream area), but even so anoxic conditions over the entire water column developed very fast once the reservoir was filled (see Sect. 24.1).

Kafue Gorge Lower, Zambia: this is the third (lowermost) in a cascade of three reservoirs. The total volume of the reservoir is 83 million m^3, and the average inflow is 276 m^3/s. This means that the yearly inflow is 105 times the reservoir volume, water residence time in the reservoir, on average, being only 3.5 days. First, it was planned to operate it as a ROR plant. This would have meant that during the wet season, when inflow is higher than average, there would have be a large amount of spilling, with the reservoir level remaining constantly at its maximum, while during the dry season, when inflows are lower, the rule would have been a daily drawdown and refilling. It was then decided to increase the installed capacity and to operate the plant in a peaking regime. This will lead to less spilling, but a very heavy drawdown, almost emptying the reservoir every day and filling it again during the night, will occur throughout the year under this regime. Stratification of the reservoir is not possible under such conditions, and with such a small reservoir and the rather small amount of submerged vegetation there is no risk of oxygen depletion. The large water turnover rate will constantly replenish oxygen in the reservoir, and under this regime biomass will be flushed out. Vegetation in the reservoir area was Miombo forest, which as an open-canopy forest with mostly not very large trees. The calculations show three possibilities, assuming low, medium and high amounts of biomass (whereby it has to be pointed out that even "high" would be much lower quantities than what can be found e.g. in a moist, dense forest). Even with the highest possible biomass density the amount of oxygen in the water would be largely sufficient for breaking down all the biomass in very little time. The possible impacts caused by stratification as described above will therefore not arise in this case, even if the Froude number suggests stratification. However, the effect of the peaking process will be very marked (Pöyry and Zenith 2016; Zwahlen et al. 2019).

Therefore the question: is it worthwhile to apply a complex model, or will a simpler approach be sufficient? The answer to that may depend, as is often the case, on the specific situation. First of all: for an ESIA in Laos, where the application of a specific model is mandatory, it obviously has to be done, and this might also be the case in other countries. One more reason for using such a model is the fact that it does not only answer the most basic question—whether there will be a problem or not—but also provides quantitative indications—however exact they may be—on one very important issue, namely, the amount of GHGs to be expected. Finally, and this might be a kind of a Machiavellian reason: people tend to believe results of an investigation more readily if they are being reached by means of a sophisticated computer model.

17.5 Mitigation

Impact assessment, obviously, is done for deciding whether mitigation measures need to be taken, and then for deciding what these measures are going to be, and finally planning them; the latter step, the detailed planning of the measure, is not necessarily a part of the ESIA as such, but is included in the ESMMP. A first, draft version outlining the measures is usually included in the ESIA, either as a chapter or as a separate report, but detailed planning of the measures in most cases is only done once the project—and with it the ESIA—has been accepted and the environmental authority in charge has decided what measures to include as a condition for project implementation.

The most serious issue related to reservoir stratification is usually the problem of submerged biomass and the oxygen depletion and GHG emission resulting from it. As was shown above, each reservoir created in a moist tropical environment, unless it is very small and shallow (as was e.g. the case for Memve'ele in Cameroon), is faced with this risk. There are basically three ways for addressing this issue:

- One possibility is not to do anything, to deal with the problems as good as possible as they arise (as was done in the case of Brokopondo), and to wait for the situation to improve all by itself. This will certainly happen, however, it may take a number of years, and it can be associated with some serious issues like water quality deterioration in the reservoir and in the downstream area, problems for fish and other aquatic organisms, and potentially high GHG emissions during this time. In the case of Brokopondo, one additional issue was the proliferation of hydrogen sulphide (H_2S), which is also produced under anoxic conditions, and which is highly toxic and corrosive. Apparently, the situation there was so bad that power house staff had to wear gas masks for some time.

- The other option is to eliminate as much as possible of the soft biomass before impoundment with the aim of reducing the risk of anoxic conditions, or at least the aim of reducing the time span required for the development of a new equilibrium.
- The third option is an entirely technical or engineering solution: not to eliminate biomass, but to artificially re-aerate the reservoir or the water released from it, by means of re-aeration schemes of different types which have the aim to destabilise reservoir stratification and in this way bring oxygen to deeper layers of the reservoir, or to re-aerate the water released from the reservoir in order to bring oxygen content up to a level which is conform to legal standards (Illinois Environmental Protection Agency 1977; Moore et al. 2015).

I shall not say much about these technical solutions, since I do not really know much about them. At least some of these devices seem to work quite well in reservoirs for drinking water (as the ones described in Moore et al. 2015), but so far I have not yet come across a reaeration scheme that would really work well in a large reservoir. Artificial aeration has been used rather extensively in some of the smaller Swiss lakes, natural lakes where water quality had deteriorated due to eutrophication, and where this eutrophication was due mainly to phosphorous input stemming from treated waste water and runoff or drainage water from agricultural areas. The situation has improved with this measure, but it had to be maintained, and a return to natural conditions can only be achieved if the phosphorous content—including what had been fixed in the sediment over the entire period of heavy eutrophication—can be reduced much further that what was achieved already (Wüest 2009; BAFU 2016); aeration is only an accessory tool for improving the situation, but does not really lead to a sustainable solution. The situation is very different from what we have in a newly impounded reservoir. Here, oxygen consumption is not (or at least not mainly) triggered by primary production within the water body, but by the submerged biomass, and looking at the sheer amount of oxygen required to break this down (see Table 17.3) it seems doubtful whether this can be done effectively by artificial aeration.

One more concern over these technical solutions is of an entirely different nature. It could be imagined that the possibility of applying such a method if and when the problem arises could be used as an excuse for not taking any measures aiming at preventing this from happening. One example where this, in a way, has happened, is Lom Pangar HPP in Cameroon. There, it was decided not to remove any biomass, but to build a re-aeration sill downstream of the dam (Savatier et al. 2016). This resulted in a fast re-oxygenation of the effluent water, but the BOD load was so high that at least for a time after impoundment oxygen added in this way was consumed very rapidly, causing very low oxygen concentrations, and therefore problems for the fish stock, in the river downstream of the dam (see Sect. 24.1 for more details).

With or without such a system, the situation will improve. Anoxic conditions and related problems, depending on the conditions in the respective reservoir, will cease to exist after period of about 5–10 years. The most relevant parameters, in addition to the amount of biomass submerged, are size and depths of the reservoir, in- and outflow, and the configuration of water intakes, outlets and spillways.

So, while doing nothing might be an option, it is certainly better to take measures which will, if not completely prevent the problems from arising, at least contribute to improving the situation; improving the situation here means to reduce the duration during which there will be water quality problems in the reservoir as well as in the downstream area, and, very importantly, to reduce the GHG emissions, mainly the amount of methane, from the reservoir. This can only be done by pre-impoundment biomass clearing. Here, the BioREM documents (MONRE 2010) mentioned above provide a very good guideline for the preparation of a biomass clearing plan. Its application certainly does not have to be limited to Laos, where however it is mandatory. An example of a biomass removal plan based on BioREM is the one that was prepared for the Nam Ngiep HPP (Earth Systems 2015).

One thing that is often done before impoundment is salvage logging. This activity aims at making use of the timber in the reservoir area, which might otherwise be lost. This is certainly a good thing to do, since timber is a valuable resource, and even trees that do not produce high value timber can still be used for fuelwood or for making charcoal. This, however, is feasible only in cases where settlements where fuel wood and charcoal is used are rather close by, and where sufficiently good roads are available for taking the wood to those places where there is a market for it. If this condition is not given, transport costs get prohibitive very quickly. This issue had to be faced, even for good quality timber, in the case of Nangbéto, Togo (EWI 1984). One solution that I proposed was to send in people on foot to cut the really valuable trees and then, after impoundment, pick them up from the surface of the reservoir. I was told that the really valuable trees, like ebony (*Diospyros* spp.), are "sinkers", their wood being so dense and heavy that they do not swim; some good ideas just don't work. Still, while it makes sense to take out and use as much of the wood as economically possible, it has to be stated very clearly that this activity does not contribute in any way to a reduction of fine biomass in the future reservoir. Loggers only have one aim: to go in, and to get as much timber as they possibly can in as little time as possible. Everything that cannot be used, and this includes all the fine biomass, is left where it is, and for oxygen consumption it does not really matter whether it is already dead at the time of impoundment or not. Salvage logging can even contribute to another problem: since not only leaves and twigs, i.e. the soft biomass, is left on site, but also the lager branches of the trees as well as the occasional smaller tree which had to be felled to give access to the valuable trunks, this can contribute greatly to the floating debris on the reservoir.

A few considerations may be helpful when preparing such a biomass removal plan.

Yes or no? The first question is obviously whether the preparation of such a plan would be required or not. It is certainly always required when the impact assessment—whether based merely on looking at the situation, a simplified calculation or the results of a sophisticated model—comes to the conclusion that water quality problems in the future reservoir will have to be expected or will arise with certainty.

As a first step, a simple calculation may be used for deciding whether or not to pay any attention to the issue. In a second step, it can still be a good approach to apply a model, in order to get more confidence in the necessity of the measure and for calibrating it.

Will it do any good? The effectiveness of the measure is another important consideration. It has to be quite clear that in probably every case where a large area of dense forest will be impounded, it will be as good as impossible to prepare the reservoir area in a way that there will be no problem of this type at all. The figures provided for Lom Pangar (Table 17.3), e.g., show that even if only 10% of the soft biomass is left, meaning that 90% of it would have been eliminated before impoundment, there would still be a problem. This may lead to the conclusion that it might be better not to do anything, or to try with other approaches. Still, every reduction in the pre-impoundment phase will contribute to the faster return to an acceptable situation. However, one should not be surprised if even after a thorough clearing the results would still not be as expected (or, for that matter, calculated!). Some of the reasons why this is so are discussed shortly in the following paragraphs.

How much? Here, the model mentioned above, or possibly even my simplified approach, give an answer on how much of the soft biomass would have to be eliminated in order to achieve a satisfactory result. However, easy as it is to calculate this, as difficult can it be, or usually is, to achieve this goal in the field. It cannot be expected, therefore, that there is a simple answer to this question; in most cases it probably has to be just: as much as possible and feasible under the specific circumstances. For Nam Ngiep 1 HPP, with a reservoir area of 66.9 km^2, the area identified for being cleared was 19.1 km^2, for NT1, with a reservoir area of 93.6 km^2, it was 57 km^2.

How to decide what and where to clear? For a large reservoir, to clear the entire surface is probably as good as impossible. Different points have to be considered when deciding what areas should be cleared and what areas should be left out. In the case of NT1, the two most important criteria for this decision were effectiveness (i.e. work done per unit of time) and safety (i.e. reducing the risk of accidents for the work force). Focus was on areas which were easily accessible, covered with dense vegetation, and with a slope of <30%, as work on steep slopes is more dangerous. Safety considerations for the work force were not limited to this; equipment (type of tools to be used, personal protection equipment) was defined, as well as planning of emergency measures. One issue which is high on the agenda in Laos is "UXO": unexploded ordnance, a leftover, sadly all too real, of the Indochina wars. Before any work can start, a so-called UXO-clearance has to be done.

What other conditions to consider? One of the constantly repeated points of this book is that every project is different from every other, and that the specific conditions have to be taken into account. This is certainly the case here. In NT1, e.g., one important point was that the largest part of the reservoir is located within a protected area. This was also the part that was mainly covered with dense (although not necessarily always primary) forest, and it was the deepest part of the future reservoir, i.e.

the part that most needed clearing. In this case, a few additional conditions had to be met, the most important ones being the following:

- Access: like many protected areas in Laos, this one suffered from la lack of management and control and, directly linked to it, an excess of poaching. One preoccupation therefore was access. For the salvage logging as well as for the biomass clearing access was required. Since additional access roads also open the way for illegal hunting, it was not allowed to open any new roads, and where the existing ones needed to be upgraded this had to be done to the absolute minimum standard required. Furthermore, there was a condition for effectively blocking all these roads after the end of the work (for the problems related to access roads see Chap. 15). Furthermore, in order not to facilitate access to the core areas of the protected zone via the reservoir itself, it was decided to leave a band of 10 m below FSL without clearing.
- Disturbance: impacts on wildlife had to be minimised. For this reason, any activities above the FSL as well as hunting in the entire area were forbidden; this had to be monitored constantly by supervising personnel stationed on site with the work force carrying out the clearing.
- Protection of wildlife: one issue when submerging natural habitats is the threat to wildlife by drowning during impoundment. In order to minimise that risk, biomass clearing was planned and carried out in a way as to start from the river banks, wherever feasible, and work uphill, in this way driving out animals to higher and therefore safe places. At the same time, the work was done in a way as not to leave island of forest in an otherwise cleared area, since these could provide shelter to animals, which then might turn into deadly traps during impoundment.

How to get rid of the biomass? If salvage logging is being done, a potentially considerable part of the biomass, the valuable tree trunks, will be removed. However, this is the part of the biomass that is of least concern in terms of water quality. A very large, and presumably by far the largest, part of the biomass will still be left on site. This can again be separated in two broad categories: one is the remaining hard biomass, i.e. the larger branches of the harvested trees as well as the tree trunks of smaller and/or not commercially interesting trees. The problem associated with this category is that most of it, if left back, will swim up at the surface of the reservoir and will ultimately be swept to the dam, where it can cause problems with floodgates and other structures. The soft biomass will affect the water quality. The amount of both categories, therefore, has to be reduced. The only practical way to do it, at least in a large reservoir and in areas not in immediate proximity to settlements, is to burn it. I have come across—mainly theoretical—considerations for actually removing this biomass from the reservoir area, but looking at the sheer amount of it makes it clear very quickly that this is not feasible. In the example of Nam Ngiep 1 HPP mentioned above, the estimated amount of soft biomass in the vegetation was 40 t/ha; finally, an area of somewhat over 1600 ha was cleared, which amounts to roughly 64,000 t of biomass. Considering that an estimated half of the hard biomass of 200 t/ha was not taken out as timber, one would have to add to this another 160,000 t. Costs for transporting this out of the area to be submerged would be prohibitive, and in addition to that space would be needed for disposing of this enormous volume of biomass. Burning it in a biomass-fuelled thermal power plant might seem a good idea, however, the transport issue would still not be solved, and in addition to that the fact had to be considered that huge amounts of fuel would be made available, but only for a very short time; this is no sustainable way for operating such a plant. Burning, as said, is the only way. This, of course, turns the carbon in the plant biomass into CO_2, which is a GHG, but the aim here is to prevent that large amounts of it will be decomposed under anaerobic conditions, which would lead to large quantities of CH_4, a GHG which is about 20 times more climate efficient than CO_2. In the case of smaller reservoirs which are not located too far from human settlements, some of the low-value timber and wood can potentially be used, e.g. as timber for constructions in the area (fencing of fields or pastures, sheds for animals, or even parts of houses). In the case of Bujagali HPP, in Uganda, larger branches and smaller tree trunks were piled up in suitable places, to be used by the local population as fire wood. In the rehabilitation of Mount Coffee HPP, Liberia, inhabitants were encouraged to cut suitable wood prior to impoundment and use it for charcoal production. However, this only works where distances are not too long, otherwise transport costs are prohibitive. Burning the biomass means basically to apply the system used for slash-and-burn agriculture: the vegetation is chopped down and left to dry, and it is burned as soon as it is dry enough. However, for slash-and-burn agriculture, usually smaller trees are cut, the large ones are left standing—a process which is made easier by using regrowth areas rather than primary forest with dense stands of large trees. Only the dry soft biomass, leaves, stems of herbs, twigs, will burn, tree trunks and lager branches will be left lying on the ground (see Fig. 17.7a). In the case of reservoir area preparation for impoundment, however, the aim will be to burn as much as possible of this coarse fraction of the biomass as well. In order to do that, the material will have to be stockpiled after drying, and then burned. This leads to much hotter fires than just burning the soft parts, and this in turn means that measures have to be taken to prevent forest fires, which could spread beyond the limits of the reservoir area.

What is the best timing for this work? With the condition described above much about timing is said: cut—let dry—burn: this does not leave much more than cutting during the dry season, early enough to allow for sufficient time for drying, and then burning before the rainy season starts. One thing that has to be considered is the fertilising effect of the ash. Many tropical soils are very poor in nutrients. This is the main reason why slash-and-burn agriculture is being done in rotation with fallow: after cutting, the land is cultivated, in South-East Asia usually with upland rice. The crop in the first year is good, but then productivity decreases rapidly. After 3–4 years a field is abandoned since fertility by then is too low and weeds have taken over; the field is left fallow, in order to "restore soil fertility", as is usually said. This, however, is not really correct. Levang (1993: 186) has monitored soil nutrients, in

Fig. 17.7 Biomass clearing. **a** (top left): Preparation of a field for swidden (slash-and-burn) cultivation; dry vegetation will be burned, tree trunks left on the ground (Sarawak, Malaysia). **b** (top right) and **c** (bottom left): Biomass clearing after burning; bigger trees can be left standing, felled trees should be removed or burned, since otherwise they will float on the reservoir surface after impoundment, presenting a risk for boats and potentially causing problems at the dam. **d** (bottom right): Regrowth on a site cleared a year before; this can be a problem, because large amounts of soft biomass are produced after one or two rainy seasons, and therefore regrowth may have to be cut and burned again. **b–d**: Pre-impoundment biomass clearing for NT1, Laos

this case Ca, Mg and K, over a fallow period of 20 years, and there was no increase in soil content in these elements during this entire period, concentrations remaining at a very low level throughout. However, after burning, there was a sharp peak, caused by the ash from the burnt plant material; one year later, after the harvest, concentrations were down again. This clearly shows that nutrients are not stored in the soil, but in the plants. Burning the plants releases these nutrients and serves as fertilizer. For pre-impoundment biomass clearing this means that in the rainy season following burning there will be a considerable regrowth of plant material, all of it soft biomass. Therefore, ideally biomass clearing would be done in the last dry season before impoundment, and the first rains should not yet be retained. This would allow, at least in reservoirs with rather steep flanks, for flushing out at least some of the ash lying on the soil, ash which otherwise would be retained in the reservoir and encourage growth of algae, i.e. biomass again. Such an ideal case, however, will rarely if ever come into being. One dry season is usually not enough for doing all the work, since it would require an incredibly large work force, which would be very difficult to mobilise and to manage. This means that the difficult balance has to be made of starting early enough for doing the work, but not too early in order to minimise regrowth. The biomass removal has to be coordinated with the construction schedule of the dam. If however, for whatever reason construction suffers a delay—and in my experience this seems to be the rule rather than the exception—this immediately means that there will be one more rainy season with regrowth. If the dry season is dry enough, it might be possible to reduce biomass from regrowth at least to some extent by burning it before the onset of the next rainy season. Cutting the entire areas again will be a costly undertaking. One good approach was chosen in the case of NT1 for areas that were suitable for this: they were sowed with upland rice in the rainy season after clearfelling and burning. This means that much of the nutrients were taken up by and subsequently taken out with the crop, and after the harvest the dry rice stalks can be burned.

Who should do the work? Given the large areas involved, it will usually have to be companies with experience in that kind of work which are contracted to do it. It involves not only the actual process of cutting. The entire organisation—mobilising the work force, bringing it to the sites, providing food and shelter, equipping it for the work, which in addition to suitable tools includes the required safety equipment, and planning for emergency cases, i.e. first aid on site and evacuation, if required, in case of accidents—needs considerable planning and experience. In the case of Houay Ho HPP, located on the Boloven plateau in southern Laos, an at least theoretically good solution was found: two companies were given the right to take out the timber from the reservoir area, of which there was a considerable quantity, at no cost to them, under the condition that in addition to this salvage logging they would also do the biomass clearing. Unfortunately, and probably for cost reasons, there was no monitoring of this work. A post-impoundment check done about a year or so after impoundment revealed a somehow mixed picture. The good part was that apparently the FSL, which had been marked on the ground, had been respected, there was no indication that trees would have been cut above that elevation. On the other hand, the very dense stands of dead bushes and small trees in the drawdown area (see Fig. 17.8) implied that biomass clearing had not really been done as it should have. Monitoring is always required.

Fig. 17.8 Hoay Ho reservoir, Laos. Reservoir area during drawdown; the shoreline of the reservoir is clearly visible. The density of dead small trees and bushes indicates that pre-impoundment clearing has not been done as required. Picture taken 2001-02-10

References

Abril, G., A. Boudou, P. Cerdan, R. Delmas, C. Erard, P.M. Forget, P. Gosse, F. Guerin, L. Guillemet, J.M. Pons, A. Vaquer, Y. Vaquer, and R. Vigouroux. 2008. *Petit Saut: Environmental report after 12 years running*. Report by the Comité Scientifique du Petit Saut, French Guyana.

BAFU. 2016. *Der Greifensee: Zustand bezüglich Wasserqualität*, 8 pp. Faktenblatt. *Der Hallwilersee: Zustand bezüglich Wasserqualität*, 10 pp. Faktenblatt. https://www.bafu.admin.ch/bafu/de/home/themen/wasser/fachinformationen/zustand-der-gewaesser/zustand-der-seen/wasserqualitaet-der-seen.html.

Boehrer, B., and M. Schultze. 2010. Density stratification and stability. In *Lake ecosystem ecology. Encyclopedia of inland waters*, ed. G.E. Likens (2010b), 96–106. Oxford: Elsevier.

Clarke, R., and J. King. 2004. *The atlas of water. Mapping the world's most critical resource*, 127 pp. London: Earthsacn.

De Angelis, D.L., R.H. Gardner, and H.H. Shugart. 1981. Productivity of forest ecosystems studied during the IBP. The woodlands data set. In *Dynamic properties of forest ecosystems*, ed. D.E. Reichle, 567–673. International Biological Programme 23. Cambridge, UK: Cambridge University Press.

De Santo, R. 1978. *Concepts of applied ecology*, 310 pp. New York: Springer.

Deas, M.L., and C.L. Lowney. 2000. *Water temperature modeling review*, 113 pp. Central Valley, CA: California Water Modeling Forum. http://cwemf.org/Pubs/BDMFTempReview.pdf.

Earth Systems. 2015. *Biomass removal plan (BRP) for Nam Ngiep Power Company (NNP1)*. Prepared by Earth Systems for Nam Ngiep 1 Power Company. https://namngiep1.com/wp-content/uploads/2014/10/NNP1_Bio-Mass-Removal-Plan_July2015.pdf.

EAWAG. 2006. *Independent review of the EIAR 2005 on the future Ilisu dam*. Switzerland: EAWAG.

EWI. 1984. *Aménagement hydroélectrique Nangbéto: Deboisement de la cuvette du Barrage.* Projet de développemend de la pêche. Faisabilité Technique et Economique. EWI et SOGREAH. Comunauté Electrique du Bénin.

EWI. 1987. *Projet de surélévation du barrage de Mauvoisin. Analyse des effets sur l'environnement.* Forces Motrices de Mauvoisin.

Fearnside, P.M. 1995. Hydroelectric dams in the Brazilian Amazon as sources of "greenhouse" gases. *Environmental Conservation* 22 (1): 7–19.

Ford, D.E., and M.C. Johnson. 1983. *Assessment of reservoir density currents and inflow processes*, 81 pp. Tech. Rept. E-83-7. Vicksburg, ME: US Army Engineer Waterways Experiment Station, CE, 81 pp. https://apps.dtic.mil/dtic/tr/fulltext/u2/a137303.pdf.

Garzon, C.E. 1984. *Water quality in hydroelectric projects. Considerations for planning in tropical forest regions*, 33 pp. World Bank Technical Paper No. 20. Washington, DC: World Bank.

Gauquelin, T., G. Jalut, M. Iglesias, F. Valle, F. Fromard, and J.J. Dedoubat. 1996. Phytomass and carbon storage in the Sitpa tenacissima steppes of Baza basin, Andalusia, Spain. *Journal of Arid Environments* 34 (3): 277–286.

Havas, P., and E. Kubin. 1983. Structure, growth and organic matter content of an old spruce forest in Northern Finland. *Annales Botanici Fennici* 20: 115–149.

Illinois Environmental Protection Agency. 1977. *Lake aeration and circulation.* http://www.epa.state.il.us/water/conservation/lake-notes/lake-aeration.pdf.

Imai, N., K. Kitayama, and J. Titin. 2010. Distribution of phosphorous in an above-to-below-ground profile in a Bornean tropical rain forest. *Journal of Tropical Ecology* 26 (6): 627–636.

Koç, A., and A. Gökkuş. 1996. Annual variation of above ground biomass, vegetation height and crude protein yield on the natural rangelands of Erzurum. *Turkish Journal of Agriculture and Forestry* 20: 305–308.

Lehmusluoto, P., B. Machbub, N. Terangna, et al. 1999. Limnology in Indonesia. In *Limnology in developing countries. Vol. 2. International Association for Limnology*, ed. R.G. Wetzel and B. Gopal, 119–234. New Delhi, India: International Scientific Publications.

Levang, P. 1993. Jachère arborée et culture sur brûlis dans les îles extérieures de l'archipel indonésien. In *La jachère en Afrique de l'Ouest*, ed. C. Floret and G. Serpantié, 179–192. Paris: ORSTOM.

Mishra, N., S.E. van Wieren, P. Ketner, I.M.A. Heitkönig, and H.T. Prins. 2004. Competition between domestic livestock and wild bharal *Pseudois nayaur* in the Indian Trans-Himalaya. *Journal of Applied Ecology* 41 (2): 344–354.

Mol, J.H., B. de Mérona, P.E. Ouboter, and S. Sahdew. 2007. The fish fauna of Brokopondo reservoir, Suriname, during 40 years of impoundment. *Neotropical Ichthyology* 5 (4). https://doi.org/10.1590/S1679-62252007000300015.

MONRE. 2010. *Environmental guidelines for biomass removal from hydropower reservoirs in Lao PDR, Lao People's Democratic Republic. A biomass removal model user's guide. A biomass removal model technical reference.* Lao People's Democratic Republic, Strengthening Environment Management-Phase II. Vientiane Capital, Lao PDR: Swedish International Development Cooperation Agency, Ministry of Natural Resources & Environment. http://monre.myqnapcloud.com/2017/emsp/images/doc/c2/bioremusersguide.pdf.

Moore, B., M. Mobley, J. Little, B. Kortmann, and P. Gantzer. 2015. *Aeration and oxygenation methods for stratified lakes and reservoirs.* Lakeline, Spring 2015. z0ku333mvy924cayk1kta4r1-wpengine.netdna-ssl.com.

Nykvist, N. 1996. Regrowth of secondary vegetation after the 'Borneo fire' of 1982–1983. *Journal of Tropical Ecology* 12 (2): 307–312.

Odingo, R.S. 1979. An African dam. Ecological survey of the Kamburu/Gtaru hydro-electric dam area, Kenya. *Ecological Bulletins* 29: 183 pp. Stockholm: Swedish Natural Science Research Council.

Orlob, G.T. (ed.). 1983. *Mathematical modeling of water quality: Streams, lakes and reservoirs*, 518 pp. International Institute for Applied Systems Analysis. Chichester, UK: Wiley.

Parker, F.L., B.A. Benedict, and C. Tsai. 1975. *Evaluation of mathematical models for temperature prediction in deep reservoirs*. Corvallis, Oregon: National Environmental Research Center, US Environmental Protection Agency.

Pöyry, and Zenith. 2016. *Kafue Gorge Lower HPP ESIA: Environmental and social impact assessment*, 287 pp. Report prepared for ZESCO and Kafue Gorge Lower Power Development Corporation, Ltd., Zambia.

Pöyry. 2014. Environmental and social impact assessment for Rogun Hydropower Plant. In *ESIA*, vol. I: Text; vol. II: Annexes. Report prepared for OSHPC "Barki Tojik", Republic of Tajikistan.

Pöyry. 2015. *Biomass clearance plan. Environmental and social studies*, 65 pp. Nam Theun 1 Hydropower Project. Report prepared for Phonesack Group, Lao PDR.

Reeves, M. 2002. Quantifying herbaceous biomass in a rangeland ecosystem using MODIS land products. http://www.ntsg.umt.edu/projects/rangeland/.

Savatier, J., M. Lino, S. Erpicum, M. Pirotton, S. Descloux, C. Daux, G. Gwet, and A. Emadak. 2016. *Aeration weir at Lom Pangar dam. "Congrès SHF "HydroES 2016""*. Grenoble. https://www.researchgate.net/publication/306403250.

Schwoerbel, J. 1980. *Einführung in die Limnologie*, 4. Aufl, UTB 31, 196 pp. Stuttgart: Fischer.

Straskaba, M., and P. Mauersberger. 1988. Some simulation models of water quality management of shallow lakes and reservoirs and a contribution to ecosystem theory. In *Wetland modelling*, ed. W.J. Mitsch, M. Straskraba, and S.E. Jorgenson, 153–176. Amsterdam: Elsevier.

Van der Heide, J. 1982. Lake Brokopondo. Filling phase limnology of a man-made lake in the humid tropics, 427 pp. Diss, Universität Amsterdam.

Walter, H., and S.-W. Breckle. 1983. *Oekologie der Erde Band 1: Ökologische Grundlagen in globaler Sicht*, 238 S. Band 2 (1984): *Spezielle Ökologie der Tropischen und Subtropischen Zonen*, 461 S. Band 3 (1986): *Spezielle Ökologie der Gemässigten und Arktischen Zonen Euro-Nordasiens*, 587 S. Band 4 (1991): *Gemässigte und Arktische Zonen ausserhalb Euro-Nordasiens*, 586 S. Stuttgart: UTB Grosse Reihe, Fischer.

Whitmore, T.C. 1985. *Tropical rain forests of the Far East*, 352 pp. Oxford, UK: Oxford University Press.

Whitten, T., R.E. Soeriaatmadja. and S.A. Afiff. 1996. *The ecology of Java and Bali. The ecology of Indonesia series*, vol. II, 969 pp. Singapore: Periplus Editions (HK), Ltd.

Wüest, A. 2009. *Belüftung Greifensee Sommer 2009*, 25 pp. EAWAG, Swiss Federal Institute of Aquatic Science and Technology. https://www.dora.lib4ri.ch/eawag/islandora/object/eawag%3A14898/datastream/PDF/W%C3%BCest-2010-Bel%C3%BCftung_Greifensee_im_Sommer_2009-%28published_version%29.pdf.

Zauke, G.-P., R.-G. Niemeyer, and K.-P. Gilles. 1992. *Limnologie der Tropen und Subtropen. Grundlagen und Prognoseverfahren der limnologischen Entwicklung von Stauseen*, 171 S. Landsberg/Lech: Ecomed.

Zwahlen, R. 2003. Identification, assessment and mitigation of environmental impacts of dam projects. In *Modern trends in applied aquatic ecology*, ed. R.S. Ambasht, 281–370. New York: Kluwer Academic Publishers.

Zwahlen, R., L.W. Kahongo, and V. de Genot de Nieukerken. 2019. Kafue Gorge Lower HPP—A low-impact addition to an existing HPP cascade, Zambia. Paper presented at the Africa 2019 Hydropower Conference, Windhoek, Namibia, Apr 2019.

Chapter 18
Downstream Impacts

Si le quitamos agua al río es porque se van secando los bofedales, las vegas y todo eso.
(If we remove water from the river, this is why the wetlands are drying up, the meadows and all that.)
Luisa Huanuco; cited after Berenguer (2017: 100)

Dams with retention and regulation capacity will invariably have an effect on the characteristics of the river in the downstream area. The natural difference in river discharge between dry (low flow) and wet (high flow) season will be reduced. This is very often the main purpose of a dam: retaining water during the high flow season, in order to make it available for the generation of electricity in the low-flow season, which is often the period of the year where energy demand is highest. While this makes perfect sense from an energy market point of view, it invariably has repercussions on the situation in the downstream area. The effects are felt mainly in—sometimes very vast—floodplains of rivers which are highly dynamic habitats depending entirely on the dynamics of the river, which includes not only seasonal differences in the amount of water, but also flood events. One specific habitat type often affected by HPPs is the spray zone of water falls.

18.1 Main Types of Impacts

Depending on type and layout of an HPP, the effects on the downstream area can be of different nature and intensity. The main groups of these impacts are shortly listed and described below.

- ROR plant: since the flow regime of the river is not changed, or only marginally, the downstream impacts are usually rather marginal.
  - Seasonal flow patterns: not changed.
  - Sediment transport: can be influenced by accumulation of sediment in the reservoir; however, since there is no large storage area, the capacity of sediment

R. Zwahlen, *Assessing the Environmental Impacts of Hydropower Projects*, Environmental Earth Sciences, https://doi.org/10.1007/978-3-030-91185-0_18

retention is also limited, and possibilities for sediment routing of flushing are rather good, and certainly in the interest of the owner of the power plant.
  - Fish: migration can be interrupted or at least hindered already by rather small structures, and this in turn can negatively affect fisheries as well as the quality of the river as natural habitat. However, options for fish ladders and other structures enabling fish migration are rather good, and the use of fish-friendly turbines, ideally in combination with other structures to allow drift and downstream migration, can help reducing mortality among fish migrating downstream.
  - Flood protection: depends on design and operation of the structure, but usually limited.

- Storage plants: intensity of impacts on the downstream area will be greater that for ROR plants, and as a rule will increase with an increase in dam height and reservoir volume.

  - Seasonal flow pattern: the change can be very marked, and is actually usually the main reason for creating large reservoirs: to store water during the high-flow season and using it for electricity generation during the low-flow season; the effect then is a decrease of high flows and an increase of low flows, but in an extreme case the pattern can even be reversed (see examples in Figs. 9.8 and 24.1).
  - Sediment transport: this can be greatly affected, and the reduction of the sediment load can have serious impacts on the downstream area, from increasing erosion to receding deltas (see Sect. 8.6). Sediment flushing is possible, but may be more difficult in the case of high dams and deep reservoirs.
  - Fish: a high dam is an obstacle that will invariably interrupt upstream fish migration, and at the very least be a serious obstacle for organisms moving downstream.
  - Flood protection: large storage reservoirs usually have a potential for flood protection, and in many cases this is one of the roles such reservoirs are planned for, in addition to energy generation. How effective a specific reservoir is in this sense depends mainly on design, operation and suitability of the flood forecast system in place, mainly on the question of how much storage room is available in the reservoir in the moment a flood peak reaches it. Flood protection reduces the risk for people and infrastructure in the downstream area; on the other hand, it reduces or eliminates the role extreme floods can play in a natural floodplain dynamic (see the following section).

- Deviation scheme: depending on its design, a deviation scheme can be a ROR structure or a storage plant, and will therefore have the impacts described above for these two groups. In addition to that, it will reduce flow in the river stretch between weir/dam and tailrace outlet to the amount of water determined as residual flow (or ecological flow, see Chap. 20), with additional volumes during spilling. Since in such a part of the river the difference between normal flow (in this case the residual flow) and flood peaks (corresponding more or less to flood peaks in the undisturbed rivers) will be greater than it was under natural conditions, such floods can be more destructive. The effect may be more complex in the case of interbasin projects, whether these are ROR or storage HPPs.

- Peaking operation: this will usually be related to a high dam with an important storage volume, although in some cases this volume may just allow for daily peaking (as e.g. KGL, see Sect. 9.1.1.2); the peaking capacity of ROR structures is usually very limited. In addition to the impacts listed for storage plants, there is the additional impact caused by the large short-time variations in discharge from the power plant, which can go, in just a few seconds, from 0 to several hundred m^3/s. From an energy system management point of view, this is one very relevant advantage of hydropower, since it gives the flexibility to adapt production to demand on very short notice. Directly downstream of such a plant, however, the negative impacts can be rather severe: the surge wave caused by a rapid start of the turbines poses a high risk to people who are within its reach, and it will flush out the river bed and organisms within it, greatly reducing, in extreme cases to almost zero, any populations of plants and animals living there under natural conditions.

Many of these impacts have been described in previous chapters, mainly those dealing with the main topics to be covered in an EIA, i.e. Chaps. 8–10. The one system which can and will be affected by these impacts are the floodplains of rivers in the downstream area of weirs and dams. This is discussed in some detail in the following section, whereby some duplication with points that have been mentioned before are inevitable.

For obvious reasons: gaining head, hydropower plants often use water falls, the dam and intake structure being located above, the power house below the falls. This will reduce the water flowing over the falls, and this in turn can negatively affect one very specific habitat: the spray zone. This is highlighted in Sect. 18.3.

18.2 Floodplains

18.2.1 Main Characteristics

The floodplain is the land on either side of the river which is directly—although, depending on elevation above the river bed, to various extents—influenced by the river dynamics, and mainly by the seasonal difference in river flow. Form and extent can be very different, mainly depending on the local topography. Its width can vary from a narrow strip of land on either side of the river to areas stretching over several km, up to 100 km in the seasonally flooded forest called várzea (floodplains of nutrient-rich white rivers) or igapó (floodplains of nutrient-poor black- and clearwater rivers) in the Amazon basin (Walter and Breckle 1984: 195; Junk 1989). In the latter case, i.e. in areas where the land surrounding the river is flat and the gradient very low, rivers will normally show typical flow patterns, either meandering, i.e. flowing

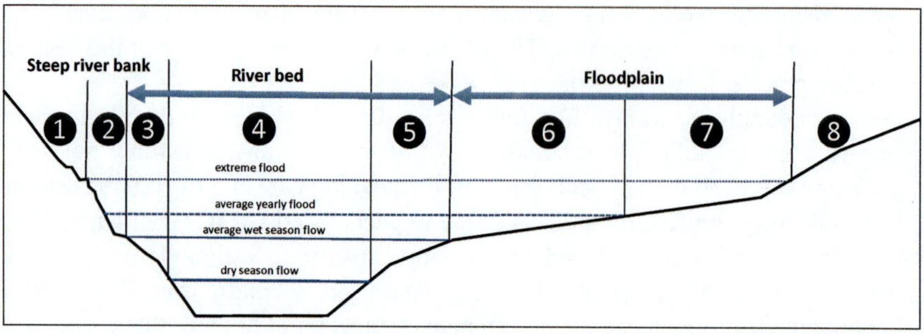

Fig. 18.1 Cross-cut through a hypothetical floodplain

in long, wide bends, or braiding, i.e. splitting up in many arms (see Bravard and Petit 2010, for a definition and physical characteristics, and Fig. 6.2). The main parts of a floodplain are shown in a diagrammatic way in Fig. 18.1 and commented on shortly here.

Numbers 1–3 are the main levels on a rather steep river bank; while they can largely correspond to the levels 5–8, these steep slopes are often located on the outside of river bends, where flow velocity is higher, and therefore subject to rather high erosion; the river, in a way, "eats into" these slopes. 4 is the part of the river bed which is constantly covered by water. It is therefore the only permanent aquatic habitat. Some species, especially sessile ones (water plants, but e.g. also some molluscs) are restrict to this part of the river bed, and it is the part where highly mobile aquatic organisms, like e.g. fish, will concentrate during the low-flow season. 5 (and 3) is the part of the river bed which is regularly covered by water during the high-flow season, meaning that this part will be regularly covered during some weeks or up to some months every year. Its extent depends entirely on steepness of the river bank, and it can be very extensive. The biome here depends largely on the physical characteristics of the site: size, flow speed of the water, duration of submergence, water and soil chemistry, and temperature. Narrow and steep banks under this regime (3 in the graph, which will usually not be considered as a "floodplain"), which are exposed to erosion, are very often bare of any vegetation (as shown in the photo of the Mekong in Fig. 18.2). Flat areas, climate permitting, will usually be covered more or less by a specific vegetation, like e.g. reeds or other types of vegetation, including certain forest types, consisting of plants which are tolerant to or require being covered by water for extended periods, but which can also sustain more or less dry periods during the low flow season. Whether this strip of land is considered as part of the river bed or as part of the floodplain is a matter of definition and certainly depends on the local conditions. 6 indicates the land that will be affected by frequent flows, i.e. by the average yearly flood or other recurrent flood events. The main characteristic here is the fact that the vegetation is made up by species tolerant of regular floods of short duration (occurring yearly or every few years, but usually lasting only a few days). 7 is a transition zone: influenced by the river only rarely, only during extreme floods, which may last usually only a few hours or in the

Fig. 18.2 Mekong shoreline zonation. Low-flow situation. The bare strip along the river is covered with water during the high-flow period; since then flow velocity is high, most of the fine soil material is carried away, no plants grow there. The strip above it is covered with water during higher than average floods, fine material is deposited there and annual plants grow easily; this zone is often used as river bank gardens. Note that the hills above the first steep part of the slope are almost completely denuded due to extensive slash-and-burn cultivation

extreme a few days. While damage can be caused by such floods, they usually do not have much influence on the vegetation, although some species of perennial plants (mostly trees) which do not tolerate to have their roots covered with water even for short periods, will not grow there; otherwise, under natural conditions the vegetation in zone 7 will be very similar to the one found in zone 8, the land which is outside the influence of the river. Plants specifically adapted to floodplain conditions as such, especially perennial ones, i.e. mostly trees, will usually not be found here, since the upland plants will normally outcompete them. This leads to a specific zonation of the vegetation, directly determined by frequency, height and duration of flooding. For a zonation of vegetation e.g. in central European floodplains see Ellenberg (1982: 335), or Walter and Breckle (1986: 125 ff.) for a more general discussion. This zonation also affects the fauna living in the different habitat types (Geier and Best 1980).

18.2.2 Floodplain Dynamics

Floodplains are ecosystems depending on a meandering river and its dynamics. These dynamics are conditioned by the variation in river flow, by regular seasonal variations as well as by rare and extraordinary flood events, and this causes the specific structure of such landscapes (see Fig. 18.3). The following main conditions determine the composition of the species living in such an ecosystem:

1. Low flow season: a large part of the soil is dry, water only in the river itself and in lakes (oxbows); in cold climates, soil and vegetation is exposed to frost. This eliminates all plants which cannot tolerate longer dry periods and frost.
2. High flow season: a large part of the floodplain is under slow flowing or almost stagnant water. These inundations can last for a few days or for longer periods, up to several months. The water brings sediments (fine sediments in the form of suspended solids in moving water, which will slowly sink to the bottom when the water is stagnant). These floods eliminate all plants that are not tolerant of prolonged inundation (i.e. dryland species). At the same time, normal high

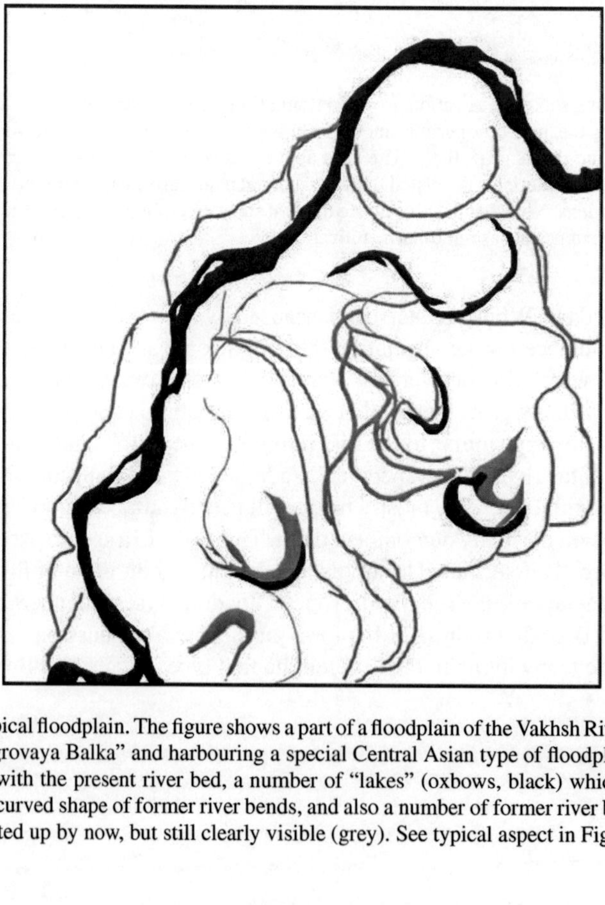

Fig. 18.3 Typical floodplain. The figure shows a part of a floodplain of the Vakhsh River, Tajikistan, known as "Tigrovaya Balka" and harbouring a special Central Asian type of floodplain vegetation called Tugai, with the present river bed, a number of "lakes" (oxbows, black) which all have the characteristic curved shape of former river bends, and also a number of former river beds which are completely silted up by now, but still clearly visible (grey). See typical aspect in Fig. 18.4

flows tend to erode the river bank on the outside of bends, while at the inside, where flow is less strong, sediments are being deposited.

3. Extreme floods: "catastrophic" events characterised by large amounts of water and high flow speed. Under such conditions, the meandering river can change its bed, destroying the vegetation in its course. An old river bend, which by this process will be cut off and no longer be part of the river, will then be transformed into one of the floodplain "lakes" or oxbows.

This means that such a floodplain is a system which is highly dynamic and is maintained by the ever changing conditions dominated by the river: seasonal flow variation, successions of dry and wet years, and extreme flood events. The development at the local level is governed by different processes, some of them slow and gradual, other short and violent:

- An oxbow which is cut off from the river will develop a typical pond vegetation, with a succession of reeds and other water plants. Occasional marginal flooding will still bring some sediment from time to time. This, and the dead and decaying plant material, leads to a gradual filling in of this oxbow. The free water surface diminishes, and over time it disappears altogether, even if the structure as such may still be visible for a long time given a difference in vegetation. This is a very slow process, which can take tens of years. The same basic process happens in any lake. Over time, a lake will be slowly filled in with soil and other materials carried by inflowing waters, and eventually become a marsh, and, ultimately, a terrestrial system (Ryding and Rast 1989: 1; Ellenberg 1982: 78393; Schwoerbel 1980: 11;49).
- The meandering river gradually changes its bed by the process of erosion and sedimentation in the bends described above. This is usually a slow, gradual process, which however can be accelerated momentarily during higher than average floods. This process can also lead to the formation of oxbows, in cases where the river erodes away the remaining narrow land between two bends.
- Exceptionally high floods, of the order of a one in 50 years flood or higher, can occasionally bring about very considerable changes in very short time (hours): the river might deposit large amounts of sediments in some places, possibly blocking its own bed, and then breaking through and excavating an entirely new bed, or else following again an ancient, long abandoned bed, destroying the vegetation on its way and mobilising large amounts of soil, to deposit them somewhere further downstream. If a river bend is cut off in the process, a new oxbow is created, and the river has a new bed surrounded by smaller or larger areas of bare soil (gravel, sand or silt depending on the situation). These are sites where so-called pioneer plants (plants able to grow rapidly on bare soils often poor in nutrients) and their associated fauna can develop.

All this means that a set of specialised plants which are tolerant of these often extreme conditions—or, more exactly, adapted to the specific conditions in specific parts of this ecosystem—will grow here; many of these plants are limited to such sites, i.e. they will not grow in other places. Likewise, the vegetation formed in this way will be the habitat for a set of animal species which are adapted to—or which depend on—the conditions of this special habitat, characterised mainly through seasonal changes in water level and flow velocity. Since habitats of this type are limited in extent and number, many of the specialised species living there are rare, and as a consequence endangered if their habitat is negatively affected. A sudden and in specific cases dramatic shift in the course of the river bed, called avulsion, can present a considerable risk to human populations living in the affected area, especially when this happens in large rivers flowing through densely populated areas. Human influence can also be a main driver of such developments, e.g. by increasing the silt load of a river through deforestation in the catchment area (Pearce 2021), or by channelling the river in a part of its floodplain, leading to higher flow and flow velocity further downstream.

18.2.3 Effects of River Regulation on Floodplain Ecosystems

River regulation, as caused e.g. by a large storage reservoir like, in the case of Tigrovaya Balka shown in Fig. 18.3, Nurek upstream of such a floodplain, has the following main direct effects on floodplain dynamics:

- Shift of water from the wet (high flow) season to the dry (low flow) season. This means that normal summer flooding will on average reach lower levels than it did before, and with higher winter flows less land will be dry during this period.
- Reduction in frequency and severity of high floods: a large reservoir has, among other effects, also an effect of reducing flood peaks, in this way providing flood protection to downstream areas. This means that the effects of such flood events as described above will be less marked.

These direct impacts on the floodplain dynamics influence the vegetation mainly in the following way:

- Areas no longer flooded regularly in summer become gradually colonised by dryland species which would not be able to grow under flood conditions, but which replace floodplain species in the absence of such floods. It is also possible that people start cultivating areas which are flooded only very exceptionally (Dister 1983; Fig. 26.5).
- Erosion as well as sediment accumulation processes are slower, and events where the river actually destroys vegetation occur more rarely or not at all any longer; this means that pioneering plants find no suitable habitats any longer.
- The "lakes", i.e. the old oxbows, continue to be filled in and disappear over time; if no new such lakes are formed due to the river changing its bed, this means that over time there will be no such lakes left any longer.

The entire process can lead to a more uniform habitat where many of the very specialised species of plants and animals cannot live any longer. In addition to that, the inflow of drainage water from irrigation schemes in the surroundings leads to the input of salts, fertilisers and agrochemicals, which all contribute to changing the situation in this ecosystem. Channelling a formerly meandering or braiding river, with the aim of increasing flow speed and protecting parts of the floodplain from regular flooding will have the same effect, maybe even more marked, since the effects of the natural river dynamics will be completely eliminated.

18.2.4 Case Studies

This section shortly describes two cases of floodplains under very different climatic conditions. Another, highly important case is the Great Lake/Tonle Sap in Cambodia; see Sect. 22.5 for a discussion of HPP effects on this system.

18.2.4.1 Tigrovaya Balka

The floodplain called Tigrovaya Balka (Fig. 18.3) is located in the lowermost part of the Vakhsh River basin, Tajikistan, down to its confluence with Pyanj river (where these two rivers form Amu Darya), close to the border with Afghanistan. It was declared a protected area on November 4, 1938. Its main objective was the conservation of the unique Tugai complex and of the animals living in it. Tugai is the name of a specific type of floodplain habitat in desert areas of Central Asia; it is characterised by a groundwater level close to the surface, which conditions a specific vegetation type composed of a number of specialised tree species, reeds etc. The reserve has an area of 49,786 ha. It is of great importance for the conservation of the Tugai ecosystem and its unique fauna and flora. It was the last place where the Caspian tiger lived, which finally disappeared around 1950. Human pressure on this area reached its maximum after the collapse of the Soviet Union and the ensuing civil war. Illegal and uncontrolled logging, hunting and fishing led to a sharp decline of many species of animals—the Bukhara deer, black and golden pheasant, gazelles, the striped hyena and other species.

The climate is continental and arid. The average annual temperature is +14 to + 17 °C, the temperature of the coldest month (January) is +2 °C, the one of the hottest (July) 32–38 °C. Temperature in July sometimes reaches up to 48°, and length of the frost-free period is 250–310 days. Winters are short and mild, which is typical for dry subtropical zones. Precipitation is distributed unevenly throughout the year, up to 70% falling in winter and spring months, usually in the form of rain. The entire system depends on river water and its dynamics, which originally was characterised by high flows in spring and summer (snow melt in the high mountains in the upper catchment) and low flows in winter.

The reserve contains about 20 lakes of different size (Fig. 18.4); their water is poorly mineralised, with about 1.92–4.67 mg/l. Besides 438 species of vascular plants, within the reserve there are about 30 species of reptiles, 34 species of mammals, 2 species of amphibians and 150 species of birds. It is one of the few

Fig. 18.4 Aspect of Tigrovaya Balka in winter. Shows one of the larger oxbow lakes surrounded by reed vegetation. See also Fig. 18.3

remaining habitats of the Bukhara subspecies of the red deer or Hangul (*Cervus elaphus bactrianus*).

Main threats for the reserve are the massive development of land adjacent to its borders, the lack of buffer zones, poaching, regular forest fires, illegal logging and, since the start of operation of Nurek HPP, reduced seasonal fluctuations in water levels in the Vakhsh. This latter effect largely eliminated natural floodplain dynamics in this area and encourages invasion of upland plants; see Chap. 23; one of the concerns related with Rogun HPP was that the negative impact on this area might increase.

18.2.4.2 Kafue Flats

The Kafue Flats are a vast wetland area in the Southern part of Zambia along a stretch of the Kafue River and conditioned by the dynamics of this river. At the lower end of this plain, the river enters a gorge, which brings it down to the Zambezi valley. A dam and hydropower plant, Kafue Gorge Upper (KGU) were built in the upper part of this gorge. KGU has a small reservoir with very limited storage capacity and is operated as a ROR plant. Since the Kafue River is highly seasonal, power generation was very limited in the low-flow season, while a large amount of water had to be spilled during the high-flow season. For this reason, a second dam, Itezhi-Tezhi, was built further upstream, upstream of the Kafue Flats, forming a large reservoir capable of seasonal storage; this was originally the only purpose of this dam, a power plant was added to it only much later (see Table 9.1 for details on the Kafue cascade). The direct result—and obviously the aim of this project—was that a considerable amount of water could be stored during the high-flow season for being released during the low-flow season, which led to a considerable increase of power generation at KGU.

The effect of this second dam on the Kafue Flat wetlands was as described above: less seasonal fluctuations, which led to a gradual shrinking of the typical wetland vegetation. A number of studies have investigated the effects, with emphasis on two issues. One was the effect on the population of the Kafue lechwe (*Kobus leche kafuensis*), a subspecies of an antelope endemic to this wetland and very well adapted to it, which came under pressure due to its shrinking habitat (Sayer and van Laviedren 1975; Hagen and Hagen 1979). The other concern was fisheries, an important resource for the local population (Chipungu 1981; Godet and Pfister 2007; Deines et al. 2013); the latter study came to the conclusion that the fish stock had suffered more from overfishing, mainly due to an uncontrolled influx of fishermen from outside, rather than from the impacts of the dam.

18.2.5 Measures

The single most decisive impact of a storage dam on floodplains in the downstream area is certainly the change in seasonal flow variation. The increase in discharge during the low-flow season, at least as long as this is done in a continuous way and not in the form of surge waves and short term fluctuations due to peaking, may not really change the situation very much. However, the reduction of high flows, i.e. the reduction of normal high water levels as well as of frequent flooding, will have the effects mentioned above: moving in of dryland vegetation, in this way reducing the extent of the typical wetland vegetation, and slowing down or completely eliminating the dynamics of the river, will have profound effects. These adverse effects on the floodplain as a natural habitat, or as an important fishing ground, can be exacerbated further by encroachment of forestry or agriculture into these areas.

A first step for impact mitigation will be a careful analysis of the situation, and of the likely changes to be brought about by the project. One approach for maintaining the ecosystem functions of a floodplain may consist in releasing periodical flood pulses from the dam; in order to have the intended effect, these must be timed adequately, i.e. be released during the period where natural floods would occur, and should be of an adequate magnitude, i.e. about the size of an average yearly flood, and if possible higher every few years (Staehly et al. 2019). Such a regime may be adequate for maintaining the floodplain ecosystem. However, with this it will usually not be possible to maintain the dynamic part of the system as caused by a meandering river, i.e. shifting the bed and in this way forming new oxbow lakes, and occasionally creating new pioneer habitats caused by the destructive force of an exceptionally high flood. If the aim is, for reasons of nature conservation, to maintain the entire range of such habitats, from the early stage where vegetation and top soil were removed to a complete siltation of such lakes and a gradual takeover by upland vegetation, the application of artificial measures will be required. This is especially required where the original floodplain was cut off from the river by longitudinal dikes, i.e. river training, as would be the case for the Bosna River cascade described in Sect. 26.4. In a study preparing a management concept for the former Aare River floodplain upstream of Bern, Switzerland, two possible approaches were described. There, the river had been straightened and put between such longitudinal dykes, and the floodplain was completely isolated from the river and was therefore no longer subject to a natural dynamic. Oxbow lakes are important as habitats for different semi-aquatic organisms like a large number of insect, amphibian and bird species. During the natural process of siltation and changing vegetation, the suitability of the place for different species will change, but in the end, when the lakes are silted up completely and no new ones are formed, this habitat type will be lost (Mazzucchi 1971). To maintain these different stages, two different strategies were proposed (Krüsi et al. 1987). The first one was a so-called maintenance approach: periodically taking rather small measures to re-open a number of places by eliminating patches of vegetation, and by occasional dredging of some of the oxbows. The other one was termed rotation strategy; it would have consisted in accepting natural succession in the existing oxbows, to the point where they would finally disappear, but then moving in with heavy machinery from time to time, completely removing a part of the forest and digging up the soil, in this way forming depressions (imitating new oxbows) and heaps of gravel and other material (imitating new gravel and silt banks). While this second approach would, by imitating the effect of a rare, but massive flood event, come closer to the natural dynamic of the site, it was clear that in this situation, the entire area also being an important recreation area for the nearby city, it would not be applicable: a measure like this would be conceived as "making a terrible mess in such a nice place" and would provoke massive protests from the local population. Such a measure, even if it might be preferable from a conservation point of view, is politically not feasible.

18.3 Spray Zones

One specific habitat type which can be influenced by a dam project are the spray zones below waterfalls. For one obvious reason, the gain of head for the power plant, waterfalls are preferred places for HPPs, with the dam placed above and the power house below the fall, in order to make use of this drop.

The spray zone is the area that is constantly wet due to the spray caused be the waterfall. Once the flow over the fall is reduced due to an HPP, so is the extent of the spray zone, and this can in an extreme case mean that such a zone, which was dripping wet all the time under natural conditions, now is permanently or temporarily completely dry. This must have a negative impact on the specific habitats that became established in such special zones. Even if a residual flow over the falls is maintained, the result would still be a reduction in size of the spray zone, and with it a reduction of the populations of organisms depending on such conditions. This, as for other types of habitats, can lead to the local extinction of some species.

There is one case which very well illustrates the potential impact of such a development: the Kihansi spray toad (*Nectophrynoides asperginis* or spray zone toad), a species of toad restricted to the spray zone of Kihansi Falls, southern Tanzania, where the reduction of its habitat due to a dam upstream of the falls led to the extinction of the species in the wild. In addition to the change in habitat conditions a parasitic fungus, *Batrachochytrium dendrobatidis*, causing a deadly disease called chytridiomycosis, may have had an important part in this (see Sect. 6.3.1). There are indications that in the case of the spray toad this fungus may have been accidentally introduced by the scientists who came there to work on saving the species. This tiny toad was then bred successfully in several US zoos. A sprinkler was installed in the habitat to recreate the spray zone, and in 2012, reintroduction of the species to its natural habitat started and seems to be successful (Thomas et al. 2013).

Figure 18.5 shows a number of waterfalls and their spray zone. Rusumo Falls (top left) is on Akagera River which there forms the border between Tanzania and Rwanda; an HPP is presently (2020) under construction. Here, at the request of the

Fig. 18.5 Waterfalls with spray zone. Top left to bottom right: **a** Rusumo Falls, Rwanda. **b** Memve'ele, Cameroon. **c** Amaila, Guyana. **d** Grand Poubara, Gabon. **e** Kongou, Gabon. **f** Kaieteur, Guyana

World Bank, a Spray Zone Study had been prepared, which showed that, while plants depending on permanently wet conditions are growing there, there are no species of a high conservation concern; it is probably also too small as to be a likely candidate for harbouring endemic species. Memve'ele (top centre), Cameroon, is—or was—a site with numerous smaller waterfalls; an HPP has in the meantime been built there (Pöyry 2010), and the situation must have changed fundamentally. Amayla (top right), Guyana, is the site for the planned HPP of the same name, where, with the exception of basic investigations (and, unfortunately, the partly construction of an access road, see Chap. 15), nothing has happened so far. Grand Poubara (bottom left), Gabon, is a rather small, although fairly wide waterfall, with a drop of only about 2–3 m, and part of the water from this fall was already used by a ROR structure at the time the studies for a dam project at that site were carried out (Pöyry 2008). Kongou (bottom centre), Gabon, is a special case in a number of ways. At this site the Ivindo River drops over a series of steps with a total head of 50 m, forming a complex of smaller and larger waterfalls and cataracts, "the most beautiful and largest waterfalls in African forests" (Anonymous 2002; translated from the French original; see also Vande Weghe 2006: 23; and Fig. 2.5 for a different part of the same falls). The falls are located in Ivindo National Park. The site had been identified as one of four potential HPP sites along this river and was considered as a part of a large development project (mining, railway link, harbour and HPP; Pöyry 2009). To the best of my knowledge, the project has not moved on since then (see Sect. 26.1.3). Kaieteur Falls (bottom right), Guyana, is a 226 m high drop into a narrow gorge (Kok and Kalamandeen 2008), and it has a spray zone of considerable size; this fall, located within the Kaieteur National Park, is the only waterfall shown here which is not an HPP site.

References

Anonymous, 2002. *Les parcs nationaux du Gabon. Stratégie pour le troisième millénaire*. Libreville, Gabon: Multipress Gabon.

Berenguer, R.J. 2017. *Taira, el almanecer del arte en Atacama. Taira, the dawn of art in Atacama*, 115 pp. Santiago de Chile: Museo Chileno de Arte Precolombino.

Bravard, J.P., and F. Petit. 2010. Geomorphology of streams and rivers. In *River ecosystem ecology: a global perspective*, ed. G.E. Likens, 44–52. Elsevier: Academic Press.

Chipungu, P.M. 1981. The impact of the Kafue Gorge dam on the Kafue floodplain fishery. *CIFA/T* 8: 120–129.

Deines, A.M., C.A. Bee, C. Katongo, R. Jensen, and D.M. Lodge. 2013. The potential trade-off between artisanal fisheries production and hydroelectricity generation on the Kafue River, Zambia. *Freshwater Biology* 58: 640–654.

Dister, E. 1983. Anthropogene Wasserstandsänderungen in Flussauen und ihre ökologischen Folgen—Beispiele vom Oberrhein und vom Rio Magdalena (Kolumbien). *Verhandlungen der Gesellschaft für Ökologie (Festschrift Ellenberg)* XI: 89–100.

Ellenberg, H. 1982. *Vegetation Mitteleuropas mit den Alpen*, 3. Aufl, 989 S. Stuttgart: Ulmer.

Geier, A.R., and L.B. Best. 1980. Habitat selection by small mammals of riparian communities: Evaluating effects of habitat alterations. *Journal of Wildlife Management* 44 (1): 16–24.

Godet, F., and S. Pfister. 2007. Case study on the Itezhi-Tezhi and the Kafue Gorge dam. *The Science and Politics of International Water Management*. https://www.researchgate.net/publication/259533453.

Hagen, H., and U. Hagen. 1979. Störung des Ökosystems der Kafue-Litschi-Moorantilope, Kobus leche kafuensis Haltenorth 1963, durch Errichtung von Staudämmen im Kafue (Sambia). *Säugetierkdl Mitt.* 27: 161–168.

Junk, W.J. 1989. Flood tolerance and tree distribution in central Amazonian floodplains. In *Tropical forests. Botanical dynamics, speciation and diversity*, ed. L.B. Holm-Nielsen, I.C. Nielsen, and H. Balslev, 47–64. *Proceedings of the symposium on tropical forests: Botanical dynamics, speciation and diversity*, Aarhus University, 8–10 Aug 1988. London: Academic Press.

Kok, P.J.R., and M. Kalamandeen. 2008. Introduction to the taxonomy of the amphibians of Kaieteur National Park, Guyana. *ABC Taxa* 5: 287. http://www.abctaxa.be/volumes/volume-5-introduction-taxonomyamphibians.

Krüsi, B.O., R. Zwahlen, and P. Furrer. 1987. Naturschutzgebiet Aarelandschaft Thun-Bern: Bisherige Entwicklung, heutiger Zustand, Schutzkonzept. In *Naturschutzinsp. des Kt. Bern, Bericht 1986*: 51–70. Sonderdruck aus Mitt. Naturf. Ges. Bern, N.F., Bd. 44.

Mazzucchi, L. 1971. Beitrag zur Nahrungsökologie in der Nähe von Bern überwinternder Krickenten (Anas crecca L.). *Der Ornithologische Beobachter* 68 (4–5): 161–178.

Pearce, F. 2021. When the levees break. *Science* 372 (6543): 676–679.

Pöyry. 2008. Projet Hydroélectrique du Grand Poubara. Volume I: Etude d'Impact sur l'Environnement et Etude d'Impact Social. Rapport préparé pour MMPHERHPEN (Ministère des Mines, du Pétrole, des Hydrocarbures, de l'Energie, des Ressources Hydrauliques et de la Promotion des Energies Nouvelles), Gabon.

Pöyry, 2009. Projet Minier de Belinga: Etude d'Impact Environnemental et Social pour le Projet Hydroélectrique de Kongou. Volume III: Projet Hydroélectrique de Kongou. Report prepared for Compagnie Minière de Belinga (COMIBEL).

Pöyry et Alpha Technology, 2010. Projet Hydroélectrique de Memve'ele. Etude d'Impact Environnemental et Social. Rapport Final EIES. Cellule d'Appui à la Maîtrise d'Ouvrage du Projet d'Aménagement Hydroélectrique de Memve'ele, Cameroun.

Ryding, S.O., and W. Rast (eds.). 1989. The control of eutrophication of lakes and reservoirs. In *Man and the biosphere series*, vol. 1, 314 pp. Paris and Park Ridge: UNESCO and Parthenon.

Sayer, J.A., and L.P. van Laviedren. 1975. The ecology of the Kafue lechwe population of Zambia before the operation of hydro-electric dams on the Kafue river. *East African Wildlife Journal* 13: 9–37.

Schwoerbel, J. 1980. *Einführung in die Limnologie*, 4. Aufl, UTB 31, 196 pp. Stuttgart: Fischer.

Staehly, S., M.J. Franca, D. Tonolla, et al. 2019. Increase of hydraulic habitat diversity downstream of dams with sediment replenishment and artificial flooding. Paper presented at the Africa 2019 hydropower conference, Windhoek, Namibia, Apr 2019.

Thomas, P.R., J.J. Breheny, and D.M. Boyer. 2013. Return of a rare Tanzanian native. The reintroduction of the Kihansi spray toad. *The Wildlife Professional* 7 (1): 42–45.

Vande Weghe, J.P. 2006. *Ivindo et Mwanga. Eaux noires, forêts vierges et bais. Les parcs nationaux du Gabon*, 272. Libreville, Gabon: Wildlife Conservatrion Society.

Walter, H., and S.-W. Breckle. 1983. Oekologie der Erde Band 1: Ökologische Grundlagen in globaler Sicht. 238 S. Band 2 (1984): Spezielle Ökologie der Tropischen und Subtropischen Zonen, 461 S. Band 3 (1986): Spezielle Ökologie der Gemässigten und Arktischen Zonen Euro-Nordasiens, 587 S. Band 4 (1991): Gemässigte und Arktische Zonen ausserhalb Euro-Nordasiens, 586 S. UTB Grosse Reihe. Stuttgart: Fischer.

Walter, H., und S.-W. Breckle. 1986. *Oekologie der Erde Band 3: Spezielle Ökologie der Gemässigten und Arktischen Zonen Euro-Nordasiens*, 587 S. Stuttgart: UTB Grosse Reihe, Fischer.

Chapter 19
Transboundary and Cumulative Effects

Rivers have a perverse habit of wandering across borders, ...
nation states have the perverse habit of treating whatever
portion of them flows within their borders as a national resource
at their sovereign disposal.
Waterbury (1979) cited after Lagendijk (2016)

In this chapter, two specific requirements of many ESIAs are highlighted: the TBIA (Transboundary Impact Assessment), i.e. the identification, analysis and description of impacts caused on international rivers and their potential effects on other countries, and the CIA (Cumulative Impact Assessment), i.e. the impact the project to be investigated will add to already existing effects stemming from other projects, or the impacts caused by such other projects will add to specific project-caused impacts, or in which way such effects may interact. One important difference these two analyses have in comparison with those parts of the ESIA dealing with impacts stemming from the "own" project is the fact that, while impacts and their consequences can be analysed, it is here not possible to formulate mitigation measures which then can be turned into firm obligations; they can only be seen as recommendations. The importance of including the other stakeholders, i.e. specifically the other countries potentially affected, is emphasised.

19.1 Transboundary Impacts

19.1.1 Definition of International Waterways

The assessment of transboundary impacts deals with this habit of rivers to cross borders, and specifically with the impact an activity on a river within one country will or may cause in another country, usually one in the downstream area. In order to do so, a definition is required, although at first sight the meaning of "transboundary or international river" may be very clear. The World Bank, in its Operational Policy 7.50, Projects on International Waterways (2013), provides the following definition:

R. Zwahlen, *Assessing the Environmental Impacts of Hydropower Projects*,
Environmental Earth Sciences, https://doi.org/10.1007/978-3-030-91185-0_19

(a) any river, canal, lake, or similar body of water that forms a boundary between, or any river or body of surface water that flows through, two or more states;
(b) any tributary or other body of surface water that is a component of any waterway described in (a).

The specification given in (b) above is very important: it could be argued that a tributary which is entirely located within the country, but joins a river which at some point crosses the border, is an exclusively national river, giving the country full jurisdiction over it and over the use of its water. Quite clearly, activities like the construction of storage dams on tributaries, can have serious consequences for the other countries in the downstream area. In addition to his, there could be a discussion on the exact meaning of "River XY". This is exemplified in the block diagram in Fig. 19.1: Pyanj River originates in the extreme North-Eastern part of Afghanistan, near the border with China, and then forms the border between Afghanistan and Tajikistan. At the point where it meets its largest North bank tributary, Vakhsh River, the name changes to Amu Darya (which is, or historically at least was, the largest tributary of the Aral Sea, see Sect. 23.5). Vakhsh itself is formed at the confluence of Obikhingu and Surkhob, and the name changes again further upstream, the part coming from Kyrgyzstan being called Kizil-Su. It could therefore be argued that Vakhsh River is not an international river, since it flows exclusively within Tajik territory over its entire length, and that therefore the impacts of Rogun HPP have not to be considered under the aspect of a transboundary impact. This point of view was brought up in discussions on impacts of this project (see Chap. 23; Pöyry 2014), but it was then made very clear, also based on the WB definition above, that obviously this was not a view that could be sustained. One other example of downstream impacts

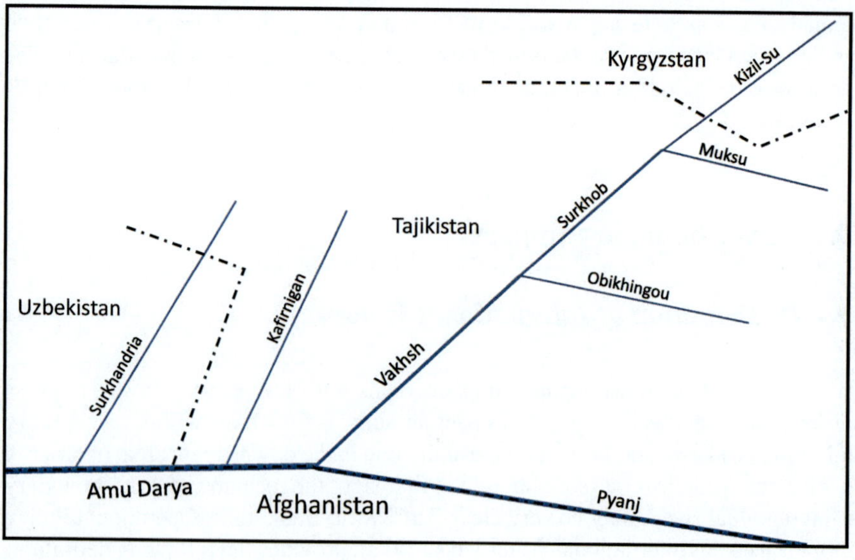

Fig. 19.1 The origin of the Amu Darya River

mainly caused by structures on tributaries rather than by those on the mainstream river is the Mekong—Tonle Sap system (see Sect. 22.5).

19.1.2 The Problems

For the case of the Aral Sea basin in Central Asia Mukhammadiev (2014) gives a short overview of the main issues related to impacts of projects (see Sect. 23.3, Rogun HPP case study). In this case the most important issue is the question of allocating water for hydropower, mainly in the upstream parts of the river basins, and for irrigation, mainly in the downstream parts. Even if it can be argued that water used for hydropower will still be available for irrigation further downstream, since hydropower is not a consumptive use of water, the problem can be very severe due to the fact that the seasonal requirements of the two water uses are different: retention of water in summer (= high flow season, but also growing season) would reduce water availability for irrigation, releasing it in winter (low flow season with high electricity demand) would not be of any use for irrigation in the downstream areas. In addition to this issue of seasonal distribution of flows there can be some other problems like handing down flood peaks (this however more related to river training rather than to storage dams, see Sect. 26.4 on the Bosna cascade), water pollution (this as well not primarily related to hydropower development), interrupting fish migration, concerns over dam safety, etc. Most of these impacts will be of concern for the downstream riparian countries; impacts on fish migration, however, can also affect upstream regions. One impact in the upstream area which has been caused by a number of hydropower projects, as e.g. in the case of the Aswan Dam in Egypt, is the creation of a reservoir which extends to upstream of the border, in that case into Sudan. One other case like this is the Lago di Lei, formed by a dam in Switzerland, the reservoir being almost entirely in Italy. If the "high dam option" would have been selected for the Rusumo HPP in the border between Rwanda and Tanzania, the largest part of the reservoir would have submerged land in Burundi (see Sect. 14.1.2).

19.1.3 TBIA: Transboundary Impact Assessment

Whenever a hydropower project is planned on an international river, transboundary effects have to be assessed, either in a specific chapter of the ESIA report, or as a separate document. The issues to be covered will have to be largely the ones shortly described above, as always considering the specific situation of the project at hand. The approach will basically be the same as for other project related impacts, by providing a description of the pre-project situation, identification of potential impacts and formulation of mitigation measures if required.

If impacts of a certain relevance are identified, a consultation process with the affected or potentially affected riparian countries may be required. One case of potential impacts and proposed mitigation measures is described in Chap. 23, for Rogun HPP, where the World Bank had initiated such a consultation process, which ultimately proved to be very useful.

19.1.4 Solutions to Transboundary Issues

While it was pointed out above that transboundary issues have to be dealt with in an ESIA, it has also to be seen clearly that such a study can only analyse the situation, show the problems and propose solutions, but it cannot really solve the problems, if there are any. Several cases have shown that such problems can be very severe, if not on the purely scientific and technical level, but rather on the political level. When Tajikistan started resuming works on the Rogun site with the aim of ultimately building this dam, there was strong opposition from Uzbekistan, since there was a concern that this dam might severely reduce water availability for irrigation. There were even threats from Uzbekistan that they might bomb the construction site for stopping this project. Luckily, in the meantime the situation has changed, and the two countries have come to an agreement in this matter; it could well be that the process initiated by World Bank, mentioned above, had started to pave the way for this. Presently, there is another project, with similar issues, in the public discussion: the Grand Ethiopian Renaissance Dam (GERD) in Ethiopia (Stockstad 2016; Roussi 2019, 2020). The issues are similar to what was or is experienced with Rogun: fear from downstream riparian countries, in this case Egypt, that control of water flow in the upstream area could cause problems for them, possibly even shortage of water during sensitive periods, limiting the control over a resource that was so far considered as their own. The solution is the same as well: reaching an agreement between the affected parties (Ethiopia, Sudan and Egypt) which would guarantee an operation of the entire system in a way as to maximise the benefits for all. One crucial point is the filling time of the reservoir (which should be 5 years according to Ethiopia, but 7 years according to Egypt; in Rogun, it is assumed that complete filling will be a process stretching over 16 years, with the aim of not reducing water availability for downstream irrigation). Mutual benefits of the GERD, if operated in a coordinated fashion, in addition to electricity generation, would be a better management of river flows, more storage available for coping with droughts, reducing flood risks in Sudan and reducing sediment input in the downstream reservoirs (Wheeler et al. 2016).

19.1.5 River Basin Organisations

Transboundary issues clearly demonstrate the importance of looking at the river basin as a unit, while the ESIA for a specific HPP mostly has a much more reduced scope. While, as mentioned, a Transboundary Impact Assessment can identify the issues and propose mitigation measures, the implementation of such measures, and earlier on discussions on them and reaching an agreement, cannot be done by the developer and the country in charge alone, it has to involve the other stakeholders as well, i.e. the other riparian countries which are or might be affected by a project. This can be done in the form of an ad hoc process, but in many cases, especially when a number of projects are developed, or are likely to be developed and the management of water resources is recognised as an important ongoing issue, it is certainly preferable to have a standing organisation, with clearly defined ways of information sharing, communication and decision taking. A number of such organisations exist, as e.g. the following ones:

- Rhine River: ICPR—International Commission for the Protection of the Rhine, https://www.iksr.org/en/. Founded in 1950, this organisation includes Switzerland, France, Germany, Luxemburg, the Netherlands and the European Commission and successfully co-operates with Austria, Liechtenstein, the Belgian region of Wallonia as well as Italy.
- Mekong River Commission (MRC, https://www.mrcmekong.org/): this commission was established in 1995 in the Lower Mekong River Basin and comprises Lao PDR, Thailand, Cambodia and Viet Nam. Every project which affects the Mekong, and especially projects on the Mekong mainstream as the HPPs mentioned in Chap. 22 needs to go through an evaluation and communication process with MRC as the coordinating body. The shortcoming of this organisation is the fact that China, in whose territory the upper and already heavily used part of the Mekong basin is located, does exchange information with MRC, but is not a member.
- Nile Basin Initiative (NBI; https://nilebasin.org/): established in 1999, this is an intergovernmental partnership of 10 Nile Basin countries, namely Burundi, DR Congo, Egypt, Ethiopia, Kenya, Rwanda, South Sudan, The Sudan, Tanzania and Uganda.
- Aral Sea Basin: In 1987, based on an analysis of water shortages in 1974–1975, and especially in 1982, two Basin Water Organizations: BVO "Amu Darya", and BVO "Syr Darya" (http://www.icwc-aral.uz/) were created. The BVO Amu Darya includes the four formerly soviet countries Kyrgyzstan, Tajikistan, Turkmenistan and Uzbekistan, and its main obligations are the maintenance and operation of the hydrological network for data collection and the yearly allocation of water to the member countries, based on a yearly forecast of expected river flows deducted from data on snow cover in the basin and long-term river flow data. This organisation remained functional after the end of the Soviet Union (Pöyry 2014).
- Senegal River: the Senegal River Basin Development Organisation (Organisation pour la Mise en Valeur du Fleuve Sénégal, OMVS, http://www.omvs.org/) includes the four states Guinea, Mali, Mauritania and Senegal. It was formally founded in 1963, with prior organisations starting during colonial times, in 1934.

19.2 Cumulative Impacts

As a rule, ESIA guidelines ask for assessing "direct, indirect, cumulative and induced environmental impacts" (ADB 2009: 16; WB 2017: 18). Definitions or descriptions of cumulative impacts are provided by WB (2017: 18, 23) and IFC (2012: 4). Shortly, cumulative impacts are the incremental impacts the project at hand adds to the effects of already existing projects, developments and activities in the same area, or such affects that will be added at the implementation of other projects in the planning stage. One important point here is that "cumulative impacts can result from individually minor but collectively significant activities taking place over a period of time" (WB 2017: 18). One example of drastically showing the potential cumulative impact is certainly the overall irrigation in the Amu Dara basin: while every single one of the projects for water abstraction from Amu Darya River (with the exception of the Karakum Channel, which caused certainly a major impact by itself) might have been rather minor, in their total the cumulative impact led to the demise of the Aral See (see Sect. 23.5).

Cumulative impacts of an HPP can be assessed either in the form of a chapter in the ESIA, or as a stand-alone report; this will often be specified in the country-specific guidelines issued by the competent authority. For assessing such cumulative impacts, two main points will have to be defined, namely:

1. The area to be considered; in the case of HPPs, this will usually have to be the river basin, where up- as well as downstream projects will have to be included in the analysis. With this, the Cumulative Impact Assessment (CIA) will include—or will be included in—the TBIA shortly described in the previous section.
2. The types of projects which could have cumulative impacts and which therefore will have to be included in the evaluation; this will be, first of all, any other HPPs, planned, under construction or in operation, on the same river, and possibly also those on tributaries in case of the main river in a basin. One such example, to some extent, is discussed in Chap. 22 for the case of the Mekong River. However, other types of projects may have to be considered as well, namely those which might have an impact on the project (as e.g. major logging in the catchment area, producing a risk of increased siltation of the reservoir) or projects which might be required as a consequence of HPP impacts (as e.g. flood protection measures required in the downstream area due to channelling a river as described in Sect. 26.4).

In a sense, every ESIA is necessarily also a CIA: it has to analyse the prevailing situation in the moment of preparation of the study, and this will necessarily have to be the situation as it may be already changed by other previously implemented projects. A good example here is the NT1 HPP in Laos, where right from the start the planning had to be done taking into account the fact that there were hydropower structures already present in the upstream area which deviated a considerable part of the water in the main river to another river basin, water which therefore would not be available for NT1 (see Sect. 22.3). A formal CIA will add to this mainly three additional elements, namely (i) other projects which may not affect the river as such, but could lead to cumulative effects of any type, (ii) projects in the pipeline or "reasonably foreseeable developments", as the guidelines say, and (iii) projects in the downstream area which might be affected, or where impacts could accumulate.

There is one important limitation of a CIA: while for the direct project impacts, the ESIA can formulate mitigation measures which will then be turned into conditions to be complied with on implementation and operation of the project, for many of the identified potential cumulative effects it will usually only be possible to give recommendations, which however cannot be turned into conditions on the level of the individual project. This is the same limitation as a Watershed Management Plan, prepared within the framework of planning an individual HPP, has: it can point out problems, actual or potential, but the project owner, and often the authorities involved in planning and authorising such a project as well, will have no authority for implementing or enforcing implementation of such measures. This would have to be done on a higher level, in the form of a comprehensive watershed or river basin management plan, an endeavour which goes far beyond the scope of any ESIA for any type of project.

References

ADB. 2009. *Safeguard policy statement*, 92 pp. Policy Paper. Asian Development Bank.

IFC. 2012. Performance standards on environmental and social sustainability. Performance Standards PS1-8, Guidance Notes GN1-8. http://www.ifc.org/performancestandards.

Lagendijk, V. 2016. Europe's Rhine power: Connections, borders, and flows. *Water History* 8: 23–39. https://doi.org/10.1007/s12685-015-0140-z.

Mukhammadiev, B. 2014. Challenges of transboundary water resources management in Central Asia, Chap. 9. In *The Aral Sea*, ed. P. Micklin, N.V. Aladin, and I. Plotnikov, 233–251. Berlin: Springer.

Pöyry. 2014. Environmental and social impact assessment for Rogun hydropower plant. In *ESIA*, Vol. I: Text. Vol. II: Annexes. Report prepared for OSHPC "Barki Tojik". Republic of Tajikistan.

Roussi, A. 2019. Nations clash over giant Nile dam. *Nature* 574 (7777): 159–160.

Roussi, A. 2020. Row over giant Nile dam could escalate, experts warn. *Nature* 583 (7817): 501–502.

Stokstad, E. 2016. Power play on the Nile. *Science* 351 (6276): 904–907.

Waterbury, J. 1979. *Hydropolitics of the Nile Valley*, 301 pp. Syracuse: Syracuse University Press.

WB. 2017. *The World Bank environmental and social framework*, 106 pp. Washington, DC: International Bank for Reconstruction and Development/The World Bank. http://documents.worldbank.org/curated/en/383011492423734099/pdf/114278-WP-REVISED-PUBLIC-Environmental-and-Social-Framework.pdf.

Wheeler, K.G., M. Basheer, Z.T. Mekonnen, et al. 2016. Cooperative filling approaches for the Grand Ethiopian Renaissance Dam. *Water International* 41 (4): 611–634. https://doi.org/10.1080/02508060.2016.1177698.

Chapter 20
Residual or Ecological Flow

Environmental flow assessments (EFA) can be done at several levels of detail, from a simple statement of water depth to provide wetted habitat for a particular fish species to a comprehensive description of a flow regime with intra-annual and inter-annual variability of low flows and floods in order to maintain complex river ecosystems.
WCD (2000: 294)

Dams and similar structures have the potential not only to reduce downstream flow to some extent, but to actually reduce it to zero. The cases where this may happen are mainly the following:

- Intake of lateral rivers (see the example of Shnogh HPP in Sect. 9.1.1.1), deviating the water to the reservoir.
- All storage plants, including peaking plants with sufficient storage capacity to retain the entire inflow, even if only for short periods of a few hours.
- Plants with river deviation, where the stretch of the river between retaining structure (weir, dam, intake) and tailrace outlet can be left without water.

In such cases, the river affected by the structure would run dry, at least temporarily, down to the point where the next lateral tributary would join it. This was actually done in many cases, causing very considerable ecological damage. Even short periods without any water in a river bed will eliminate fish and a large number of other aquatic organisms. In cases where the affected river is large enough for retaining some water in stagnant pools during the time of shutdown of the power plant, some species may survive, but certainly others will not. In order to prevent such a situation from happening, a minimum flow to be released at all times, also called ecological, environmental or residual flow, needs to be determined. As with a number of other issues related to impacts from hydropower, only bad experiences led to pressure and finally to regulations. Today, many countries have a legal requirement for defining residual flows, as well as more or less specific rules on how to do that. All major international environmental guidelines include such conditions. Since this is a field which is evolving quite rapidly, with regulations and procedures that get ever more

R. Zwahlen, *Assessing the Environmental Impacts of Hydropower Projects*, Environmental Earth Sciences, https://doi.org/10.1007/978-3-030-91185-0_20

sophisticated and complex, it is not intend here to go into the details on how to do it. Still, some background information and considerations may be useful.

20.1 A First Step

Whatever the legal framework and/or the guidelines to be followed in defining the residual flow for a given project, a first step should always be to get a clear picture on actual requirements for such a flow. In order to answer this very first question, namely, whether a residual flow may be needed or not, a good look at the project and the local situation is necessary. The design of the project, however preliminary it may be, will indicate whether or not a situation can arise in which flow reduction would be such as to require a determination of a minimum flow. The next point is then to get an idea on the possible justification for such a flow. Here, there can be two main reasons, with possibly a few specific points in each of them.

- Biological reasons, the river as habitat. The focus here will usually and foremost be on fish, but depending on the situation other types of aquatic animals or plants may be of importance.
- Socio-economic reasons. Here, the main question will be whether, for what purposes and to what extent water from the river is being used by the human population living along the affected stretch of river. The major use types are drinking water, irrigation or dilution of inflowing waste water (see the example of Shnog HPP, Sect. 9.1.1.1).

It may often be easier to find a suitable solution for the socio-economic requirements, since this is mostly just a question of water quantity at given times, e.g. during the season when water is required for irrigation. In addition to that, alternative solutions may be available, as e.g. reduction of the inflow of waste water, installing waste water treatment plants, or providing drinking water from other sources. In such cases, the issue might turn away from a simple quantification of water to the question of whether and to what extent the project (assuming it is a hydropower project) is capable of and has the obligation to provide such alternative solutions.

The issue can be more complex when a residual flow has to be granted for biological reasons. First of all, it must be clear that a river influenced in such a way will, in any case, no longer be the habitat it was before. Reduction in flow is one point, but the other project impacts, like interruption of migration routes, changes in seasonal flow variation, or short-term flow variation due, e.g., to occasional spilling or peaking plant operation, will have to be considered. Even with a good solution concerning the residual flow, this part may no longer be a suitable habitat for some of the species, and potentially some of the key species to be considered in this area. In any case, this first step should provide an answer to the most basic question, namely, whether under the circumstances given by the site and the project a residual flow is required or not; note that in many cases this question may already be answered by the legal requirement that a residual flow is required in any case, which is certainly justified by the fact that even a highly influenced river section will be a habitat for some species, even if possibly not for the ones which lived there originally. The next important step is then to determine the amount required, and this is a crucial point which often has no easy answer. From a biological point of view the answer may often be rather simple: the residual flow must guarantee that there is water flowing all the time, and the amount and flow pattern should be as close to the natural situation as possible. The project proponent, whose aim is to produce electricity, will often have a different view: water that has to be released is lost for energy production, and this can sum up to quite considerable amounts. The critical period here will often be the dry season, when there is less water available than could be used for production.

But even here, special cases may have to be considered: providing a minimum flow at all times may not always be the best solution. Datry et al. (2018) point out that intermittent streams can and do play important ecological roles, and that they often contain a biocenose which is necessarily different from that of permanent streams. Under such conditions, building dams for retaining water during the high flow season, for having a sufficient storage for being used during the dry season, be it for energy, drinking water or irrigation, may be highly desirable, but providing flowing water at all times may not the be best solution from an ecological point of view in every case. Still, the second part of the "rule" as mentioned above would still hold: as close as possible to natural conditions, which, in such a case, may include periods without flowing water.

20.2 Rules for Calculation

As mentioned, the first concerns that finally led to regulations concerning a minimal residual flow were those aimed at maintaining a continuous flow in affected streams and rivers in order to assure the survival of fish populations. A large number of different rules have been developed in various countries for defining the required minimum flow. Some of them have legal status, some are just more or less firm recommendations. A number of these rules are briefly described here:

- Swiss legislation: based on Q_{347}, defining the minimum flow to be released in any case; this means that the minimum flow to be released has to correspond to the flow which, under natural conditions, would be exceeded during 347 days per year; Fig. 20.1 provides an example for this approach. The lower threshold value is 50 l/s (for streams with a Q_{347} below 60 l/s), increasing to 10 m^3/s (for rivers with a Q_{347} of 60 m^3/s or more). The legislation also indicates a number of conditions under which a higher residual flow can be fixed. As in many other cases, this

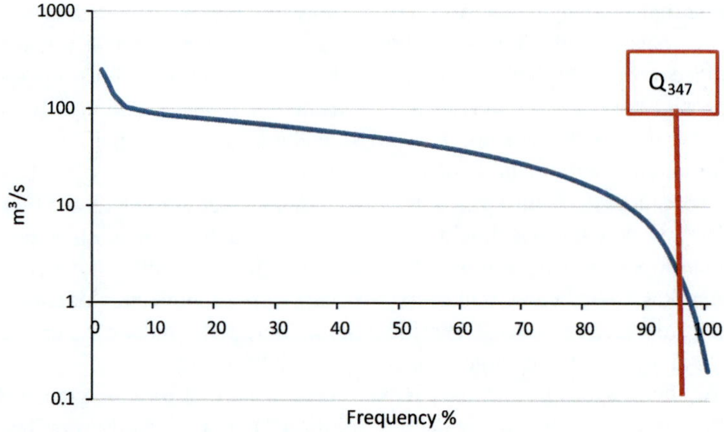

Fig. 20.1 Frequency distribution of flow values. Results in a residual flow of 1.6 m^2/s

rule was defined mainly for maintaining fish (especially salmonid) populations in mountain streams. It does not provide e.g. for maintaining floodplain dynamics.

- UK: Q95% (plus a possibility for correction); this corresponds to the Q_{347} rule.
- USA: minimum flow based on mean annual flow (MAF) and the role of the river as fish habitat: 10% of MAF for low quality habitats, 30% for moderate and 60% for excellent habitats.
- 10% of natural flow: this imitates seasonal dynamics, but is difficult to implement unless based on average monthly or seasonal flows.
- 15% of mean seasonal flow (mainly aiming at maintain fish populations).
- Minimum flow corresponding to at least the minimal monthly flow.
- Minimum flow equal to smallest ever observed flow.

Note: this is by no means a complete list. It just shows that there is a rather large variation in such rules. It also has to be kept in mind that such rules can change, based on new scientific data and/or on political pressure.

20.3 Modern Scientific Approach

Newer publications on residual flow (e.g. Arthington et al. 2006; Krchnak et al. 2009) do not give rules for defining residual flows. The main points are rather statements like: "In the past decade, environmental flow science has progressed considerably, and scientists now warn that maintaining minimum low flows is necessary but insufficient to maintain healthy river ecosystems. Instead, a naturally variable pattern of water flow is needed to sustain biodiversity and ecosystem services provided by rivers" (Krchnak et al. 2009: xi). One important conclusion from this is that the focus has widened from a concentration on fish populations to the entire river ecosystem.

This then leads to the following conclusion: "Even with the best-available expert knowledge and analysis, dam developers, governments, and stakeholders need to understand that the environmental consequences of hydropower dam development and operations cannot be predicted with complete certainty. To be ecologically and socially sustainable, water and energy development and management need to be perpetually informed by monitoring, carefully targeted data collection and research, and further analysis to address new uncertainties or surprises. Therefore, a program of monitoring, evaluation, and adjustment—commonly referred to as adaptive management—should be fully and explicitly integrated into any hydropower development or reoperation plan so that management approaches can be continually modified in response to increased understanding or changes in human or ecosystem conditions. The economic risks from uncertain future environmental constraints need to be addressed explicitly as part of an adaptive management strategy" (Krchnak et al. 2009: xii).

As a consequence, these "modern" approaches, e.g. Arthington et al. (2006), or the ELOHA (Ecological Limits of Hydrologic Alteration: A Framework for Determining Regional Environmental Flow Needs; http://conserveonline.org/workspaces/eloha), do not provide any rules or mathematical models for determining residual flow. Rather, they give indications for determining the need for residual flows, including seasonal flow modifications, flood events etc., based on the following 4 steps:

(1) Develop classification for reference streams;
(2) Develop frequency distributions for each flow variable in each class;
(3) Compare frequency distributions from flow-modified streams with reference conditions in the same class; and
(4) Develop flow-response relationships for ecological health data from reference and flow modified streams for each flow variable.

This is in line with the WCD Report which states "At least twenty-nine countries seek to minimise ecosystem impacts from large dams by using an EFR [environmental flow release] to meet predetermined ecosystem maintenance objectives. The practice of EFRs began as a commitment to ensuring a 'minimum flow' in the river (often arbitrarily fixed at 10% of mean annual runoff). It has since grown to include a definition of ecosystem requirements and a planned flow release programme, which may vary annually or seasonally, to meet downstream needs for both the environment and people. The level of EFR required is determined by the need to maintain particular ecosystem components downstream, often with reference to national legislation. The countries that use this method have recognised that a short-term reduction in financial returns from a project often leads to improved long-term sustainability and attainment of broader societal objectives for a healthier environment. Still, this represents a redistribution of the benefits of a dam project and thus existing beneficiaries such as irrigators and operators of hydropower facilities may resist EFRs" (WCD 2000: 81). The same basic approach is used by IHA (Table 20.1).

Table 20.1 Environmental flow according to IHA

Issue	Mitigation strategies
Downstream hydrology and environmental flows Changes to downstream hydrology impact on river hydraulics, instream and streamside habitat, and can affect local biodiversity. Operating rules should not only consider the requirements for power supply, but also be formulated, where necessary and practicable, to reduce downstream impacts on aquatic species and human activities	Operating schedules should, where necessary and practicable, incorporate environmental water release patterns (including environmental flows) within the operational framework for the supply of power Downstream regulating ponds and other engineering solutions may provide cost-effective alternatives to environmental flow releases directly from power stations It is important that the environmental objectives of any flow release are identified in a clear and transparent manner. These releases need to be developed within the context of environmental sustainability and also take into account local and regional socio-economic factors. It is desirable that the environmental flow objectives be agreed with local communities

Source IHA Sustainability Guidelines, 2004

In a World Bank study, Hirji and Davis (2009a, b) discussed the issue and provided case studies, pointing out the importance of environmental flows in a number of cases, without however providing any "rules" on how to determine residual flows in a specific case. Table 20.2 lists a number of cases indicating the residual flow determined for each of these projects.

20.4 Guidelines and Models

IFC (2018) has published a comprehensive handbook on environmental flows. In its Table 3 (pp. 50–54) it also provides a commented list of models used for residual flow assessment.

20.4.1 How Much Good?

The definition and maintenance of an environmental flow is now a procedural or legal requirement in most countries and according to most if not all ESIA guidelines specifically referring to hydropower, and in this way the question whether or not a minimum flow should be defined will be obsolete in most cases. A short and necessarily incomplete account on approaches etc. is provided above. Still, it is worthwhile at least to think about the usefulness of a residual flow, among other reasons also because releasing such a flow comes at a cost, this water not being

Table 20.2 Examples for residual flow

Project	River	Country	Env. flow	Comments	Source
Houay Ho	Houay Ho, Se San tributary	Laos	0.05 m^3/s	In the absence of a regulation in Lao PDR defining required minimum flow in rivers the Swiss regulation was applied, which aims at maintaining a river as a suitable habitat for aquatic species by guaranteeing a minimum discharge tat is close to the minimum natural flow in this river. The basic figure for calculation of this minimum flow is the Q_{347}, i.e. the flow that is exceeded on 347 days of an average year. In absence of a detailed flow curve, the average discharge of the month with the least discharge can be taken. In this case, this would mean that a minimum flow of 0.05 m^3/s (= 50 l/s) immediately downstream of the dam would have to be maintained. In periods when this flow does not occur naturally from runoff below the dam, this amount of water would have to be released from the reservoir. Application of the pertinent US code would lead to similar values The Houay Ho is a highly seasonal river with an average discharge of 10 m^3/s and a minimum monthly discharge of 0.04 m^3/s	EWI (1996)

(continued)

Table 20.2 (continued)

Project	River	Country	Env. flow	Comments	Source
Nam Ngum 2	Nam Ngum	Laos	0	Tributary with substantial flow a few 100 m d/s from dam site; project discharges almost directly into the next reservoir (Nam Ngum 1), which is already existing	EWI (1998bb)
Murum	Sungai Murum	Sarawak, Malaysia	8 m^3/s	Sungai Murum is a river in a moist area without a very marked dry season; therefore, the discharge pattern is not very distinctly seasonal. Average annual flow is 248 m^3/s. In the EIA, two cases had to be considered To maintain river stretch of Murum between dam and power house (12 km): compensation flow of 10% of minimum monthly average flow, corresponding to 8 m^3/s With Bakun dam in place: FSL of Bakun will reach power outlet of Murum, no need for residual flow	EWI (1998aa)
Nam Theun 1 (NT1)	Nam Theun/Nam Kading	Laos	10 m^3/s	Nam Kading is a tributary of the Mekong. The power house, located at the foot of the dam, discharges into the Mekong floodplain (33 km u/s from Mekong confluence), considerable backwater effect from the Mekong when discharge in Nam Kading is low. The project is for peaking production. 10 m^3/s as residual flow were defined to prevent the river stretch immediately d/s of the dam from falling dry during off-hours	Electrowatt-Ekono, (2005)
Allain—Duhangan	Beas	India	0.231 and 0.162 m^3/s	2 × 96 MW high head ROR scheme using the water from two tributaries of Beas river. Mean flow 1.5–22 m^3/s (Allain) resp. 1.4–7.3 m^3/s (Duhangan). The proposed environmental flow corresponds to 15% of the available flow	ERM (2009)

(continued)

Table 20.2 (continued)

Project	River	Country	Env. flow	Comments	Source
Nadarivatu	Sigatoka	Fiji	200 l/s	42 MW installed capacity, 40 m high dam. Minimum natural flow 0.2 m^3/s, high seasonal and short term variability of discharge. Maximum turbine output 15 m^3/s	SKM (2007)
Lajitan	Youshi	China	"a certain volume"	Average flow 115 m^3/s, reservoir with 10 day regulating capacity. No indication on amount of environmental releases	CRAES (2000b)
Dongping	Zhongjian	China	0.1 m^3/s	Average river discharge 45.6 m^3/s. Proposed environmental flow assumed to be provided by leakage through dam	CRAES (2000a)
Rogun	Vakhsh	Tajikistan	10 m^3/s	335 m high dam located 15 km upstream of the upper end of the existing Nurek HPP reservoir. Planned for peaking operation. Affected part of Vakhsh river, average discharge of 631 m^3/s, is very short, no inhabitants, limited if any importance for fish. Recommended to release a residual flow of 10 m^3/s, not required in case the additional dam between Nurek reservoir and Rogun dam would be built	Pöyry (2014)

(continued)

Table 20.2 (continued)

Project	River	Country	Env. flow	Comments	Source
Batoka Gorge	Zambezi	Zambia /Zimbabwe	Scenarios: 94 m^3/s 180 m^3/s 216 m^3/s 255 m^3/s	177 m high dam, located in the Batoka Gorge 47 km downstream of Victoria Falls, reservoir surface 23 km^2. Mean annual flow 1268 m^3/s. HPP planned to be for peaking. The environmental flow assessment showed a very significant impact of peaking, which could mostly be prevented by a ROR operation pattern. Evaluated effects of different EF releases, with decreasing impact with increasing amount of EF	ERM (2019)

available for the production of energy. A few points concerning this issue are briefly discussed here.

First of all, it has to be stated very clearly that a residual flow is just that: a kind of left-over, severely reduced flow of water in a part of the river which is affected by the project; it will not be the original river any longer, and in most cases will not even come close to the original; this is quite clearly illustrated by Fig. 9.12. Some of the species that originally lived in this part of the river may be able to survive under these condition, others may not (Bogan et al. 2017: 366), and in some cases additional species may be able to newly colonise this site. It has to be considered that flow reduction is not the only condition that changes. In the case of a deviation scheme average flow will be greatly reduced between intake structure and tailrace outlet. However, during high and exceptionally high flows, which cannot be stored and used for energy generation, the "surplus" flow will be spilled, i.e. left to flow either over a sill provided for that purpose, or through special flood evacuation structures. In this way, seasonal peak flows and any type of flood events can lead to a discharge which is comparable to what was flowing there under natural conditions. However, now the difference between average flow (= residual flow) and peak discharge will be much more important than it had been before, and it will have a tendency to happen much more abruptly than before. This will likely have a much greater impact on the river bed than it had under natural conditions. One additional problem arises in the case of peaking power plants. Here, such abrupt changes in discharge are no longer a seasonal and/or rather exceptional phenomenon, they can and do occur much more frequently, sometimes several times a day. The surge wave which will be formed by a sudden start of the turbines will literally flush out the river bed, often over a considerable distance, and the sudden recession of flow poses a risk for water organisms to find themselves stranded in a suddenly dry part of the river bed. No aquatic life can persist under such extreme conditions. As Datry et al. (2017) show, intermittent streams can be very valuable habitats, but only as long as there is a more or less reliable seasonal pattern of presence and absence of flow. A stretch of river under a peaking regime can certainly not be qualified as such.

This should not be understood as an argument against defining and maintaining a minimum flow, but rather as a request for clearly defining the usefulness of such a measure. Two main aspects can be seen as justifying such a measure. The first is the maintenance of aquatic life. If the altered river can still fulfil, at least in part, its functions as an aquatic habitat, e.g. for the maintenance of some fish species, even if this might not be at the original population size, then the measure is justified. The other point consist in the requirements of a human population along the affected part of the river, be it for irrigation, as a source of drinking water, for diluting waste water or any other use. This, finally, means that in the end the residual flow will have to be defined in a site-specific way, considering requirements and usefulness of the measure, rather than just the application of a, however sophisticated, formula or model.

To finalise this line of thought, one could imagine the following situation: a basin comprising of three similar rivers (similar in terms of morphology, underlying geology, climatic and biotic conditions). Would it, from an environmental (or, more precisely, ecological) point of view, be better to use two of them, literally squeezing every last drop out of them, and leave one untouched, or alternatively to use all three of them, but leaving appropriate residual flows everywhere? This is a highly hypothetical example, no real case can be presented here where this was actually discussed and implemented, but it could well be that if such a planning would be done somewhere option 1 might turn out to be preferable. One big problem here is that in the real world, rivers in a region are most often not treated as a unit, but as separate entities which can (and must) be "developed" independently from each other. Even if in an ESIA it is proposed (and accepted) that as a mitigation or offset measure for impacts caused in one river, another one in the same area should be kept intact, the next thing to happen is another project being studied and implemented in this "compensation" river; this, unfortunately, is a very "real world" situation.

References

Arthington, A.H., S.E. Bunn, N.L. Poff, and R. Naiman. 2006. The challenge of providing environmental flow rules to sustain river ecosystems. *Ecological Applications* 16 (4): 1311–1318.

Bogan, M.T., E.T. Chester, T. Datry, et al. 2017. Resistance, resilience and community recovery in intermittent rivers and ephemeral streams, Ch. 4.6, in *Intermittent rivers and ephemeral streams: Ecology and management*, ed. T. Datry, N. Bonada, and A.J. Boulton, 349–376. London: Academic Press. https://doi.org/10.1016/B978-0-12-803835-2.00013-9.

CRAES. 2000a. *Dongping hydropower project in Xuanen County of Hubei Province*. Environmental Impact Assessment Report, First Draft. Chinese Research Academy of Environmental Sciences, World Bank.

CRAES. 2000b. *Lajitan hydropower project in Laifeng County of Hubei Province*. Environmental Impact Assessment Report, First Draft. Chinese Research Academy of Environmental Sciences, World Bank.

Datry, T., N. Bonada, and A. Boulton. 2017. *Intermittent rivers and ephemeral streams. Ecology and management*, 597 pp. London: Academic Press, Elsevier.

Datry, T., A.J. Boulton, N. Bonada, et al. 2018. Flow intermittence and ecosystem services in rivers of the Anthropocene. *Journal of Applied Ecology* 55 (1): 353–364. https://doi.org/10.1111/1365-2664.12941.

Electrowatt-Ekono. 2005. *Nam Theun 1 hydropower project, Laos: Environmental impact assessment*. Malaysia: GAMUDA Berhad.

ERM. 2009. *Environmental and social impact assessment: 192 MW hydroelectric power project, Tehsil Manaly, District Kullu, Himachal Pradesh*. Revised and updated report prepared for AD Hydro Power Limited. http://adhydropower.com/pdfs/adhpl_revised_updated_esia_2009.pdf.

ERM. 2019. Proposed Batoka Gorge hydro-electric scheme (Zambia and Zimbabwe) on the Zambezi River. In *Environmental and social impact assessment (ESIA) for the project area of inundation, staff villages and quarries (V4.0)*. Report prepared for Zambezi River Authority. Specifically: Annex J: Environmental Flow Report.

EWI. 1996. *Houay Ho hydroelectric power project. Environmental impact assessment*. Seoul, Korea: Daewoo Corporation.

EWI. 1998a. *EIA for Murum HPP, Sarawak, Malaysia*. Report prepared for Sarawak Electricity Board.

EWI. 1998b. *Nam Ngum 2 hydroelectric project, Lao P.D.R.: Environmental impact assessment (EIA)*. The Shlapak Group.

Hirji, R., and R. Davis. 2009a. Environmental flows in water resources policies, plans and projects. In *Case studies. Environment department papers, natural resource management series*, Paper No. 117, 159 pp. Washington, DC: World Bank.

Hirji, R., and R. Davis. 2009b. *Environmental flows in water resources policies, plans and projects: Findings and recommendations*, 212. Washington, DC: World Bank.

IFC. 2018. *Environmental flows for hydropower projects. Good practice handbook*, 136 pp. The World Bank Group. http://www.ifc.org/sustainabilitypublications.

IHA. 2004. *Sustainability guidelines*, 24 pp. International Hydropower Association.

Krchnak, K., B. Richter, and G. Thomas. 2009. *Integrating environmental flows into hydropower dam planning, design and operations*. Water Working Notes 22, Nov 2009, 44 pp. World Bank.

Pöyry. 2014. Environmental and social impact assessment for Rogun hydropower plant. In *ESIA*, Vol. I: Text. Vol. II: Annexes. Report prepared for OSHPC "Barki Tojik", Republic of Tajikistan.

SKM. 2007. *Nadarivatu hydropower scheme*, Fiji. Consolidated EIA. FEA.

WCD. 2000. *Dams and development. A new framework for decision-making*. The Report of the World Commission on Dams, 404 pp. London: Earthscan.

Chapter 21
Dam Projects: Pro and Contra

Complex problems have simple, easy-to-understand wrong answers.
Bloch (1980: 17)

As mentioned in the introduction, dam projects tend to be discussed mainly from two points of view, one side stressing the advantages, the other insisting on the disadvantages of such undertakings. This Chapter looks into these discussions; to do so it is subdivided into three parts. In the first one, two controversial projects are presented, the Aswan High Dam and the Three Gorges Dam. The second part deals with the interesting and highly relevant question of sustainability of hydropower, while the third gives a few examples of some wrong arguments which have been used against hydropower or a specific project. The aim here is to show, once again, that the situation does in no way correspond to a simple black and white picture, but that there are many graduations in between, and that simple answers often tend to be wrong.

21.1 Controversial Projects

A number of hydropower projects have led to extensive discussions, during the planning stage as well as far into the operation period. The two probably most famous projects in this respect, the Aswan High Dam in Egypt and the 3 Gorges Dam in China, are presented here shortly. The third in this list, which is not included here, is most probably Narmada in India (see e.g. Khagram (2004) for a description of the situation around this project.

R. Zwahlen, *Assessing the Environmental Impacts of Hydropower Projects*,
Environmental Earth Sciences, https://doi.org/10.1007/978-3-030-91185-0_21

21.1.1 The Aswan High Dam

> *Experts agree that this dam has led to a steady deterioration in the fertility of Egypt's soil—the Nile no longer deposits silt traditionally used as fertilizer by Egyptian farmers. In addition, the bed and banks of the dam are becoming eroded downstream of the dam; the coastline of the Nile delta is retreating; vast amounts of pesticides and fertilizers have killed fish; and the rising water table has weakened the foundations of many buildings.*
>
> WWF (1999: 29)

> *Based on the study carried out by the Third World Centre for Water Management, it is evident that the High Aswan Dam is one of the most successful hydraulic structures of the world, and not one of the worst as is often presently contended at the international level.*
>
> Biswas and Tortajada (2012: 394)

Without any doubt, the High Aswan Dam is one project which has received—and certainly deserves—a lot of attention. Descriptions of the project, its history and negative effects can be found, e.g. in Lavergne (1986), WCD (2000:103, 117), Hughes (2009:175), Wohl (2011:94ff.), who also emphasise the political issues that were involved in the construction of this dam, which was commissioned in 1969, at the height of the cold war. Among the most important points raised are the resettlement of 110,000 people (Lassally-Jacob 1996; Scudder 2005:73), the loss of water due to evaporation in the reservoir and seepage from the lake as well as from the irrigation systems downstream of the dam, in total estimated to be about 20%, and the increasing river bed and bank erosion in the downstream area. The reservoir, Lake Nasser, has a total storage volume of 164 km^3, which corresponds almost exactly to twice the yearly discharge of the Nile at Aswan, which is about 80–90 km^3 per year. This large storage volume allowed for a control of the yearly floods of the Nile; besides flood control, making water available continuously for irrigation and for electricity generation were the main aims of the project. As can be seen in Fig. 21.1, river

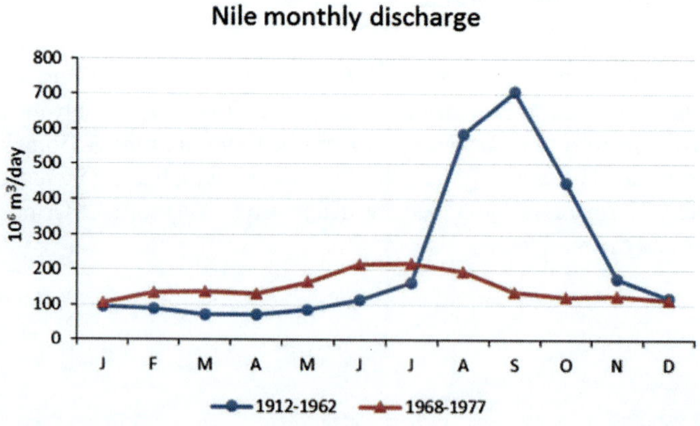

Fig. 21.1 Average monthly discharge of the Nile without and with Aswan high dam. Based on data from Shalash (1980)

discharge downstream of the dam changed completely, from a very seasonal pattern characterised by a few months with very high flows and a prolonged dry season to a more or less continuous flow the year round. The difference is even more striking when only maximum yearly flows are compared (see e.g. Wohl 2011:74). Ezzat (1993) mentions an increase in total flow in the period of 1904–1964, and then a decrease; this could be at least part of the explanation for the fact that there is a marked difference in the recorded flows between the two periods shown in Fig. 21.1 (yearly average 228 $\times$ 10^6 m^3 per day before *versus* 151 after commissioning of the dam), but the retention of a large quantity of water for the filling of the reservoir, which took several years, certainly explains a large part of this difference; evaporation in the reservoir may have contributed.

Retention of the water in the large reservoir did not only allow for an almost complete elimination of the yearly floods, but also causes retention of almost all the sediments in the reservoir. As all the descriptions of the Aswan High Dam mentioned above point out, this, in addition to the fact that at the present rate the reservoir will be filled with sediments in about 500 years, also causes the two possibly most severe impacts on the downstream area, namely the fact that the fields along the Nile no longer receive the fertilisation by deposits of silt brought by the floods, and that the Nile delta started eroding due to the lack of sediment. The lack of sediments fertilising the fields led to a drop in fertility and made the increasing use of artificial fertiliser inevitable. However, while this is emphasised by all the descriptions of the project listed above, only one even mentions another issue of high importance: "By the end of the (19th) century, however, the population of Egypt increased beyond the capacity of already irrigated lands to provide food. Partly in consequence of this desperate need to increase irrigation, projects to measure and control Nile flows increased substantially with the British occupation of Egypt in 1882". (Wohl 2011:94); this had already been pointed out by Biswas (1995), when he mentioned that the high population pressure and the scarcity of land and water forced Egypt to use its resources to the extent possible, and that the Aswan dam had improved the availability of water and allowed for an increase in irrigated areas and agricultural production. Without pretending to say that Lake Nasser allows for a sustainable use of agricultural land and water it is certainly still correct to point out that the old (biblical) situation with the Nile fertilising the fields of Egypt and thus maintaining the wellbeing of its people as well was no longer sustainable, and that alternative solutions had to be sought. Just to illustrate this point: the population of Egypt was 10 million in the 1960s, when the dam was built, increased to 63 million in 1995 and 93.4 million in 2017 (PRB 2017), and might increase to 97.3 million in 2025 (Benazzou 1997). Concerning delta erosion, opinions and information about this issue vary widely, from alarmist statements to ones that consider the situation as manageable, as discussed briefly by Scudder (2005:229).

Just two other points have to be mention here: it is certainly true that a considerable amount of water is lost due to evaporation from the reservoir. However, if this water would continue to be just flushed through Egypt during the 3 months of the yearly flood season, this would certainly not help diminishing the water scarcity in this country. On the other hand, as Afifi (1993) pointed out, the reservoir saved Egypt

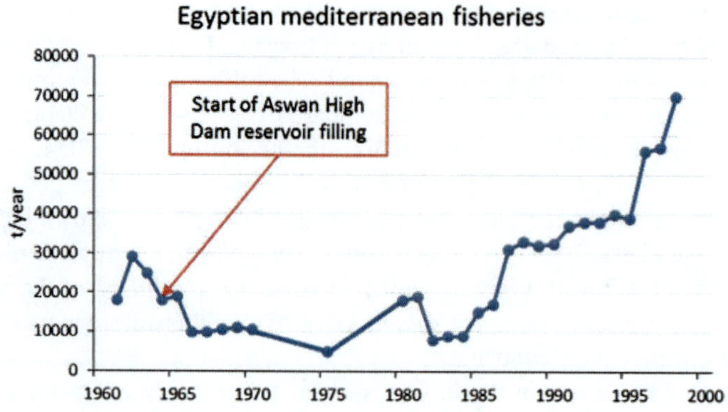

Fig. 21.2 Development of Egyptian Mediterranean fisheries 1960–2000. Based on data from Nixon (2003)

from the most severe effects of the drought North Africa experienced in the years from 1979 to 1988. The other point is fisheries: Awan dam is blamed for the crash of fisheries in the Mediterranean Sea in the region of the Nile delta; when it is mentioned e.g. that after the closure of the Aswan High Dam sardine catches dropped from 18,000 to less than a thousand tons per year (McCully 1996:45). In a careful analysis of the data Nixon (2003) showed that the decline in fish catches had started before the closure of the dam, i.e. before it could have affected nutrient input into this part of the Mediterranean in any way; it is very likely that overfishing in this period caused the drop in yield (Fig. 21.2). The paper also shows that catches have increased again, and this was probably caused, to a large extent, by the massive input of fertiliser from the intensified Egyptian agriculture.

All this does not necessarily mean that the Aswan High Dam was definitely the best possible solution for the burning problems of Egypt, or that it could be considered as a sustainable solution without taking additional measures. The impacts listed and discussed in all these publications are, without any doubt, to be taken seriously, and the way this project was conceived and pushed through, mainly also by rigorously silencing any criticism, especially from an environmental and social point of view, was far from what should have (and might have) been done. It might have been possible to come up with a better solution, one that would have had most or all the benefits of this project, but with considerably less negative impacts; the solution would certainly not have been as simple as the one that was proposed by Lavergne (1986), who suggested to stop using Lake Nasser as a storage reservoir to control river discharge and instead of that do storage in Lake Albert. Doing so might actually be difficult considering the fact that 86% of the water flowing into Lake Nasser stems from Ethiopia and not from the White Nile (e.g. Batisha 2012). However, two things have to be pointed out:

- When discussing any project, we should always try to consider positive and negative aspects before coming to a conclusion.

- In a changing world, just conserving the status quo is not a guarantee for a sustainable solution.

In their analysis of the effects of the High Aswan Dam, Biswas and Tortajada (2012) come to very similar conclusions. They point out that, while Egyptians consider the dam as an important improvement of the situation of the country, many voices from outside described it—and continue to do so—as a major ecological problem. They point out that there is a "danger of drawing simplistic conclusions because they often prove to be wrong".

21.1.2 Three Gorges HPP

The most extensively discussed and criticised HPP of the last 30 years or so, the one probably the largest number of people have heard of, is certainly the Three Gorges HPP on the Yangtze Kiang in China. The intention here is not to provide a thorough analysis of this truly gigantic project, it is mentioned here mainly to emphasise the issue of selecting and using information in the discussion around such a project.

This discussion at an international level took place mainly in the 1990s. Construction started in 1994, and the World Bank had pulled out of the project due to potential major environmental and social problems. Construction at the dam itself started in 1997 and was finalised in 2006, full capacity was reached in 2012. It is often called "the largest hydropower plant in the world", and it is certainly exactly that in terms of installed capacity (22,500 MW), and in terms of water passing through the structures (on average 14,300 m^3/s, 451 km^3/y) and of annual electricity output (95 TWh/y). On the other hand, with 181 m the dam is by no means among the highest; there was a number of far higher dams, and the highest, Nurek, in Tajikistan, with 300 m, had already been in operation for about 30 years at that time. The reservoir is very long, about 660 km, but very narrow, about 1.5 km on average, with a surface are of about 1000 km^2; this is certainly not small, but there are far larger reservoirs.

One interesting aspect of the discussions around the project was the fact that a large amount of wrong information—fake news as this would be called now—was circulating, illustrated here with the example of a newspaper article (SOZ 1997, citing information from International Rivers Network). In this article, among graphs illustrating the dam and some background information, there were the following points:

- Endangered animal species: Yangtze River Dolphin, Chinese Sturgeon, Yangtze Finless Porpoise, Giant Panda, Siberian Crane.
- Endangered cropland: "The fields going to be submerged today produce 40% of the Chinese wheat harvest and 70% of the Chinese rice harvest".

Some comments concerning the endangered species. All the species mentioned are listed as critically endangered by IUCN (https://www.iucnredlist.org/), meaning that a threat to any of these species is a serious issue. The Yangtze Sturgeon (*Acipenser dabryanus*), as well as another species, the Chinese Paddlefish (*Psephurus gladius*), which was not mentioned, are endangered. The sturgeon may by now be extinct. The Three Gorges dam has certainly contributed to the decrease in the populations of these species, together with other stress factors like e.g. overfishing, and other dams upstream of Three Gorges increasing fragmentation of the habitats of numerous fish species (Qiu 2012). The Siberian Crane (*Leucogeranus leucogeranus*) is threatened mainly through loss of wintering habitats, and at least some of this loss is caused by impacts of dam projects on wetlands. The main treat to the Yangtze Finless Porpoise (*Neophocaena asiaeorientalis*, the subspecies living in the river), seems to be fisheries, where animals of this species, like other small dolphins, die as "bycatch" (a euphemism if there ever was one) in fishing nets. An especially prominent case is—or was—the Chinese River Dolphin (*Lipotes vexillifer*). It was extremely rare before the construction of the Three Gorges dam started, affected by fishing activities ("bycatch"), water quality deterioration etc., and concerns were raised that the dam might lead to its extinction. It is now considered as extinct or probably extinct (IUCN). Although some sightings have been announced, there are no confirmed sightings since 2001, and this in spite of some effort focusing on finding it. It is always difficult to find a rare, seclusive species, especially in a habitat which does not favour observation, and it is nearly impossible to prove that such a species does no longer exist, but even such species just happen to be observed on rare occasions, so there is a high probability that it is actually extinct. The dam might have played a role in this, but it is certainly not helpful, also for the conservation of other species, just to put the blame on one project—or one impact—even if it is one of considerable extent. Remains the Giant Panda (*Ailuropoda melanoleuca*). Here, the situation is totally different. Figure 21.3 shows the present range of this species in comparison with

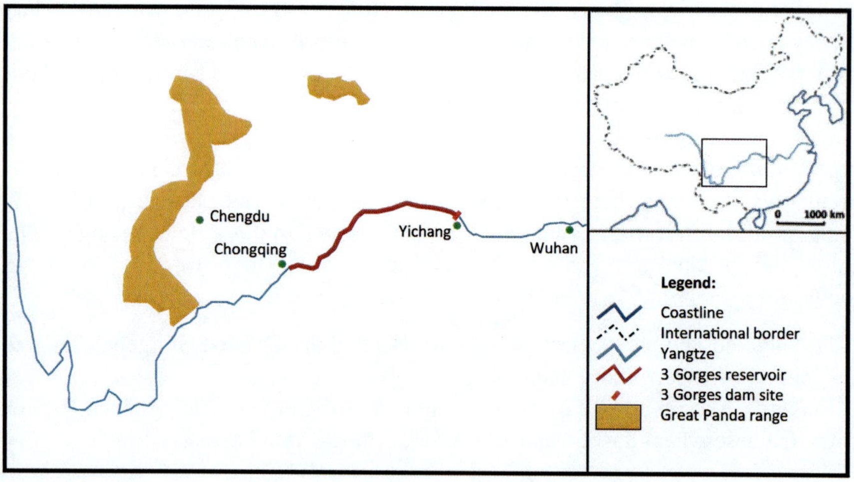

Fig. 21.3 Range of the great panda and 3 gorges HPP reservoir. For a more detailed map of the great panda range see https://www.iucnredlist.org/species/712/121745669

the Three Gorges reservoir. It is difficult to imagine how the project could possibly affect the Panda. Even its historical range, which was considerably larger (Chorn and Hoffmann 1978) did not, or just very marginally near the city of Chongqin, reach the area affected by the reservoir. So why mention the Panda in this context? If the focus is on endangered species, there seems to be sufficient matter of concern with the ones mentioned previously. It could just be that whoever put this information together was aware of the fact that the other species are not very prominent (in the sense of being known by the general public), while there is probably no other animal which is so easily identifiable and stands so much for nature protection like the Giant Panda, and the occasion of having this animal so "close" to the project area was just too good to let it pass. This does not make the argument any better, and certainly does not lend credibility to the anti-project campaign.

It is even worse with the information on agricultural land which was—or would have been—terrible news indeed, and it is difficult to imagine that any country would even consider to risk such losses. But it seems almost as incredible that a serious newspaper would write a thing like that without apparent quality control. The general thrust of the assertion was taken up in a letter from a group of NGOs addressed to the Swiss government (since the latter had considered participating in the financing of the project), stating that "tens of thousands of square kilometres of fertile land will be submerged". This was, most likely, just mixing up km^2 and ha: figures about lost farmland vary, depending on the source, between 16,000 and 28,000 ha; in fact, the 660 km long reservoir is, as mentioned, very narrow and covers a total area of about 1000 km^2, which corresponds to 100,000 ha. A figure of 28,000 might then qualify as "tens of thousands". The figures on lost crops most likely stem from a misinterpretation of a report which criticised the feasibility study for the project (Ryder 1990), which says: "The Yangtze River Valley, encompassing an area roughly one-fifth that of Canada's, is China's agricultural and industrial heartland. Supporting roughly 400 million people, one-third of China's population, the valley produces 40 percent of the nation's grain, 70% of its rice, and 40% of China's total industrial output". Obviously, there is a big difference between the river basin and the reservoir area. Most of this production is in the Yangtze floodplain downstream of the dam, and this latter, in addition to producing electricity and improving navigation, also has an important role in flood protection for the downstream area. In this area, four major floods in the twentieth century have caused close to 300,000 deaths, left millions homeless (Edmonds 1992; Liu and Xu 1999), and caused immense costs in terms of loss of harvests and damage to infrastructure.

This, by all means, is not intended to say that the project did not have any negative impacts; it certainly did. As described above, several critically endangered species were affected, and two of them are probably extinct by now, although it is not easy to say exactly what the part of the Three Gorges Dam was in this. But in any case, whether it is the largest hydropower plant in the world or not, it certainly holds one world record, namely, in the number of displaced persons. Like with other parameters related to this project, figures differ widely depending on the source consulted, covering a range of about 1–1.6 million; actual numbers were most likely between 1.2 and 1.4 million. This was an enormous task, mainly considering the fact that approximately half of that number was a rural population with a livelihood based on agriculture. The challenge, as always in such cases, is to find adequate areas of land which should be not too far away (same region, of specific importance where ethnic minority groups are concerned; same climatic conditions, and, as the relevant guidelines state, "at least of the same quality as the submerged land"). Obviously, as always, the farmers cultivated mostly the fertile alluvial soils in the vicinity of the river, and these are exactly the soils that are lost to submersion. Moving the people uphill, as was done to a great extent, is not a good idea, because land there is too steep, soils are not deep enough, and the land is by far not as productive. Problems were known right from the start, and the outcome finally was not satisfactory (Fearnside 1994; Chau 1995; Duan and Steil, 2003).

21.2 Sustainability of Hydropower

21.2.1 The Concept of Sustainability

The land belongs to the countless numbers wo are dead, the few who are living, and the multitudes of those yet to be born.

Sarawak, Borneo, native elder (Hong 1987: 80)

The discussion on whether hydropower could or should be considered as a sustainable form of resource use is ongoing, and different views are expressed (discussing it: Dorcey et al. 1997; WCD 2000; IHA 2010; Goldemberg and Lucon 2010:46; Jia 2010; Zwahlen 2011; Moran et al. 2018; affirmative: Frey and Linke 2002; negative: McCully 1996:140; Khagram 2004:211). Sometimes, small hydro is considered as sustainable, while "large hydro" is not (e.g. Goldemberg and Lucon 2010:47). One organisation, the International Hydropower Association (IHA) underlines the importance of sustainability, and the role hydropower can or has to play in this respect, e.g. in its 2020 Charter for Sustainable Hydropower (https://www.hydropower.org/who-we-are/iha-charter-for-sustainable-hydropower), without however attempting to give a simple definition of what is sustainable hydropower and what is not.

The original development of the concept stems from forestry and fisheries, and was first used under the term of "sustained yield", a concept first developed under the German term of "Nachhaltigkeit" (Leibundgut, 1970:190; FRIM 1990; Rietbergen 1993:4; Bächtold 1998), which meant the use of a resource, in this case forests and fish stocks, in a way that allows for an "endless" use of the resource, by harvesting only as much a will regrow in the same period, paying attention to not depleting the resource. While this concept is easy to understand and based on common sense, *Homo sapiens*, as we all know, does not seem to be very good in applying it: forests as well as fish stocks have been and continue to be ruined by overexploitation, and it is the same for other natural resources. Elephants are a good example: if used wisely, which for a species that calls itself "*sapiens*", wise, should be the normal way of doing things, ivory could be used in a sustainable way forever. Ivory is a wonderful material, and especially in Asia admirable skills have developed in the art of working this material. Unfortunately, greed and prospects for high gains over a short period led to a degree of exploitation which would eradicate all elephants in a rather short time, and the only way of hoping to prevent this from happening is a total ban on ivory trade, with seized ivory being burned. In this way, a potential source of revenue which could be generated from carefully managed elephant populations can no longer be used, and, maybe even more regrettable, the skills of working with ivory, which have been developed over centuries, may be lost. There are numerous examples showing that it is possible to sustainably hunt populations of large and slow-growing animals, as demonstrated e.g. for alligators (*Alligator mississippiensis*; Joanen et al. 2021).

Sustainability was defined in the so-called Brundtland Report: "Sustainable development is development that meets the needs of the present without compromising the ability of future generations to meet their own needs" (WCED 1987:43). Costanza et al. (1991:8) give the following definition: "Sustainability is a relationship between dynamic human economic systems and larger dynamic, but normally slower-changing ecological systems, in which (1) human life can continue indef initely, (2) human individuals can flourish, and (3) human cultures can develop; but in which effects of human activities remain within bounds, so as not to destroy the diversity, complexity and function of the ecological life support system". The definition of Dasgupta (2021:508) is: "Sustainable—A situation is sustainable if it can persist indefinitely. An unsustainable state of affairs cannot persist indefinitely". Otherwise, definitions are difficult to find, and one often has the impression that everybody talks about it, assuming that everybody knows what is meant. This was expressed in a perfect manner, almost ironically, although probably not meant to sound that way, allegedly by former British Prime Minister John Major in 1994: "Sustainable development is difficult to define. But the goal of sustainable development can guide future policy" (Irwin 2001:43). A good example for this type of using the term is presented by Razmjoo et al. (2020), who in a short paper mention "energy sustainability" about 20 times and point out its importance for "water sustainability", without actually saying what the importance is and what these two terms mean. A recent, useful overview of the sustainability concept (even with the strong focus on the situation in the USA, and also not proposing a definition) is provided by Portney (2015).

A useful description of the concept is the following: "Sustainability as traditionally used by natural resource specialists implies maintaining the status quo. Sustainable development implies improving human livelihoods while maintaining options to allow for adaptation to change. So our objective is not sustainability, it is sustainable development and that means evolution and adaptation, at all levels, from the individual cropping system to the entire planetary environment" (Sayer and Campbell 2004). This definition is satisfactory, but it should not make a difference between "sustainability" and "sustainable use": there is no such thing as a sustainable resource, it is only the use of the resource which is sustainable or not. Let's take the example of coal, or of any fossil fuel: as long as we do not use it, it will just remain where it is, whereas use, however small, will deplete the stock, which does not regenerate, at least not within a time frame that could be of any practical meaning to us: sustainability or sustainable use in the strict sense, in such a case, cannot be achieved, although e.g. for coal it might be argued that stocks will last for a very long time as long as use remains on a low level (Zwahlen 1995).

The situation is different for hydropower: using water for generating electricity does not destroy the resource, the water, once used, is still available in the downstream area and can basically fulfil all functions like irrigation, use as drinking or service water, elimination of waste, aquatic habitat etc. Water is an "undepletable resource" (Zwahlen 1995), meaning that it may be available in limited quantities at any given time and place, but will constantly be re-supplied. In this sense hydropower can be classified as "sustainable".

This, of course, would be a too simplistic approach. There are a few problems which at least make this reasoning questionable. For one, there is a limited number of suitable dam sites, and most of the very good ones have probably been developed already. This is certainly the case for Switzerland, where dam construction for generating power started over a hundred years ago. Hardly any sites suitable for medium-sized or large new dams are left, and where there are sites that might have a potential there is strong opposition against such ideas.

A recent case shows this quite clearly: the Trift glacier dam project. Trift is the name of a glacier in the Swiss Alps. In recent years, it has diminished dramatically, and where a few decades ago there was still a massive glacier there is now a small valley which contains a newly formed small lake. A plan was prepared to build a dam there for creating a reservoir which would allow to use the water flowing from the small catchment, and mainly to use it in winter, when demand for electricity is greatest. A 177 m high dam would form a reservoir with a volume of 85 M m^3 and an FSL of 1767 m asl. It would use a head of about 425 m, and the 80 MW turbine would produce 145 GWh of electricity per year (KWO, 2021; Stalder 2020). A group has now formed opposing this project on the ground that this untouched alpine landscape should be preserved (Häne 2019). One can certainly see it like this. On the other hand, it can also be argued that this would be a good project for different reasons. It is a landscape that changed dramatically due to the retreat of the glacier. There would be a visual impact in this landscape, the dam being clearly visible, but otherwise the impact would presumably be small: no destruction of a well-established lake ecosystem, no interruption of fish migration, no submerging of valuable alpine vegetation etc. And what is even more important: Switzerland has decided to abandon nuclear energy and to change to renewable electricity generation. Small hydro, which is sometimes considered as sustainable regardless of project-specific impacts, is developed, affecting a large number of small streams which so far have not been exploited. It takes a lot of small hydro for producing the amount of electricity that could be generated at Trift, and each of these small streams is a potentially valuable ecosystem. Seen like this, the choice seems to be obvious.

Another issue is the fact that dams and HPPs are not built for eternity. If well planned, built and maintained, they will last approximately 100 years; after that, structures may no longer be safe and will have to be removed or replaced; see e.g. the example of the Rheinfelden HPP on the border between Switzerland and Germany, described in Sect. 13.2. Even before that, there can be the problem that the reservoir is filled with sediments and will therefore no longer be productive; this was an issue in the case of Rogun HPP, discussed in Sect. 23.2.

An increasing number of dams has been removed recently, mainly in North America (see Sect. 13.4). In a number of cases, it has been observed that the river reverted to a situation rather close to the original conditions in a surprisingly short time. One could therefore argue that the river as well as the site are still there, and could possibly be used for a new structure or for any other purpose. The resource was not destroyed by use, which would indicate that the use was sustainable.

It might also be possible to define hydropower sustainability in relation with specific issues. In this sense, IHA has developed a tool (G-Res tool) for determining greenhouse gas emissions to be expected from a specific project (https://g-res.hydropower.org/). The definition applied there is that if lifetime emission, determined for a lifetime of an HPP of 100 years, is lower than 100 g CO_{2eq}/kWh, then it is a sustainable project. This value is certainly much lower than emission values for power plants using fossil fuel; on the other hand, if we compare it to average values for hydropower plants (see comparisons provided in Fig. 8.8 and in Table 14.3), the aim of 100 g may not seem to be too ambitious. Still, this value could be accepted as a guideline for sustainability with respect to GHG emissions, but obviously sustainability cannot be based on one parameter only.

Overall, if sites are chosen carefully, if unsuitable sites (unsuitable, here, not from an engineering, but from and environmental and social point of view) are left alone, if projects are being optimised and well maintained, and if water resources are being used in a way as to take into consideration not only hydropower, but all other relevant aspects as well, then hydropower can and should be seen as a sustainable form of resource use.

21.2.2 Combination with Other Renewable Energy Sources

Renewable energy is an important way forward in the efforts to limit and reduce GHG emissions (which need to be, as Gates 2021, repeatedly stresses, reduced to zero), and here solar and wind are the most relevant alternative sources. As every publication discussing pros and cons of these energy sources points out, they have one major drawback: they produce energy when they are available, not when energy is needed. Hydropower, or at least storage plants, do not have this problem, they allow for production whenever there is a requirement (at least as long as the reservoirs are full). One major requirement for wind and solar, therefore, is storage of energy. On a small scale, this can be done by batteries, but these are, at least for the time being, not sufficiently efficient for storing large amounts of energy, as is e.g. required for supplying a large city during a night, during a period of several days without wind or for allowing a reliable supply for peak consumption hours, let alone for an entire winter season. A battery which can have a capacity in this order of magnitude is a pump storage plant, where surplus energy is used for pumping water up into a reservoir, to use it later for generating electricity in exactly the moment when it is needed. In this way, hydropower and other renewables can form a very useful combination. This can be developed even further, e.g. by placing floating solar panels on a reservoir. In this way, electricity produced by the solar panels cold be used directly, as long as it is sufficient, saving the water for the time without sunshine, and whenever the solar panels produce a surplus of energy, this can be used for operating the pumps. A number of other advantages of such a combination of solar and hydro are mentioned in the literature: no additional space (i.e. terrestrial habitats; Leitner 2009) is required for placing the panels; the cooling effect of the water increases efficiency of the solar panels; and shading the reservoir can reduce algal growth and with it eutrophication of the reservoir (Alvoeiro and Allen 2019; Reithe and Grossmann 2019; McKuin et al. 2021).

21.2.3 The Times They Are A'Changin'

> *... for the times they are a'changin'*
> Bob Dylan 1964

One additional dimension has to be included in the discussion: it is not only the way a resource is being used, but also the intensity of the use, which can change over time. One can read descriptions of how carefully, and indeed sustainably, traditional societies were using their environment and its resources, an how we do not do that any longer (Titilola 1994; Aumeerruddy 1995; Basiago 1995; Mathias 1995; IUCN 1997). However, as e.g. Eibl-Eibesfeldt (1997:38) has pointed out, "harmony with nature" was mainly a consequence of small populations and, one could add, the absence of means enabling people of really causing harm; so e.g. it has been observed that at least some members of traditional societies were "in contradiction with some principles of their traditional ecological philosophy" once they became involved in the management of their forests and were given the means to exploit them (Charest 1995). The view that traditional resource use is or has been sustainable for extended periods is certainly true for any number of systems, but it is definitely not true universally; Diamond (2005) describes a number of cases where traditional use turned out not to be sustainable at all. The problem here is usually development over time, and related to demographic development of the societies in question.

This is an important fact to consider: times have changed, and they continue to do exactly that. This will be exemplified shortly for the issue concerning fisheries in the Mekong basin in Chapter 22. The basic reason for that, and the main issue, is population increase. Malthus (1798), Ehrlich (1971), Goldsmith and Allen (1972) and Meadows (1972) may have been wrong, but maybe rather they have just not been exactly right: even if what they prophesised did not occur in that way, population increase is still a problem, or certainly even the most basic and overarching problem (Dasgupta 2018; O'Sullivan, 2020).

Still, evoking the "good old times" doesn't make anything any better, and is usually wrong anyway. Wallace-Wells (2019:269) writes: "We are still now, in much of the world, shorter, sicker, and dying younger than our hunter-gatherer forebears, who were also, by the way, much better custodians of the planet on which we all live. And they watched over it for much longer—nearly all of those 200,000 years". In my view, all of this is wrong. While apparently human health decreased in general with the onset of agriculture (and according to Bodsen and Paine 1995, more so for women than for men), supposedly due to the fact that the hunter-gatherer diet had been much more varied and therefore healthier than the food of farmers mainly based on one or two main crops, average life expectancy at birth of the hunter-gatherers was still just about 25–40 years. In just the last two centuries, life expectancy has increased from this to about 67 worldwide, and to over 80 in the developed world (Ehrlich 2002:629; Oeppen and Vaupel 2002; Wahl and Heyl 2004:23; Harari 2011:57, 299; McCauley 2019). While it is true that modern hunter-gatherers, the few that still follow this lifestyle, can live to the age of 70 (Gurven and Kaplan 2007; McCauley 2019), this is true only for the relatively few who survived childhood, child mortality still being very high. And I doubt very much that humans, at any time than just very recently, had considered themselves as "custodians" of the planet: people were just not numerous enough, and they did not have the technical means, to inflict too much damage on the earth, but even so, looking at the human-aided if not human-provoked extinction of the megafauna in Australia and America, they definitely did leave traces. The "noble savage" approach, in the end, does not bring us any further (Zwahlen 1996).

Looking at traditional practices one can find numerous examples which over time turned into a problem. The so-called "caterpillar fungus" (*Ophiocordyceps sinensis*) is a fungus living in the Himalayan region as a parasite in the caterpillars of a moth which lives in the soil, feeding on plant roots. Once the caterpillar is dead, the fungus grows into an above-ground fruiting body, which is used as a traditional medicine. This medicine has turned into a much sought-after commodity, and exploitation of this resource leads to severe damage to mountain pastures as well as to conflicts between local herders and collectors coming from outside (Jamtsho and Katel 2019; Tan 2020). In addition to fisheries, there is another traditional practice which illustrates this problem very well: swidden or slash-and-burn cultivation. The process is simple: a farmer cuts a bit of forest, burns the vegetation after having it left to dry, and then plants his crops there, e.g. upland rice, usually planting a number of other crops mixed into it (intercropping). In the following year, he cuts an additional plot for his rice. Last year's field might provide one more harvest, depending on the situation and mainly on soil conditions, and some other plants grown there, biannual or perennial ones, will continue to produce. In the third year, soils will in most places be too exhausted for another good crop, and weeds will have taken over the plot to a great extent, but some plants will still provide at least a little bit of fruit or tubers. The first plot will then be abandoned, forest plants will take over, and regrowth will turn into secondary forest. The cycle will continue. After a suitable fallow period, the farmer will return to a plot used earlier. Fifteen years of fallow seem to be, at least in many cases, a suitable fallow period: in this time soil fertility will have been restored, the vegetation will be sufficient as to produce enough fertilising ash when burned, and it is easier to fell secondary rather than primary forest (as described e.g. by Coomes et al. 2021). This practice was used in probably all tropical forest areas for millennia, without negatively affecting the forests: a truly sustainable form of use. However, over time—and often under changing conditions, one of which was access to medical services, be it even at a relatively modest level—population increased, while agriculture often remained the same. This led first to an increase in the area under cultivation, and finally, when simply suitable land became scarce, to a decrease in the length of the fallow cycle. Floret and Serpantié (1991) provide a good overview of the importance of the fallow cycle, with many case studies mainly from Africa. With maybe only two to five years of fallow remaining, soil fertility cannot regenerate to a satisfactory level, and harvests decreased, forcing people to increasingly use land of less suitability, and to further reduce the fallow cycle. According to Teng (1990), swidden cultivation with 20 years fallow can produce 1 t of rice per ha without requiring any fertiliser, and this drops to 300–350 kg with a fallow cycle of only 3–5 years. Under changing conditions, a very traditional and very sustainable way of resource use had turned into a very destructive and totally unsustainable one. Figure 21.4 illustrates the two conditions. The picture on the left shows a few houses and their fields in a vast area of primary forest in Long Malim, a small settlement located at the upper end of the Murum HPP in Sarawak (Malaysian part of Borneo) in 1993. It was a settlement of Penan people, a small ethnic minority living in the forests of Borneo. This group was no longer following the traditional life style as forest nomads, whose livelihood was entirely based on hunting and gathering, as was described in detail by Manser (1992). They lived in a type of longhouse, a housing stile that had been taken over from the much more numerous Dayak, and they practiced agriculture. Still, they lived isolated and far away from the other, dominant ethnic groups inhabiting the downstream area. With the intensity shown there, slash and burn is certainly sustainable. The picture on the right shows a situation in Laos: people encroaching on forest in a protected area, continuously advancing more into

Fig. 21.4 Slash-and-burn cultivation: from sustainable to destructive. Left slash and burn cultivation in Long Malim, Sarawak. Right: slash and burn cultivation encroaching on a protected area in Laos

the forest and using the soil, including increasingly steep, i.e. more erosion prone land; note that this encroachment was in no way related to any hydropower project.

This may, at first sight, not have anything to do with hydropower. However, hydropower projects can play a very decisive role in this context. As mentioned in Sect. 11.1, drowning a valley in a reservoir will in very many cases submerge a considerable area of alluvial soils. These are usually the most productive soils, much less prone to erosion and exhaustion than soils on hill slopes, and often the only soils available which can be used continuously, without requiring a prolonged fallow period. In many places, these soils are being cultivated in the form of paddy fields (irrigated rice), providing not only staple food, but very often also cash income from selling the surplus. Today, if such soils are lost, e.g. to submersion, it is as good as impossible in most places to find adequate in-kind replacement, since all such soils are already used by someone. If people affected in such a way are simply moved to higher elevations, there is a high risk that this will force them into slash-and-burn cultivation, with all the negative consequences just outlined, and with a major risk of driving these farmers into poverty.

Today, mankind definitely has a very marked effect on the environment on a global scale, and in a way which goes far beyond the two aspects of unsustainable agriculture and fisheries. Therefore, we actually should be "custodians of the planet", since we are causing the problems, we know what exactly causes the problems, and we have the means to do something about it. However, in most cases we do not really act as custodians, or at the very least we make a bad job of it (see Syvitski et al. 2020).

Returning to the "good old times", as mentioned, is not an option, and neither is continuing with business as usual. We need to make efforts for reaching sustainability, but in doing so we have to take reality—that of today and not that of sometime in the past—into account and look for adequate solutions. As Diamond (1999) puts it, although he considers the development of agriculture as the "worst mistake in human history", archaeology is "offering no lessons for the present": we are simply too numerous to go back to the life of hunter-gatherers! Ellis et al. (2021) point out the fact that already 10,000 years ago there was hardly any "wildlands" left on earth, whereby this term is understood as a terrestrial habitat which is completely free of any human interference, and that land inhabited and used during millennia by traditional societies have been very rich in terms or biodiversity. They deplore the loss of traditional land use knowledge and practices and point out the importance of such practices for biodiversity conservation. However, as they do not say explicitly but show clearly in their illustrations: the real problems were closely linked to population increase. Today, there are just too many of us, there is no way back. A possible way forward, with a very important focus on energy, is outlined by Gates (2021).

21.2.4 Towards Defining Hydropower Sustainability

Translating this into hydropower means to make wise choices about what potentially feasible projects to implement and what not, and to make the implemented ones as good—technically, economically, but also environmentally and socially—as we possibly can. This would make hydropower really sustainable, and this also includes looking for and finding alternatives where these may be the better option. For specific applications or cases, general definitions of sustainability as those mentioned in Sect. 21.2.1 are not very useful. More precise and applicable criteria have to be defined for determining the sustainability of an action or a project. The EU is defining a classification system, "the so-called Taxonomy—to determine whether an economic activity is environmentally sustainable" (EU TEG 2019:15). On page 19, this report provides the following definition:

"For an action to meet the definition of an 'environmentally sustainable economic activity' and thus be considered Taxonomy-eligible, it must:

1. Contribute substantially to one or more of the environmental objectives
2. Do no significant harm to any other environmental objective
3. Comply with minimum social safeguards (under the draft regulation, these are defined as ILO core labour conventions).
4. Comply with the technical screening criteria".

Here, an attempt is made to list the relevant criteria to be met to make hydropower sustainable:

- Reaching the economic goals; this is not an environmental requirement, but building a hydropower plant which does not fulfil its economic goals can certainly not be considered as sustainable; WCD (2000) lists and discusses a number of such projects.
- Substitution of fossil fuel; this is a point that can be considered as fulfilled by any hydropower project by just assuming that otherwise the energy produced there would be generated by using fossil fuel.
- Emission, over the lifetime of the project, of less GHGs than would be emitted from fossil fuel for generating the same amount of energy; here, the value of 100 g CO_{2eq}/kWh for a lifetime of 100 years as proposed by IHA could be taken as a threshold value. EU TEG (2019:245): "Facilities operating at life cycle emissions lower than 100 g CO_{2eq}/kWh, declining to 0 g CO_{2eq}/kWh by 2050, are eligible".
- Maximising the life span with the aim of reaching 100 years, which is feasible provided it was planned and built according to a state of the art level, and well maintained throughout.
- Maximise energy density, expressed as the relation of W per m^2 of land submerged; a threshold of 5 W/m^2 is proposed (Climate Bonds 2021).
- Planning and implementing a project in a way which will allow to continue to use the site after the plant has reached the end of its useful life; continued use does not necessarily mean an ongoing use in the same way; basically, it can be a replacement with another structure, with one of the same type, or removal of the structure for restoring it to its previous condition.
- If for a project resettlement of people is inevitable, then compensation has to be made according to the recognised standards, the most important requirement there being that after resettlement people should be better off than they were without

the project. Specifically, no farmers should be forced into adopting slash-and-burn cultivation. A project cannot be considered as being sustainable if it pushes people into an unsustainable way of living (this is an example of adopting point 2 of the EU TEG 2019, definition given above).

- No net loss in biodiversity; this may require, in many cases, specific measures for fish migration, and/or biodiversity offset measures, as well as measures in the downstream area.
- For existing plants, whenever an upgrade is required, increase efficiency to the extent possible by using latest technology, and reduce existing impact, e.g. by providing an adequate ecological flow or by integrating fish pass facilities into an existing structure.
- Combine hydropower, where feasible, with other sources of renewable energy, e.g. solar and wind.
- Carefully plan and implement reservoirs for a combination with other forms of use (as water supply, irrigation, transport, fisheries, leisure, downstream water uses).
- Take efforts, in combination with all stakeholders involved, for protecting the watershed for reducing sedimentation of the reservoir and deterioration of water quality.

This list does not necessarily cover all aspects which might have to be considered in every specific case, and it does not allow for a clear separation of unsustainable and sustainable projects. However, the better the performance of any project in any of these criteria is, the closer it will be to reaching sustainability.

21.3 Wrong Arguments

Doch d Wält isch so perfid, dass si sech sälte oder nie
nach Bilder, wo mir vore gmacht hei, richtet.
Mani Matter, Swiss singer-songwriter
(However, the world is as mean as to seldom, if ever,
conform itself to images we made of it.)

Before going into this, it is necessary to emphasise very clearly once again what has been said several times already: while in general hydropower can be considered as a good and environmentally friendly form of generating electricity, and while dams and reservoirs play a decisive role in this, one has to be fully aware of the fact that every dam and hydropower project inevitably also has negative impacts. And, just as important: not every technically and economically feasible dam project is also a good project, and there are projects that should be (or should have been) abandoned. Some of these points were expressed in a paper entitled "How to maintain the bad reputation of dam projects" (Zwahlen 2011). Resistance against "bad" and controversial projects has to be expected and can be fully understood.

This section lists a number of "wrong arguments" that I have come across, be it in the literature or in my own experience. These are arguments which were brought up in the discussion about hydropower and dams, usually with the aim to show what kind of adverse effects such project have, but which never became reality. I call these "cry-wolf-arguments". I am of the opinion that they invariably backfire: when the public realises that some extremely dramatic effects are, at best, largely exaggerated or do not materialise at all, this can only have the effect of reducing the overall credibility of the author who brought them up.

21.3.1 Cry Wolf

On a larger scale, not limited to, and not even focussing at, dams and hydropower, there were, and there continue to pop up, doomsday prophecies. Here, it is not about the foretelling of the end of the earth, examples of which are too numerous to keep track of, like the one predicting that the apocalypse would happen on July 14, 1960, at 13.45 h, and that only those people gathering at that time on the Mont Blanc, in the French Alps, would be saved. Not surprisingly, a few hundred (or maybe a few thousand) actually did climb the mountain on that day, and many more waited in terror for the terrible things to happen. Well, as we know, and as was the case on so many other similar occasions, it did not happen, but it is always amazing to see that, however absurd a theory is, there are always some people ready to believe it. However, there are serious studies who came to more or less doomsday-like conclusions, some of which are listed and commented here shortly.

- Malthus 1798: in a way, he could be seen as the "father of the doomsday prophets", at least of the modern ones. In his essay, he set forth the theory that population growth is only checked by limited supply, mainly of food, and that therefore every increase in food production would inevitably lead to a stronger increase in population, which would more than compensate the better supply.
- Ehrlich 1971: in his famous book "The population bomb", he went in the same direction as Malthus, predicting mass famines in the near future, with hundreds of millions of people dying from hunger.
- Meadows 1972: this Club of Rome report focussed on the risks of depleting resources of all kind, consumption increasing due to population growth, and depletion ultimately leading to a crash in the human population.

These three publications, which I all consider as very important, all basically reflect the same situation: there is a problem—increasing human population as the most pressing issue—and the scenarios developed were more or less simple extrapolations of the existing situation into the future. All the potential alternative development paths like new technologies, improved crop varieties, development of alternative energy sources etc., were not—and could not have been—taken into consideration.

There are others. In the 1970s, before it became increasingly clear that global warming would be the trend, we were warned about an imminent new ice age.

Shortly afterwards, there was the forest dieback ("Waldsterben"); while there was definitely a problem, mainly increasing acidity of rain and soils due to air pollution, some alarmist voices at that time sounded as if very soon, at least in Central Europe, there would be no forests left at all. There was desertification, and the ozone hole. The latest I have heard of are the warning that in 60 years agricultural soils will have been completely lost worldwide, and the book by Wallace-Wells (2019) entitled "The uninhabitable earth", which starts with the sentence "It is worse, much worse, than you think".

Most of these publications and movements reflect a serious concern, and one which is very real and obvious: no unlimited growth is possible within a closed and limited system as our earth undoubtedly is (Mohr 1995:65). This basic concern receives continuous attention (concerning questions around "Malthusian" and "non-Malthusian" approaches see e.g. Ayres 1993; McNicoll 1998; Jackson and Victor 2019; Rees 2020; Vesco et al. 2020). However, the individual topics made it into the news, very prominently for most of them, for some time, and then just disappeared as if they had never been there.

The problem here is the alarmist nature of many of these publications etc., their "cry wolf" attitude. This may seem required in order to attract the due attention to the serious issues raised, but it inevitably leads to the reaction described in the original tale of the boy who got bored when he had to look after the cows of the village and thought that it would be fun to see the reaction of the villagers to an alarm cry: one gets tired of it and stops to pay attention. As Shermer (2018:155) very aptly put it: "Beware the prophet who claims the end of the world, the apocalypse, doomsday or judgement day is upon us".

One note in addition to that: I would not call the 1972 Club of Rome report as alarmist. The authors were very conscious about the limits of their model and of their data, and already in the introduction they make a clear statement that what they calculate is likely to happen "if present conditions continue", but that there are alternatives. It was then turned into just another doomsday story mainly by press articles, first pointing out the great danger humankind is facing, and later on blaming the study for "not having been right". I am quite sure, however, that at least to the same extent it was a self-destroying prophecy as mentioned in Sect. 1.2.

There is a new addition to this series of books, which points out more than once the seriousness of the situation, and specifically of climate warming, e.g. by saying things like "Yet if nothing else changes, the world will keep producing greenhouse gases, climate change will keep getting worse, and the impact on humans will in all likelihood be catastrophic" (Gates 2021:4). However, this book is not about the end of the world, but about what we need to do to prevent this from happening.

21.3.2 The Risk in Using Models

The examples mentioned above, in one way or another, reflect the outcome of model calculations, whereby such models can be mere thought models or ones based on sophisticated calculations using a large amount of data.

As I used to say to my students when we talked about applying computer models in the context of impact assessment: if the model predicts the end of the world, there certainly is something wrong with the model or, more probably, with the input data used, and not with the world.

This is not meant to say that models should not be used in our approaches. Many such models, although not necessarily reflecting the development of the entire world ecology over the next 100 years, are well-tested and very useful tools for assessing probable outcomes in the future, when some conditions will have changed. This applies mainly to models for calculating noise emissions, the distribution of contaminant plumes in air or water, or, of high importance in our context, hydraulic models. Even in biology, models based on comparative data on other species can be used to determine such highly relevant parameters as the population density of Loch Ness monsters (Sheldon and Kerr 1972) or the total abundance of *Tyrannosaurus rex* (Marshall et al. 2021). I have, admittedly, much less confidence in ecosystem models, see above, or in econometric models. With the latter I had an experience which might shed some light on my reluctance to believe in their results. I participated in a comprehensive study, which had the objective to assess the potential effects of an adherence of Switzerland to the European Union (Liniger and Zwahlen 1992; Hanser et al. 1995). Three scenarios had to be compared: integration into the European common market, full membership in the European Union, or continuing independently (which, ultimately, was the way that Switzerland chose). One important point was the different economic development within these three scenarios and their consequence for the environment. Our assessment of this was done on the basis of an econometric model which had been developed and applied by a university institute. This model predicted development of different sectors of the economy in terms of percentage of the total work force in the three scenarios, and the outcome was that, e.g., in 1988 2.2% of the total work force were employed in the chemical industry, and this would change till the year 2000 – 2.0% (independent way), 2.1% (adherence to the common market), and 2.3% (full EU integration). This was a kind of a "black box" model, nobody with the exception of the authors new what the model did and what kind of data were used to feed it, and I was really very surprised that with all the discussions we had with all kinds of experts during the preparation of our study, nobody ever questioned the results of this econometric model. This is why I told my colleagues, when we submitted the final report, that in fact our conclusions should end with the sentence "However, the outcome could be very different".

Overall, I think that I have confidence in results reached by the application of a (sophisticated) model when I have the impression (gained on the basis of the document where its application and results are described) that the person(s) applying the model (i) understand what the model is doing, and what its limitations are, (ii) have the required knowledge about and understanding of the system, e.g. a river, to which the model is applied, and (iii) have the data required for actually using the model; in fact, on several occasions I had the impression that condition (ii), but especially condition (iii) were not fulfilled, this latter meaning that a model was applied using a largely insufficient database. But mainly, I tend to accept the results of a modelling when I have the impression that they are plausible. If this is not the case, if a model provides highly surprising results, this does not mean that it is necessarily wrong, but a really good explanation why this is or will be as described would be required in such a case. Therefore: whenever you use a model, on looking at the results, lean back for a moment and ask yourself: is this plausible? And if you have doubts, check again!

21.3.3 Earth Rotation

In the prologue to his book entitled "Deep Water: the epic struggle over dams, displaced people and the environment" Jaques Leslie writes: "The world's dams have shifted so much weight that geophysicists believe they have slightly altered the speed of the earth's rotation, the tilt of its axis, and the shape of its gravitational field" (Leslie 2005: 4). This topic is then not discussed any further, but the fact that it is mentioned at all, in the very introduction to the book, and even repeated on its back cover, seems to point out that here we are facing a potentially very serious problem. A similar, even more alarming statement can be found somewhere else, which says: "Billions and billions of tons of what had been free-flowing water is now sequestered in set locations around the middle latitudes, moving it away from the equator. Furthermore, most of the dams are at high altitudes, not sea level. This shift of mass towards the axis has fundamentally changed the wobble of the earth. It's not a drastic shift, something on the order of one tenth of a microsecond, but it has been proved that we, in our hubris, have the power to change even the motion of the earth itself". https://theglyptodon.wordpress.com/2012/04/21/whoa-dam-water-reservoirs-shift-earths-rotation/. A similar statement, although less alarmist, is made by Nilsson (2009:215), who mentions the influence of reservoirs on earth rotation and concludes that "no other human activity has been big enough to cause any appreciable alteration in these global phenomena".

These statements are based on the following: geophysical research in the mid-1990s has shown that since the early 1950s 88 large dams had been built with a storage volume corresponding to 10 billion tons each at least. Tidal drag caused by the moon slows down earth rotation, while moving large amounts of water closer to the earth's axis would tend to speed the planet's spin. "Without lunar tidal drag, the reservoir effect would have reduced the length of a day by 0.2 millionths of a second a day for the last 40 years. ... These effects are several hundred times smaller than natural variations in Earth's motion" (Browne 1996, in a NYT article citing Benjamin Fong Chao, the geophysicist who made these calculations). This means that in 60 years, i.e. until approximately today, day length would have been reduced by 0.0044 s. Two things I was wondering: to what degree has this effect been counterbalanced by the shift of weight from high mountains in mid to high latitudes back to the seas due to the melting glaciers, and what was the situation during the height of glaciation, when really a lot of water had been shifted from the seas to just such places especially in the Northern hemisphere? For the first question one can find an answer in Zemp et al. (2019), who estimate that between 1961 and 2016, without including Greenland and Antarctica, major glaciers have lost a total of 9.6 Gt (gigatons, 1 Gt = 10^{12} kg) of water, and mention that about as much again was lost from the Greenland ice sheet and approximately a third again of that amount from Antarctica. According to Nilsson (2009:218), the water stored in reservoirs in 1997 corresponded to 13.6 mm of sea level equivalent, while at the maximum of the last glaciation the sea level was up to 150 m lower than it is today (Whitten et al. 1987:16; Reich 2018:160; Corlett 2019:5), and the rise in sea level due to climate warming has been 20 cm in the period 1901–2018 (IPCC, 2021:6). For me, identification of the "impact" of reservoirs on earth rotation mainly shows the degree of exactitude with which geophysicists can make measurements, and the sophistication of their calculations, but I do not think that we have to worry about dam effects in this respect, maybe even in the contrary: Frederikse et al. (2020:369) state that "Reservoir impoundment caused a substantial, although temporary, slowdown of GMSL (global mean sea level) rise during the 1970s". Voosen (2020) reports that between 1940 and 1960 the rise in sea level slowed down from 2 to about 0.5 mm per year, a development to which "dam building may have contributed", but might now be up to 4.8 mm annually. No amount of dam building will ever be able to counterbalance that effect.

21.3.4 The Kingfisher

A coalition of several NGOs had formed to organise the opposition against the Ilisu Dam Project in Turkey (see Sect. 26.3 for more details on this project). Here, one point of overall minor importance is picked out.

Among the wide variety of topics investigated and discussed in the framework of the project, there was also its impact on fauna. In one of the meetings held with the presence of the project proponents and the ECAs (which were involved in the project for a potential financing), representatives of one of the participating NGOs presented a new argument against this project: implementation of the project could threaten and potentially extinguish the entire European population the pied kingfisher (*Ceryle rudis*). Like all kingfishers, this is a bird species which lives in the vicinity of water, mainly lakes, ponds and other stagnant water bodies, and also along usually larger, slow flowing rivers, where it finds its prey, small fish that are caught by diving. In recent years, one or two pairs of this species had established themselves in the vicinity of the Dicle River (as the Tigris is called in Turkey) in the area of the future reservoir. The concern was now that these two pairs, who represented this "entire European population", could lose their habitat and disappear. And this was not announced as a

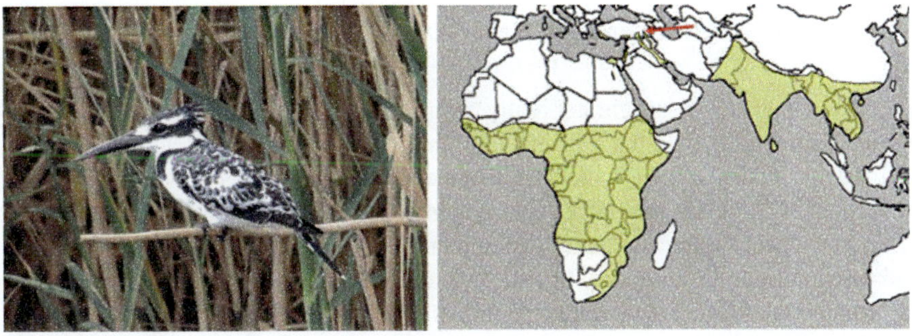

Fig. 21.5 The pied kingfisher (*Ceryle rudis*) and its geographic range. Map based on: https://www.iucnredlist.org/species/22683645/110626699#geographic-range. The red arrow indicates the Ilisu site. Photo of a pied kingfisher in Akagera National Park, Rwanda

risk to "two breeding pairs of the pied kingfisher", but as a risk to "its entire European population".

I had—and still have, a number of problems with this argument:

- These were two pairs breeding at an extreme edge of their area of distribution, as far north-west as they ever went. Every biologist dealing with the distribution of species knows that populations at the edges of their occurrence are very vulnerable and unstable: if there is a series of a few years with favourable conditions, they can become established and maybe even reproduce; however, one single or a few less favourable years may lead to the disappearance of the species from this area.
- Ilisu is located in South-East Anatolia, and this is usually not identified as being a part of Europe, neither geographically, politically or zoogeographically.
- The most important point, however: the pied kingfisher is probably the most numerous of all kingfisher species, and certainly the one with the widest distribution: from Senegal to Viet Nam, and from Northern India to South Africa (see Fig. 21.5).

It is possible that a dam project actually threatens a species, or an important population of a species, and this can happen mostly in the case of endemic species, i.e. species which have a very restricted distribution and are often dependent of very specific habitats, as e.g. the spray toad, briefly discussed in Sect. 18.3. In such a case every possible measure has to be taken to preserve an affected species, and it could be (or at least: should be considered as) a no-go for the project if measures cannot grant the survival of such a species. In the case of the pied kingfisher, however, it was certainly a wrong argument, and the consequence of such wrong arguments is just that in similar cases, but real ones, this will no longer be taken seriously.

21.3.5 The Jaguar and the Reservoir

The following example might not really be alarmist, but is just another example of a wrong argument, this time not against, but in favour of a dam project.

I was asked once by a colleague to give him some comments on a report he was about to finalise. He had been on a mission to Belize for a short appraisal of the prospects for hydropower development, and especially for one existing dam and a second one in the feasibility stage, both of which were contested by a local environmental NGO. There were two things that I especially noticed. The first one was the fact that he did not mention the NGO in question at all. I asked him whether he had talked to them, and what the outcome was; his answer was very short and clear: "You cannot talk to these people". This is an argument I have heard several times, expressed more or less in this way, and with which I do not agree at all: you can always talk "to these people", even if you may not always agree. I'll elaborate on this a bit more (see Sect. 25.1). The second was a photograph in the report showing footprints of an animal. The legend to this illustration said "footprints of a young jaguar in the drawdown area of the existing reservoir". The implicit message of this, obviously, was clear: the jaguar (*Panthera onca*) is a rare, threatened and usually very shy and seclusive species. If this shy animal likes the surroundings of the reservoir, this means that the reservoir is good for the environment, and especially for the fauna of the project area, and if one reservoir is good, then two reservoirs must even be better. However, looking at the picture it was immediately clear to me that these were by no means the footprints of a jaguar: it was undoubtedly a dog. I pointed this out to him, and needless to say I was not really pleased to notice that in the final, submitted version of his report there was a picture showing the "footprints of a young jaguar". However, there was a point that upset me more than the refusal of my colleague to believe me. When it comes to check on hydropower projects and the possibility to implement additional ones, I fully appreciate that a dam or hydropower engineer (which he was) has to participate in this exercise. What I do not appreciate at all is the fact that an organisation, which had "sustainability something" in its name, and in the case of projects criticised for environmental reasons, apparently was of the opinion that an engineer could also, of course, pronounce himself competently on environmental issues. I am firmly of the opinion that in such cases, and this especially also when it comes to first identification of potential dam and hydropower sites (see Sect. 5.5), in addition to the engineer an expert in environmental (and/or, depending on the situation, social) issues must be included in the team.

References

Afifi, A.K. 1993. Role of Asswan High Dam in safeguarding Egypt from floods and droughts. Internat. Symposium on High Asswan Dam vital achievements, fully controlled. ICOLD: 139–159

Alvoeiro, J., and Allen, S.M., 2019. Mitigation of drought impacts on reservoirs using hybrid solar/hydropower. Paper 22.01 presented at the Africa 2019 Hydropower Conference, Namibia: Windhoek.

Aumeerruddy, Y. 1995. Perceiving and managing natural resources in Kerinci, Sumatra. *Nature and Resources* 31(1): 28–37.

Ayres, R.U. 1993. Cowboys, cornucopians and long-run sustainability. *Ecological Economics* 8 (3): 189–207.

Bächtold, H.G. 1998. Nachhaltigkeit: Herkunft und definition eines komplexen Begriffs. *Schweizer Ingenieur und Architekt* 116(13): 4–7

Basiago, A.D. 1995. Sustainable development in Indonesia: Case study of an indigenous regime of environmental law and policy. *International Journal of Sustainable Development and World Ecology* 2 (3): 199–211.

Batisha, A.F. 2012. Hydrology of the Nile River basin in the era of climate changes. *Irrigation and Drainage Systems Engineering* S5: e001. https://doi.org/10.4172/2168-9768.S5-e001.

Benazzou, C. 1997. Le profil démographique des pays méditerranéens à l'horizon 2025. Dans: *L'Annuaire de la Méditerrannée 1997*, ed. El Malki, H., 148–157. Rabat/Paris: GERM-PUBLISUD.

Biswas, A.K. 1995. Environmental sustainability of Egyptian agriculture: Problems and perspectives. *Ambio* 24 (1): 16–20.

Biswas, A.K., and Tortajada, C. 2012. Impacts of the high Aswan dam. In: *Impacts of large dams: A global assessment*, 379–395. Chapter 17, ed. Tortajada, C., Altinbilek, D., and Biswas, A.K., Berlin: Springer. https://doi.org/10.1007/978-3-642-23571-9_17.

Bloch, A. 1980. *Murphy's law book two: More reasons why things go wrong*, 96. Magnum Books, London.

Bodsen, J.L., and Paine, R.R. 1995. The evolution of human longevity from the Mesolithic to the middle ages: An analysis based on skeletal data. In: *Exceptional longevity: From prehistory to the present*, 169. ed. Jeune. B., and Vaupel, W., Monographs on Population Ageing 2. Odense University Press. https://www.demogr.mpg.de/Papers/Books/Monograph2/start.htm

Browne, M.W. 1996. Dams for water supply are altering earth's orbit, expert says. New York Times, March 3, 1996. https://www.nytimes.com/1996/03/03/news/dams-for-water-supply-are-altering-earth-s-orbit-expert-says.html.

Charest, P. 1995. Aboriginal alternatives to megaprojects and their environmental and social impacts. *Impact Assessment* 13 (4): 371–386.

Chau Kwai-Cheong. 1995. The three-gorges project of China: Resettlement prospects and problems. *AMBIO* 24(2): 98–102.

Chorn, J., and Hoffmann, R.S. 1978. Ailuropoda melanoleuca. *Mammalian Species* (110): 1–6.

Climate Bonds. 2021. Hydropower criteria. Development of eligibility criteria for the climate bonds standard and certification scheme. Background paper. https://www.climatebonds.net/standard/hydropower.

Coomes, O.T., Cheng, Y. Takasaki, Y., and Abizaid, C. 2021. What drives clearing of old-growth forest over secondary forests in tropical shifting cultivation systems? Evidence from the peruvian amazon. *Ecological Economics* 189. https://doi.org/10.1016/j.ecolecon.2021.107170.

Corlett, R.T. 2019. *The ecology of tropical East Asia*, 3rd ed., 320 pp. Oxford, UK: Oxford University Press.

Costanza, R. (ed.) 1991. *Ecological economics: the science and management of sustainability*, 525 pp. New York: Columbia Universtiy Press.

Costanza, R., Daly, H.E., and Bartholomew, J.A. 1991. Goals, agenda, and policy recommendations for ecological economics. In: *Ecological Economics*, 1–20. ed. Costanza, R., New York: Columbia University Press.

Dasgupta, P. 2018. Global population size and the United Nations' sustainable development goals. *Ecology, Economy and Society—the INSEE Journal* 1 (1): 11–16.

Dasgupta, R. 2021. The economics of biodiversity: the Dasgupta review. London: HM Treasury, 604 pp. http://www.gov.uk/official-documents

Diamond, J. 1999. The worst mistake in the history of the human race. *Discover*. https://www.discovermagazine.com/planet-earth/the-worst-mistake-in-the-history-of-the-human-race.

Diamond, J. 2005. *Collapse: How societies chose to fail or survive*, 575 pp. London: Penguin.

Dorcey, T., Steiner, A., Acreman, M., and Orlando, B. (eds.) 1997. *Large dams: Learning from the past, looking at the future. Workshop proceedings, Gland, Switzerland, April 11–12, 1997*, 145 pp. Gland and Washington: IUCN/The World Bank.

Duan, Y.F., and S. Steil. 2003. China three gorges project resettlement: Policy, planning and implementation. *Journal of Refugee Studies* 16 (4): 422–443.

Edmonds, R.L. 1992. The Sanxia (three gorges) project: The environmental argument surrounding China's super dam. *Global Ecology and Biogeography Letters* 2 (4): 105–125.

Ehrlich, P. 1971. *Die Bevölkerungsbombe*, 191 pp. München: Hanser.

Ehrlich, P.R. 2002. *Human natures: Genes, cultures and the human prospect*, 531 pp. New York: Penguin.

Eibl-Eibesfeldt, I. 1997. *Die Biologie des menschlichen Verhaltens: Grundriss der Humanethologie*, 1118 pp. 3. Auflage, Weyarn: Seehammer.

Ellis, E.C., Gauthier, N., Goldewijk, K.K., et al. 2021. People have shaped most of terrestrial nature for at least 12,000 years. *PNAS* 118(17). https://doi.org/10.1073/pnas.2023483118

EU TEG. 2019. Taxonomy technical report. Financing a sustainable European economy. https://ec.europa.eu/info/sites/default/files/business_economy_euro/banking_and_finance/documents/190618-sustainable-finance-teg-report-taxonomy_en.pdf.

Ezzat, M.N. 1993. Nile water flow, demand and development. International Symposium on High Asswan Dam vital achievements, fully controlled. ICOLD, 77–89.

Fearnside, P.M. 1994. The Canadian feasibility study of the three gorges dam proposed for China's Yangzi River: a grave embarrassment to the impact assessment profession. *Impact Assessment* 12(1): 21–57.

Floret, C., et Serpantié, G., (eds.) 1993. La jachère en Afrique de l'Ouest. Paris: Atelier International, Montpellier, du 2 au 5 Décembre 1991. ORSTOM, 494 pp.

Frederikse, T., Landerer, F., Caron, F., et al. 2020. The causes of sea level rise since 1900. *Nature* 584(7821): 393–397. https://www.nature.com/articles/s41586-020-2591-3.pdf

Frey, G.W., and D.M. Linke. 2002. Hydropower as a renewable and sustainable energy resource meeting global energy challenges in a reasonable way. *Energy Policy* 30 (14): 1261–1265.

FRIM. 1990. ITTO guidelines for the sustainable management of natural tropical forest. *ITTO Technical Series* 5: 18 pp.

Gates, B. 2021. *How to avoid a climate disaster*, 257 pp. New York and Toronto: Knopf.

Goldemberg, J., and O. Lucon. 2010. *Energy, environment and development*, 2nd ed., 448 pp. London: Earthscan.

Goldsmith, E., and R. Allen. 1972. *Planspiel zum Ueberleben*, 115 pp. Stuttgart: DVA.

Gurven, M., and H. Kaplan. 2007. Longevity among hunter-gatherers: A cross-cultural examination. *Population and Development Review* 33 (2): 321–365.

Häne, S. 2019. Grüner Bruderkampf am Gletscher. *Tagesanzeiger* 2019–08–09:4

Hanser, C., Kuster, J., Liniger, H.U., und Zwahlen, R. 1995. Europäische integration: Ökologische Konsequenzen für den Industrie-und Dienstleistungsstandort Schweiz? In: *Schweizerische Umweltpolitik im internationalen*, 14–41. Herausg. *Kontext* Jochimsen, M., und Kirchgässner, G. Basel: Birkhäuser.

Hanser, C., J. Kuster, J.P. Porchet, and R. Zwahlen. 1995. Incidences de l´intégration européenne sur l´environnement. *La Vie Economique* 68 (12): 20–28 (in French).

Harari, Y.N. 2011. *Sapiens: A brief history of humankind*, 498 pp. London: Vintage.

Hong, E. 1987. *Natives of Sarawak. Survival in Borneo's vanishing forests*, 265 pp. Malaysia: Institut Masyarakat.

Hughes, J.D. 2009. *An environmental history of the world: Humankind's changing role in the community of life*, 2nd ed., 306 pp. New York: Routledge.

IHA. 2010. Sustainability assessment protocol. New edition 2020. London: International Hydropower Association, 213 pp. https://www.hydrosustainability.org/assessment-protocol

IPCC. 2021. Summary for policymakers. In: *Climate Change 2021: The Physical Science Basis. Contribution of Working Group I to the Sixth Assessment Report of the Intergovernmental Panel on Climate Change*, ed. Masson-Delmotte, V., P. Zhai, A. Pirani, S.L. Connors, C. Péan, S. Berger,

N. Caud, Y. Chen, L. Goldfarb, M. I. Gomis, M. Huang, K. Leitzell, E. Lonnoy, J.B.R. Matthews, T. K. Maycock, T. Waterfield, O. Yelekçi, R. Yu and B. Zhou. Cambridge University Press. In Press. https://www.ipcc.ch/report/ar6/wg1/downloads/report/IPCC_AR6_WGI_SPM.pdf

Irwin, A. 2001. *Sociology and the environment*, 210 pp. Cambridge: Polity Press.

IUCN. 1997. *Indigenous peoples and sustainability. Cases and actions. IUCN indigenous peoples and conservation initiative, IUCN inter-commission task force on indigenous peoples*, 364 pp. Utrecht: International Books.

Jackson, T., and P.A. Victor. 2019. Unraveling the claim for (and against) green growth. *Science* 366 (6468): 950–951.

Jamtsho, Y., and Katel, O. 2019. Livestock depredation by snow leopard and Tibetan wolf: Implications for herders' livelihoods in Wangchuck centennial national park, Bhutan. *Pastroalism* 9(1): 1–10. https://doi.org/10.1186/s13570-018-0136-2

Jia, J. 2010. Rethinking the role of dams and hydropower in sustainable development. World Atlas 2010. *The International Journal of Hydropower and Dams* 8–9.

Joanen, T., M. Merchant, R. Griffith, J. Linscombe, and A. Guidry. 2021. Evalutation of effects of harvest on alligator populations in Louisiana. *Journal of Wildlife Management* 85 (4): 696–705. https://doi.org/10.1002/jwmg.22028.

Khagram, S. 2004. *Dams and development: Transnational struggles for water and power*, 270 pp. Cornell University Press, Ithaca.

KWO, 2021. Neubau Speichersee und Kraftwerk Trift. Brochure, Kraftwerke Oberhasli. www.grimselstrom.ch

Lassally-Jacob, V. 1996. Land-based strategies in dam-related resettlement programmes in Africa. In: *Understanding impoverishment: The consequences of development-induced displacement*, 187–199. ed. McDowell, C. Refugee and Forced Migration Studies Vol. 2. Oxford: Berghahn Books.

Lavergne, M. 1986 The seven deadly sins of Egypt's Aswan high dam. In: *The social and environmental effects of large dams*, 181–183. ed. Goldsmith, E., and Hildyard, N. Vol. 2: Case Studies.

Leibundgut, H. 1970. Der Wald. Eine Lebensgemeinschaft. Frauenfeld: Huber, 197 pp.

Leitner, P. 2009. The promise and peril of solar power. *The Wildlife Professional* 3 (1): 48–53.

Leslie, J. 2005. *Deep water: The epic struggle over dams, displaced people, and the environment*, 352 pp. New York: Picador.

Liniger, H.U., und Zwahlen, R. (eds.) 1992. Europäische integration und ökologische folgen für die schweiz. *Schriftenreihe Umwelt Nr.* 177: 364 pp. Bern: BUWAL.

Heng, Liu, and X. Zikai. 1999. Les inondartions du Yiang-Tsé-Kiang en 1998. *Nature Et Resources* 35 (3): 14–21.

Malthus, T. 1798. *An essay on the principle of population.* Digital: J. Johnson, St. Paul (Word version of the text, 91 pp).

Manser, B. 1992. *Stimmen aus dem Regenwald. Zeugnisse eines bedrohten Volkes.* Bern: WWF/Zytglogge, 299 S.

Marshall, C.R., Latorree, D.V., Wilson, C.J., Frank, T.M., Magoulick, K.M., Zimmt, J.B., and Poust. 2021. Absolute abundance and preservation rate of Tyrannosaurus rex. *Science* 372(6539): 284–287. https://doi.org/10.1126/science.abc8300.

Mathias, E. 1995. Framework for enhancing the use of indigenous knowledge. *Indigenous Knowledge and Development Monitor* 3 (2): 17–18.

McCauley, B. 2019. Life expectancy in hunter-gatherers. In: *Encyclopedia of evolutionary psychological science*, ed. Shackelford T. and Weekes-Shackelford, V. Cham: Springer. https://doi.org/10.1007/978-3-319-16999-6_2352-1

McCully, P. 1996. *Silenced rivers: The ecology and politics of large dams*, 350 pp. Zed Books, London.

McKuin, B., A. Zumkehr, J. Ta, R. Bales, J.H. Viers, T. Pathak, and J.E. Campbell. 2021. Energy and water co-benefits from covering canals with solar panels. *Nature Sustainability*. https://doi.org/10.1038/s41893-021-00693-8.

McNicoll, G. 1998. Malthus for the twenty-first century. *Population and Development Review* 24 (2): 309–316.

Meadows, D. 1972. *Die Grenzen des Wachstums. Bericht des Club of Rome zur Lage der Menschheit*. Stuttgart: DVA, 180 pp.

Mohr, H. 1995. *Natur und Moral: Ethik in der Biologie. Dimensionen der Modernen Biologie Bd*, 4: 191 pp. Darmstadt: Wiss. Buchges.

Moran, E.F., Lopez, M.C., Moore, N., Müller, N. and Hyndman, W. 2018. Sustainable hydropower in the 21st century. *PNAS* 115(47): 11891–11898. https://doi.org/10.1073/pnas.1809426115

Nilsson, C. 2009. Reservoirs. In: *Lake ecosystem ecology. Encyclopedia of inland waters*, 211–219. ed. Likens, G.E. 2010b. Oxford: Elsevier.

Nixon, S.W. 2003. Replacing the Nile: Are anthropogenic nutrients providing the fertility once brought to the Mediterranean by a great river? *Ambio* 32 (1): 30–39.

Oeppen, J., and Vaupel, J.W. 2002. Broken limits of life expectancy. *Science* 296(5570): 1029–1031. https://science.sciencemag.org/content/296/5570/1029.full.

O'Sullivan, J.N. 2020. The social and environmental influences of population growth rate and demographic pressure deserver greater attention in ecological economics. *Journal of Ecological Economics* 172. https://doi.org/10.1016/j.ecolecon.2020.106648.

Portney, K.E. 2015. *Sustainability*. MIT Press Essential Knowledge Series Massachusetts Institute of Technology, 234 pp.

PRB. 2017. 2017 World population data sheet. www.worldpopdata.org

Qiu, J. 2012. Trouble on the Yangtze. Upriver habitats—including a critical refuge created when construction began on the three gorges Dam—are now at risk from a series of new projects. *Science* 336(6079): 288–291.

Razmjoo, A., N. Khalili, M.M. Nezhad, N. Mokhtari, and A. Davarpanah. 2020. The main role of energy sustainability indicators on the water management. *Modeling Earth Systems and Environment* 6: 1419–1426. https://doi.org/10.1007/s40808-020-00758-1.

Rees, W.E. 2020. Ecological economics for humanity's plague phase. *Ecological Economics* 169: 106519. https://doi.org/10.1016/j.ecolecon.2019.106519.

Reich, D. 2018. *Who we are and how we got here: Ancient DNA and the new science of the human past*, 335 pp. Oxford University Press, Oxford, UK.

Reithe, G., and Grossmann, J. 2019. Floating PV on reservoirs: a threshold analysis by means of a cost-benefit appraisal referencing various potential sites across Africa. Paper 22.06 presented at the Africa 2019 Hydropower Conference, Namibia: Windhoek.

Rietbergen, S. (ed.) 1993. *The Earthscan reader in tropical forestry*, 328 pp. London: Earthscan.

Ryder, G. 1990. *Damming the three Gorges: What dam builders don't want you to know*. Second Edition. 135 pp. Tronto: Probe International.

Sayer, J., and B. Campbell. 2004. *The science of sustainable development*, 268 pp. Local Livelihoods and the global environment: Cambridge University Press, Cambridge.

Scudder, T. 2005. *The future of large dams: Dealing with social, environmental, institutional and political costs*, 389 pp. London: Earthscan.

Shalash, S. 1980. The effect of the High Aswan Dam on the hydrological regime of the Rive Nile. Proceedings of the Helsinki Symposium, June 1980. http://hydrologie.org/redbooks/a130/iahs_130_0244.pdf.

Sheldon, R.W., and S.R. Kerr. 1972. The population density of monsters in Loch Ness. *Limnology and Oceanography* 17: 796–798.

Shermer, M. 2018. *Heavens on earth: The scientific search for the afterlife, immortality and utopia*, 305 pp. London: Robinson.

SOZ. 1997. *Das Monstrum der Pekinger Machtelite*. Zurich: Sonntagszeitung, 1997-02-02.

Stalder, H. 2020. Stauseeprojekt am Triftgletscher nimmt Form an. NZZ (Neue Zürcher Zeitung), 2020-08-20. https://www.nzz.ch/schweiz/das-projekt-fuer-einen-stausee-am-triftgletscher-nimmt-form-an-ld.1572154?reduced=true.

Syvitski, J., C.N. Waters, J. Day, et al. 2020. 2020: Extraordinary human energy consumption and resultant geological impacts beginning around 1950 CE initiated the proposed Anthropocene

Epoch. *Communications Earth and Environment* 1: 32. https://doi.org/10.1038/s43247-020-00029-y.

Tan, G.G. 2020. *Pastures of change: Contemporary adaptation and transformations among nomadic pastoralists of Eastern Tibet*. Studies in Human Ecology and Adaptation 10: 200, Springer. https://doi.org/10.1007/978-3-319-76553-2.

Teng, C.S. 1990. Shifting cultivation of hill padi—a traditional method of utilizing and managing natural soil fertility for rice production in Sarawak, Malaysia. Sarawak: Soils Division, Research Branch, Department of Agriculture.

Titilola, T. 1994. IKS and sustainable agricultural development in Africa: Essential linkages. *Indigenous Knowledge and Development Monitor* 2 (2): 18–21.

Vesco, P., Dasgupta, S., De Cian, E., and Carraro, C. 2020. Natural resouces and conflict: a meta-analysis of the empirical literature. *Ecolocial Economics* 172: 106633. https://doi.org/10.1016/j.ecolecon.2020.106633.

Voosen, P. 2020. Seas are rising faster than ever. *Science* 370 (6519): 901.

Wahl, H.W., and V. Heyl. 2004. *Gerontologie—Einführung und Geschichte*, 247 pp. Stuttgart: Kohlhammer-Urban.

Wallace-Wells, D. 2019. *The uninhabitable earth*, Life after warming, 493 pp.: Tim Duggan Books, New York.

WCD. 2000. *Dams and development: A new framework for decision-making: The report of the world commission on dams*, 404 pp. London: Earthscan.

WCED. 1987. *Our common future: world commission on environment and development*, 383 pp. Oxford: Oxford University Press.

Whitten, A.J., M. Mustafa, and G.S. Henderson. 1987. *The ecology of Sulawesi*, 777 pp. Yogyakarta, Indonesia: Gadjah Mada University Press.

Wohl, E. 2011. *A world of rivers: Environmental change on ten of the world's great rivers*, 359 pp. Chicago: University of Chicago Press.

WWF. 1999. A place for dams in the 21st century? A WWF International Discussion Paper. Gland: WWF International, 108 pp.

Zemp, M., M. Huss, E. Thibert, et al. 2019. Global glacier mass changes and their contributions to sea-level rise from 1961 to 2016. *Nature* 568 (7752): 382–386. https://doi.org/10.1038/s41586-019-1071-0.

Zwahlen, R. 1995. The sustainability of resources versus the sustainability of use: A comment. *International Journal of Sustainable Development and World Ecology* 2 (4): 294–296.

Zwahlen, R. 1996. Traditional methods: A guarantee of sustainability? *Indigenous Knowledge and Development Monitor* 4 (3): 18–20.

Zwahlen, R. 2011. How to maintain the bad reputation of dam projects. In: *Dams and reservoirs under changing challenges* eds. Schleiss, A., and Boes, R. Proceedings of the 79th Annual Meeting of ICOLD, 761–768. Switzerland: Lucerne.

Part IV
Case Studies

This part presents a number of case studies. These are of quite different type and volume, from a focus on one specific project to a description of an entire river basin. The main purpose of these case studies, however, is not to provide detailed, let alone complete, information on one or another of the projects mentioned. Rather, these cases were chosen in order to illustrate one or several of the specific topics addressed earlier on. And here once more it has to be pointed out that if a specific study is criticised, a point raised in it or a conclusion reached, the purpose is not to criticise this study or its author(s), but to point out aspects that may have been overlooked, or simply interpreted in a way with which I do not agree.

Chapter 22
Mekong Basin

He tells me how local fish catches are in decline and how recently the Mekong's water level will rise unpredictably in the middle of the dry season. Both situations are related to China's dams located hundreds of kilometres upstream from Chiang Khong.
Eyler (2019)

The Mekong is often called "one of the great rivers of the world", ranking 10th in terms of mean annual flow at the mouth (e.g. Rainboth 1996:1; MRC 2005:1). It originates on the Tibetan plateau, in the Quinghai province (China) at an altitude of about 5000 m asl. From there it flows over a distance of 2200 km through the Yunnan province, in a generally South to South-Eastern direction to the border with Myanmar and Laos, leaving China at an elevation of about 500 m asl. This is the part generally called Upper Mekong Basin (UMB), however, the river there is called Lancang Jiang. The catchment here is very narrow, three major rivers (Yangtze, Salween and Lancang) all flowing through narrow parallel gorges. It enters the Lower Mekong Basin (LMB) and continues its flow, first forming the border between Myanmar and Laos, then trough Laos, along the border between Laos and Thailand, into Cambodia, for after 2600 km reaching the delta and finally the sea in Viet Nam (Fig. 22.1). This case study does not focus on one specific HPP. Hydropower development was—and continues to be—very extensive throughout both parts of the basin, with two types of developments, namely, dams on the mainstream Lancang/Mekong (with ROR as well as storage dams) on the one hand, and projects on Mekong tributaries, mostly storage dams in Laos, on the other hand. The focus of this case study, therefore, is on cascade effects, impacts on floodplains and on fisheries.

22.1 Short Overview

One very important organisation active in this river basin is the Mekong River Commission (MRC). It is an inter-governmental organisation that works directly with the governments of Cambodia, Lao PDR, Thailand and Viet Nam; MRC's

R. Zwahlen, *Assessing the Environmental Impacts of Hydropower Projects*, Environmental Earth Sciences, https://doi.org/10.1007/978-3-030-91185-0_22

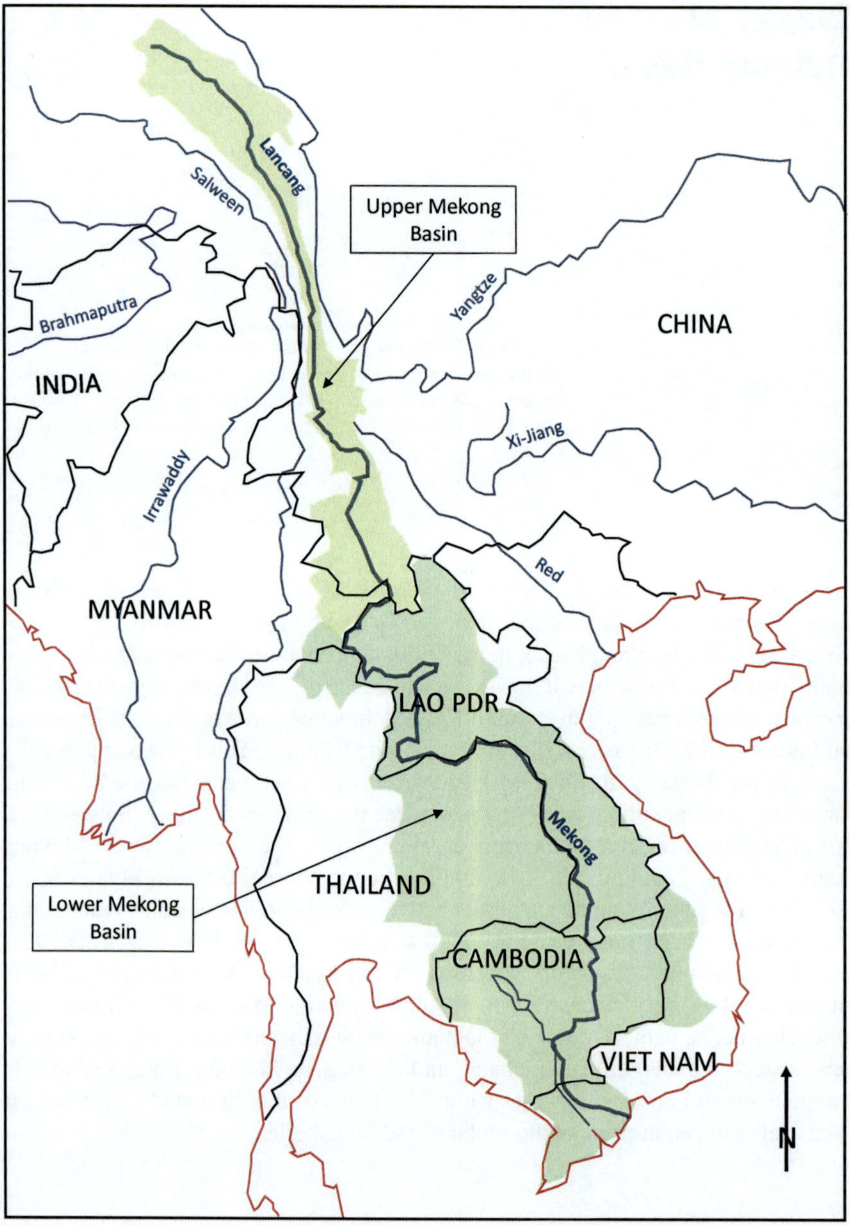

Fig. 22.1 Mekong basin

Table 22.1 Catchment area and flow from Mekong River basin countries

Country	Catchment area [km^2]	Catchment % of MRB	Flow % of total flow
China	165,000	21	16
Myanmar	24,000	3	2
Lao PDR	202,000	25	35
Thailand	184,000	23	18
Cambodia	155,000	20	18
Viet Nam	65,000	8	11
Total	795,000	100	100

Source MRC (2005)

home page (http://www.mrcmekong.org/about-mrc/) contains a wealth of information, specifically on hydropower development and issues related to it in the basin. Its activities cover mainly the Lower Mekong Basin (LMB), the Upper Mekong Basin (UMB) is not directly covered, since Myanmar, which has only a very small part of the total catchment, and China, on whose territory basically the total UMB is located, and which is also very active in hydropower development, are not members of this commission. The repartition of the Mekong Basin between the states that share it is shown in Table 22.1.

The Mekong is also famous for its diversity in fish species; with at least 1200 species (Rainboth 1996) it is considered as ranking third after the Amazon and the Congo.

22.2 The Mekong Cascade

According to ICEM (2010:64), until 2030 water storage by hydropower projects on the Mekong mainstream and its tributaries will increase from 9.9 to 69.8 km^3, which corresponds to 14% of the Mekong's mean annual flow. 80% of the reservoirs, by number, will be in Laos and Viet Nam, but the Yunnan or Lancang cascade, i.e. the dams on the upper Mekong, in China, will account for 23.7 km^3 of this storage volume, 94% of it by only two reservoirs (Xiaowan and Nuozhadu). This, according to the same source, "represents the first time that a development sector will threaten to significantly alter the hydrological regime of the entire Mekong Basin". Figure 22.2 illustrates this effect.

The graph shows a comparison of two series of measured values at Luang Prabang, 1960–2008 and 2015–2019; the difference between these two periods is that in the meantime two large storage plants (Xiaowan with a total storage of 14,560 million m^3 in 2009, and Nuozhan with 23,703 million m^3 in 2012) were commissioned as part of the Lancang cascade; the other HPPs already in operation there are ROR plants with very little storage. The graph clearly illustrates the general effect of storage plants: higher flows in the dry season and lower flows in the wet season. However, it has

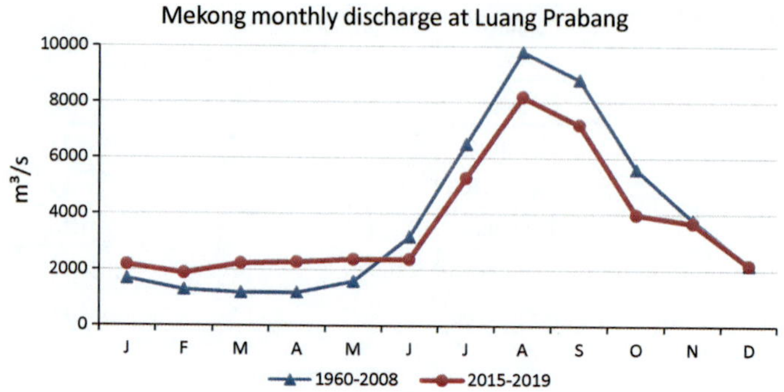

Fig. 22.2 Change of Mekong flow by storage hydropower projects. *Source* based on data from Pöyry (2020)

to be mentioned that there is a possible source of misinterpretation: the 2015–2019 period is too short to provide statistically significant results. While the higher flow in the dry season is most probably due to the effect of the upstream storage plants, the fact that the average yearly flow in this shorter period is lower than the one of the longer period (3670 as compared to 3910 m^3/s) is most probably not, but is rather caused by lower than average precipitation during the period 2015–2019; in fact, 2017 was an exceptionally dry year (see Fig. 22.3). Longer monitoring periods are required for finally assessing the effect of the projects as such.

The main point here, however, is not to illustrate the development at one specific site, but to show a general development: storage means a shift in overall water volume from the high-flow to the low-flow season. This is a desired effect in terms of energy production, and as a matter of fact the mainstream hydropower plants in the Lower Mekong Basin (LMB) will benefit from this effect of the storage volume of the large Chinese projects. However, the ecological impact of this development has to be considered, an impact which becomes manifest mainly in the floodplains in the downstream area (as was described in Sect. 18.2). It will be discussed in Sect. 22.5

Fig. 22.3 Mekong in Vientiane, dry and wet year. Two pictures from the high-flow season, September: left 2017, right 2018. While 2018 the flow was higher than average, in 2017 it was far lower than average

for the specific situation of the Tonle Sap. Figure 22.4 and Table 22.2 provide an overview of the Mekong cascade.

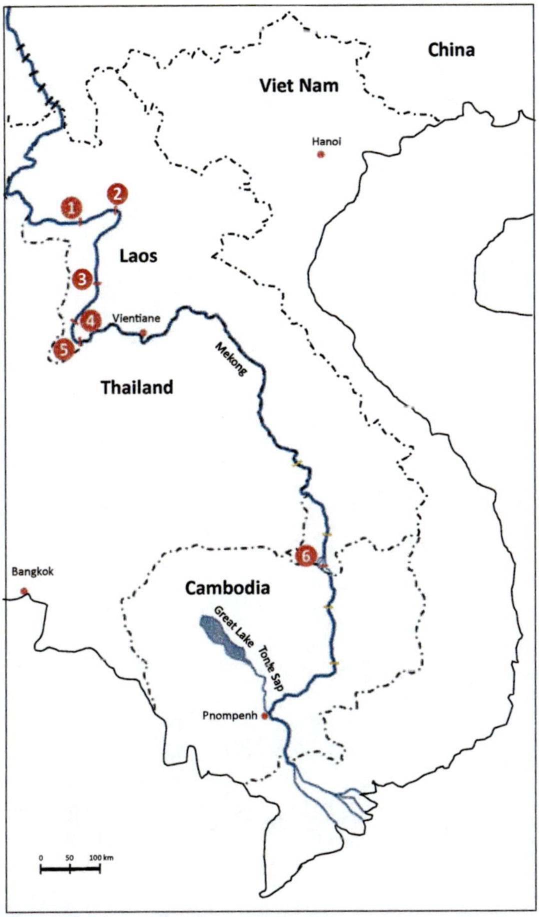

Fig. 22.4 The Mekong cascade. *Legend* red bars: dam sites mentioned in the text, 1: Pak Beng HPP, 2: Luang Prabang HPP, 3: Xayaburi HPP, 4: Pak Lay HPP, 5: Sanakham HPP; 6: Don Sahong HPP. Black bars: cascade on the Lancang (= Mekong), China, from upstream: Xiaowan HPP (4200 MW), Manwan HPP (1550 MW), Dachaoshan HPP (1350 MW), Nuozhadu HPP (5850 MW), Jinhong HPP (1750 MW), Ganlanba HPP (155 MW). Orange bars: other dam projects in lower Mekong, from upstream: Ban Koum HPP (1820 MW), Phu Ngoy HPP (651 MW), Stung Treng HPP (880 MW), Sambor HPP (2600 MW). 2018 *Source* MRC 2010:187; MRC:181. (https://www.internationalrivers.org/sites/default/files/attached-files/lower_mekong_dams_map_june2017.jpg)

Table 22.2 HPPs in the Mekong cascade

Name	Country	Commissioned	Type	River km	Cap	Energy	Dam height	FSL	Reservoir volume		Res. area
									full	live	
					MW	GWh/y	m	m asl	10^6 m^3	10^6 m^3	km^2
Gongguoqiao	China	2012	ROR		900	4,041	130	1319	510	10	–
Xiaowan	China	2010	o.a		4200	18,990	295	1236	14,560	9990	194
Manwan	China	1995	Seas		1570	6,710	132	994	920	257	415
Dachaoshan	China	2003	Seas		1350	5,900	111	895	940	367	26
Nuozhadu	China	2014	o.a.		5850	23,912	262	807	23,703	12,300	320
Jinhong	China	2009	Seas		1750	5,570	110	602	1140	249	510
Ganlanba	China	Plan	ROR		155	1,777	60		557		58
Pak Beng	Laos	Plan	ROR	2188	1230	5,517	85	345	280	442	87
Luang Prabang	Laos	Plan	ROR	2036	1460	6,622	50	312	1332	12	49
Xayaburi	Laos	2019	ROR	1930	1285	7,400	32	275	1300	225	49
Pak Lay	Laos	Plan	ROR	1818	1320	6,460	35	240	904	58	59
Sanakham	Laos	Plan	ROR	1772	570		38	215		106	81
Pak Chom	Laos	Plan	ROR		1079	5,318	53	192	100	12	
Ban Koum	Laos	2030	ROR		1820	8,434	53	115	630	0	133
Phu Ngoy	Laos	2029	ROR		728	2,751	27			530	13
Don Sahong	Laos	2020	ROR		260	2,000	22.5	75	25		2.2
Stung Treng	Camb	Plan			880	5,052	55	55	1550	70	80

(continued)

Table 22.2 (continued)

Name	Country	Commissioned	Type	River km	Cap	Energy	Dam height	FSL	Reservoir volume		Res. area
									full	live	
					MW	GWh/y	m	m asl	10^6 m^3	10^6 m^3	km^2
Sambor	Camb	Plan			2600	11,740	56	40	3794	465	620

Main sources https://en.wikipedia.org/wiki/Hydropower_in_the_Mekong_River_Basin; Yu et al. 2019 (for Lancang cascade projects); ICEM (2010) (for Lower Mekong projects).

Types: ROR = run-of-river (where active storage > 0, some limited peaking is possible); seas. = seasonal storage, o.a. = over-annual storage.

Note concerning the Lancang cascade: Yu et al. (2019:5) list a total of 23 HPPs on the Lancang: 6 operating and one planned as listed above, 15 upstream of Gongguoqiao, out of which the 7 lowermost are under preparation, and one (Mengsong), located directly on the border to Laos, is listed as cancelled (situation of 2016).

General observation: depending on the source consulted, the values for some of the parameters (e.g. installed capacity, storage volume, reservoir area) can be significantly different, probably due mainly to the fact that during project optimisation in the planning process significant changes can be made; the figures indicated above cannot necessarily be interpreted as "true" values, however, the order of magnitude is important, as well as the operation mode (ROR, with possibly some margin for peaking, or large storage). Likewise, not all sources necessarily list exactly the same projects.

22.3 Nam Theun 1

This case is mentioned mainly because it is a good example for illustrating effects of a combination of a cascade of dams along one river in combination with trans-basin projects. The situation is illustrated in Fig. 22.5.

The Nam Theun is one of the major tributaries of the Mekong in Laos. It drains a large highland area between the border with Viet Nam and the Mekong floodplain, and it changes its name to Nam Kading after its confluence with Nam Mouan. Three power companies share the hydraulic resources of this river basin, namely Theun-Hinboun, Nam Theun 2 and Nam Theun 1 (listed in the order in which they were built; see key data in Table 22.3).

The Theun-Hinboun (TH) scheme is operating since 1998; its main components are a weir on the Nam Theun (No. 1 in Fig. 22.5), a headrace structure which deviates water from the plateau to the power house in the plain, from where it is then directed into Nam Hinboun, another, smaller tributary of the Mekong; head- and tailrace structures are indicated by solid black lines. TH is a run-of-river structure with very limited storage. In order to increase production of electricity, another dam was built

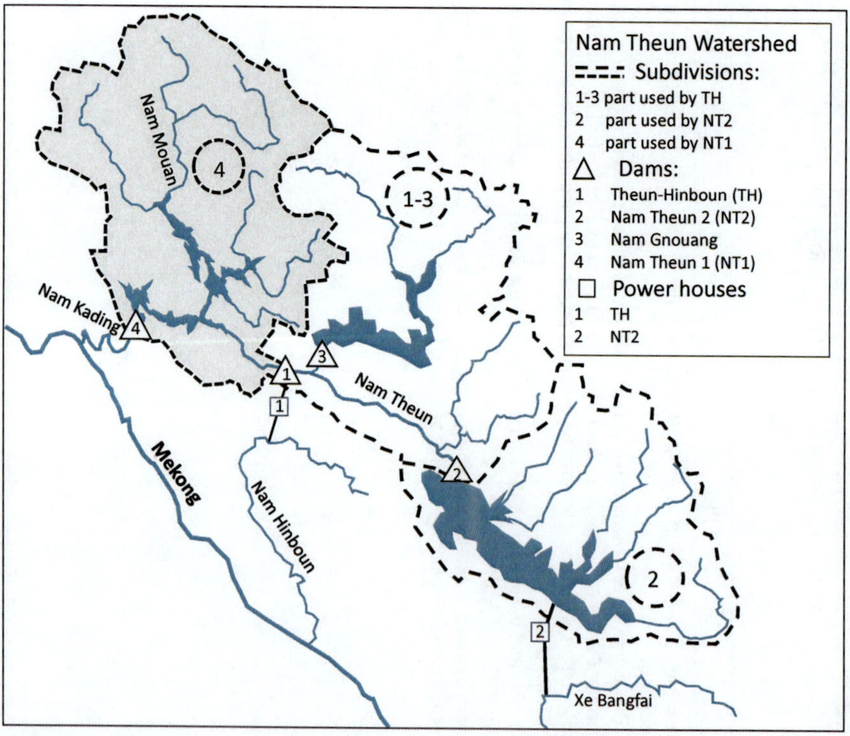

Fig. 22.5 Nam Theun—Nam Kading watershed. Redrawn mainly based on Pöyry 2015; further explanation in the text

Table 22.3 Key features of the Nam Theun HPP cascade

Feature	Unit	NT2[1]	TH[2],[3]	NGN[3]	NT1
Year of commissioning		2020	1998/2013	2013	u.c. (2022)
Dam type		Gravity	Concrete gravity	RCC	RCC
Height	m	39	25	65	177
Crest length	m	439	265	480	771
Reservoir					
FSL	m asl	538	399	455	292
MOL	m asl	525.5	396	420	250
Area at FSL	km^2	489		105	93.6
Area at MOL	km^2	86			31.6
Tot. volume	km^3	3.9	0.04	2.45	3.01
Live storage	km^3	3.6		1.26	2.02
Max. depth	m	39			160
Drawdown	m	12.5			42
Catchment	km^2	4390	4903	2942	5085
av. inflow	m^3/s	238	180	95	315
Turbine discharge	m^3/s	333	110		513.7
Energy					
Installed capacity	MW	1070	240/500	60	300
Annual output	GWh	6000	1400/2950		2,595
Head	m	350	235	56	140
Operation type		Peaking inter-basin	ROR inter-basin	Peaking	Peaking
Resettlement					
Villages	N	17	–	12	7
Persons	N	6′300	–	4367	3240

Sources [1]Descloux et al. (2014), [2]http://globalenergyobservatory.org/geoid/41640; [3]THPC (2014)
TH: before/after expansion (Nam Ngouam Dam)

on the Nam Gnouang (No. 3 in the figure), a major tributary to this part of Nam Theun, which created a reservoir with a surface of 105 km^2 and a total volume of 2.45 km^3; it is also equipped with a power house with an installed capacity of 60 MW, but its main purpose is to provide storage and regulation capacity for TH.

In 2010, Nam Theun 2 (NT2, No. 2 in the figure) went into operation. During a period of about 10 years, this project was studied over and over again, which makes it probably the best studied hydropower project ever; these were the 10 years during which the World Bank refrained almost completely from being involved in hydropower development. I often have the impression that I am probably the only environmental specialist involved in hydropower who worked in Laos quite often during this period without having ever been directly involved in NT2. It uses the upper part of the Nam Theun watershed, by means of a 39 m high dam, a large and rather shallow reservoir on the Nakai Plateau with a surface area of 489 km^2 and a volume of 3.9 km^3. From the reservoir, a headrace structure takes water to the power house (1070 MW) at the foot of the escarpment, the tailrace channel of which delivers the water to Xe Bangfai, yet another tributary of the Mekong; its confluence with the Mekong is outside of the map shown in Fig. 22.5.

Nam Theun 1 (NT1, No. 4 in the figure) is presently (2020) under construction, start of operation is planned for 2022. It also has a rather long history. We were involved in feasibility studies in 1995, then again in 2005, for a different developer, and finally since 2015 for the entity who is now building the dam and power scheme. The dam is located at the point where the river—here already called Nam Kading—leaves the highland to enter the Mekong plain. The scheme is characterised by a 177 m high dam, creating a reservoir of 93.6 km^2 and a volume of 3.01 km^3. Unlike the other two schemes, it has a power house directly at the foot of the dam, with an installed capacity of 600 MW

In a situation that could be called a "normal cascade", as e.g. in the case of the dams along the Kafue River in Zambia (Sect. 9.1.1.2), in the Sanaga basin in Cameroon or in the Mekong mainstream, the lowermost hydropower plant would usually benefit from the regulating capacity of the dams upstream of it. This is not the case here: from the parts identified as 1–3 and 2 in Fig. 22.5, NT1 receives only what is provided by TH dam as residual flow or what is being spilled, since the remaining part of the water is deviated by the two upper dams, from part 2 by NT2 dam to the Xe Bangfai, and from TH to Nam Hinboun; only part 3 of the natural catchment of the Nam Kading, shaded grey in the Figure, is actually available for this hydropower plant, water stemming mainly from Nam Mouang and its tributaries, and only to a small extent from Nam Theun. This effect is illustrated in Fig. 22.6.

NT1, as can be seen from this graph, therefore has only about half the water available as it would have without the two dams further upstream; of course, this did not come as a surprise, planning of NT1 was done taking the other structures and their effects into account. The other, indirect conclusion of this is that the two

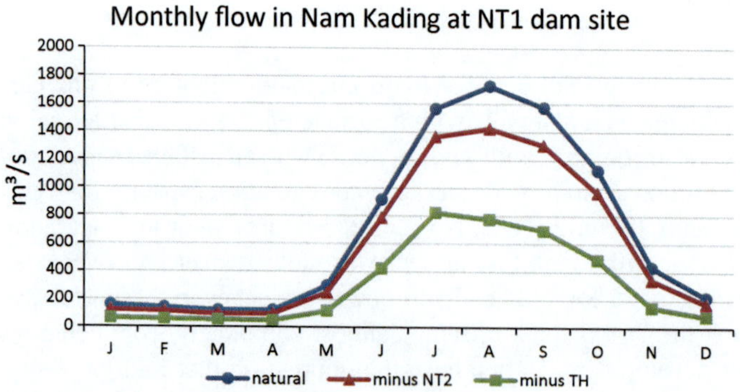

Fig. 22.6 Change in river discharge by interbasin deviation

receiving rivers, Nam Hinboun in the case of TH, and Xe Bangfai in the case of NT2, now have much more water to cope with than they did under natural conditions, and water which is not just added to the rivers, but which comes in pulses influenced by the operation pattern of the two power houses.

22.4 Don Sahong

Don Sahong HPP is a comparatively small dam project presently (2019) under construction on the Mekong mainstream, in Southern Laos, near the border to Cambodia. Probably mainly due to the fact that it is located "on the Mekong mainstream" it has received a rather high degree of attention, at least some of which is probably due to a fundamental misunderstanding of this project. One example which shows what leads to this misinterpretation is a study on the impact of Don Sahong on Mekong fisheries (Baird 2009). In this study, the positon of the project is illustrated in a figure showing the lower Mekong Basin, approximately in the same way it is shown as No. 6 in Fig. 22.4. Without wanting to suggest that the intention was to mislead the readers, this is actually what this illustration does. In his thesis, Baker (2015:153) cites this study and mentions "… the commencement of the construction of the Xayaburi Dam and the more recent initiation … for the Don Sahong Dam—both projects being Mekong mainstream dams". Another report based on Baker states that "the dam will cut across the major part of the river". All this gives the impression (although not really saying so) that Don Sahong is similar to Xayaburi dam, which is located much further upstream. However, there is a very significant difference between the two: Xayaburi, as the dams on the Mekong already built in China (where the river is called Lanchang) is a dam that was built across the river, and in this way blocks it completely. Don Sahong is not. And there is more misleading information to be found: the IRN webpage on the Don Sahong Project (https://www.internationalrivers.org/campaigns/don-sahong-dam) shows, first of all, a picture of a waterfall which is in the Khone or Siphandone region, but which is not at all the site of the Don Sahong Dam; it shows the Tak Ho Fall and was probably chosen because it shows a much more spectacular scenery.

The Siphandone (meaning four thousand islands) is a place where the Mekong encounters a major obstacle: tectonic activities caused a conspicuous rupture in this area, which interrupted the Mekong, caused it to break up in a large number of wider and narrower channels, which fall over a step of about 30 m in a large number of waterfalls of different sizes, the Khone Falls, which in their entirety are the largest waterfall in Asia. This is illustrated in Fig. 22.7.

A short evaluation of the Don Sahong HPP was done in 2012. A small group visited the area for a few days in order to prepare a rapid assessment including technical, environmental and social issues. It became immediately clear that in spite of its comparatively small dimensions (in terms of dam height, area occupied, reservoir size and installed capacity), Don Sahong HPP wold be a project with a very high visibility, mainly due to the fact that it is located on the Mekong main stream, and

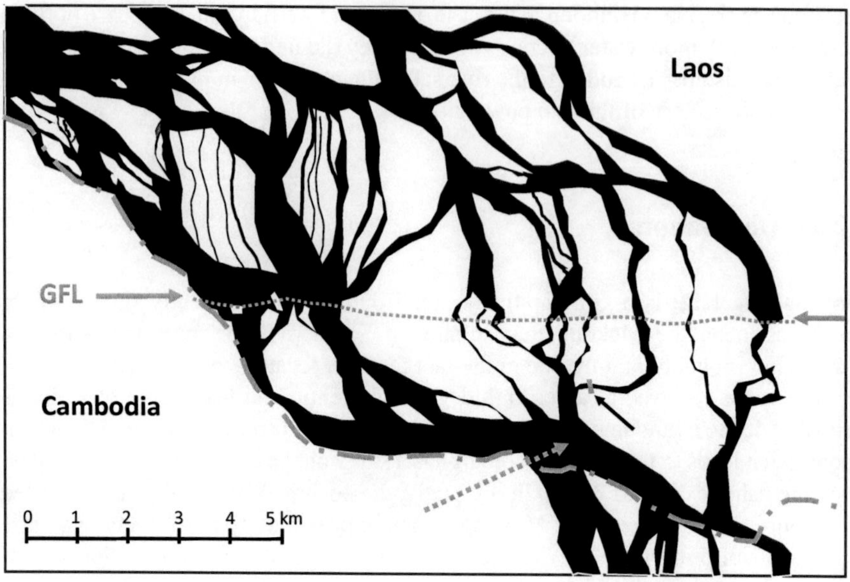

Fig. 22.7 Siphandone with Don Sahong dam site. Grey arrows and grey dotted line indicate the Great Fault Line (GFL) which causes a backwater effect on the Mekong and makes it to split up into a large number of arms which then fall down over the about 30 m drop forming numerous rapids and water falls. At the arrow on the right side of the picture: the Tak Ho waterfall, on the main channel on the Lao side, easily accessible and probably the largest and most spectacular waterfall here. The black arrow indicates the site of the Don Sahong dam, construction of which started in 2016, and which will block the channel leading to it. The dotted grey arrow indicates the wide and deep area of the Mekong where the different channels meet again; this is the habitat of the local population of the Mekong dolphin (*Orcaella brevirostris*)

that therefore it would inevitably be criticised and subject to close scrutiny by its opponents. Having said that it is on the main Mekong, it is however important to clearly indicate that it does in no way block the entire river. As can be seen from Fig. 22.7, it is located on one of the numerous channels, the Hou Sahong channel. It uses on average about 15% of the total Mekong flow, or about 3% during the peak flow in the wet season. With an installed capacity of 260 MW it is rather small, if compared e.g. to the 1250 MW of Xayaburi, which can use the entire water of the Mekong.

The GFL (Great Fault Line) is an important natural barrier and a zoogeographical limit for a number of fish species occurring in the lower Mekong, i.e. down-stream of the falls. Still, out of the about 200 species of fish present in this part of the river, 90 are thought to be carrying out migrations, and 58 species of significant commercial and subsistence importance migrate upstream across the GFL. One species of migrating fish which can be considered as especially prominent is the endemic Mekong giant catfish (*Pangasianodon gigas*), reaching a length of up to 3 m and a weight of 200 kg; it is now bred in captivity, but has become very rare in the wild. The migration pattern is not well known. Lorenzen et al. (2006:9) indicate two locations where this fish was caught in recent years, namely in the Tonle Sap river in Cambodia and at Chiang Khong/Huay Xai in Northern Thailand and Laos (just North of the reservoir of Pak Beng), and this latter site was identified as the probable area of reproduction; this would imply long range migrations, although it does not seem to be clear whether the species is capable to master the Khone Falls when migrating upstream. There is also an important downstream migration, and mainly a drift, of eggs, larvae and juvenile fish. Not all of the numerous arms of the Mekong in this area are of the same importance for fish migrations. The Hou Sahong Channel is the most important one, as it provides the major dry season and to a lesser extent wet season migratory pathway, being the only one (or the only of the more important ones) which, although there are some rapids, is not blocked by a high waterfall which constitutes a barrier for fish migration. Don Sahong dam will block (or already blocks, construction having started in 2016) this channel and will pose a serious obstacle to fish migration from down- to upstream of the GFL (MRC 2010:31). This is the main criticism raised against this project, and a point that has to be taken seriously.

Publications mentioned above point out that the Mekong supports one of the most important inland fisheries, and that a very high number of people depend on it, as their main source of protein in their daily diet, but also as a source of income (Poulsen et al. 2002; Hortle 2007). This is undoubtedly true and needs to be taken into consideration as an important fact for any development project in the wider area which could influence this resource in any way. Many of the papers also mention the importance of the Mekong fishery for the traditional way of life of populations along the river. Most if not all of them also state that fishermen are complaining about a decline in fish catches in recent years, and they all agree on the reason for this: dams. Just two statements to illustrate this point: "If all 11 lower Mekong dams slated for completion by 2030 are built, annual fish hauls from the lower Mekong could decline by up to 880,000 tons, or 42%" (Stone 2016); "This type of subsistence fishing supports about 60 million people living along the 4350 km of the mighty Mekong—but maybe not for long. Laos and other countries—China, Myanmar, Cambodia, Thailand and Viet Nam—are on a dam-building spree, which will reduce the water flow in the Mekong and prevent the passage of fish, threatening the livelihoods of these subsistence fishers, diminishing biodiversity and flooding cultural sites" (Gies 2017).

This clearly states as a fact that 60 million people depend directly on Mekong fisheries, and this is confirmed by numerous statements one can find on this issue; to mention just a few here, the following short list:

- http://www.fao.org/3/ad089e/ad089e09.htm (FAO, 2003/04): The food security of 60 million people living in the Lower Mekong Basin (LMB) is based on rice and fish.
- https://www.grida.no/resources/5630 (2006): More than a third of the population of Cambodia, Lao PDR, Thailand and Viet Nam—some 60 million people—live in the Lower Mekong Basin, using the river for drinking water, food, irrigation, hydropower, transportation and commerce.

- https://theaseanpost.com/article/whats-stake-mekongs-fishery (n.d.): Originating in the Tibetan highlands and running through China, Myanmar, Thailand, Lao PDR, Cambodia, and Viet Nam, the Mekong and its tributaries provide water, food and income for 60 million people.
- https://www.internationalrivers.org/where-we-work/asia/mekong/ (n.d.): Continuing down the current path would mean death by a thousand cuts to the river's rich fisheries and ecology and the 65 million people in the Lower Mekong Basin who depend upon them.
- http://www.mekongwatch.org/platform/bp/english1-5.pdf (n.d.): The population of the Mekong River Basin exceeds 60 million .... Approximately 85% are agricultural communities. Agriculture, fishing, and collection of aquatic life and vegetation are the main economic activities of the people in the basin, and about 62.6% of the working population in the region make their living primarily from water resource related work.
- https://fishbio.com/about-the-mekong (n.d.): More than 60 million people, from nearly 100 distinct ethnic groups, depend on the fisheries of the Mekong Basin for subsistence and livelihoods.
- https://greatermekong.panda.org/discovering_the_greater_mekong/landscapes/mekongriver/ (WWF, n.d.): With over 1300 species of fish, the Mekong River is the world's most productive inland fishery. Providing livelihoods to 60 million people, this fishery accounts for up to 25% of the global inland catch, providing up to 80% of all animal protein to the people of the Mekong River basin. Fisheries in the Mekong River are worth 2.5 billion USD per year to the countries of the Mekong.
- https://www.jsps.go.jp/english/e-astrategy/date/07_asiahorcs_03_p1/keynote/keynote5_%20Baran.pdf (Baran, n.d., pdf of PPT presentation): "Mekong Population: 60 million people in the watershed (76 million by 2020) .... High economic dependence on fisheries in rural households (can exceed 80% in rural households) ... Fish consumption in the Mekong Basin: 24–34 kg/person/year = 49–82% of animal protein consumed ... The Mekong River produces around 21.6 million tonnes of fish per year.
- https://www.nationalgeographic.com/environment/2019/07/mekong-river-lowest-levels-100-years-food-shortages/ (2019): MekongRiver at its lowest in 100 years, threatening food supply .... The river basin is home to the largest inland fishery in the world and more than 60 million people depend on it for their livelihoods.

Looking at this list (which could easily be expanded) two things are striking. I have failed in finding out who was the first to come up with the 60 million. It might stem from MRC (2010:32), which provides figures for the population in the LMB in 2007 (in millions): Cambodia 13.0, Laos 5.2, Thailand 23.1, Viet Nam 18.7, which sums up to 60 million. However, FAO already mentions the same figure in 2003/04, meaning that it must have been around earlier than that. Anyway: the only one in this list who considers an increase in population is Baran, indicating a number of 72 million for 2020. The second surprise is that while everybody agrees that dams

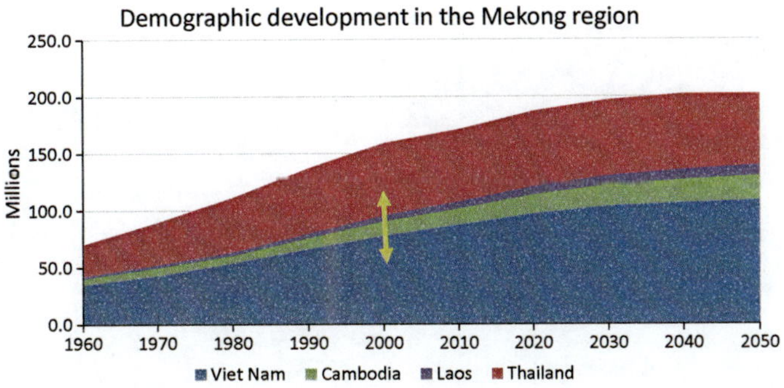

Fig. 22.8 Demographic development in four countries of the Mekong region. Figures for 2020–2050 based on PRB 2017

are to blame for ruining fisheries in the Mekong, hardly anybody even mentions the possibility of overfishing. None of these sources asks the question of sustainability of the situation as it would be without dams, and what is completely left out of the picture is the problem of demographic development.

Figure 22.8 shows the demographic development in the four countries in the Lower Mekong Basin (LMB). It clearly shows that since 1960 the population in this area has increased dramatically. From 1960 to 1990, the population in these four countries has doubled, from roughly 70 to 140 million, and it continues to increase, and is expected to reach 200 million in 2040 (almost three times the numbers of 1960), which will hopefully be the maximum. The 60 million, according to the MRC source cited above, would roughly correspond to the situation in 2000, assuming that the entire populations of Cambodia and Laos are included there, as well as about one third of the population of Thailand and one quarter of the population of Viet Nam, as indicated by the yellow arrow in the Figure. If we further assume that this part of the population would increase at the same rate as the population of the four countries in total, this would mean that the LMB population would be roughly 74 million in 2020 and reach 76 million in 2050.

With this development, the least one would have to ask is whether Mekong fisheries as practiced currently are sustainable, and the answer would most probably be no. After having seen the intensity of fishing in the Siphandone area, albeit just during a short visit, one can very well imagine that there is massive overfishing—or more to the point, it is difficult to imagine that fishing as done presently on the Mekong is sustainable—and this at a time when there was no dam on the lower Mekong mainstream as yet (Fig. 22.9). Just blaming the decrease in fish stocks on dams is too easy, and too simple to be true. More importantly: it does not help. The problem does not go away by pointing the finger at a reason which may not be the real causing agent, maybe not the most important, but certainly not the only one. The fisheries reports prepared as part of the ESIA for the project (DSHPP 2010, 2011) mentioned a decrease of fish catches reported by fishermen of 50–90% since the 1990s, and overfishing was mentioned as the main threat to fish stocks in this area. Baird (2009:15)

Fig. 22.9 Fish traps in the Siphandone area. Left: lee traps for fish being swept downstream. Right: traps for fish migrating upstream. At the time of the visit (June 2012) water level was too low for most of those being active

makes an interesting statement on one fish species, *Cirrhinus microlepis*, saying that this species "has demonstrated its ability to rebound from over-fishing if the conditions are right". This, at least, is an indication that overfishing is a problem, and that there might be solutions for it. However, he does not push this issue any further. Only one of the other available studies (Mather et al. 2009) mentioned overfishing as one of the causes for the decreasing fish stocks, while there as well dams are identified as the main reason.

What would be required is a careful and comprehensive monitoring of fish stocks and fisheries in this area, which then could lead to a proper management of the

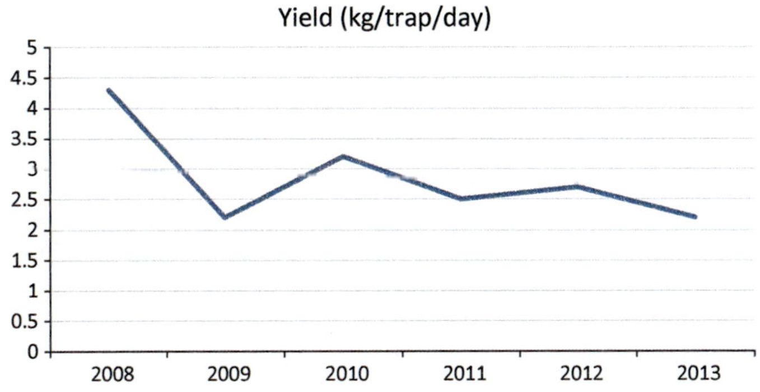

Fig. 22.10 Fisheries monitoring at Khone Falls 2008–2013. Data *source* MRC 2018

resource. One effort to monitor fishing in this regions was carried out by MRC (2018; Fig. 22.10). Monitoring was carried out by recording fish catches in so-called lee traps, a type of trap used in rapids in the Khone Falls area for catching fish wandering or being swept downstream (see Fig. 22.9).

The results from this monitoring suggest a declining trend in yields; unfortunately, when the use of such traps was banned in 2013, this monitoring could not be continued.

It has, as Dawkins and Wong (2016:41) state, "… become fashionable to regard hunter-gatherers … as more in balance with nature than us" (whereby they clearly include traditional ways of fishing in the term "hunter-gatherer"). This, as they state, and as has been pointed out by many other studies, has often not been the case, as even traditional societies in many places all over the world tended to overexploit the very resources they depended on (e.g. Diamond 2005:105). A demographic development and resulting pressure on fisheries as described above for the Mekong cannot be sustainable in the long run, and changes are happening one way or the other. This is not meant to say that dams as such will be a solution of the problem, but just assuming that "business as usual" in the Lower Mekong Area would be a solution, and would not cause major problems in itself, is just too much of a simplification.

As mentioned, the channel to be blocked by the Don Sahong dam has been identified as the most important route for fish migration across the GFL, and possibly the only one where fish could pass during the dry season. To mitigate this impact, the main mitigation measure proposed by the ESIA was to make two channels passable for fish in the dry season to replace the one which will be blocked by the project, one to the left and one to the right of the Hou Sahon; work for doing this was already under way during our site visit. If done properly, this could even be an improvement in comparison with the present situation, because if really only one channel is available, this can be considered a high risk situation for fish populations depending on it. Even so, it is difficult to imagine that only one channel open for migration could ever have been sufficient for maintaining mass migrations across this formidable obstacle in the long term, given the dynamics of this entire system, where any channel could be blocked, or the water could open a different way, as happens often in floodplain situations. It would now be important to closely monitor fish migrations along these alternative routes (and, simultaneously, looking for others which might be there). One additional point makes this "one channel theory" even more unlikely: about 100 years ago, when the French were colonising Indochina, they modified this channel in an attempt to make it navigable. This attempt failed, but it has probably facilitated fish migration through this channel. However, fish migration across the GFL must have taken place before that already, which would certainly suggest the presence of other routes.

The other main environmental issue there is the fact that a small population of Irrawaddy or Mekong Dolphins, *Orcaella brevirostris*, one of the few species of freshwater dolphins, lives in the area below the GFL, where the different arms come together to form a broad, deep and rather calmly flowing river again (indicated by the dashed arrow in Fig. 22.7); during the short visit mentioned above a few of them have been seen surfacing shortly, impossible to say how many individuals, but probably at least four to five. Like the other freshwater dolphins, this is also an endangered species. The subpopulation in Laos and Cambodia is considered as critically endangered. Reliable data on its population are difficult to obtain, but all relevant sources report a considerable decline over the past years. This small population, which may consist of fewer than 10 individuals, is extremely threatened. Like others, it is declining, also due to accidental drowning in gill nets and to (illegal) fishing with explosives. The reduction in fish stocks might have led to a reduction in its food basis. Concerns were expressed that construction and operation of Don Sahong HPP might put additional pressure on this population, which could lead to its complete extinction, mainly by deteriorating water quality (increase in suspended solids and accidental water pollution with fuels, lubricants and concrete, by underwater blasting, and finally by reducing the fish stocks and therefore the food basis for the dolphins).

If the information that there are only about 10 individuals left and that the population is isolated from others further downstream is correct, one can certainly say that the prospects for its future are bleak. Small populations at the edge of the area of distribution of a species are always at a risk of disappearing. They are vulnerable because of the small numbers of individuals, and if then an additional mortality factor arises (in addition to the inevitable natural mortality) this can easily lead to the extinction of such a population, as losses cannot be compensated. The main such risk, according to available sources, is accidental drowning of dolphins in gill nets. That the risk is very real can be demonstrated with the following simple calculation, which is done assuming that the key parameters of the dolphins' life history (sexual maturity at the age of seven, one offspring per female every two years, maximum life span 30 years) as indicated in the literature are correct (Walker 1968; Lekagul and McNeely 1988; Corbet and Hill 1992; Kreb 1999).; Stacey and Arnold 1999).

Assuming a population of 10 animals (5 males, 5 females) evenly distributed over the life span of the species, this would mean one male and one female per 6-year cohort, and this again would mean 4 females of reproductive age (one being a juvenile). In this case, there would be a maximum of two new-borns per year.

This again means that a mortality of more than 2 animals per year would inevitably mean a shrinking of the population which, with such a small number of individuals, could easily lead to its extinction. How much additional mortality can arise due to accidentally getting caught in in fishing nets has been emphasised by the case of the vaquita (*Phocoena sinus*), the smallest porpoise species and an endemic to the Gulf of California which is critically endangered and at risk of becoming extinct mainly due to this (Dalton 2004; Brownell 1983; Würsig et al. 2021).

The risk for DSHPP here is that, if this population should disappear after construction has started, the project would very certainly be blamed for this (note e.g. that Mather et al. 2009, p. 73), make the following statement: "The (Don Sahong) dam would be sited less than 2 km from the trans-boundary dolphin pool on the Lao-Cambodian border. This would lead to the demise of this dolphin group, reducing the remaining range of the highly endangered Mekong dolphin by up to one third, and making it extinct in Laos". With the population being so small, the likelihood that it disappears is very high. It would be impossible to really know what had been the ultimate cause, but the obvious thing to do would be to blame it on Don Sahnog dam. The same has happened with the Yangtse Dolphin (*Lipotes vexillifer*) and the Three Gorges Dam.

According to Brownell et al. (2017), the entire population of the Mekong Dolphin downstream of Khone Falls numbered about 85 individuals in 2010, distributed over 200 km downstream into Cambodia, and even this is a very small population. They point out another, certainly very real threat to this population: if a mainstream dam should be built across the Mekong, as the Sambor Dam envisaged in Cambodia roughly 100 km downstream of the falls, this would lead to a fragmentation of the population, and would compromise its chances of survival.

A very thorough management program for this population would be required in order to maintain it. One immediate and urgently required measure would be to completely ban gill net fishingwithin the range of this species. This would, obviously, require offering an alternative livelihood to fishermen affected by this measure. Such a management program cannot be implemented within the framework of Don Sahong HPP, nor by any other dam project. Other river dolphins face a similar situation, and their populations are declining or even being driven to extinction, as was most probably the case for the Yangtse Dolphin (*Lipotes vexillifer*) due to human pressure of all types on their habitat, dams being one of them (Reeves and Leatherwood 1994).

22.5 Mekong—Tonle Sap

The Tonle Sap is a large wetland system connected to the Mekong in Cambodia. It consist of two main parts, the Great Lake and the Tonle Sap River (whereby Tonle Sap is often being used for the entire floodplain system of lake and river), the latter connecting the lake to the Mekong at Pnom Penh, the capital of Cambodia. The water regime of this floodplain is determined by the seasonal flow variation of the Mekong. During the low flow season (January to May), the Tonle Sap River drains the lake towards the Mekong, the lake reaching its minimum surface area of approximately 3500 km^2 and a depth of around 0.5 m. When the water level of the Mekong rises, the flow direction in the river turns, and the lake is gradually increasing in surface, reaching up to 14,500 km^2 in September/October (MRC, 2005). At its peak size, the lake stores up to 80 km^3 of water; 12.5% of the water inflowing to the lake stems from rainfall, 34% from 11 river tributaries, and as much as 53.3% from the Mekong (Arias et al. 2014).

The Tonle Sap wetland is one of the most productive fishing grounds worldwide. The yearly fluctuations in lake area and depth plays an important role in this productivity, and is also essential for the livelihood of the human population living in this area (see e.g. Hortle and Bamrungrach 2015). Like other floodplain systems, the Tonle Sap area comes under increasing pressure (Rainboth 1996; Corlett 2019). Storage reservoirs in the Mekong basin are one major concern in this respect. It has to be expected, as shown for other rivers, that these storage reservoirs will act here as they usually do, namely, by reducing the flow during the wet (or better: the high flow) season, when the reservoirs are being filled, and increasing it during the dry (low flow) season, when the reservoirs are being drawn down for generating electricity.

A recent study (Arias et al. 2014) has analysed this effect, focussing on the influence of the reservoirs (in operation, under construction or planned, 42 in total) in the so-called "3S system", i.e. in the basins of the three rivers Sesan, Srepok, and Sekong, which are shared by Laos, Viet Nam and Cambodia and join the Mekong in Cambodia, contributing nearly a quarter of the total flow. This publication modelled a situation including the storage in the Lancang cascade (23.2 km^3 existing storage) plus the expected future storage in the 3S catchment (26.3 km^3) in comparison to the baseline situation, i.e. the Mekong basin before the operation of major storage plants. This clearly illustrates the expected effects: a raise in water level in the Tonle Sap in the low flow season and a reduction in the high flow season. It has to be expected that this effect will become more pronounced in the future, since every storage project that will be added will act in the same direction. Here this effect is evaluated shortly by adding storage plants on Mekong tributaries in Laos, including data for 10 such reservoirs (Nam Ngum 1, 2 and 3, Nam Leuk, Nam Bak, NT1, TH/Nam Gnouam, NT2 and Houay Ho), for a total additional storage volume of 16.8 km^3; this is live storage, although in one or two cases the source was not really specific. Still, the volume will finally certainly be larger, since this is by no means a complete list of HPPs in this area. It has to be pointed out that, unlike in Arias et al. (2014) here no sophisticated model was applied, this volume was just added under the assumption that monthly distribution would be the same as for the scenario they analysed. The results are shown in Fig. 22.11. It is also important to note that the HPP cascade on the main Mekong in Laos will not contribute to this in any major way, since these are all ROR schemes which will change flows, if at all, only very marginally, and will not lead to a seasonal shift.

The graph shows that the effect of these storage plants is clearly visible, and much more so in the dry than in the wet season. According to these results, the dry season water level will be 0.76 m higher than originally, and 0.44 m lower during the wet season, meaning that the amplitude between the two seasons, on average,

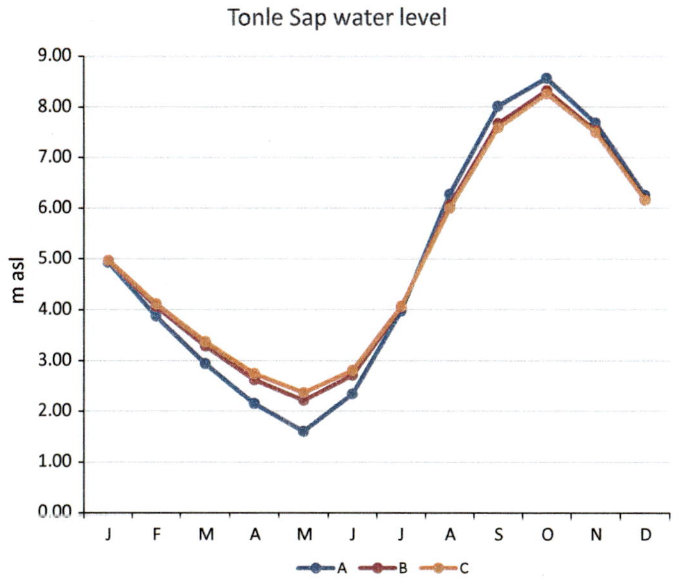

Fig. 22.11 Effects of storage plants on the Tonle Sap water level. A: Baseline B: scenario analysed by Arias et al. (2014). C: 10 storage reservoirs in Laos added (see explanation in text). Values = median values of Tonle Sap water level

will be reduced by more than 1 m. In terms of surface this means that on average, the minimum surface of the Great Lake in the dry season will increase by about 1200 km^2 (from now 3500 to 4700 km^2), while the maximum size will shrink by 500 km^2, from now 14,500 to 14,000 km^2. This effect will be masked, to some extent, by year-to-year variations, but the overall effect will be very noticeable. This will certainly have negative effects on the ecosystem as such, but specifically also on the land use pattern (at least part of the natural drawdown area being presently used for agriculture) and on fish, most likely by reducing fish stocks due to a general reduction in productivity of the entire system.

The Tonle Sap is probably the best known, and best investigated, such system of connection of a river to a wetland which is seasonally regulated by a backwater effect from the river during the high-flow season. A similar situation is presented by the swamp area in the lower Sinú River area, Colombia (ciénaga grande del Bajo Sinú, ciénaga grande de Lorica). As in the case of the Mekong-Tonle Sap system, river water flows back into the swamp during the rainy season, during which time the flooded area expands greatly, while it recedes again during the dry season. The area is used for agriculture and, recently, also for cattle ranching, and it supports an important inland fishery; it has been used for over 2000 years already before the arrivals of Europeans in this area. The Urrá HPP, located some 100 km upstream from the swamp on the Sinú River, has changed the situation by reducing the seasonal differences in flows, as described above for the Tonle Sap. Here as well, other human influences, like draining and filling in parts of the swamp for gaining pasture land, are changing the system, and there are concerns about overfishing (Salazar-Mejía, 2008; Olaya-Nieto et al. 2015). Another, somewhat smaller such system is the Rawa Aopa (Aopa swamp) in South-East Sulawesi, Indonesia, a swamp area alimented by Konaweha River. Drainage for an irrigation scheme on this river was threatening to drain this swamp as well, and it was saved by the fact that it had been declared a National Park, and that it was used also as the main source of drinking water for Kendari, the provincial capital (EWI 1989; Zwahlen 1992a, b). Additional irrigation schemes as well as a planned multipurpose dam project (Pelosika dam project for irrigation and hydropower, which was in its detailed design stage in 2018) further upstream on the same river are likely to have an impact on this wetland as well.

References

Arias, M.E., T. Piman, H. Lauri, T.A. Cochrane, and M. Kummu. 2014. Dams on Mekong tributaries as significant contributors of hydrological alterations to the Tonle Sap Floodplain in Cambodia. *Hydrology and Earth System Sciences* 18: 5303–5315. https://doi.org/10.5194/hess-18-5303-2014.

Baird, I.G. 2009. The Don Sahong dam: potential impacts on regional fish migrations, livelihoods and human health. *Critical Asian Studies* 43: 211–235. https://doi.org/10.1080/14672715.2011.570567. https://www.researchgate.net/publication/232894456.

Baker, C.G. 2015. The problem with environmental security: challenging the either/or approach of national versus human security in the context of the Mekong river basin. Doctoral Thesis, 290. University of Sydney: Centre for International Security Studies.

Brownell, R.L., Jr., Reeves, R.R. Thomas, P.O., Smith, B.D., and Ryan, G.E. 2017. Dams threaten rare Mekong dolphins. *Science* 355(6327): 805

Brownell, R.L., Jr. 1983. Phocoena sinus. *Mammalian Species* 198: 3

Corbet, G.B., and J.E. Hill. 1992. *The mammals of the Indomalayan region: a systematic review*, 448 pp. Oxford University Press, New York: Natural History Museum Publications.

Corlett, R.T. 2019. *The ecology of tropical East Asia*, 3rd ed., 320 pp. Oxford, UK: Oxford University Press.

Dalton, R. 2004. Net losses pose extinction risk for porpoise. *Nature* 429: 590.

Dawkins, R., and Wong, Y. 2016. The Ancestor's tale. A pilgrimage to the dawn of life, 771 pp. London: Weidenfeld and Nicolson

Descloux, S., Guedant, P., Phommachanh, D., and Luthi, R. 2014. Main features of the Nam Theun 2 hydroelectric project (Lao PDR) and the associated environmental monitoring programmes. *Hydroécology Application*. https://doi.org/10.1051/hydro/2014005.

Diamond, J. 2005. Collapse. How societies chose to fail or survive, 575 pp. London: Penguin.

DSHPP. 2010. Fisheries study in Hou Sahong, Hou Sadam and Hou Xangpheuk.

DSHPP. 2011. Report on supplementary fisheries study in Hou Sahong, Hou Sadam and Hou Xangpheuk.

EWI. 1989. Wawotobi drainage improvement study. Annual 5: environmental evaluation study, Jakarta, Indonesia: EWI for Ministry of Public Works.

Eyler, B. 2019. *Last days of the mighty Mekong*, 365 pp. London: Zed Books.

Gies, E. 2017. The real costs of energy. *Nature* 551 (7682): S145–S147.

Hortle, K.G. and Bamrungrach, P. 2015. Fisheries habitat and yield in the lower Mekong basin. MRC technical paper no. 47, 80. Cambodia: Mekong River Commission, Phnom Penh, ISSN: 1683–1489.

Hortle, K.G. 2007. Consumption and the yield of fish and other aquatic animals from the lower Mekong basin. MRC technical paper no. 16, 87. Vientiane: Mekong River Commission.

ICEM. 2010. Strategic environmental assessment of hydropower on the Mekong Mainstream. Report prepared for MRC, 198 pp.

Kreb, D. 1999. Observations on the occurrence or Irrawady dolphin, Orcaella brevirostris, in the Mahakam River, East Kalimantan, Indonesia. *Zeitschrift für Säugetierkunde* 64 (1): 54–58.

Lekagul, B., and McNeely, J.A. 1988. Mammals of Thailand. 2nd edition, 758 pp. Thailand, Bangkok: Association for the Conservation of Wildlife.

Lorenzen, K., N. Sukumasavin, and Hogan, Z. 2006. Development of a conservation strategy for the critically endangered Mekong giant catfish: quantitative assessment report. Mekong Giant Catfish Working Group Report 3. All MGCWG Reports are available on: http://www.aquaticresources.org/mekongcatfish.html

Mather, R., S. Kritsavanarin, G. Pangare, and Weerapong, D. 2009. Siphandone: the Mekong under threat. Mekong Region Water Dialogues Series Publication No. 1, 180 pp, IUCN, Gland.

MRC SEA. 2010. Fisheries baseline assessment, Working Paper Vol. II.

MRC. 2005. Overview of the hydrology of the Mekong basin, 73 pp. Vientiane: Mekong River Commission.

MRC. 2010. State of the basin report 2010. Mekong River Commission, 232 pp. Laos: Vientiane

MRC. 2018. Lee trap fishery monitoring in Southern Lao PDR reveals fishery resources degradation from 2008–2013. Digital: http://www.mrcmekong.org/assets/Publications/Fishery-Monitoring-in-Lao-PDR-final.pdf

MRC. 2019. State of the basin report 2018. Mekong River Commission, Vientiane, 226 pp.

Olaya-Nieto, C., Segura-Guevara, F., and Vergara-Paternina, A. 2015. Growth and mortality of Cachana (*Cynopotamus atratoensis*) in the cienaga Grande de Lorica, Colombia. Rev. MVZ Córdoba 20(Supl.): 5028–5036, ISSN: 0122–0268. https://doi.org/10.21897/rmvz.16

Poulsen A.F., Ouch Poeu, Sintavong Viravong, Ubolratana Suntornratana and Nguyen Thanh T. 2002. Fish migrations of the lower Mekong river basin: implications for development, planning and environmental management. MRC Technical Paper No. 8, Mekong River Commission, Phnom Penh, 62. ISSN: 1683–1489

Pöyry. 2015. Nam Theun 1 hydropower project: environmental and socail studies. environmental impact assessment (EIA). Report prepared for Phonesack Group.

Pöyry 2020. Luang Prabang HPP: environmental impact assessment. vol. 4.1 of the feasibility study. Report prepared for Luang Prabang Power Company Limited.

PRB. 2017. 2017 world population data sheet. www.worldpopdata.org

Rainboth, W.J. 1996. Fishes of the Cambodian Mekong. FAO, Rome, 265 pp. 27 color plates

Reeves, R.R., and S. Leatherwood. 1994. Dams and river dolphins: can they coexist? *Ambio* 23 (3): 172–175.

Salazar-Mejía, I. 2008. La economía de la Ciénaga Grande del Bajo Sinú: lugar encantado de las aguas. Documentos de Trabajo Sobre Economía Regional 102. Banco de la República. Centro de Estudios Económicos Regionales CEER. Cartagena, Colombia. https://repositorio.banrep.gov.co/bitstream/handle/20.500.12134/469/?sequence=1

Stacey, P.J., and Arnold, P.W. 1999. Orcaella brevirostris. *Mammalian Species* 616: 8.

Stone, R. 2016. Dam-building threatens Mekong fisheries. *Science* 354 (6316): 1084–1085.

THPC. 2014. Theun-Hinboun plant summary. http://www.thpclaos.com/images/stories/pages/documents/theun-hinboun_summary_report.pdf

Walker, E.P. 1968. Mammals of the world. 2nd edition. Baltimore: The Johns Hopkins Press, Vol. I: 1–646, Vol. II: 647–1500.

Würsig, B., T.A. Jefferson, G.K. Silber, and R.S. Wells. 2021. Vaquita: Beleaguered porpoise of the Gulf of California México. *Therya* 12 (2): 187–206. https://doi.org/10.12933/therya-21-1109.

Yu, X., He, D., and Phousavanh, P. 2019. Balancing river health and hydropower requirements in the Lancang River basin. Springer Nature Singapore, 285. https://doi.org/10.1007/978-981-13-1565-7

Zwahlen, R. 1992. Failure of irrigation projects and consequences for a different approach: a case study. *Ecological Economics* 5: 163–178.

Zwahlen, R. 1992. The ecology of Rawa Aopa, a peat-swamp in Sulawesi, Indonesia. *Environmental Conservation* 19 (3): 226–234.

Chapter 23
Rogun: Water Scarcity in Central Asia

Les responsables des collectivités savent que leur premier devoir est d'assurer l'approvisionnement en eau. Sans elle, pas de vie possible.
(Community leaders know that their first duty is to secure the water supply. Without it, no life is possible.)
Cans (1997: 6)

The Aral Sea basin is a closed Central Asian river basin, with two main rivers, the Amu Darya and the Syr Darya, flowing into the Aral Sea, without any natural outflow from this basin. The Aral Sea tragedy, the shrinking and almost disappearing of this large water body in the second half of the twentieth century, is one major example of negative effects of water management measures. The situation in the basin is of high relevance for a large hydropower project in this area, Rogun HPP. The chapter describes and discusses this project, but it also exemplifies the fact that a hydropower project is not just an isolated undertaking serving a specific purpose. Rather, is has—or can have—far reaching consequences beyond the project site, the region of its implantation, and even the country undertaking this development. Inclusion of all stakeholders in order to reach an optimal outcome is required.

23.1 Background

The Rogun Hydropower Project (Rogun HPP) was first conceived in the Soviet Union in the 1950s and 1960s as part of the regional Central Asian development, mainly for providing irrigation water for the cotton producing region in the lower Amu Darya basin, and in addition, like the already existing Nurek HPP just downstream of Rogun, for generating peaking power for the Central Asian grid, while baseload would be provided by thermal power plants in other parts of the area. Construction of Rogun HPP started in 1982. By 1990, around 100 km of tunnels had been built, which consisted of the river diversion tunnel, various access tunnels, head- and tailrace tunnels and the cavern for the underground power house. The coffer dams were built and the river was deviated, construction of the dam as such was about to start.

R. Zwahlen, *Assessing the Environmental Impacts of Hydropower Projects*, Environmental Earth Sciences, https://doi.org/10.1007/978-3-030-91185-0_23

The entire situation changed with the end of the Soviet Union, which led to the independence of Tajikistan and the other Central Asian states in 1991. The change to a market economy, also for electricity, resulted in severe problems for Tajikistan, since now it had to pay in hard currency for imported energy, which it required mainly in winter, and it could not make up for this with hydroelectricity, which produced a surplus in summer, but suffered from a shortfall in winter. The main producer of electricity was Nurek HPP, with an installed capacity of 3000 MW, with an additional number of smaller hydropower stations with a total of 1032 MW (EWE and SMEC 2000). Nurek was the only one with large storage, while the others were mostly ROR schemes, which meant that their production in winter, when demand was highest, was very low. With the exception of some coal, all fossil fuels had to be imported, in addition to electricity produced in foreign thermal power plants. All this posed a heavy burden on the economy of Tajikistan, a country which on top of that had problems due to its lack in natural resources (with the exception of a considerable hydropower potential) and its remoteness, which requires long transport ways for all goods to be exported or imported. In addition to that, soon after independence a civil war broke out, which officially ended with the agreement on peace reached in 1997 (Curtis 1997: 267; UNDP 1999: 25; Breu et al. 2003: 12) and caused heavy damage to the infrastructure of the country. The consequences for the country can be exemplified with Fig. 23.1, which shows development of energy consumption (statistical data 1985–1999, forecast till 2004).

In this situation, due to a lack of funds and its location in an area where factions opposing the government were still active, Rogun HPP had come to a complete stop. In 1993, debris brought by a flood blocked the diversion tunnel, which led

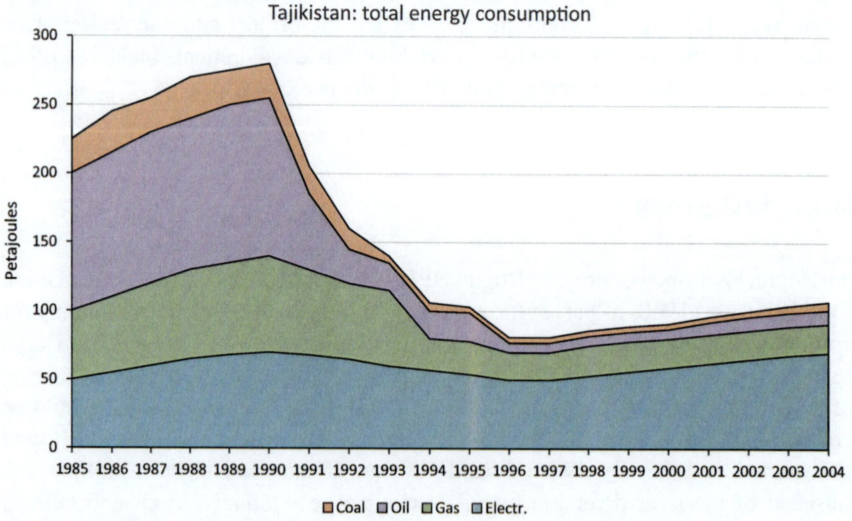

Fig. 23.1 Energy consumption in Tajikistan 1985–2004. Based on date from EWE and SMEC (2000)

to damage in some of the already built structures and to the coffer dams being swept away. Further damage was caused by an earthquake in 1995. Work on site was not resumed before 2008, but basically only care and maintenance activities were carried out. In 2011, a Technical and Economic Assessment Study (TEAS) and an Environmental and Social Impact Assessment (ESIA) started, with funding provided to the Government of Tajikistan by the World Bank. The agreement was that until the results of these studies would be available, no further investment in the project would be made, and only work required for maintenance and for preventing additional damage to already existing structures would be carried out. I was the team leader for the ESIA, which was originally planned to take one year to prepare, but which was then finalised only in September 2014.

23.2 The Project

The immediate reason for reactivating the project was the fact that Tajikistan suffered from a critical lack in winter electricity, with demand exceeding supply by up to 25%. This led to load shedding even in the capital Dushanbe, and to some regions being completely cut off from electricity supply in winter. Demand-side changes such as improvements in energy efficiency were needed as part of the solution, but the shortfall could only be overcome in the mid- and long term with additional generation capacity. Rogun HPP has been identified as part of the long-term solution, long-term not least due to the time span required for construction, estimated at 16 years. The Rogun HPP site lies about 110 km east of Dushanbe (Fig. 23.2), and the HPP is the uppermost of a planned and partly built hydropower cascade on the Vakhsh river, the most important of which so far is Nurek HPP, 70 km downstream from Rogun, which started operation in 1980. The Nurek HPP includes a high dam—at 300 m the highest in the world—and a reservoir with a storage capacity of 10.5 km^3. Nurek is operated in such a way as to shift about 4.2 km^3 of water from summer to winter (see Fig. 23.9) to meet Tajikistan's electricity demand in the cold season. The other HPPs in the cascade are run-of-river (ROR) schemes with little or no storage capacity. Still, the ones downstream of Nurek benefit from water released from this large storage reservoir.

In the meantime, the original purpose of the Rogun project had evolved from supporting regional irrigation and hydropower generation, to the present plan, which calls for Rogun to serve as a multi-purpose project providing power, flow regulation, flood control and sediment retention. In addition, two alternatives had been developed for being compared with the original design (Table 23.1).

The primary purpose of the Rogun HPP is to generate electricity to help Tajikistan overcome critical shortages of electricity in winter and meet future demand, while respecting the water needs of downstream riparian countries. In addition, the project as planned can provide other significant benefits, including:

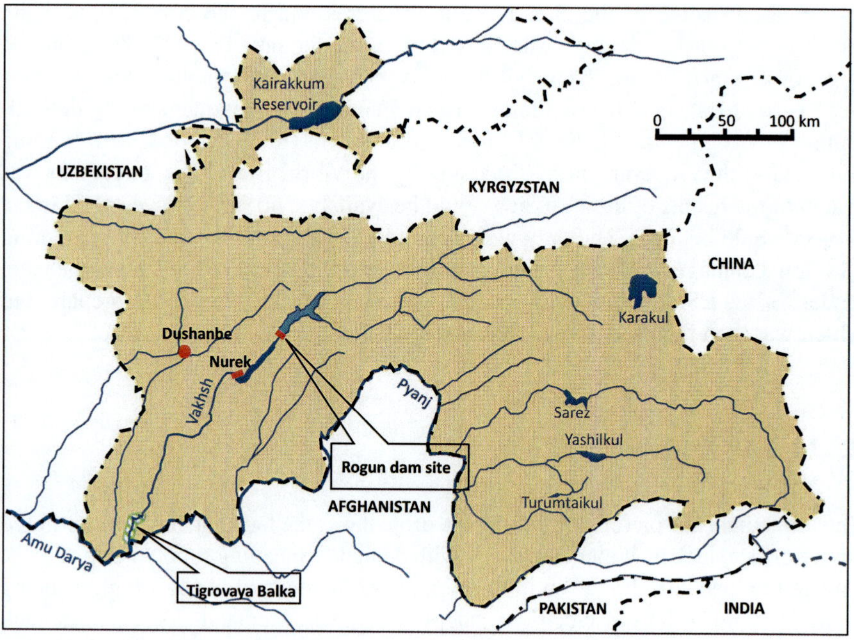

Fig. 23.2 Map of Tajikistan indicating Rogun HPP dam site

Table 23.1 Rogun HPP: main parameters of studied alternatives

Alternative*	Unit	FSL 1290	FSL 1255	FSL 1220
Dam height	m	335	300	265
Total reservoir capacity	km^3	13.3	8.55	5.22
Reservoir active storage	km^3	10.3	6.45	3.93
Reservoir area	km^2	170	114	68
Investigated installed capacities	MW MW MW	3600 3200 2800	3200 2800 2400	2800 2400 2000
Construction time	years	14	12	10
Time for reservoir filling	years	16	13	9
Annual energy production (for the three installed capacities)	TWh TWh TWh	14.4 14.3 14.1	12.4 12.3 12.1	10.1 10.0 9.8
Resettlement	Villages Households Persons	77 6035 42,000	36 2433 18,000	25 1825 13,000
Expected minimum life span	years	115	75	45

Source Pöyry (2014)
Alternatives: labelled by FSL (m asl)

- **Downstream flow regulation potential**: the added storage capacity provided by Rogun can help overcome shortages that are now felt in downstream areas in dry years.
- **Electricity production for export**: Rogun HPP would allow power sales to neighbouring countries, including in particular Pakistan and Afghanistan.
- **Flood routing**: the existing Vakhsh cascade, including Nurek dam but without Rogun, is not designed to handle the Probable Maximum Flood (PMF). The construction of Rogun HPP would provide flood routing capacity and protect Nurek and the lower cascade. The fact that the FSL 1220 alternative (265 m high dam) would not have assured safety of the cascade in the case of a PMF was one of the reasons to eliminate it from further investigations.
- **Retention of sediments**: by retaining the high sediment load which otherwise would reach and fill Nurek reservoir, Rogun (FSL 1290 alternative, 335 m high dam) would effectively extend the life of Nurek HPP and the Vakhsh cascade by over 100 years.

The TEAS studies recommended a 335 m high earth core rock filled dam with its Full Supply Level (FSL) at 1290 m asl, with an installed capacity of 3200 MW; this alternative also resulted as the best in the economic analysis.

The ESIA came to the following conclusions: While the FSL 1220 alternative has lower resettlement and a lower capacity for shifting water from summer to winter, it has a reduced energy output, a relatively short life span and it does not mitigate the Vakhsh cascade's inability of handling PMF. Overall, it offers no comparable advantages to the two higher dam alternatives. FSL 1290 and FSL 1255 alternatives differ considerably in the main environmental and social impacts and risks. In addition to the best economic results, the longer life span and the better flood and drought mitigation potential are strong arguments in favour of the FSL 1290 alternative. On the other hand, the lower potential for adversely influencing the downstream flow conditions, and mainly the fact that resettlement would be reduced by more than half are similarly strong arguments for the FSL 1255 alternative. Thus from an environmental and social point of view, there were good arguments for and against each of these two alternatives, and important trade-offs had to be considered. Given the considerably longer life span of the FSL 1290 alternative, the higher electricity production and the fact that the incremental environmental and social impacts can be appropriately mitigated, the ESIA consultant recommended to take the FSL 1290 alternative forward for detailed consideration.

23.3 Transboundary Impact

The main issue to be considered in the framework of the Rogun HPP was its potential transboundary impacts, meaning the possibility to influence the downstream flow, and with it water availability for Tajikistan's downstream neighbours, mainly Uzbekistan.

The potential conflict here consists in the fact that a dam with a large storage can—and usually does—retain water during the high-flow season, to use it for producing electricity in the low-flow season. In the case of Rogun, this would mean that water would be retained in summer, when river discharge is high, for it to be used to produce electricity in winter, when natural river flow is low and electricity demand is at its peak. Uzbekistan, however, is in a different situation: it covers its electricity demand mainly with fossil fuel fired thermal power plants, and it uses the water for irrigation, on which a very considerable proportion of its population directly depends. A reduction of river discharge in summer, therefore, would directly lead to a shortage of water for irrigation, while additional water in winter would be of no use. Although Rogun had originally be planned for regulating flows mainly in the interest of irrigation in the downstream area as the primary goal, under the changed political conditions after the end of the Soviet Union there was no longer a guarantee that this would still be so. For this reason, Uzbekistan was strictly opposed to the project, to the point that for a while there were even threats that it would send its air force to bomb the construction site, should construction activities be resumed.

The World Bank was well aware of this situation and therefore had organised right from the start a consultation process which included all riparian states. In a number of meetings, the delegations of all riparian states were informed about the progress of the two studies, and all pending issues were discussed with them. The meetings were usually held in Almaty, in Kazakhstan, which was neutral ground, this country largely being located in the Aral Sea basin, but not in the Amu Darya basin. Unfortunately, Uzbekistan refused to participate in the first ones of these meetings, and it was certainly a success for the efforts taken by the World Bank that finally Uzbekistan decided to participate as well. It also had been very clear right from the start of the studies that a solution would have to be found which respected the interests of all riparian countries, and mainly for the two most affected ones, Tajikistan with the main aim to improve its electricity requirements, and Uzbekistan with its interest in assuring sufficient water supply for irrigation. The text on the following pages is an attempt to describe this rather complex situation and the solution which was finally recommended.

The main issue to be addressed here is the management and sharing of transboundary water resources in the Amu Darya basin. Mukhammadiev (2014) provides a very clear-cut analysis of the situation in his text entitled "Challenges of Transboundary Water Resources Management in Central Asia" (in Micklin et al. 2014). The abstract of this text is reproduced here:

> Central Asian major river basins link the countries of Afghanistan, Kazakhstan, Kyrgyzstan, Tajikistan, Turkmenistan and Uzbekistan. Water management in Central Asia continues to be the most important transboundary environmental issue and the biggest problem remains how to allocate water for upstream hydropower production and downstream irrigation. Disagreements between the upstream and downstream states have increased regional tensions and slowed development plans. National responses to existing cooperative opportunities are essentially driven by a policy of national self-sufficiency in energy and water. While it is reasonable to be concerned about water and/or energy security, it is also critical to understand that a policy of self-sufficiency incurs substantial costs for all. As long as self-sufficiency dominates the policy agenda, the benefits of cooperation will not materialise. International

water law could provide a rational avenue toward achieving international consensus on both use and allocation of water resources in the basin, with international legal agreements to reinforce the consensus. Incentives to cooperate through the application of the benefit-sharing concept as a development model in the basin would include decreased costs and increased gains in many dimensions of regional cooperation, including the benefits that stem from better agricultural practices and its competitiveness, joint development of the region's energy resources, and better management of regional environmental risks.

Source Mukhammadiev (2014: 233)

23.4 The Amu Darya Basin

The origin of the Amu Darya (which has this name only from the point of the confluence of its main tributary, the Pyanj, with Vakhsh, 1445 km upstream of the lake) is in the Pamir, in Afghanistan, close to the Chinese border, at an altitude of 4900 m asl (Fig. 23.3). The total length of Pyanj-Amu Darya is 2540 km. Most of its water stems from snow and glacier melt. Two major tributaries come from Afghanistan, the Kochka (upstream from Pyanj-Vakhsh confluence) and the Kunduz (between Vakhsh and Kafirnigan), bringing about 7–8% of the total flow. These two

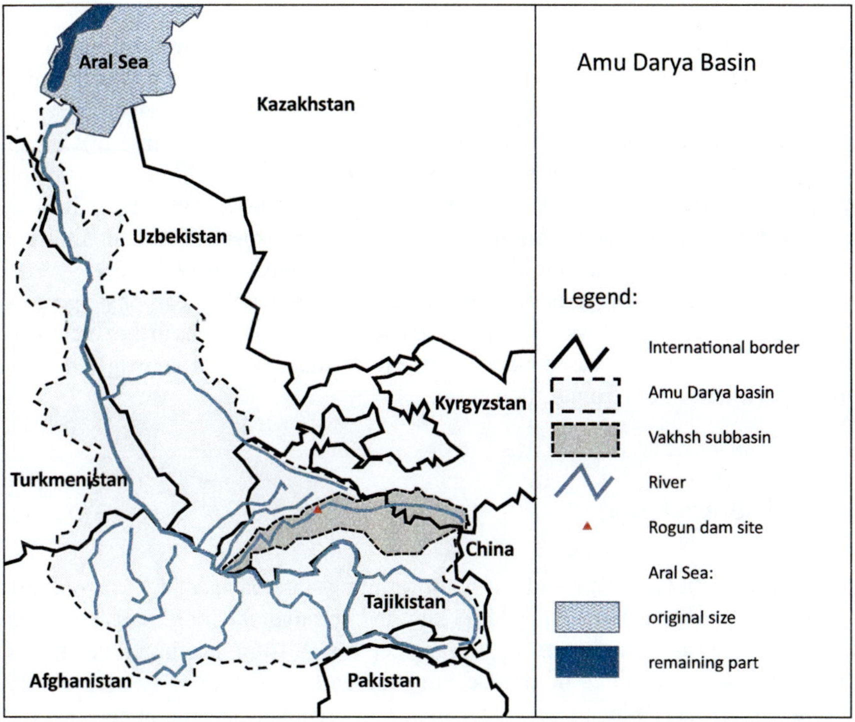

Fig. 23.3 Amu Darya basin

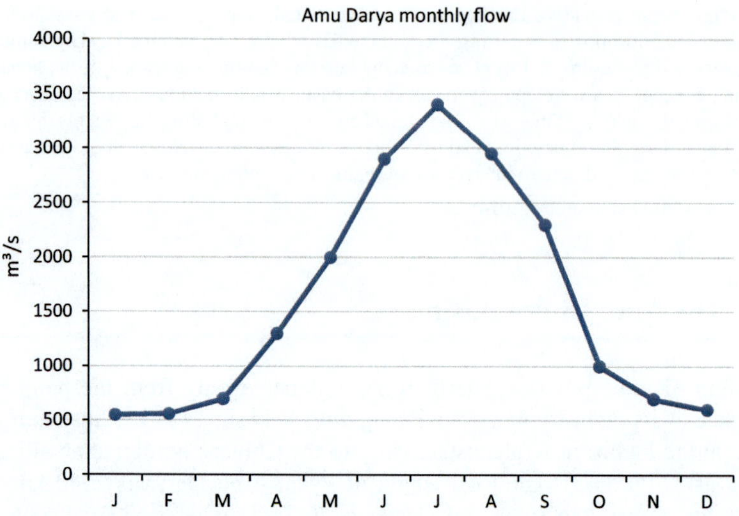

Fig. 23.4 Monthly discharge of Amu Darya at Kerki (1959, 50 years average). *Source* Based on Létolle and Mainguet (1993: 58). Kerki is located in Turkmenistan, about 250 km downstream from the Tajik-Uzbek border

rivers form extensive floodplains near the city of Termiz, which marks the end point of the mountainous section of Amu Darya. Further west, there are no other major tributaries. The Amu Darya has a highly seasonal flow pattern, as shown in Fig. 23.4.

In its publication on shared rivers, ICOLD (2007: 233), for the case of Central Asia, under the heading International Shared Rivers does not list individual rivers, but the entire Aral Sea basin as a unit. As for the Amu Darya basin, depending on the source, information on the size of the basin and the respective parts in different countries vary greatly, as exemplified by the figures provided in Table 23.2.

The water of Amu Darya has a high concentration of calcium carbonate, and its content in organic matter is very low, due to the low content of humus in the catchment area. Salt content has increased markedly due to drainage water from irrigated areas and waste water inflow from settlements.

23.5 The Aral Sea

The Aral Sea is a large lake without outflow, inflows are balanced by evaporation. Around 1960, it had a surface of 66,458 km^2 and was then the fourth largest lake in the world in terms of surface area. Average depth was 16 m, maximum depth was 68 m in the western and 29 m in the central and eastern part; these two parts were separated by islands which today form an almost complete separation between the two parts.

Table 23.2 Country areas in the Aral Sea basin (km^2)

Country	ICOLD (2007)	FAO (2013)	
		Aral Sea basin	Out of which Amu Darya basin
Kazakhstan	124,739	345,000	
Uzbekistan	382,286	425,030	364,630
Turkmenistan	69,671	461,740	359,730
Kyrgyzstan	112,093	118,370	7800
Tajikistan	135,942	141,130	125,450
China	1583		
Afghanistan	100,691	246,000	166,000
Pakistan	3349		
Aral Sea basin total	1,230,408	1,773,270	1,023,610

In historical and prehistorical times, the lake level has known rather important fluctuations, the amount of which is still a matter of debate. However, for a time span of at least two centuries before 1961, it has been very stable, with a variation in the order of magnitude of 1–2 m.

The Aral Sea has only two tributaries of importance, the Syr Darya in the North and the Amu Darya in the South, both originating in the high mountain areas to the east of the lake, in Kyrgyzstan, Tajikistan and Afghanistan. Records from the twentieth century show that the overall runoff of the Amu Darya at the foothills of the mountains (i.e. in the region of Termiz, a short stretch downstream of the Tajik-Uzbek border) was 78.6 km^3/yr, with year to year fluctuations between 91.5 and 41.9 km^3/yr, however without showing a downward trend (Létolle and Mainguet 1993: 50). Out of this, originally (i.e. before about 1960) around 60 km^3/yr reached the Aral Sea (see Fig. 23.6). The fact that in spite of the high annual variation of inflow the lake level remained rather stable might have been due to the mitigating effect of the groundwater reserves in the soils along the river plains and the lake.

23.5.1 Development in the Twentieth Century

After the end of World War II, and especially towards 1960, large irrigation projects were implemented in the lower parts of the Aral basin, along the Amu as well as the Syr Darya (Micklin 2014c). One major step in the development of the water resources which was decisive for the Amu Darya and the Aral Sea was the construction, in the 1950s (Létolle an Mainguet 1993: 151) of the Karakum canal, which feeds the irrigated areas in Turkmenistan and also provides drinking water to Ashgabat, the capital of the country. It derives 13 km^3 of water annually from the Amu Darya, i.e. about 16% of the total flow of the river, and leads it along the Karakum desert towards the Caspian see (where rarely if ever any water from the canal actually arrives). Large quantities of water are lost due to infiltration and evaporation (Fig. 23.5).

Fig. 23.5 Karakum canal near Ashgabat

This canal, together with the other irrigated areas, lead to the immediate effect of massively increasing water abstraction from Amu Darya, thus reducing inflow to the Aral Sea (see Fig. 23.6). This led to a very marked reduction in the total amount of water reaching the Aral Sea, since the losses from evaporation were no longer compensated. Thus the level of the lake decreased, (from around 53 m asl before 1960 to 42 m asl in 1985), and the surface shrunk from 65,000 to 45,000 km^2 in the same time span.

Since then, the Aral Sea has continued to shrink; see Fig. 23.3, which indicates approximately what is left of the larger Southern part of the Aral Sea; the Northern part, which is located in Kazakhstan and which is alimented by the water from the Syr Darya, was separated be a dam which prevents water flowing from there to the Southern part. After the construction of this dam this part, now called the Small Aral Sea, has recovered at least to some degree.

Figure 23.7 shows the water balance of the Aral Sea for the last 100 years. The graph shows the following:

- **Inflow**: this is the total inflow (i.e. from Amu and Syr Darya combined). It was stable (around 56 km^3/yr) for the period 1911–1960. After that, with the construction and operation of the major irrigation and diversion schemes, inflow started to drop. In recent years, annual inflow was between 0 (in dry years) and about 8 km^3 (in wet years). The last two series from 1990 to 2010 are for the South Aral Sea only, since by then the northern part (Small Aral Sea) got separated from the (originally much larger) southern part.

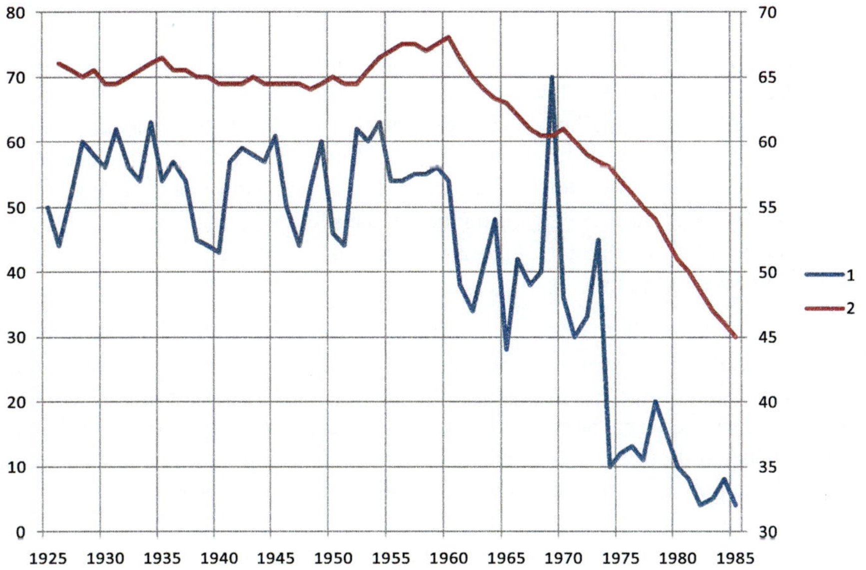

Fig. 23.6 Inflow to and surface of the Aral Sea. 1: Inflow to Aral Sea (in km^3/yr, left axis). 2: Surface of Aral Sea (in '000 km^2, right axis). Based on data from Létolle and Mainguet (1993: 186)

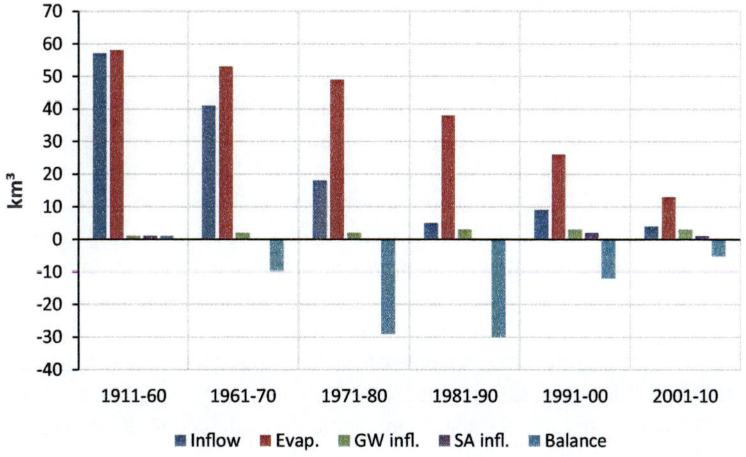

Fig. 23.7 Aral Sea water balance 1911–2010. Based on data from Micklin (2014a) (Tables 5.1 and 5.2). Explanations see text

- **Evaporation**: before 1960, there was a balance between total inflow and evaporation. Since the early 1960s, evaporation, although it also decreased in line with the shrinking surface of the sea, was always considerably larger than total inflow.

- **Groundwater (GW) inflow**: estimated at 2–3 km^3/yr. Lower sea level has led to a slightly larger inflow of groundwater, which however did by far not compensate the decrease of surface inflow.
- **Small Aral Sea (SA) inflow**: since the separation of the two parts, the Large Aral Sea received inflow from the northern part, the Small Aral Sea. This has decreased since the construction of the dam between the two parts. However, this inflow was (and is) too small to make any difference in the southern part.
- **Balance**: with evaporation being higher than inflow, the water balance was (and continues to be) negative.

Around 1990, northern and southern parts of the Aral Sea became separated. The northern part, the Small Aral Sea, which receives the inflows from Syr Darya, was then isolated from the southern part (Large Aral Sea) by the construction of the North Aral Dam between the two parts. This lead to a recovery of the Small Aral Sea; salinity there has dropped from around 30 g/l to pre-shrinking values of 10 g/l or less. The southern part, however, continues to shrink and salinity is rising accordingly.

23.5.2 Impacts on Aral Sea Biology

Before 1960, salinity of the Aral Sea was around 1% (10 g/l). With the proportional increase of evaporation, due to the shrinking water volume salt content started to rise and reached approximately 3% (30 g/l) in 1990. This development led to a gradual extinction of a large number of invertebrates and of most freshwater fish species in the lake, partly due to the detrimental effects of the shrinking lake on spawning areas, but mostly due to mortality caused by increasing salinity. Natural reproduction of freshwater fish had stopped completely at a salinity of 14 g/l, which was reached in the mid-1960s. To compensate this, some marine invertebrates and fish were introduced into the lake, and some of these species developed very well, allowing even for a considerable commercial fishery as a replacement for the original freshwater fisheries (see Plotnikov et al. 2014 for details). Hypersaline conditions (i.e. salt contents of >35 g/l, the content of sea water) then led to a gradual decrease in the number of the introduced species of marine invertebrates and fishes as well. Since the late 1990s, when salinity reached 60 g/l, the Large Aral Sea is completely without fishes, and only a few crustaceans, which are adapted to conditions of such extreme salinity, survive.

23.5.3 Prospects for Saving the Aral Sea

As mentioned above, the Small Aral Sea has recovered since its separation from the southern part of the lake. The drop in salinity led to the re-establishment of populations of some fresh water fish which had disappeared before, and some fisheries can be maintained there today.

However, the situation for the "Large" Aral Sea is completely different (Micklin 2014b:361 ff.). To actually restore the Aral Sea to its former condition over a period of 100 years, an annual inflow of 54 km^3 would be required. The following options were considered:

- Improving irrigation: the irrigation systems along the Amu Darya are highly inefficient, and losses through seepage and infiltration are very high. It would be possible to save considerable amounts of water by improving these systems, albeit at a high cost. A 1996 estimate for a complete rehabilitation of all irrigation systems in the Aral basin (about 8 million ha) came up with costs of 16–22 billion USD, and this would save 9.2 km^3 of water annually. At the most, and obviously at still much higher costs, it might be possible to increase inflow to the Aral Sea by up to 20 km^3/yr. Even this, however, would not be sufficient for restoring it.
- During the Soviet period, a project was developed to transfer water from Siberian rivers to Central Asia and to the Aral Sea. This project was stopped in the 1980s, due to its negative environmental impacts, but probably even more so due to prohibitive costs. In fact, in addition to the construction costs there would also arise very high operation costs, since this water would have to be pumped over the continental divide which has an elevation of about 120 m asl. Under this scheme, it was planned to divert 27 km^3/yr to Central Asia, out of which an estimated 15 km^3 might have actually reached the Aral Sea, not really sufficient to save it.
- Another more recent project envisages the transfer of water from Lake Zaysan (source of the Irtysh) in Kazakhstan to the Aral Sea. Here again, detrimental environmental impacts (lack of water in the Irtysh) and high costs (e.g. for a 100 km long tunnel to convey the water) would have to be faced. In an extreme case, directing all the outflow from the lake towards the Syr Darya, this would amount to 18 km^3/yr, out of which about 12–13 km^3 would reach the Aral Sea.

With this, there is no prospect of restoring the Large Aral Sea in the foreseeable future. Salinity is increasing, and the eastern (shallow) part has in the meantime dried out almost completely, while the deeper western part "could turn into a lifeless water body akin to the Dead Sea" (Plotnikov and Aladin 2014: 355).

The Aral Sea is probably the most well-known case of extreme shrinking and associated increasing salinisation of a lake due to human activities, mainly water abstraction for irrigation, but there are several other inland lakes which face the same fate, as e.g. Lake Balkhash in Kazakhstan, whose inflow is increasingly reduced mainly due to irrigation schemes in China (Stone 2012; Ussenalieva 2020), Lake Urmia in Iran (Stone 2021), Lake Poopó in Bolivia, which receives its water from Lake Titicaca (Torres-Balló et al. 2019) and Lake Chad (Bossard 2009: 116; Vincent

2010: 66). Presently, this seems to be a common fate of endorheic lakes, i.e. of lakes without effluent, which are generally saline (Wurtsbaugh et al. 2017).

23.5.4 Problems with Water Allocation

According to UNDP (2005: 92), in Central Asia "water shortages are predominantly a management and incentive problem, not a resource problem". In spite of the political tensions between the countries of the region, the intergovernmental organisation which decided on water distribution among the member countries continued to work. As Khudaiberganov (2007: 40) describes, this worked generally rather well, although in dry or extremely dry years there were problems. These problems were mostly not due to disagreements between the countries, but resulted from the fact that upstream water users within the two most affected countries, Turkmenistan and Uzbekistan, used more than the share allocated to them. This resulted in the fact that downstream water users under such conditions suffered most of all, since for them there was simply not enough water left.

The UNDP report mentioned above also identifies annual losses in agricultural production in the Central Asian republics due to inefficient water management at 1.75 billion USD, as shown in Table 23.3. These losses are due to inadequacies in water availability, salinity, poor weed control and cultivation practices, soil compaction, leaching water, drainage problems and land abandonment.

These figures clearly show the importance of an improved water resources management. It is also pointed out that, since many consequences of inefficient water management are shared with neighbouring countries, regionally concerted action by all countries would achieve maximum benefits for everyone. This certainly also includes management of hydropower.

Table 23.3 Estimated losses in agriculture due to inefficient irrigation (M USD/yr)

Country	Syr Darya basin	Amu Darya basin	Aral Sea basin	% of 2003 GDP
Kazakhstan	206	0	206	0.7
Kyrgyzstan	81	0	81	4.3
Tajikistan	58	112	170	10.6
Turkmenistan	0	378	378	6.1
Uzbekistan	390	529	919	9.3
Total	735	1019	1754	3.6

Source UNDP (2005: 93)

23.6 The Role of Dams on Vakhsh River

Figure 23.8 shows flow parameters for a number of relevant stations. The topmost curve shows the largely natural flow of the Amu Darya at Chatly, 240 km upstream of the former inflow into the Aral Sea, in the period before the massive water abstraction; it can be assumed that most of this water reached the Aral Sea. The difference between this and the lowermost curve, inflow to the Aral Sea 2000–2010, indicates the amount of water derived from the system for irrigation. The two intermediate curves show the flow in Vakhsh River at two stations, namely Darband (upstream of Nurek dam) and Tigrovaya Balka (downstream of Nurek). The difference between these two curves is in part due to water used in summer for irrigation in Tajikistan (mainly derived from Nurek reservoir) and the seasonal redistribution due to the operation of Nurek HPP: water retention in summer, used for generating in winter. The graph clearly shows that the contribution of Vakhsh to the flow of the Amu Darya is rather substantial, and that the part of this water flowing in summer, i.e. available for irrigation further downstream, was clearly influenced by Nurek HPP. The concern of the countries in the downstream area, and mainly Uzbekistan, was that Rogun would significantly increase this effect, since its storage volume is even larger than the one of Nurek. An additional risk is the filling phase of the reservoir, which can only be done by retaining water, most of it in the summer months. Effects of Nurek are shown in Fig. 23.9. One clear effect is the shift from summer to winter; less obvious, but just as important, is the fact that during the filling phase average yearly flow was reduced from about 560 to 500 m^3/s.

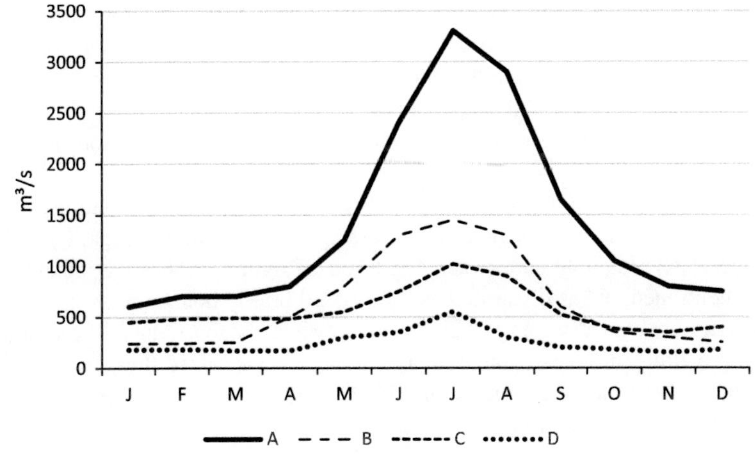

Fig. 23.8 Mean hydrographs of Vakhsh and Amu Darya. A: Amu Darya, Chatly, 1931–1941 (original flow situation); B: Vakhsh, Darband, 2000–2010 (upstream of Nurek, unregulated); C: Vakhsh, Tigrovaya Balka, 1984–1990 (downstream of Nurek, regulated); D: Release to Aral Sea, 2000–2010 (all major irrigation systems installed). Based on data from Pöyry (2014)

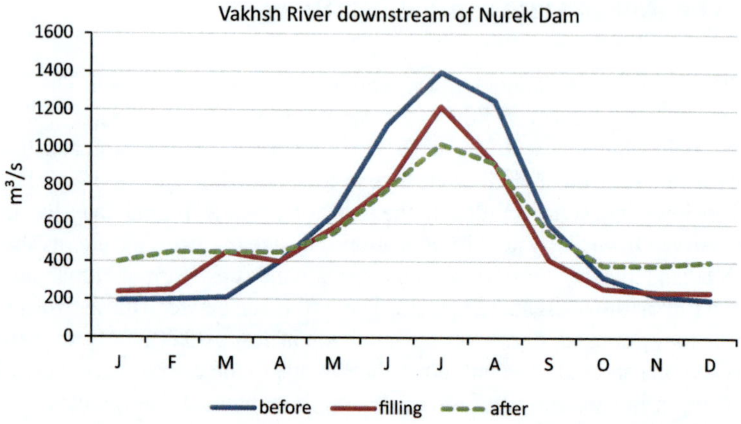

Fig. 23.9 Influence of Nurek dam on Vakhsh river flow. Based on data from Pöyry (2014)

Present water use systems are based on water availability in a normal year, i.e. in a year with approximately average flow. This inevitably leads to difficult situations in dryer than average years, and especially in extremely dry years, since there is no buffer for coping with this situation. An obvious solution for this problem, or at least a possibility to make it less severe, would be to create storage capacity in the upstream areas, retaining water in normal and wet years for being used in dry years. Given their topography, neither Uzbekistan nor Turkmenistan have good possibilities for such storage reservoirs; they would cover large areas of potentially arable land, and such large and shallow reservoirs would increase losses due to evaporation. Tuyamuyun reservoir, a large storage reservoir in Uzbekistan, is located far downstream and therefore of use only for a fraction of the irrigated areas. Tajikistan, however, has the potential for creating large and deep reservoirs. It was an important task of the ESIA for Rogun HPP to clearly show the problems and the potential solutions related to water use and allocation.

As mentioned above, Rogun was originally planned for adding regulating capacity to the entire system, and especially for reducing problems in dry years by making additional water available under such conditions. Rogun HPP has this potential, as already mentioned and as illustrated and discussed briefly below.

Problems arise in dry years. "Dry year" always means reduced precipitation (mainly snowfall) in the catchment area, and therefore reduced flow of water from snow melt in the following vegetation period. When talking about dry years here, three possible cases have to be considered, namely:

1. reduced precipitation in the Pyanj catchment;
2. reduced precipitation in the Vakhsh catchment; and
3. reduced precipitation in both catchments.

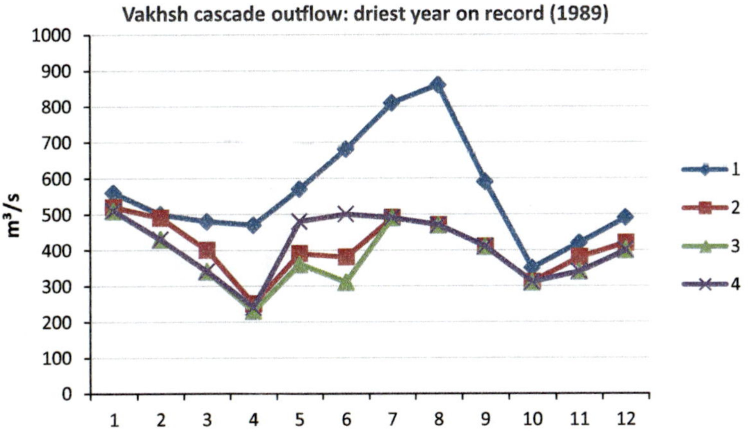

Fig. 23.10 Vakhsh river flow, driest year on record. Cases shown: 1 = Flow in average year; 2 = Flow in driest year, present situation[a]; 3 = Flow in driest year, future situation (Tajikistan fully using allocated water); 4 = Flow in driest year, with Rogun HPP in place. All cases include Nurek HPP. [a]At that time, 2014, Tajikistan did not make full use of the amount of irrigation water allocated to it by ICWC, the International Coordinating Water Commission

The most severe drought has to be expected in case 3; however, Rogun can only influence the flow in Vakhsh River, therefore only this part was taken into consideration. Figure 23.10 shows the situation for the year 1989, the driest year on record (when the total Vakhsh cascade outflow was 12.2 km^3 as compared to the average value of 16.3 km^3).

The figure shows that in such a dry year, while winter flow changes only marginally, summer flows are reduced by a very considerable amount, which then translates directly into water shortage in the downstream areas. Figure 23.11 illustrates, to what extent Rogun HPP could provide compensation in such a situation. The basis for this are the flow figures from the same year as above, i.e. 1989.

Calculations for this figure were made assuming that the additional amount of water made available by Rogun would be evenly distributed over the entire vegetation period April–September, with the same amount each month. Obviously, the system would allow for a better fine tuning according to requirements.

The figure shows that Rogun HPP, if operated accordingly, would allow for a compensation of the water shortage in the Vakhsh catchment, either in part or fully, depending on the additional amount of water made available.

Two conditions need to be considered:

- This can only be done if at the beginning of such a dry season enough water is stored in the reservoir.
- It can compensate the effect of a drought in the Vakhsh catchment, but it would not be sufficient for completely compensating water shortage in the case of a similar drought in the (much larger) Pyanj catchment.

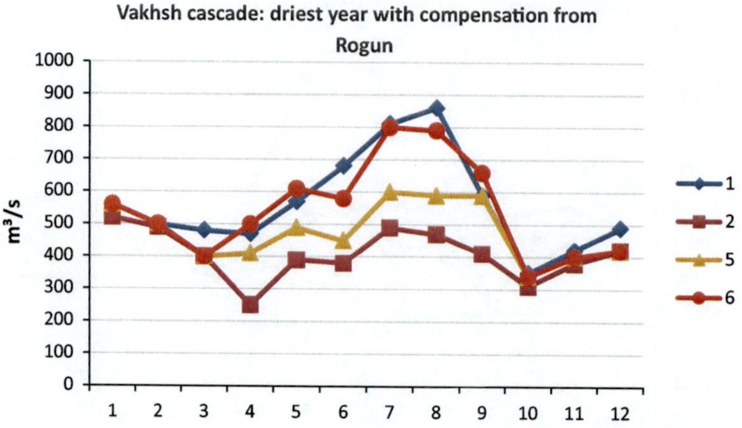

Fig. 23.11 Dry year, compensation from Rogun HPP. Cases: 1 = Vakhsh flow in average year; 2 = Vakhsh flow in dry year; 5 = Vakhsh flow in dry year, additional 2 km^3 made available from Rogun reservoir; 6 = Vakhsh flow in dry year, additional 4 km^3 made available from Rogun reservoir

Even so, the figure shows that Rogun HPP has the capacity of reducing negative impacts of a dry year to a considerable extent. However, it has to be seen clearly that this could be disadvantageous to Tajikistan for the following reasons: making available more water during one summer would require a different operation pattern, and it could potentially reduce winter energy generation; in addition, it would also require the possibility for refilling the reservoirs after such a dry year, which might require an amount of water which would be above of current allocations. This clearly shows that the Vakhsh cascade with Rogun, if operated accordingly, could have considerable advantages for all parties concerned, but that this would require modifications of existing agreements and practices between the states in the Amu Darya basin to efficiently manage flow between the countries. This seems all the more necessary in the light of a probable future development where, under changing climatic conditions, years with extremely high and extremely low flows are likely to occur more frequently than this was the case in the past. It was therefore recommended that the ICWC member states, Kyrgyzstan, Tajikistan, Uzbekistan and Turkmenistan, should modify existing agreements and practices to include operation of Rogun HPP in a way as to maximise benefits for all parties, like flood protection, additional water releases during dry summers and additional hydropower generation during exceptionally cold winters. Such agreements would have to specify the use of the regulating capacity of the Vakhsh cascade for optimising downstream flows under extraordinary conditions. One additional point here is the fact that Afghanistan, although it has a part of the Amu Darya catchment, is not part of ICWC, but has plans to use part of the water from its Amu Darya tributaries for irrigation, and possibly to develop hydropower in this area. It was, however a participant in the consulting group of riparian states which accompanied the preparation of the Rogun ESIA. One recommendation of the ESIA was to include Afghanistan in this process.

Managing reservoirs and hydropower in the Amu Darya catchment does not reduce the need for undertaking efforts for improving the efficiency of the irrigation systems, e.g. by reducing seepage and infiltration losses, increasing efficiency of field irrigation, improving drainage for re-using drainage water, increasing the use of ground water, switching to less water consuming cultures, etc., as also recommended by other authors like UNDP (2005), Khudaiberganov (2007), FAO (2013), Jalilov (2010), and Jalilov et al. (2011).

With all this, one point has to be clearly mentioned: none of the measures discussed above will have an effect of sufficient magnitude as to have any beneficial impact on the Aral Sea. But even so, the demise of the Aral Sea and the efforts to improve the use of shared water resources among riparian countries might serve as an example for managing difficult international issues related to shared water resources. There are many situations of shared rivers and conflicting interests, as e.g. the Great Renaissance Dam under construction on the Blue Nile by Ethiopia and contested by Egypt (Stokstad 2016; Wheeler et al. 2016; Roussi 2019, 2020), the case of the HPP cascade on the Mekong (see Sect. 22.2), or the rather recent information on plans of China to build a major dam in the headwaters of the Brahmaputra (Perras 2020). And even within Central Asia, there is another lake, Lake Balkhash in Kazakhstan, which is threatened by the same fate as the Aral Sea: it is also a lake in a basin without effluent, and is fed mainly by rivers from China. Development of irrigation along these rivers has already led to a reduction of inflow and a shrinking of the lake, which raises fears for its future (Stone 2012; Ussenalieva 2020).

References

Bossard, L. (ed.). 2009. *Regional atlas on West Africa.* West African Studies, 284 pp. Paris: OECD.

Breu, T., H. Hurni, and A. Wirth Stucki (eds.). 2003. *The Tajik Pamirs: Challenges of sustainable development in an isolated mountains region*, 80 pp. Centre for Environment and Development (CDE), University of Berne.

Cans, R. 1997. *La bataille de l'eau*, 232 pp. Paris: Le Monde Editions.

Curtis, G.E. (ed.). 1997. *Kazakhstan, Kyrgyzstan, Tajikistan, Turkmenistan and Uzbekistan country studies*, 565 pp. Area handbook series. Washington DC: Federal Research Division, Library of Congress.

EWE and SMEC. 2000. *Power sector development project.* ADB 3207 TAJ. Vol. 1: Energy sector review and policy formulation. Vol. 2: Diagnostic analysis of Barki Tajik. Vol. 3: Rehabilitation and investment program. Vol. 4: Environmental and socio-economic analysis. Vol. 5: Feasibility report. Reports prepared by Electrowatt Engineering and SMEC International for Asian Development Bank and Government of Tajikistan.

FAO. 2013. *Irrigation in Central Asia in figures.* In AQUASTAT survey 2012, ed. K. Frenken, 228 pp. FAO.

ICOLD. 2007. *Shared rivers: Principles and practices.* ICOLD Bulletin 132: 305 pp. Paris: International Commission on Large Dams.

Jalilov, S.M. 2010. Impact of Rogun dam on downstream Uzbekistan agriculture. Thesis, North Dakota State University, 70 pp.

Jalilov, S.M., T.M. deSutter, and J.A.L. Leitch. 2011. Impact of Rogun dam on downstream Uzbekistan agriculture. *International Journal of Water Resources and Environmental Engineering* 3 (8): 161–166.

Khudaiberganov, Y. 2007. Particular characteristics of integrated water resources management (IWRM) in the Amudarya river basin. In *Implementing water resources management in Central Asia*, ed. P. Wouters, et al., 35–43. Springer.

Létolle, R., and M. Mainguet. 1993. *Aral*, 357 pp. Paris, France: Springer.

Micklin, P. 2014a. Aral Sea basin water resources and the changing Aral water balance, Chap. 5. In *The Aral Sea*, ed. P. Micklin, N.V. Aladin, and I. Plotnikov, 111–135. Berlin: Springer.

Micklin, P. 2014b. Efforts to revive the Aral Sea, Chap. 15. In *The Aral Sea*, ed. P. Micklin, N.V. Aladin, and I. Plotnikov, 361–379. Berlin: Springer.

Micklin, P. 2014c. Irrigation in the Aral Sea basin, Chap. 8. In *The Aral Sea*, ed. P. Micklin, N.V. Aladin, and I. Plotnikov, 207–231. Berlin: Springer.

Micklin, P., N.V. Aladin, and I. Plotnikov (eds.). 2014. *The Aral Sea. The devastation and partial rehabilitation of a great lake*. Springer earth system sciences, 453 pp. Berlin: Springer.

Mukhammadiev, B. 2014. Challenges of transboundary water resources management in Central Asia, Chap. 9. In *The Aral Sea*, 233–251, ed. P. Micklin, N.V. Aladin, and I. Plotnikov. Berlin: Springer.

Perras, A. 2020. China gräbt halb Asien das Wasser ab. *Tagesanzeiger* 2020-12-31. https://www.tagesanzeiger.ch/china-graebt-halb-asien-das-wasser-ab-308409429848

Plotnikov, I., N.V. Aladin, Z.K. Ermakhanov, and L.V. Zhakova. 2014. The new aquatic biology of the Aral Sea, Chap. 6. In *The Aral Sea*, ed. P. Micklin, N.V. Aladin, and I. Plotnikov, 137–169. Berlin: Springer.

Plotnikov, I., and N.V. Aladin. 2014. The biological future of the Aral Sea, Chap. 14. In *The Aral Sea*, ed. P. Micklin, N.V. Aladin, and I. Plotnikov, 355–359. Berlin: Springer.

Pöyry, 2014. Environmental and social impact assessment for Rogun hydropower plant. In *ESIA*, Vol. I: Text. Vol. II: Annexes. Report prepared for OSHPC "Barki Tojik", Republic of Tajikistan.

Roussi, A. 2019. Nations clash over giant Nile dam. *Nature* 574 (7777): 159–160.

Roussi, A. 2020. Row over giant Nile dam could escalate, experts warn. *Nature* 583 (7817): 501–502.

Stokstad, E. 2016. Power play on the Nile. *Science* 351 (6276): 904–907.

Stone, R. 2012. For China and Kazakhstan, no meeting of the minds on water. *Science* 337 (6093): 405–407.

Stone, R. 2021. After revival, Iran's great salt lake faces peril. *Science* 372 (6541): 444–445.

Torres-Batlló, J., B. Martí-Cardona, and R. Pillco-Zolá. 2019. Mapping evapotranspiration, vegetation and precipitation trends in the catchment of the shrinking Lake Poopó. *Remote Sensing* 12: 73. https://doi.org/10.3390/rs12010073.

UNDP. 1999. *Tajikistan human development report*, 95 pp. Dushanbe: Republic of Tajikistan and UNDP.

UNDP. 2005. *Central Asia human development report. Bringing down barriers: Regional cooperation for human development and human security*, 248 pp. Bratislava: UNDP Regional Office for Europe and the CIS.

Ussenalieva, A. 2020. Save Kazakhstan's shrinking Lake Balkhash. *Science* 370 (6514): 303 Letter. https://doi.org/10.1126/science.abe7828.

Vincent, W.F. 2010. Effects of climate change on lakes. In *Lake ecosystem ecology. Encyclopedia of inland waters*, ed. G.E. Likens, 65–70. Oxford: Elsevier.

Wheeler, K.G., M. Basheer, Z.T. Mekonnen, et al. 2016. Cooperative filling approaches for the Grand Ethiopian Renaissance Dam. *Water International* 41 (4): 611–634. https://doi.org/10.1080/02508060.2016.1177698.

Wurtsbaugh, W.A., C. Miller, S.E. Null, et al. 2017. Decline of the world's saline lakes. *Nature Geoscience* 10: 816–821. https://www.nature.com/articles/ngeo3052.

Chapter 24
Impoundment and Reservoir Development

Reservoir area is one of the most crucia environmental and social variables, but has not commonly been allocated much importance nor even provided by hydro proponents.
Goodland (1997: 82)

One of the most important processes that is started at first filling of a reservoir is the breakdown of organic matter which was present on the submerged land. This is of relevance mainly in the case of reservoirs located in the moist tropics, where potentially a large amount of biomass, i.e. plant material in areas covered by forests, is present, while the generally high temperatures lead to a fast decomposition of such material through the activity of microorganisms, at the same time limiting the amount of oxygen dissolved in the water. This can quickly lead to anoxic conditions in the deeper layers of the reservoir, and this in turn can cause the formation of methane, a very potent GHG, and its emission from the reservoir. Two cases of recently impounded reservoirs, Lom-Pangar in Cameroon (without) and Nam Ngiep 1 in Laos (with pre-impoundment reservoir area clearing) are presented shortly and their development in the first stages after impoundment is discussed. Both reservoirs showed typical stratification, however, with some differences between the two cases.

Results as the ones presented here can only be obtained from monitoring campaigns once the project was built; the ESIA will have to evaluate the situation during the planning phase, with the aim of coming to a realistic picture of the probable future development, and for being able to formulate suitable mitigation measures. Comparison with the actual development in observed situations is a great help in this undertaking.

24.1 Lom Pangar

Lom Pangar dam is located in the Eastern part of central Cameroon, just downstream of the confluence of the Pangar with the Lom River; the latter joins Sanaga River, the main river of Cameroon, about 20 km downstream of the dam site. The main

R. Zwahlen, *Assessing the Environmental Impacts of Hydropower Projects*, Environmental Earth Sciences, https://doi.org/10.1007/978-3-030-91185-0_24

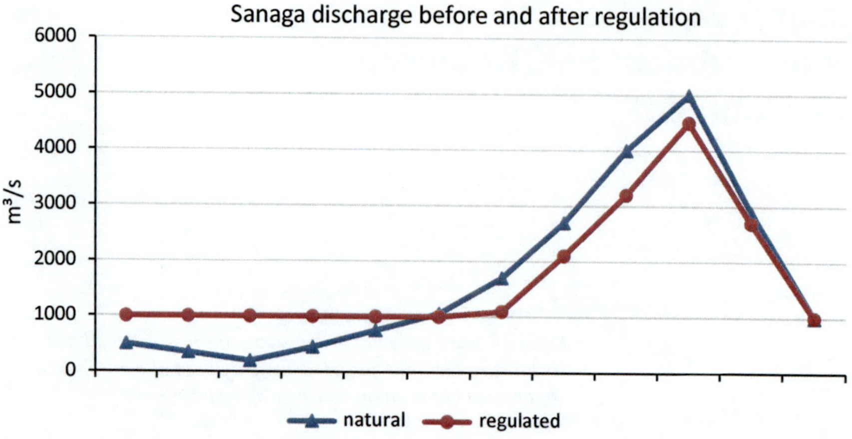

Fig. 24.1 Seasonal flow pattern of the Sanaga river without and with regulation. Based on data from Emadak et al. (2019b)

purpose of the project is not generation of electricity, but to provide seasonal storage for contributing to the regulation of the Sanaga River. Three hydropower stations are in operation along the Sanaga, and two more are planned. These are stations that are operated more or less as ROR plants, none of them has a sufficiently large reservoir as to be able to provide seasonal regulation. Since the Sanaga River is very seasonal, this means that large quantities of water have to be spilled in the wet season, while the power plants cannot be operated to their capacity during the dry season. River regulation, with several reservoirs on a basin wide scale, aims at having a constant water discharge in the Sanaga in the dry season (Fig. 24.1), which allows all these stations to run at their full capacity year round, and which in turn means a very considerable increase in overall electricity output. In order to contribute to this, Lom Pangar dam actually reverses the seasonal flow pattern in the remaining part of the Lom River before its confluence with the Sanaga (see Fig. 9.1). The main features of the project are summarised in Table 24.1.

The Lom Pangar reservoir, which submerges a considerable length of the two rivers, has a total area of 540 km^2, about 60% of which was covered by dense tropical forest concentrated in the lower part of the reservoir, the remaining part being mainly savanna covering the areas further upstream, with some water or bare surface. The amount of biomass in the forest was estimated at about 355 t/ha, while for savanna that was only about 8 t/ha. This sums up to a total of approximately 10 million t of biomass in the area to be submerged. It was clear that this had the potential to negatively impact water quality in the reservoir and in the downstream area.

Before impoundment, there was a big discussion about the necessity to carry out a pre-impoundment reservoir area clearing. While the POE, of which I was the environmental expert and the chairman, insisted on clearing as large an area as possible, other stakeholders were of a different opinion. The reasons in favour of an as thorough as possible pre-impoundment clearing were the concerns about very

Table 24.1 Main characteristics of Lom Pangar HPP

Parameter	Unit	Value
1. Dam		
Dam type		RCC
Height over foundations	m	45
Crest length	m	677
2. Reservoir		
Full supply level (FSL)	m asl	673.2
Minimum operation level (MOL)	m asl	649
Area at FSL	km^2	540
Area at MOL	km^2	136
Volume at FSL	Mm^3	7500
Volume at MOL	Mm^3	250
Active storage	Mm^3	7250
Reservoir length (Lom/Pangar)[a]	Km	179/101
Maximum depth	m	40
3. Power house		
Total installed capacity	MW	30
4. Hydrology		
Catchment area	km^2	19,700
Average discharge at dam site	m^3/s	244

Source ISL et al. (2005)
[a]See reservoir shape in Fig. 16.2

marked deterioration of water quality (anoxic conditions, emission of GHGs, largely in the form of methane) immediately after impoundment, with negative effects on the fish fauna, and the concerns about effects of the large amount of organic matter being flushed out of the reservoir, which would negatively impact water quality in the downstream area. Reasons in favour of a very limited biomass clearing, in addition to the considerable costs that this would have caused, were the facts that the reservoir is rather shallow, with a mean depth of about 14 m and a maximum depth of 40 m, and that the planned drawdown is very massive, the by far major part of the water in the reservoir being exchanged yearly, which would flush out the biomass suspended in the reservoir water. In addition, a structure for re-oxygenation was installed a short stretch downstream of the dam, and it was decided to flush the reservoir to the extent possible twice after the start of impoundment. This should lead to an improvement of water quality in the reservoir rather quickly.

Finally, it was decided to clear roughly 2500 ha in proximity of the dam (25 km^2), which corresponds to merely 5% of the reservoir area. Ironically, a few years earlier a World Bank environmental specialist had told me that the Lom Pangar project was "just a pretext for cutting the forest". In the end, only about 600 ha were really

Fig. 24.2 Lom Pangar reservoir gradually submerging the forest at first filling. Picture taken in November 2015, about 2 months after the start of impoundment. At this time, the reservoir level had reached 663 m asl, 10 m below FSL. A large amount of biomass is being drowned

cleaned, and not even the valuable timber had been taken out of the remaining part of the forest to be submerged (Ngeunga 2020) (Fig. 24.2).

Start of reservoir filling was in September 2015, and it reached a surface of about 300 km^2 in January 2016; it was then drawn down for reaching a surface of about 100 km^2 in May 2016, reached FSL and a surface of 540 km^2 in November 2016, for being drawn down again starting in the end of January 2017. Figure 24.3 shows the situation at the start and at the end of the first complete filling of the reservoir. The data show mainly that water quality deterioration was very fast, which was not surprising given the very large amount of submerged biomass; in the lower reaches of Pangar River, in November 2016 anoxic conditions were reached already at a depth of merely 2 m, and this improved only slightly till February 2017.

Effects in the reservoir, at least in the lower half of it, were quite dramatic. The fish stock literally exploded, and this in turn attracted a very large number of fishermen from different regions of the country. In 2016, the population of the village located nearest to the dam site, which had been about 200, rose to about 4000 within six months, and continued to increase. Fish yields were very high, but most of it seemed to consist of only two species, one catfish (*Clarias* sp.) and one cichlid (*Tilapia* sp.); as at that time unfortunately the fish monitoring study which should have been under way by then had not yet started, there was no reliable information on the amount of catches and on species. The two dominating species were obviously able to cope with the bad water quality, and they benefited from the availability of an enormous amount of food in form of rotting vegetation, and potentially also from the absence of predators, which had not survived under these precarious conditions. The development followed, in a somewhat almost exaggerated way, the pattern described

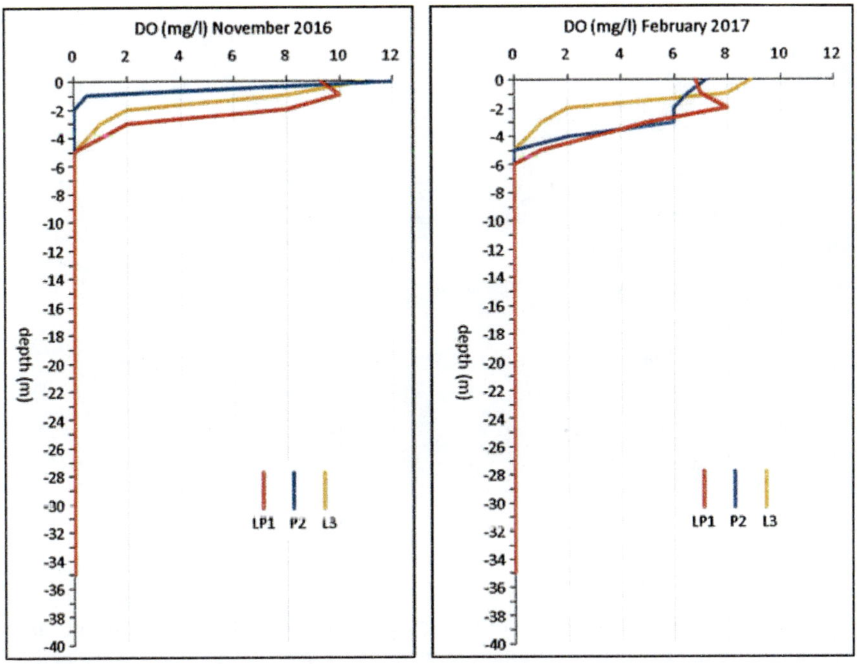

Fig. 24.3 Oxygen concentrations in Lom Pangar reservoir after first filling. LP1 = sampling station after confluence of Lom and Pangar Rivers, close to dam. P2 = sampling station in Pangar River, shortly upstream of confluence. L3 = sampling station in Lom River, shortly upstream of confluence. Based on data from EDF (2016, 2017), Emadak et al. (2019b)

by Balon (1974) for the case of Lake Kariba: massive increase in productivity, and massive decrease in biodiversity.

The situation was less favourable in the downstream area, in the remaining part of the Lom River. During filling, and then onwards at least till the start of operation of the power house, water was released from the reservoir in a way as to cause very high turbulence (see Fig. 24.4), and in addition to that there was the re-oxygenation sill a short way downstream of the dam. It was not surprising, therefore, that water quality monitoring just downstream of this sill constantly revealed a satisfactory oxygen content. However, a measuring campaign along the river down to the confluence with the Sanaga revealed a rapidly decreasing oxygen content, which after all did not come as a surprise: while water release and sill led to an immediate reoxygenation of the largely anoxic water drawn from the reservoir, BOD content in this water was so high that this oxygen was consumed almost completely immediately downstream of the sill, and here the river was flowing so calm and with so little turbulence that only very little oxygen was taken up again from the atmosphere. Unfortunately, this campaign was carried out only once, and the study, which should have closely followed the development of conditions in the downstream area, had not yet started. But fishermen in Lom 1, the only village in this downstream area, reported that fish

Fig. 24.4 Release of water during first filling for guaranteeing the environmental flow

had almost completely disappeared from this part of the river shortly after the start of impoundment, completely eliminating what had before been a very considerable part of their livelihood. In the absence of a good monitoring program it was, however, not possible to get reliable information on the development of the situation.

Greenhouse gas (GHG) monitoring during the first period after impoundment revealed a GHG emission of about 4.5 Mt CO_{2equiv} per year (Emadak et al. 2019a). Assuming that the 30 MW HPP integrated in the Lom Pangar dam would operate year-round at full capacity, and that regulation of the Sanaga would increase productivity of the Sanaga cascade by about 25%, this would roughly result in an electricity generation of 2.5 TWh/y. Producing this amount of energy in a coal fired power plant would result in the emission of about 2.5 Mt CO_{2equiv} (using the emission value from Table 14.3). This shows that in the short run, GHG emission from Lom Pangar is roughly twice that of a coal-fired power plant. However, GHG emissions from the reservoir will decrease rapidly, since the fine biomass will have been broken down within a few years. Coarse biomass, especially tree trunks which make a large part of total biomass, may remain there for a long period, and does not contribute to GHG, and especially to methane, production in to a very marked extent. Over the entire lifetime, even in this case with a rather unfavourable handling of the situation concerning biomass and GHG, hydropower is still much better in terms of emissions that fossil fuel alternatives. The ongoing monitoring will have to show when exactly the rate of emission will turn to the favourable side in this case.

24.2 Nam Ngiep 1

The Nam Ngiep 1 HPP (NNG1) is located in Bolikhamxay and Xaysomboun provinces in the central part of the Lao PDR on the Nam Ngiep River, a tributary of the Mekong. Reservoir impoundment started in May 2018, and electricity production in 2019. The reservoir is very long (73 km) and mostly very narrow, as shown in Fig. 24.5, the total surface is 66.9 km^2. Since it is largely used as a peaking plant, it was equipped with a re-regulation pond just downstream of the power house, with the aim of reducing the surge wave in the river downstream of the plant (Table 24.2).

A pre-impoundment biomass removal was carried out during the construction phase (Earth Systems 2015), and a water quality monitoring program was implemented. Some of the most relevant results of this are shown here and discussed shortly. In spite of this, oxygen consumption in the reservoir was very high, and anoxic conditions below the oxycline developed very quickly. The situation 1.5 years after reservoir impoundment is shown in Fig. 24.6. In the uppermost site, R 1, where

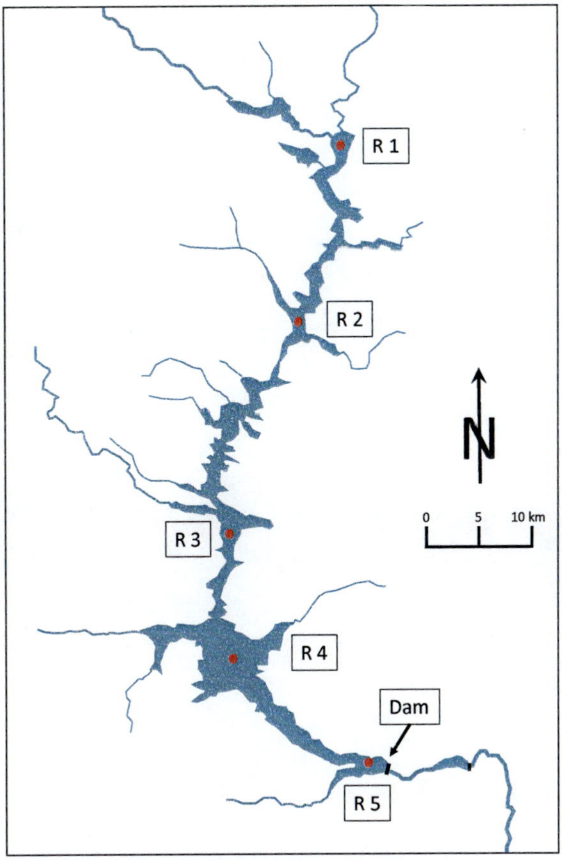

Fig. 24.5 Nam Ngiep 1 reservoir with monitoring stations. Red dots marked R 1 to R 5: water quality monitoring stations. Re-regulation basin and dam downstream of the main dam. https://www.adb.org/projects/documents/lao-41924-014-emr-7

Table 24.2 Main characteristics of Nam Ngiep 1 HPP

Parameter	Unit	Value
1. Dam		
Dam type		RCC
Height	m	167
Crest length	m	530
2. Reservoir		
Full supply level (FSL)	m asl	320
Minimum operation level (MOL)	m asl	296
Area at FSL	km^2	66.9
Volume at FSL	hm^3	2241
Active storage	Mm^3	1192
Reservoir length	km	72
3. Power house		
Total installed capacity	MW	273
Yearly output	GWh/y	1546
Maximum plant discharge	m^3/s	230
4. Hydrology		
Catchment area	km^2	3725
Average discharge at dam site	m^3/s	148.8
5. Re-regulation pond		
Dam height	m	20.6
Installed capacity	Mw	18
Yearly output	GWh/y	105

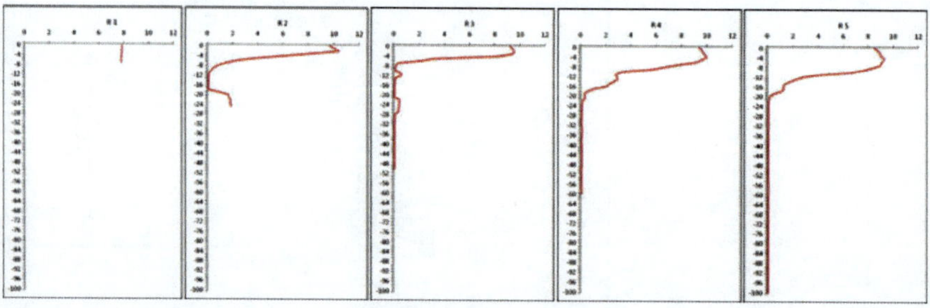

Fig. 24.6 DO in the NNG1 reservoir, 2 years after start of impoundment. R 1 to R 5: monitoring stations (see Fig. 24.5). Values for March 2020

the reservoir is still shallow and more like a slow-flowing river than a reservoir, oxygen was always present over the entire water column. But already at R 2 there was a reduction to almost zero oxygen below a depth of about 10 m, and this situation was maintained in the part of the reservoir further downstream. This means that in spite of biomass removal, there was still enough submerged biomass for a complete consumption of the available oxygen. It will now be interesting to see how this situation will develop over the next few years, and possibly to make a comparison with a similar situation where no biomass had been removed prior to impoundment.

One first indication could be what is shown in Fig. 24.7: while in April 2019, one year after the start of impoundment and about 8 months after first filling, the oxycline was very sharply defined at a depth of about 8 m, one year later it was more gradual and reached anoxic conditions only at a depth of 20 m, and in September 2020 there was some oxygen down to 40 m. This could be caused, e.g., by special weather

Fig. 24.7 DO in the reservoir near the dam. Red line: April 2019; blue line: March 2020; green line: Sept. 2020

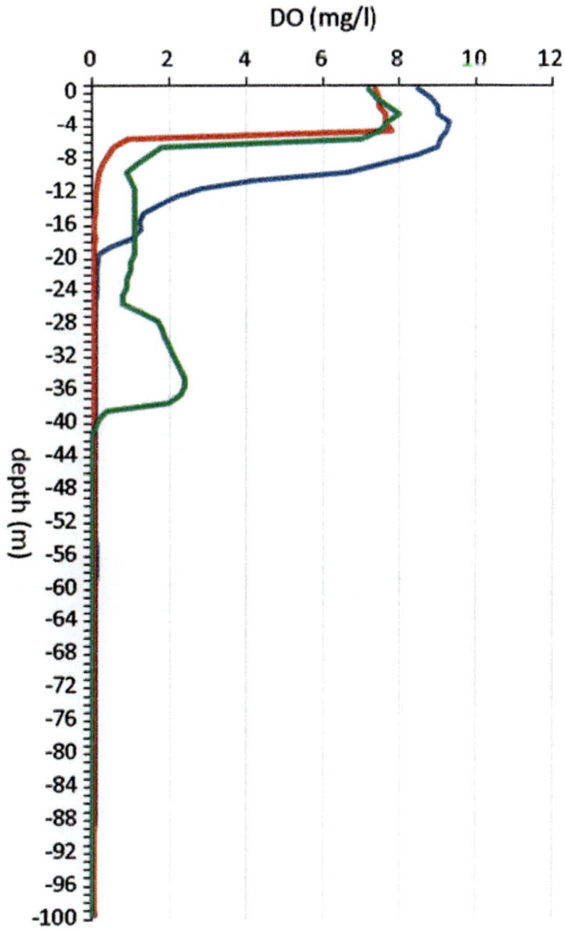

conditions, as strong winds, e.g. just before the March 2020 monitoring campaign, which might have mixed the upper layers of the reservoir to some extent, bringing more oxygen-rich water to a somewhat greater depth in the reservoir. On the other hand, it could also be a first sign of improvement of the situation. The water intake for the power house is located in the hypolimnion, below the depth of the MOL, i.e. more than 30 m below FSL. This means that anoxic water will be taken in and passed through the turbines, which is also clearly indicated by the low oxygen content of the water in the re-regulation pond. Over time, together with the ongoing breaking down of organic matter in the reservoir, this will lead to better water quality in the reservoir. Ongoing monitoring is important for documenting this development.

References

Balon, E.K. 1974. *Fishes of lake Kariba, Africa*, 144 pp. Neptune City: TFH Publications.

Earth Systems. 2015. *Biomass removal plan (BRP) for Nam Ngiep Power Company (NNP1)*. Prepared by Earth Systems for Nam Ngiep 1 Power Company. https://namngiep1.com/wp-content/uploads/2014/10/NNP1_Bio-Mass-Removal-Plan_July2015.pdf.

EDF. 2016. *Lom Pangar—Suivi de la qualité d'eau et des émissions de gaz à effet de serre*, 52 pp. Rapport de référence. EDC.

EDF. 2017. *Lom Pangar—Suivi de la qualité d'eau et des émissions de gaz à effet de serre*, 165 pp. Rapport annuel de suivi de la qualite de l'eau: année 2 de remplissage. Rapport preparé pour EDC.

Emadak, A., V. Chanudet, M. Dermaty, et al. 2019a. Greenhouse gas emissions from the reservoir of the Lom Pangar hydroelectric dam (Cameroon). Paper presented at the Africa 2019 hydropower conference, Windhoek, Namibia, Apr 2019.

Emadak, A., T. Nsangou, A. Towa, E. Bell, and H.N. Ngouyamsa. 2019b. Environmental and social management of the impoundment of the reservoir of the Lom Pangar hydroelectric dam: water quality monitoring. Paper presented at the Africa 2019 hydropower conference, Windhoek, Namibia, Apr 2019.

Goodland, R. 1997. Environmental sustainability in the hydro industry: Disaggregating the debate. In *Large dams: Learning from the past, looking at the future*, ed. Dorcey et al., 69–101. Gland and Washington: IUCN/The World Bank.

ISL, Oréade-Brèche, and SOGREAH. 2005. *Etude environnementale du barrage de Lom Pangar*. Rapport de Sythèse. République du Cameroun, Ministère de l'Energie et de l'Eau.

Ngeunga, M. 2020. Lom Pangar: Le lac de barrage aux bois morts. https://infocongo.org/fr/lom-pangar-le-lac-de-barrage-aux-bois-morts/.

Chapter 25
HPPs in High Mountain Areas

«Au grand barrage
On reprend la saison qui crache dans les bétonnières
Au milieu des grues métalliques pendues dans le vide à 3000»
"Le grand barrage", Jean Pierre Huser,
D'un millénaire à l'autre, 2007
At the big dam
They resume the season that spits in the concrete plants
Amidst the metal cranes suspended in the void at 3000

Dams and power plants in areas of high mountains may present different conditions from projects at lower elevations. On the one hand, the higher the location of such a site the smaller the catchment area will be, which means comparatively small amounts of water, water which in addition in most cases will show a very marked seasonal flow pattern. On the other hand, the topography will often be suitable for forming reservoirs with a small surface, but a large storage volume, when deep and narrow valleys can be used, and they have the potential to offer high head which can be used by means of suitable headrace structures. In this chapter, two projects are presented, Mauvoisin in Switzerland and Pamir 1 in Tajikistan. The first one, in addition to presenting options for enhancement of an existing project, also serves as a case study emphasising the issue of entering discussions with an NGO which is opposing the project.

25.1 Mauvoisin Dam Heightening and Mauvoisin II: Talking with NGOs

The Mauvoisin dam, located in the Alps in the South-West of Switzerland, was built in the 1950s, and by that time it was the highest double-arch concrete dam in the world. The reservoir is located in a narrow valley with steep flanks, and consequently has a rather small surface, but is very deep. Water stems mainly from snow and glacier melt, only a small portion is added by summer rainfall. The purpose of the plant is peaking production, mainly in winter; given the highly seasonal inflows, the reservoir

R. Zwahlen, *Assessing the Environmental Impacts of Hydropower Projects*, Environmental Earth Sciences, https://doi.org/10.1007/978-3-030-91185-0_25

is filled during summer, to be drawn down in winter. Over time, it became increasingly clear that the summer inflow exceeded the storage volume. It was therefore decided to heighten the dam with the aim of increasing storage capacity, in order to shift additional water from summer to winter, thus increasing energy generation in winter. The project envisaged a dam heightening by 13.5 m and an increase of the storage volume by 30 million m^3. Table 25.1 summarises the main characteristics of the original layout and the project (Fig. 25.1).

I was in charge of the EIA for this project, which was prepared in 1987/88. Given the fact that the increase in dam height, due to the topography of the area, would lead only to a small increase in reservoir area, and that in relation with the main river, the Rhone in Valais, the additional shift of water from summer to winter would be rather small, no major environmental impact was anticipated. Still, there had been a number of objections from NGOs (note: according to Swiss legislation on environmental protection, NGOs officially recognised as being active in the field of nature and environmental protection are entitled to make objections against projects they consider as too environmentally damaging, and in a high percentage of cases their objections are finally partly or completely accepted). However, by the time the project was ready to go into the final stage of getting construction permits, all but one had revoked their opposition. The one left was WWF Switzerland. My client had already decided to take this to court, and when I proposed that it might be worthwhile to engage in a discussion with the opponent, I was told that "one cannot talk to these people", and that in the end it would go to court anyway, a process which was expected to take three years. Still, they accepted that I could try, "if I felt like it". So I had a first meeting with the person who at that time was in charge of the local WWF section from which the opposition had originated; and of course: one could talk to them. As it turned out, their main point of objection was not the project of dam heightening as such, but the situation at the foot of the existing dam: there, in the 1950, the construction site had been installed. This part of the valley had partly been filled with excavation material from the dam foundations, and a platform for the main construction site structures had been formed, covering the original valley floor with up to 10 m of such material. In addition, excavation material from the various tunnels had been put there, and there were heaps of left-over aggregates which had been prepared for the dam construction. At the end of the construction period, all structures had been removed, but nothing else had been done there. So, almost 40 years after the end of construction works, the place still looked like a construction site, and this was also due to the fact that whoever in this region needed some gravel would easily find it there; this, given the remoteness of the site, did not happen too often, but it added to the fact that vegetation, which in this North facing valley in the shadow of the high dam and with a long-lasting snow cover had difficulties to develop anyway, was still as good as absent. And: in the first few hundred meters of the former river (which in fact had only been a small mountain stream), no water was flowing. For this reason, one major requirement of the opponent was to ensure a residual flow starting at the toe of the dam, in addition to a landscaping effort that should be done there after the end of the construction

Table 25.1 Main characteristics of Mauvoisin HPP

Parameter	Unit	Value
1. Dam		
Dam type		double curve arch concrete
Original height	m	237
Crest length	m	520
Original crest elevation	m asl	1962.5
Planned dam heightening	m	13.5
Final crest elevation	m asl	1975.0
Final dam height	m	250.5
2. Reservoir		
Original		
Full supply level (FSL)	m asl	1961.5
Area at FSL	km^2	2.08
Volume at FSL	Mm^3	270
After dam heightening		
Additional storage	Mm^3	30
New total storage	Mm^3	300
New areas at FSL	km^2	2.3
3. Water ways and power house		
Original[a]		
Upper headrace tunnel	m	5000
Upper stage head	m	469
Power house 1st step	MW	127.5
Lower headrace tunnel	km	12,500
Lower stage head	m	1016
Power house 2nd step	MW	225
Yearly output	GWh/y	820
Turbine output	m^3/s	28.75
Additional for Mauvoisin II[b]		
Headrace tunnel	m	18,380
Head	m	1485
New power house	MW	550
Overall yearly energy production	GHh/y	865[c]
Turbine output	m^3/s	46
4. Hydrology		
Catchment area	km^2	113.5

(continued)

Table 25.1 (continued)

Parameter	Unit	Value
Average discharge at dam site	m^3/s	8.2

Source EWI (1987), FFM (n.d.), Ecotec and Ecosens (1993)
[a]Existing structures, no changes required for dam heightening
[b]Mauvoisin II: additional project to increase total capacity; see explanations in the text
[c]Production of entire scheme (all three power houses); no additional water used, increase in overall energy production only due to higher efficiency of new machines and reduced losses

Fig. 25.1 Heightening of Mauvoisin dam by 13.5 m

works for the new project, requirements that seemed reasonable enough to me. We agreed to meet again, provided I could convince my client to participate.

Discussions stated, and they were not easy in the beginning. The first thing that had to be overcome was the mutual distrust. On the side of the utility, the opinion was that the main and only purpose of the NGO was to destroy their project, and on the other side there was the impression that the utility wanted to go ahead with this project with "only economic aims" in mind, and irrespective of any damage to the environment. Still, after the first few discussions it became clear on both sides that the respective opponents both had their good reasons, and that ultimately both were willing to come to a mutually acceptable solution. That was the moment when the discussion about specific measures started. As I recall it, one decisive moment was when the question of residual flow started. The argument of the WWF was that "you store 270 million (270,000,000) m^3 of water in the reservoir, so it should not be a problem to release 1 m^3/s at the foot of the dam". Quite interestingly, 1 m^3/s would sum up almost exactly to 30 million m^3 per year, the amount that was going to be retained by dam heightening. It was then agreed that a small creek should be created with 50 l/s. This water was not going to be taken from the reservoir, but from one of the lateral adductions, for a good reason: the water in the reservoir stemmed mostly from glacier melt and was extremely turbid due to the very fine silty material suspended in it, while the water from the lateral adduction was clear and therefore much more suitable for creating a small stream. In addition, it was agreed that at the end of the construction period the platform at the foot of the dam would be landscaped in a way as to better integrate it in the surroundings, and for enabling vegetation to start. One year after the first discussion an agreement was reached, and the project could go ahead.

A few years after the completion of dam heightening, there was another project, called Mauvoisin II. The plan this time was to build, in addition to the existing structures, a 18.4 km long headrace tunnel, a 2.4 km long penstock with a head of 1157 m, and a 550 MW powerhouse at the main valley floor (elevation 482 m), which would use the water from the reservoir in a single step. Since there was not more water available, this would not have increased the overall productivity (except for higher efficiency), the aim was to allow for a narrower, but higher peak production. This time, the planning approach was different: right from the start, all relevant stakeholders were included in the process: the local authorities, NGOs and, very importantly, the associations of fishermen, to take into account the fact that higher peaking would potentially have serious impact on discharge in the Rhone River, affecting the fish stocks there. The engineering work was advanced stepwise and only to the extent that it was required for having the necessary information for precisely assessing the impacts and the effectiveness of proposed mitigation measures, a procedure that was not really to the liking of some of my colleagues from the engineering department. This approach, however, lead to something that was probably a new record: no other project of similar size and importance had ever been granted all required permits in such a short time, and without any opposition. Ironically, just when construction started, the European power market was reorganised and opened, and suddenly Mauvoisin II was no longer interesting from a purely economic perspective and was abandoned.

Main conclusion from all this: you can always talk to those people; you might not always agree, and in some cases one might not come to an amicable solution, but it is worth a try (Zwahlen 2006).

25.2 Pamir 1

Pamir 1 is an HPP on the Gunt River, right bank tributary of Pyanj River, one of the two main headwaters of the Amu Darya River (see Fig. 19.1). The plant was planned, and the first stage was built, during the Soviet Union, with the aim of providing electricity to Khorog, the capital of the Gorno Badachshan Autonomous Region, the South-Eastern province of Tajikistan, and some of the surrounding villages in the main valleys of the Pamir mountains. The power station was planned for a capacity of 28 MW, and in the first stage half of that was implemented. As for other projects in that country, everything came to a halt at independence in 1991 (see Sect. 23.1). In spite of some damage caused during the civil war, the plant remained functional, but electricity generation was not sufficient for supplying the entire grid, especially in winter, when demand was highest. Before, additional demand had been covered by diesel generators which had been installed in many villages, diesel being a commodity which was supplied in sufficient amount. At the breaking up of the Soviet Union, this supply stopped immediately. In winter, the few villages which were still connected to the grid were cut off, and Khorog was supplied under a regime of load shedding; more isolated villages were left completely without any electricity year round.

In 1999, the Aga Khan Foundation, together with IFC, decided to rehabilitate the remaining structures and to bring Pamir 1 to its full capacity, and they commissioned a feasibility study, which included an ESIA (EWE 2001).

One prerequisite of this was to increase water availability in winter. The Pamir area has an extremely continental and high-mountain climate, with warm to hot summers, although with very large temperature fluctuations between day and night, at least at higher elevations, and very cold winters. Precipitation is scarce, falling mostly as snow in winter, the rivers being fed by water from snow and glacier melt, meaning that flow was rather high in spring and early summer and very low during the winter months. The plan was to put in a regulating structure at the outflow of Lake Yashilkul, a lake which had been formed by a landslide some thousand years ago (similar to the larger Lake Sarez, in the next valley to the North, which had been formed by a very large landslide in 1911, burying three villages; Alford et al. 2000). This structure would regulate the lake level, retaining some water during summer to be released as a base flow of 10 m^3/s in winter, in order to feed the power plant located some 100 river km further downstream. The lake level is at 3720 m asl, Gunt River originating from this lake drains a large part of the Pamir area, with a plateau at roughly 4000 m asl and high mountains reaching more than 6000 m asl (the highest mountain, called Peak Communism during Soviet times, whose name was then changed to Peak Ismoil Somoni, after the legendary king of Tajikistan, reaches 7495 m asl).

The original plan had foreseen to first lower the lake level by 2 m, and then to regulate it on that level. A first site visit carried out in 1999 quickly showed a potential problem of this design: people in the few villages on the Pamir plateau mainly live from animal husbandry, with yaks and sheep as the most important animals. Stocks had been raised greatly during Soviet times, since hay coming from other places had been supplied then as winter food for the herds. This supply as well was no longer available, and the herders had to make do with the fodder found in the area. At this altitude, and with the scarce precipitation, pasture productivity is very low (Breu et al. 2003). The only exception to this rule are the very few river floodplains, where plant productivity, without being high in comparison with lowland rainfed pastures, is considerably higher. One of these sites was located at the upper end of Lake Yashilkul, the delta formed by the Alichur stream flowing into the lake; this had formed a very flat area located just a few centimetres above the normal lake level, different types of grasses and sedges forming a green meadow. Lowering the lake level only a few decimetres would have dried out this plain, destroying the vegetation. This plain was a place where the inhabitants of neighbouring villages could make some hay, but mainly it was an important winter pasture for their yaks. In order to maintain this important resource, the project was then redesigned, avoiding any lowering of the lake. During design there was also a discussion of the possibility to raise the lake level by about 1 m, for having a larger storage capacity, however, this would have submerged the floodplain, destroying the pasture just as well as by drying it out. The plan was then made with the aim of keeping the lake level in summer as close as possible to its natural level, drawing it down in winter and bringing it up to normal level again as fast as possible in spring. A monitoring program was prepared and implemented with the aim of surveying pasture productivity in this area, and an obligation for the project to pay a compensation to the herders in case the construction work or the operation of the plant would lead to a lower productivity (Fig. 25.2).

Fig. 25.2 Lake Yashilkul. Above: part of the floodplain and the lake. Below: lower end of the lake, showing part of the rock fall material which had formed the lake

Fig. 25.3 Lake Turumtaikul. Left: part of the lake. Right: a small amount of water flowing through this type of river bed at very low temperatures would freeze completely

As already mentioned in Sect. 8.1.4.2, a few years later, the possibility of using another, much smaller lake for additional water supply to the power station in winter was investigated. The idea was to put in place a regulating structure at the outflow of Lake Turumtaikul, located at an elevation of 4202 m asl, and to release water from there during the winter months in the order of 1 m^3/s into the stream which reaches Gunt River about 30 km further downstream. This idea, however, was abandoned. Site visits had shown that over a stretch of several kilometres this stream flows through a rather wide river bed littered with large boulders, and a flow of this magnitude would, under the prevailing conditions of -20 °C or lower during several months, just have frozen, not leaving any water to reach the power plant (Fig. 25.3).

References

Alford, D., S.F. Cunha, & J.D. Ives. 2000. Lake Sarez, Pamir Mountains, Tajikistan: Mountain hazards and development assistance. *Mountain Research and Development* 20 (1): 20–23.

Breu, T., H. Hurni, and A. Wirth Stucki (eds.). 2003. *The Tajik Pamirs: Challenges of sustainable development in an isolated mountains region*, 80 pp. Centre for Environment and Development (CDE), University of Berne.

Ecotec and Ecosens. 1993. *Projet d'aménagement hydroélectrique de Mauvoisin II: Augmentation de la puissance de 550 MW*. Rapport d'imapct sur l'environnement. Forces Motrices de Mauvoisin.

EWE. 2001. *Evaluation of the GBAO electric networks and assistance in project preparation. Vol. 3: Environmental and social measures.* International Finance Corporation and Government of Tajikistan.

EWI. 1987. *Projet de surélévation du barrage de Mauvoisin. Analyse des effets sur l'environnement.* Forces Motrices de Mauvoisin.

FMM. n.d. *Mauvoisin II—augmentation de la puissance*, 20 p. Sion: Forces Motrices de Mauvoisin S.A.

Zwahlen, R. 2006. The value of talking to NGOs: A case study. In *Proceedings, Asia 2006, international symposium on water resources and renewable energy development*, Bangkok, 30 Nov and 1 Dec 2006.

Chapter 26
Bad Dams?

In less than a decade, hydropower and the dams associated with many developments have gone from being viewed as the most environmentally benign source of power to among the most aggressively criticized.
Churchill (1997: 105)

As mentioned earlier, I consider hydropower to be a generally environmentally friendly, renewable and important source of energy, with the clear limitation that this is not supposed to mean that every technically and economically feasible HPP is necessarily a good project. Here a number of hydropower projects are presented and discussed shortly which, in my personal view, should not be—or should not have been—built, and I'll also give the reasons why I came to this conclusion. As in the other case studies presented in this book, the examples will serve for highlighting one or a few specific aspects, it is not intended to give an overall, comprehensive account of any of these projects. One such category are projects located in a protected area, a number of which are mentioned in the first section of this chapter. A recent example of such a project, not mentioned in any detail here, is Rufiji HPP, located in the Selous Game Reserve in Tanzania, which is a UNESCO World Heritage site; this project led UNESCO in 2019 to put this important protected area on the List of World Heritage in Danger (https://whc.unesco.org/en/decisions/7503/).

26.1 Projects in Protected Areas

The only proven way to halt destabilization of the living world is to protect the largest possible reserves and the native biodiversity surviving within them.

Wilson (2016: 16)

R. Zwahlen, *Assessing the Environmental Impacts of Hydropower Projects*, Environmental Earth Sciences, https://doi.org/10.1007/978-3-030-91185-0_26

26.1.1 Ayago Nile

In the 1960s, a large HPP was envisaged to be built in Uganda, the Murchison Falls HPP, named after the major waterfall on the Victoria Nile about 50 km upstream from its mouth into Lake Albert, the head of which would have been used with a dam above and a power house below the falls. The obvious problem here was that this project was located in the centre of the Murchison Falls National Park (Anonymous 1968, 1970; Katete 1968); the falls and the gorge below it, which was the habitat of one of the largest population of the Nile crocodile left in Africa, was one of the main attractions of this national park, and the park itself was the main earner of income for the country, this at least until Idi Amin took the power and basically shut the country down.

To the best of my knowledge, this was the first hydropower project which met with a strong international opposition based on concerns about environmental protection. Besides the crocodile population, the park was also the habitat of a large elephant population, along with a considerable number of other animals. The main opponent was Prof. Bernhard Grzimek, the then director of the Frankfurt Zoo, who had become famous for his engagement for wildlife in general, East African wildlife, but mostly for his efforts to protect the Serengeti in Tanzania (Grzimek 1959, 1980; Sewig 2015). This project was eventually abandoned, but it was not entirely forgotten: an alternative site for a hydropower project was identified: Ayago-Nile. Starting in 1979, together with a Norwegian company we prepared the feasibility study for this project, and I was the one who did the environmental impact study, a subject that was still very new at that time. The main proponent of this project was the then head of the Uganda Electricity Board, an elderly English gentleman who apparently had been in this position since before the independence of Uganda. Somebody then told me that this gentleman would be going into retirement soon, and that this project was "his baby, since he wanted to leave something behind". I had the occasion to talk to him once, and I asked him why, since the original site at the Murchison Falls (which at that time were called Kabalega Falls) had been abandoned, he had not opted for a site outside the national park. To my surprise he told me that to the best of his knowledge this site was outside the park; well, it is not (see Fig. 26.1). He also told me another interesting story: Murchison falls is the place where the Victoria Nile drops from the plain into the Great Rift Valley. Since this valley had formed, the river had cut into the plain in the form of backward erosion, in this way forming the narrow gorge of roughly 10 km through which it is flowing downstream of the falls, before entering the valley floor. This backward erosion is continuing, and one day the river will reach Lake Kyoga, located further upstream, and at that time there would no longer be any Lake Kyoga and no longer any Murchison Falls; diverting most of the water from the falls through a penstock, therefore, would not destroy the falls, but actually save them in the long run. I am sure that erosion will lead to exactly that just a few million years down the road, but given the fact that such a project has a life span of maybe 100 years, I had my sincere doubts about its role in actually saving the falls! But to this day I do not know whether he really believed what he was telling me.

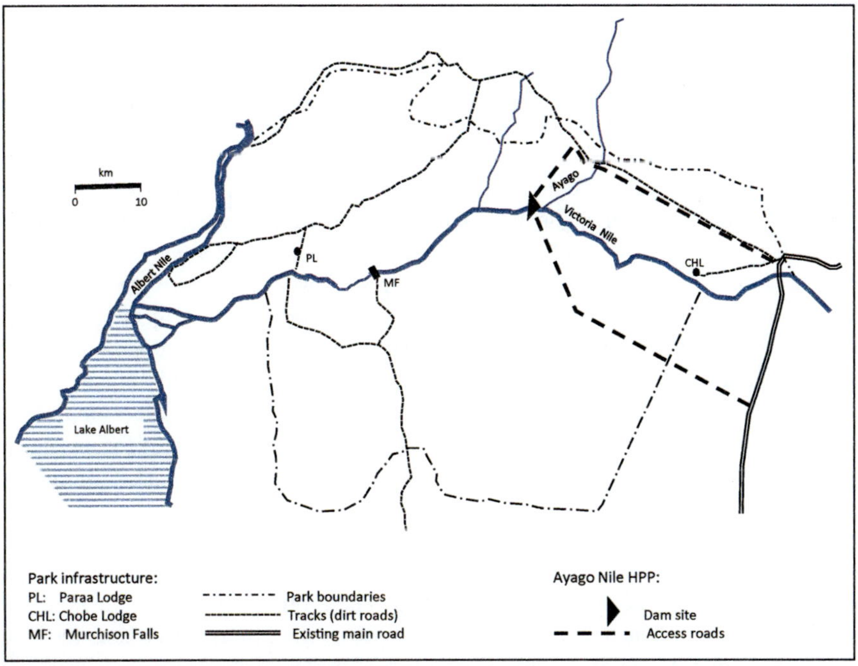

Fig. 26.1 Ayago-Nile project site in Murchison Falls National Park

It was an interesting project, and this certainly also because it gave me the occasion to walk around the project site, only accompanied by a park ranger, something one usually is not supposed to do in an African game park. The working conditions however, turned out to be somewhat difficult. Just when we started our work, Idi Amin had committed what he himself with hindsight may have considered as his greatest mistake: he had occupied a small bit of Tanzanian territory, the part between the South-Western border and Akagera River just upstream of where the latter flows into Lake Victoria, probably assuming that Tanzania would not care enough for taking any measures. He was wrong: just a few days after we had arrived on site, we were informed by radio that Tanzanian troops had started their invasion of Uganda, and that as one of the first effects the Entebbe airport had been closed. This left us in a somewhat difficult situation, because from where we were we did not have many options for leaving the country: West across Albert Nile into Zaire (now Democratic Republic of the Congo), North into the Southern part of Sudan (now South Sudan), East through the Karamoja area of Northern Uganda into Northern Kenya, a region largely characterised by unrest: none of these options looked like a good idea. Finally, we made it back to Kampala and then by road to Kenya.

One year later, I got the occasion to go back for finalising my work. A few things had changed. Soon after my arrival in the country I was advised not to use any longer the bank notes left over from my last stay, which still showed the portrait of Idi Amin. In the field, we now had a much stronger escort. In 1979, this had been two rangers, one with a spear and one with an old rifle which contained one last bullet. Now there was a group of three or four, and they were equipped with automatic weapons and apparently an ample supply of ammunition, which they were quite happy to use, e.g. once when we came down to the Albert Nile and saw a group of men intending to cross the river from the Zairian side, presumably poachers, who turned around and rowed back as fast as they could as soon as they had noticed us. I'm not so sure that this upgrade in equipment made me feel any safer, especially once, on a bumpy dust road, when the two guards in the back of our car suddenly loaded their rifles telling me that "now we were entering a dangerous area"; I was sure they were right. One other observation was made when walking along the Nile near the project site. While in 1979 I had seen several racks that poachers had used for drying hippo meat, this time I came across hippo cadavers from which only the teeth were missing, the meat having been left to rot. Hippos were numerous, and since tourism was completely down and no income could be gained from that, killing a few hippos for the meat did not seem a real problem, but killing them just for the small quantity of ebony seemed like a terrible waste.

From an environmental point of view, the main concern was not so much the dam itself, but rather the access roads which would have had to be built. In the course of the feasibility study, the project was modified. The original idea of building a high dam had been abandoned, and two ROR alternatives were investigated, one located on the north bank of the river, with an intake, a 2 km long deviation and an installed capacity of 2 × 60 MW, the other on the south bank, with a low weir across the river, a total capacity of 480 MW (2 × 3 × 80 MW) and a 9 km long deviation, which could have been implemented in stages. For these, no access road on the south bank would have been required, for the second alternative a road from the weir with river crossing "only" a road of about 10 km to the outlet of the tailrace tunnel would have been necessary. But even so, as I saw it, there would still have been major risks for the park: the fact that a large construction site lasting several years would have had to be opened there, but also the fact that at least one access roads would have had to be built into the park, opening access to areas that until then had been basically inaccessible. My recommendation was to abandon the Ayago site for good and to concentrate on Karuma Falls (a site just outside of the park, near the place where the main road crossed the Victoria Nile) and Bujagali, a site close to Jinja, both outside of protected areas and easily accessible by existing roads (EWI 1981). I'm quite happy to say that both of these sites have been developed since then, but I do not think that my report really contributed greatly to this outcome. Besides, recently I have heard that studies for developing Ayago-Nile have resumed; I still think it should not be built, but: once a site identified, it cannot be dis-identified, as already mentioned earlier.

26.1.2 Sutjeska National Park

Sutjeska National Park (NP), founded in 1962, is the oldest of the three NPs in Bosnia-Herzegovina (located within the part called Republika Srpska), with a surface of 172 km^2 and an additional buffer zone (hunting area) of 537 km^2. It contains one of the last remaining virgin forests in Europe, the Peručica forest, in addition to a wide variety of alpine and montane habitats (high mountains, alpine meadows, different forest types), lakes and rivers. It has a rich flora (2600 species recorded, among which a number of endemic and Red List species), 114 bird species and a high diversity of other animals, some rare and endangered species among them, like e.g. bear and wolf. Another NP, Durmitor, is located rather close to Sutjeska, across the border with Montenegro; Durmitor NP and Tara Gorge is a UNESCO World Heritage site (http://whc.unesco.org/en/list/100). There is a plan to link these two parks by means of a corridor along Tara River, which here forms the border between Republika Srpska and Montenegro.

In the area of this NP, on the river Sutjeska (itself a tributary of the Drina) and its tributaries, a total of 13 HPPs inside, on the border of or very near to the National Park were planned, with a total installed capacity between 43.35 and 51.8 MW, depending on the alternatives that would be chosen (Table 26.1). The locations in relation to the National Park are shown in Fig. 26.2. These were analysed as part of the SEA mentioned in Sect. 26.4.

Table 26.1 Hydropower projects on rivers in Sutjeksa NP

River	No.	Project name	MW	River-km
Sutjeska	1	S-1	5.3	7 + 500
	2	S-2 S-2a	7.4 3.8	26 + 600
	3	S3	3.2	31 + 000
Hrcavka	4	S-H-1	13.8	6 + 430
	5	S-H-2	1.7	9 + 500
Jabusnica	6	S-J-K-1 S-J-1a S-J-1b	9.2 6.8 4.4	0 + 680
	7	S-J-2	1.5	6 + 000
	8	S-J-3	1.7	9 + 000
Suški Potok (Suha)	9	S-S-1	0.8	0 + 530
	10	S-S-2	3.6	3 + 200
	11	S-S-3	1.7	5 + 700
	12	S-S-4	1	8 + 900
Kolbucarica	13	S-K-J-1 S-K-1a	0.9 0.85	2 + 000

Source Pöyry (2009b)

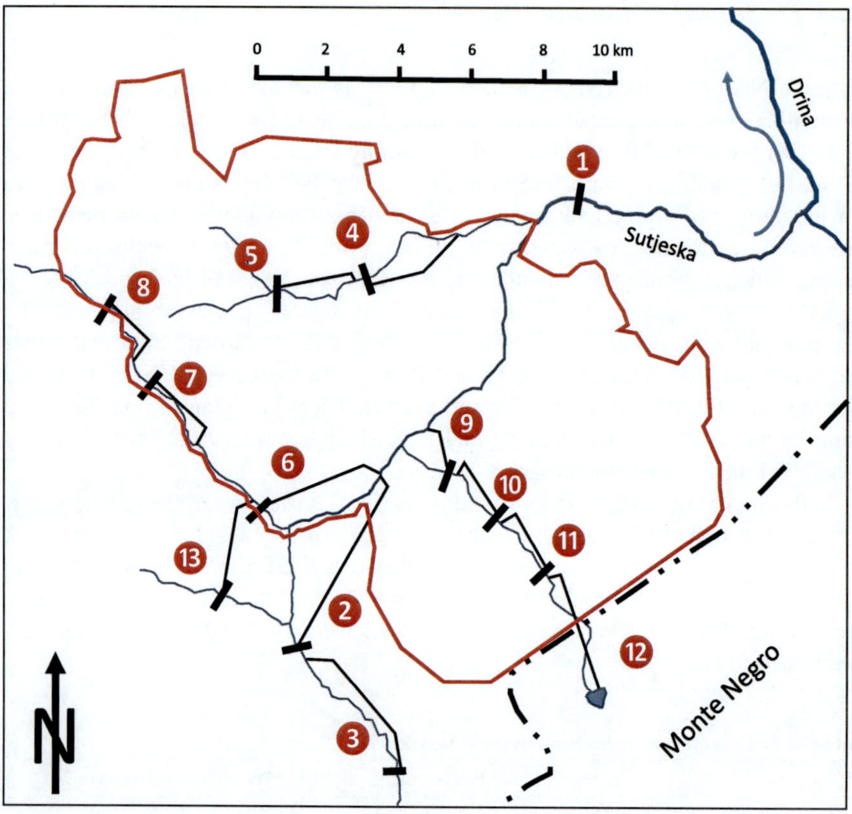

Fig. 26.2 Planned HPPs in Sutjeska National Park. Red line: NP border; dotted black line: border to Montenegro; planned dams/weirs and waterways shown in black. Numbers as in Table 26.1

Without going into details, the following main impacts that this development would have can be identified:

- Every one of the streams in this park, all of them rather small, steep and fast flowing mountain streams, would be interrupted several times, and the partly rather long deviations from intake to power house would leave considerable parts of them dry during most of the time. Fish migrations would no longer be possible. Fish are an integrating part of the Park's biodiversity.
- A major impact on the National Park would be caused by the access roads which would be required for construction, most of which would probably have to be kept for operation and maintenance. The transmission lines would constitute another serious impact.

The recommendation of the SEA (Pöyry 2009b) was not to develop any of these HPPs, each one of which would have serious consequences for the National Park.

So far, I have no information indicating that any of the projects in question would have been implemented in the meantime.

26.1.3 Ivindo National Park, Gabon: Kongou HPP

The planned Kongou HPP on the Ivindo River in Gabon was part of a much larger endeavour, which consisted mainly in an iron ore mine in the North-East of the country, a port located North of the capital Libreville for exporting the iron ore, an approximately 600 km long new railway line for transporting the ore from the mine to the port, and the HPP for providing energy for operating the entire project. Of this large undertaking, only the case of the HPP is mentioned here. The main characteristics of the project as planned at that moment are summarised in Table 26.2.

When we started working on the preparation of the ESIA for this project, the main problem became immediately apparent: it is located in a protected area, the Ivindo National Park. This park is known mainly for its large populations of two flagship species, the African elephant (*Loxodonta africana*, and here the rarer forest form of this species) and the Western gorilla (Gorilla gorilla), a critically endangered species.

In an earlier study, four sites with a potential for hydropower development had been identified (see Fig. 26.3). A short description of these four sites follows, whereby it has to be pointed out clearly that only the Kongou site (No. 1 in the figure) was evaluated in detail; the descriptions of the three others here are based only on their location in relation to the national park. One (No. 4) is close to the confluence of the Ivindo with the Ogooué, the main river of Gabon; this site is located completely outside of the park and would not affect the protected area in any way; furthermore, it is located in proximity of the existing railway line Libreville—Franceville, and this railway could most probably be used for most of the heavy transports necessary for construction. In No. 3, dam and power house are outside of the park, only a small part of the reservoir would reach into the park, affecting it only marginally. No. 2 is located completely within the park, and so is No. 1, Kongou; it can be assumed that both of them would probably affect the national park in a similar way, although no additional data are available on site No. 2. For the case of Kongou, one particular issue is the fact that the dam site is located on the Kongou Falls (hence the name of the project). These falls are often described as "the most spectacular water falls of the forested part of Africa" (Anonymous 2002; Vande Weghe 2006). Their attractiveness is not given by one major water fall, but by the fact that the river splits here over an area with a width of more than one km, where it then falls over a series of steps with larger and smaller cataracts and water falls, and this in a landscape covered by dense equatorial evergreen forest; two aspects of Kongou Falls are shown in Figs. 2.5 and 18.5e. The dam, with a length of almost 2 km, would very certainly destroy these falls completely, and the very large construction site required for the construction of the dam, waterways and power house would change the aspect of the site in a very decisive way.

Table 26.2 Main characteristics of the Kongou HPP

Feature	Unit	Value
1. Hydrology		
Catchment area	km^2	50,140
Mean annual discharge	m^3/s	823
2. Dam		
Type		Rockfill
Height	m	31
Length of dam crown	m	1935
3. Reservoir		
FSL	m asl	462
MOL	m asl	457
Size at FSL	km^2	41.97
Total storage	10^6 m^3	195
Live storage	10^6 m^3	156
4. Hydropower plant		
Total head	m	46.8 (52–41.5)
Total discharge	m^3/s	500
Residual flow	m^3/s	113
Mode of operation		ROR
Installed capacity (stage 1)	MW	150
Annual energy output	GWh	12,020
Total capacity (stage 2)	MW	500
5. Auxiliary structures		
Transmission line	km	40
Access road, existing dirt road	km	20
Access road new, within national park	km	29

Another major issue was the access road. Before the feasibility study for this project, the Kongou Falls had been accessible only by boat from Makokou: At the falls, there was a small camp with a few very simple cottages, and there was a system of foot paths for walks in the forest and for getting to see the falls from a few other places. For the study of the Belinga project, the developer had been assigned the Kongou site for developing the hydropower plant. Unfortunately, the developer then decided to open an access road to the site, mainly for bringing in the drilling rigs for the geological exploration. Unfortunately again, a logging company was mandated for opening this road, and they did it in the way logging companies do such things: fast, with heavy machinery, basically just pushing the trees and other disturbing vegetation to the sides; the result was an access road of almost 30 km inside the national park, far too wide for the purpose (see Fig. 15.1), and built without any of

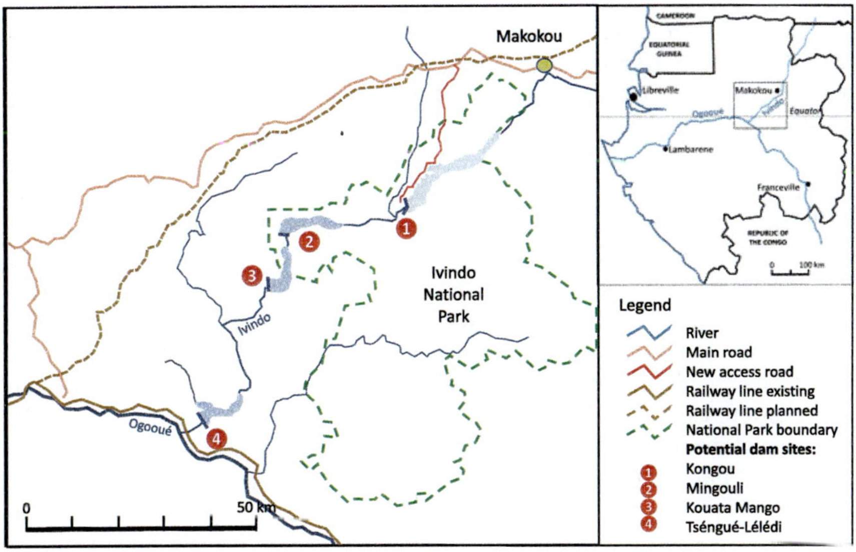

Fig. 26.3 Ivindo National Park with potential HPP sites and reservoirs

the required permits, most importantly without having informed the park authority beforehand.

The main recommendation of the ESIA (Pöyry 2009a) was to evaluate alternative dam sites and to leave the Kongou site intact. One obvious advantage of this site would have been the fact that no resettlement would have been required, since the area was not inhabited and not used in any other way than for the, at least at that moment, still very modest and low-level tourism activities and occasionally for some fishing, but the impact on this very relevant and sensitive part of the national park was considered as being too severe. Eventually, the developer pulled out, and to the best of my knowledge nothing more has happened since then in way of HPP development of this site. Still, some damage had certainly been done by opening the access road.

26.2 Palas Valley, Pakistan

A feasibility study was carried out for two cascades of hydropower plants on two left-hand tributaries of the Indus River in Pakistan (PSHC 2009, 2010). While for the Spat Gah valley no major environmental concerns had been identified, the situation was very different for the second valley, Palas, located just a short stretch further to the south. While in Spat Gah there was little forest, the higher parts of the valley slopes of Palas were covered with rather dense, intact forest, mainly oak forests up to about 1800 m asl and then coniferous forests up to the timber line at about 3300 m asl; it was supposed that the difference between these two valleys was explained at least in part by a regional gradient in precipitation, showing a gradual decrease in total annual precipitation from south to north. Palas Valley had been identified as a biodiversity hotspot (Sayer 2003), since it was a habitat for at least six locally or globally endangered species, namely, three Himalaya pheasants (Western Tragopan, *Tragopan melanocephalus*, Koklass Pheasant, *Pucrasia macrolopha biddulphi*, and Himalayan Monal Pheasant, *Lophophorus impejanus*) and three mammals (Asiatic or Himalayan Black Bear, *Ursus (Selenarctos) thibetanus*, Himalayan Musk Deer, *Moschus chrysogaster*, and Himalayan or Grey Goral, *Naemorhedus goral*). The most important of these at this site, from a conservation point of view, is the Western Tragopan, listed as vulnerable by IUCN, which has its largest remaining population in Palas valley. All these species largely depend on intact forests, and with the exception of the Goral, which can be found from approximately 1000 m asl upwards, they all live in elevations above 2000 m. With this, their habitat is well above at least the lower step of the planned cascade, which is located at about 1400 m asl.

One very important feature of Palas valley was the absence of a road. Experience from many parts of the world show that lack of access is the best protection of forest, while opening an area by building a road into it more often than not triggers logging, not seldom leading to deforestation (see Chap. 15); this is also documented for Pakistan (Ali et al. 2005; Raza 2006; both publications emphasise illegal logging as a major cause for deforestation). In the absence of a road it is simply not possible—unless helicopters are used—to take out the logs, and the few trees the local inhabitants might occasionally cut for their own use, and preferably close to their villages, had not had a major negative impact on forests. Building a dam, even if it is not a very large one, is not feasible without the construction of an access road, and there was a concern that, while the dam as such would have a minor negative impact on these important habitats, if any at all, the access road might trigger deforestation and with it destruction of the habitats of these species. Under such conditions, it would be preferable not to implement the project; in the very least, a very stringent management and protection program would have to be developed and implemented, which would have to include a strict access control.

In 2020, the WAPDA internet site indicated that the decision to build Lower Palas HPP had been taken, but did not actually mention that construction had started (http://www.wapda.gov.pk/index.php/projects/hydro-power/future/lower-palas-valley).

26.3 Ilisu

Ilisu HPP in Turkey is a special case, to which I want to give a very personal case description. To this aim, this case study starts with the following three statements:

1. In my opinion, Ilisu should not have been built.
2. I am aware of the fact that the decision on whether to go ahead with the project is not mine.

3. If it is implemented, then it should be done as well as possible, and if I can contribute to that aim I will.

The intention here is to elaborate some more on these three statements. While the first one is strictly personal, the second is quite obvious. The third however may be contested by a somewhat fundamentalist approach which would go along the line of "how can you possibly get involved in such a project".

26.3.1 The Project

Ilisu HPP is a large hydropower project located in South-East Anatolia on the Tigris River, the dam site being located approximately 50 river km upstream (N) of the border to Iraq. It is a major component of the South-East Anatolian Development Project (GAP), which aims at improving the conditions for the population in this area by providing energy, irrigation, infrastructure and improved services e.g. concerning health and education (Anonymous 2000). Ilisu as such is a single-purpose hydropower project, but it contributes to flood protection and water availability for irrigation in the GAP area. The main features of the project are listed in Table 26.3.

26.3.2 My Involvement

A consortium which had been formed for providing turbines and electro-mechanical equipment required for the construction of Ilisu had put in a demand for export risk insurance. The group of ECAs involved in a first step, then still under the leadership of the Swiss Department for Economic Development (SECO) included the USA, Germany, Portugal, Sweden, Austria, the UK and Switzerland. It had become evident at a rather early stage that there were potential problems with the ESIA, which was not really up to international standards and therefore did not meet the requirements of these countries. I had been contracted by the Swiss export risk insurance company for providing comments and input in this situation, in a first step mainly for resettlement, which had been identified as the probably most important issue, where a considerable improvement in the approach to the problem and the planning of the process to be followed was definitely required. In the ensuing discussions, the consortium changed, and so did the group of ECAs; several of the countries involved originally withdrew, and finally there was a group of three left, namely Germany, which had taken the lead, Switzerland and Austria. During this process, the requirements for ESIA-related aspects had been identified more clearly, and the three ECAs had decided to install an expanded Committee of Experts for covering these aspects. This COE finally consisted of three groups, each one chaired by an international expert and comprising several Turkish experts. The three groups were Resettlement (chair: Michael Cernea), Archaeology (chair: Margarete van Ess) and Environment (chaired

Table 26.3 Ilisu HPP: main characteristics

Parameter	Unit	Value
1. Dam		
Dam type		Rockfill with central core
Height over foundations	m	135
Crest length	m	1820
Crest elevation	m asl	530
2. Reservoir		
Full supply level (FSL)	m asl	525
Area at FSL	km^2	300
Volume at FSL	Mm^3	10,410
Active storage	Mm^3	7460
Reservoir length (Tigris valley)[a]	km	136
3. Water ways and power house		
Penstock	m	420
Total installed capacity	MW	1200
Yearly output	GWh/y	3833
Turbine discharge	m^3/s	211
Operation mode		Peaking
Powerhouse discharge	m^3/s	100–1200
4. Hydrology		
Catchment area	km^2	35,517
Mean annual rainfall	mm	814
Average natural discharge at dam site	m^3/s	490
Minimum flow[a]	m^3/s	2.5–4.9

Source IEG (2001)
[a]The minimum flow is provided by a 4.5 MW auxiliary unit

by me). Over a number of years, these three groups closely followed project development, commented on plans, and proposed mitigation and compensation measures. Finally, a long list of ECA requirements, which had to be observed as a condition for the involvement of the three countries, had been prepared. This list was discussed in a meeting with all involved parties in Ankara, and it was turned into a list of mutually agreed conditions which were to be implemented, under a continuous scrutiny of the ECAS and control by the COE, as a condition for granting the export risk insurance.

In parallel to this development, resistance against the project had formed, organised by a group of NGOs in the three supporting countries (Carle et al. 2007). There were some meetings of the project working group, and of the COE, with delegations of these NGOs, where a number of issues were discussed, but the main activities of the project opponents consisted in making political pressure in their respective home countries. The aim was to prevent an involvement of the three countries in the project, and in this way to stop the project from being built. Not surprisingly, the main arguments against the project were related to resettlement and archaeology. Environmental issues were raised, but they very clearly did not stand at the forefront of the discussions.

As mentioned in the introduction to this chapter, I am of the opinion that Ilisu should not have been built, and in this respect I agreed with the NGOs opposing the project. From an environmental point of view, the project certainly had to face the same issues as every large dam project: interruption of a river, submersion of terrestrial habitats, change from lotic to lentic conditions, change in downstream flow conditions (Turan 2000). None of these impacts had been considered as extremely serious, and none was considered as a no-go for the project; mitigation measures were available and had been agreed upon. A more difficult issue was resettlement (Koppel Maldonado 2012). As the project was going to create a large reservoir, a considerable number of villages and a population of about 12,000 people were directly affected and would have to be resettled. But here as well, this was considered as a big, but not an insurmountable task, the important issue being the way in which the resettlement was going to be implemented. For this as well, conditions had been formulated and agreed upon. The main issue, however, was archaeology. It was estimated that in the future reservoir area there were approximately 200 sites of potential interest for archaeology which had not yet been investigated; Tuna et al. (2001) provide an overview of salvage excavations done in 1999. It might have been possible to excavate a number of those before impoundment of the reservoir, but certainly not all of them. But this was not the real problem. It could be argued that a submersion of such sites would not destroy them, it would make them more or less inaccessible at least for the time during which the reservoir would be there, and it would certainly have protected them from inadvertent destruction by any type of future activity, or from excavation by amateurs or treasure hunters. It has to be seen that Turkey, with its long and interesting history, abounds in archaeological sites of all types, and that a serious investigation of all of these sites is not feasible, or will at least require a very long time.

The problem was Hasankeyf. This is a very special case of a cultural heritage site indeed. A small town in south-eastern Anatolia, on the right bank of the Tigris River, it has been inhabited without interruption since at least 5000 years, which makes it one of the places on earth with the longest history of continuous occupation. Several cultures have come and gone, and each one left its marks. One very special characteristic are the caves in the cliff which were inhabited for centuries, some of them until very recently (Arik 2001; Barchard 2009; Gies 2017). Hasankeyf has in the meantime largely disappeared in the waters of the Ilisu reservoir, and much of it, for certain all the buildings erected by mud bricks, like e.g. the tall minaret, will be destroyed. I'm not an archaeologist, but as far as I know Hasankeyf was something very special even for a country as rich in historical monuments as Turkey, and no really similar place or situation exists. Hasankeyf is not the first archaeological site affected by hydropower development, and it will certainly not be the last. The most famous example of mitigation measures taken in such a case are probably still the monumental statues of Abu Simbel which were relocated in a spectacular action in 1968 before being submerged by the raising waters of lake Nasser created by the Aswan High Dam. Other projects as well can interfere with such sites, as shown by an airport project in Peru (Wade 2019). However, for me, Hasankeyf would have been the one and only, but a very valid reason for not implementing the Ilisu project (Fig. 26.4).

This leads to a second statement: of course, the decision whether or not to implement Ilisu was not for the COE to take, and it did not depend on this group in any way. By preparing an ESIA Report, the persons in charge of this task can—or are supposed to—give advice, arguments and opinions on what to do in kind of mitigation measures, and in a way the ultimate mitigation measure certainly is a recommendation not to go ahead with the project. The ESIA for Ilisu (IEG 2001) had not come to that conclusion, however. The COE's involvement in the project had come much later, and its role here was different. The Turkish government had decided long ago to build the Ilisu dam, since this project was considered as one of the most important components of the GAP Project, a large regional project for the development of the entire south-east Anatolian region (Barchard 2009: 11). The main purpose of the

Fig. 26.4 Hasankeyf. View from the top of the cliff on the right bank. The caves visible near the right margin of the picture were partly destroyed recently by rockfall which ripped away the "front walls" of the formerly inhabited caves; caves are also visible on the left bank of the Tigris River. The red arrow shows approximately the level the water in the reservoir will reach on the high minaret. Picture taken 2007-12-06

Ilisu project, in addition to hydropower, is to provide water for irrigation in the area downstream of the dam (Turan 2000; IEG 2001).

The third point: if a project is implemented, then it should be done as well as possible. Besides the question of "what is good?", to which one will inevitably get different answers depending on whom one asks, there is also an important ethical question: if an expert thinks that a project is "not good", i.e. does not fulfil the standards he or she would like to see the project adhere to, can this expert still justify a participation in such a project? This is a question which has implications on different levels. Here, an answer is sought from the point of view of an expert who has the possibility to be involved in such a project in the way as I was involved in Ilisu. And the answer to this question is clearly yes, such an involvement is justifiable, and it might even be an obligation: if I am convinced that it is not a "good" project, but I know that it will be implemented in any case, I still think that it is worth any effort to try and make it as good as possible under the circumstances. This was certainly also the motivation of my colleagues in this Committee.

26.3.3 Project Development and Conclusions

After the meeting mentioned above, where the environmental, social and cultural conditions had been determined and agreed upon, work on the ground started. A detailed resettlement programme was developed, and work started for the relocation of the village which was located exactly in the footprint of the future dam and which therefore would have to be relocated before work on the dam as such could start. A detailed plan for salvaging a number of the historical structures, by moving them to higher ground, was being established. For the construction site as such, detailed EHS (Environment, Health and Safety) rules were determined. It became soon evident that, while all involved parties had agreed to the conditions, there was room for interpretation of some of them. As an example just the following: it had been agreed that only preparatory work would be carried out until the three supporting countries would have come to a final decision on whether or not to continue to support the project, and that up to that moment no work which would touch the river would be started. Then suddenly it became clear that the contractor had built an access road from the right bank, where all the construction site infrastructure was located, to the dam abutment on the left bank, and this access included a crossing of the river by means of a dyke with large culverts. For the ECAs, this was "work influencing the river", while for the project proponent this was "preparatory work", and it was argued that this dyke was required since the construction of a bridge which would provide permanent access required more time, and work on the left bank was not possible without access. Such differences in opinions and views are probably inevitable, but the continued follow-up of the ECAs and the COE allowed for an early detection of such discrepancies, and for finding a mutually acceptable solution.

In the meantime, the NGOs opposing the project had increased their political pressure, mainly in Germany, and this finally led to the decision of the German government to withdraw its support of the project. Since the German export risk organisation had the lead of this project setup, the other two countries, Austria and Switzerland, followed. This decision also meant that the involvement of the COE came to an abrupt end. The NGOs celebrated this as a victory, since they had succeeded in stopping Ilisu HPP. However, they soon came to realise that this was not the case at all: construction of the project continued; what they had succeeded in, however, was in stopping the support by the group of the three countries, and with this also the supervision of the works by the COE. Soon afterwards, I came across a rather disillusioned statement saying that in spite of this withdrawal the work continued, and this entirely without any external control. This, however, was a totally foreseeable development, given the determination of the county to implement this project.

I have very little information on the progress of the Ilisu project since then, or on the conditions concerning environment, resettlement and cultural heritage, and by no means I want to say that all this was badly done. However, I am rather convinced that project implementation, from this point of view, would have been better if the agreed upon type of input and control had been maintained. The success of the NGOs, in my view, was in reality a failure, and I am convinced that they would have been able to achieve more if they as well had accepted that Turkey, as a sovereign country, would not decide in their sense, and if instead had directed their effort towards helping to implement the rather far reaching measures which had already been accepted.

26.4 Lower Bosna Cascade

They say you learn from your mistakes - it's a lie.

Dani Klein, Vaya con Dios: Time flies (1992)

A Strategic Environmental Assessment aimed at comparing a total of 129 smaller to medium-sized hydropower projects (with installed capacities between 0.09 and 48 MW, the vast majority smaller than 5 MW, only 16 > 10 MW) in three river basins (Vrbas, Bosna and Drina) in the part of Bosnia-Herzegovina called Republika Srpska (Pöyry 2009b). Among the sites identified, there was a cascade of 7 power plants along the lower Bosna River, from slightly upstream of the town of Doboj down almost to the confluence with the Sava, which forms the border with Serbia, with a total planned installed capacity of 90.2 MW, the single steps being between 8 and 14.4 MW; this cascade represented a considerable part of the entire potential. All these plants were planned to be built in a similar way: lateral dikes would straighten the now widely meandering river, the dams would provide some head, but very little storage capacity, and the plants would basically be operated as run-of-river plants. In all cases the dams and parts of the dykes would be built "on dry land", i.e. not in

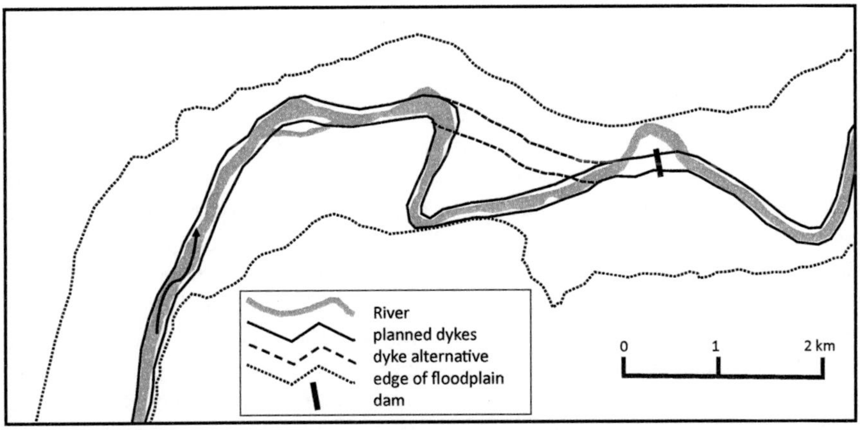

Fig. 26.5 One of the 7 HPPs envisaged for the lower Bosna cascade. Dashed lines: approximate outer limits of the floodplain, which is used mainly for agriculture. Solid black lines: dykes for channelling the river. Dashed lines: possibility for further straightening the river. Black bar: dam and powerhouse site

the present river bed, and the river would then be diverted into the new bed and over the dams (see illustration in Fig. 26.5).

At the time of our assessment, planning of these plants was far advanced, and construction was supposed to begin soon. According to a short note in Hydropower and Dams Vol. 16, Issue 4, 2009 (p. 32), EPC contractors were being sought, and construction was expected to start this same year, commissioning being planned for 2011. The assessment identified the following main impacts:

- Cutting off river bends, straightening the river and thus accelerating the flow; note that the dikes shown in Fig. 26.5 do not necessarily represent a final plan, further straightening of the river would be possible.
- Cutting off the main part of the floodplain which under natural conditions was exposed to the dynamics of the meandering river. This would, on the one hand, protect agricultural areas from flooding. On the other hand, is would change (or largely destroy) floodplain habitats in his area. According to information obtained on site, approximately 1000 ha of wetland would be affected.
- Flood protection would also encourage the construction of houses in the areas which before had been exposed to regular flooding; at that time, there apparently was no regulation on where houses were allowed to be built. This would carry the risk of greater damage in the case of extreme floods which might breach the dykes.
- The floodplain with a size of about 60 km^2 presented a large flood retention area, which would then no longer be available. According to technical information received, dykes and dams were planned for being able to evacuate a 1000 year flow. With the retention volume missing, flow speed in such a case would be

increased greatly, the floods would reach the Sava downstream of the cascade much faster and with higher peaks, increasing the risk of flooding in this area.

It had not been possible to check the ESIA which reportedly had been prepared for the cascade or for individual plants, but according to information obtained on site the risk for downstream areas had not been identified and assessed, and no mitigation measures had been envisaged. The SEA recommended to carry out these studies in any case before actually starting construction. Interestingly, planning of this cascade was done at a time when exceptional floods had affected different parts of Europe, but mainly Germany, e.g. along the Elbe River, and had caused enormous damage in the city of Dresden, among others, and a German development bank apparently was interested in financing or co-financing these projects. These floods along the Elbe River, as so often, had been "shifted downstream" due to lack of storage areas as a result of river correction similar to the one proposed for the Bosna cascade, and a lot of money was being spent for improving this situation, the measures also including widening of rivers, where still possible, for restoring some of the former retention capacity.

The map for Bosnia-Herzegovina provided in Schwarz (2020: 16) would suggest that of this cascade of 7 only one (possibly 2) are still envisaged for construction; the most recent Google Earth images do not give any indication that any of them has been constructed or would presently be under construction.

References

Ali, J., T.A. Benjaminsen, A.A. Hammad, and B. D. Øystein. 2005. The road to deforestation: An assessment of forest loss and its causes in Basho Valley, Northern Pakistan. *Global Environmental Change Part A* 15 (4): 370–380. http://www.sciencedirect.com/science?_ob=ArticleURL&_udi=B6VFV-4HCDJNH-1&_user=10&_rdoc=1&_fmt=&_orig=search&_sort=d&_docanchor=&view=c&_searchStrId=1004635318&_rerunOrigin=google&_acct=C000050221&_version=1&_urlVersion=0&_userid=10&md5=9dbc303414747342dfa2ca12f5db0b32.

Anonymous. 1968. *Hydro-electric power development threatens Uganda's Murchison Falls National Park*, 8 pp, Kampala.

Anonymous. 1970. How to destroy the Murchison Falls. *Oryx* 10: 207.

Anonymous. 2000. Water and development in southeastern Anatolia. Essays on the Ilisu dam and GAP. In *Proceedings of a seminar held at the Turkish Embassy in London*, Feb 2000, 79 p. Turkish Embassy.

Anonymous. 2002. *Les parcs nationaux du Gabon. Stratégie pour le troisième millénaire*. Libreville, Gabon: Multipress Gabon.

Arik, M.O. 2001. Hasankeyf excavation and salvage project. In *Salvage project of the archaeological heritage of the Ilisu and Carchemish dam reservoirs*, ed. N. Tuna, et al., 795–831. Activities in 1999. Ankara.

Barchard, D. 2009. *Land of the lakes. The rebirth of Southeast Anatolia*, 144 pp. Bristol Book Publishing.

Carle, C., A. Rota, and C. Eberlein. 2007. Ilisu: Umstrittenes Projekt. *TEC* 21–23 (2007): 24–31.

Churchill, A.A. 1997. Meeting hydro's financing and development challenges. In *Large dams: Learning from the past, looking at the future*, ed. Dorcey, et al., 103–110. Gland and Washington: IUCN/The World Bank.

EWI. 1981. *Ayago-Nile hydroelectric project, Uganda: Ecology*, 75 pp. Electrowatt Engineering.

Gies, E. 2017. The real costs of energy. *Nature* 551 (7682): S145–S147.

Grzimek, B. 1959. *Serengeti darf nicht sterben*, 333. Berlin: Ullstein.

Grzimek, B. 1980. Ugandas Tierwelt wird überleben! *Das Tier* 20 (5): 11–15.

IEG. 2001. *Ilisu dam and HEPP. Environmental impact assessment*. Dolsar: Ilisu Engineering Group, Consortia for Ilisu: Hydro Concepts Engineering, Colenco, Hydro-Québec International.

Katete, F.X. 1968. The mistake at Murchison Falls. *Africana* 3 (8): 18–21.

Koppel Maldonado, J. 2012. A new path forward: Researching and reflecting on forced displacement and resettlement. *Journal of Refugee Studies* 25 (2): 193–220.

Pöyry. 2009a. *Projet Minier de Belinga: Etude d'Impact Environnemental et Social pour le Projet Hydroélectrique de Kongou*. Volume III: Projet Hydroélectrique de Kongou. Report prepared for Compagnie Minière de Belinga (COMIBEL).

Pöyry. 2009b. *Vrbas, Bosna and Drina SEA. SEA of river basins of Vrbas, Bosna and Drina in Republika Srpska, Bosnia and Herzegovina*. Report prepared in association with Dekonta and University of Banja Luka for EBRD.

PSHC. 2009. *Lower Palas hydropower project. Feasibility study. Palas Valley—Spat Gah hydropower consultants (ILF, Pöyry and ACE) for WAPDA*. Pakistan: Water and Power Development Authority.

PSHC. 2010. *Lower Spat Gah hydropower project. Feasibility Study. Palas Valley—Spat Gah hydropower consultants (ILF, Pöyry and ACE) for WAPDA*. Pakistan: Water and Power Development Authority.

Raza, M.H. 2006. *A souvenir of KKH—Gilgit-Hunza & Skardu. Saga of the silk route*, 269 pp. Islamabad, Pakistan: Best Books.

Sayer, J. 2003. *Prospects and implications of promoting Palas Valley, Kohistan, as a UNESCO world heritage site/man and biosphere reserve*. Consultancy 2003.17 (No. 12), Final Report. Gland, Switzerland: IUCN.

Schwarz, U. 2020. *Hydropower projects on Balkan rivers—Update 2020*, 33 pp. Vienna/Radoflzell: RiverWatch & EuroNatur.

Sewig, C. 2015. *Bernhard Grzimek, der Mann, der die Tiere liebte*, 428 pp. Cologne: Bastei Lübbe.

Tuna, N., J. Öztürk, and Velibeyoglu (eds.). 2001. *Salvage project of the archaeological heritage of the Ilisu and Carchemish dam reservoirs*. Activities in 1999, 854 pp. Ankara: Middle East Technical University, METU Centre for Research and Assessment of the Historic Environment.

Turan, I. 2000. International aspects of water issues. In *Water and development in southeastern Anatolia. Essays on the Ilisu dam and GAP. Proceedings of a seminar*, 38–45. London: Turkish Embassy.

Vande Weghe, J.P. 2006. *Ivindo et Mwanga. Eaux noires, forêts vierges et bais. Les parcs nationaux du Gabon*, 272 pp. Libreville, Gabon: Wildlife Conservation Society.

Wade, L. 2019. Airport construction threatens Incan heartland. *Science* 363 (6427): 568–569.

Wilson, E.O. 2016. *Half earth. Our planet's fight for life*, 259 pp. New York: Liveright.

Chapter 27
Epilogue: Whither Dams?

Looking to the future, a new development paradigm is needed that can wean societies from the need for more and more large dams.
Scudder (2005: 294)

In his book entitled *The Future of Large Dams*, under the heading *Should large dams be built?*, Thayer Scudder states that "The unfortunate answer is yes, but ..." (Scudder 2005: 295). The "but" is a request for best practice in environmental and social issues related to hydropower projects, and his book provides ample evidence, mainly from a social perspective, showing that best practice is still not reached, and often by a wide margin. This is certainly also true for topics of environmental concern, as has been shown above with numerous examples. In this final chapter, after a few comments on the future of hydropower (keeping this very short, since this is not my area of expertise, and in the hope not to pass as too much of a "terrible simplifier"), the question of how to go on and hopefully to improve the ESIA process is addressed. This needs to be done on two levels: first, the ESIA as such has to be improved, and this does not only refer to carrying out the necessary studies and preparing the reports, but has to include preparation (TOR!), the administrative procedure and the evaluation process as well. Secondly, and arguably even more important, progress must be made in implementation of the measures identified, proposed and accepted. Without reaching a good level in this respect, everything else would have been in vain. A number of points are raised for aiming at improvements on both levels, being aware of the fact that a final "best practice" will never be reached, not least because, as emphasised in Sect. 21.2.3, "... the times they are a'changin"! Improvement is an ongoing process.

R. Zwahlen, *Assessing the Environmental Impacts of Hydropower Projects*, Environmental Earth Sciences, https://doi.org/10.1007/978-3-030-91185-0_27

27.1 A Future for Hydropower?

27.1.1 Growing Energy Demand

Energy demand is growing. This is a simple and unquestionable statement. According to IEA, total energy supply rose from 5800 t_{oe} (oil equivalent) in 1971 to over 14,000 t_{oe} in 2018 (https://www.iea.org/data-and-statistics/charts/world-total-energy-supply-by-source-1971-2018). Drivers of this development are numerous, but certainly population increase and technological development of all types play a role, the latter in spite of the fact that in most aspects energy efficiency is increasing, and continues to do so. One of the very important, and certainly one of the most justified, contributions to energy demand is the economic development of low- and middle-income countries. Reduction in energy use is often asked for, sometimes even as an alternative to additional generation. While just economising energy is not enough, all efforts in this sense have certainly to be taken. Speaking about electricity, this concerns the use of energy saving household appliances, but also the reduction of losses in the transmission and distribution grid. Still, even the best energy saving practice will certainly not be sufficient for satisfying the demand.

Looking at the importance of hydropower, one point to observe is the fact that of the total energy supply mentioned above, roughly 80% stems from fossil fuel (coal, oil and gas), and less than 5% from hydropower. One important aim of all measures aiming at limiting climate change is the reduction—and ultimately the total abandonment—of the use of fossil fuel (Gates 2021). In addition to that, a number of countries, Switzerland among them, have decided to get out of nuclear energy. Implementing this would mean that close to 90% of today's energy supply would have to be replaced by renewable alternatives. It is certainly not imaginable that hydropower could ever be able to fill the gap, but it will certainly have a role to play.

27.1.2 Remaining Potential

In a recent announcement of a hydropower congress the following statement was made: "It is widely acknowledged by policy-makers that vast amounts of new hydro and pumped-storage will be required in the move to achieve a more sustainable future." This raises the question on what potential for hydropower development is still available, and what the impacts would be for developing this potential, especially since we want to see this as a step towards "a more sustainable future". As mentioned above, looking at current energy consumption and the proportions the different primary energy sources have in this, it is very clear that hydropower, while it will certainly have an important role to play, will never be able to replace the portions presently contributed by fossil fuels. In addition to that, and especially under a sustainability point of view, one will always have to consider that energy is not the only, and arguably not even the most important, role that rivers have to play. There is no doubt that in many countries or regions all the best sites for hydropower have already been developed, and if a site has not been used yet there is in many cases a good reason for that. On 2020/07/08 e.g., Aljazeera announced that China had just commissioned the Wudongde HPP, which would be followed in a year by Baihetan HPP, located 170 km further downstream, on the same river, Junsha, and that this would be the last large dam ever to be built in China. The reasons for that were that all really good dam sites had been developed by now, and that alternatives like renewables and coal were so cheap that there was no reason for investing large amounts of money in further big dam projects. The future development of hydropower would largely be restricted to pump storage, which would be increasingly decentralised and smaller. This seems to be contradicted by the announcement of 2020/12/31 in newspapers that China envisaged to build a large dam in the "Grand Canyon" of Yarlung Tsangpo River, the upper part of the Brahmaputra, an announcement which apparently raised concerns about water availability in India and Bangladesh (Perras 2020).

Anyway, a good number of sites that can or could be developed is certainly still available. One potential major project that has been in the discussion a number of years ago apparently has just resurfaced: Grand Inga, to be located close to the mouth of the Congo River in the Democratic Republic of the Congo (Fig. 27.1). Ideas about developing hydropower at this site were developed first in the 1950s. Once completed, it would be the largest hydropower plant in the world, with an average discharge of 41,000 m^3/s and a total installed capacity of 45,000 MW (Green et al. 2015; Boyé and de Vivo 2016; Warner et al. 2019, the latter qualifying the project as a "fantasy").

Grand Inga, according to the available sources, would be "powering Africa". Obviously, energy generation would be too large as to be used locally, but for it to actually power Africa a number of transmission lines would be required as indicated in Fig. 27.1. Planning and building these transmission lines could easily be as much a challenge as the HPP as such; the latter could be built in several stages, which might facilitate the work, and this would probably be the case for the transmission lines as well. Still, just preparing the ESIA for one of these lines, crossing borders and therefore having to deal with a different set of legal and procedural requirements all the time, would be a real challenge. Undoubtedly, there is a large hydropower potential, as yet unused, in the Congo basin; political and social instability in this area was probably the main obstacle opposing its development.

On a much smaller scale, one newly arising potential for additional future hydropower plants is provided by the massive retreat of glaciers due to climate change. Farinotti et al. (2019) assessed the global potential for water storage and electricity generation under such conditions and came to the conclusion that developing about 40% of all the identified potential sites of this type could be feasible from an environmental, technical and economic point of view, representing a total storage capacity of 355 ± 105 km^3 and a potential to produce 533 ± 200 TWh per year. Much of this potential is located in Central Asia. They would certainly have at least three major advantages in comparison with most other potential dam sites:

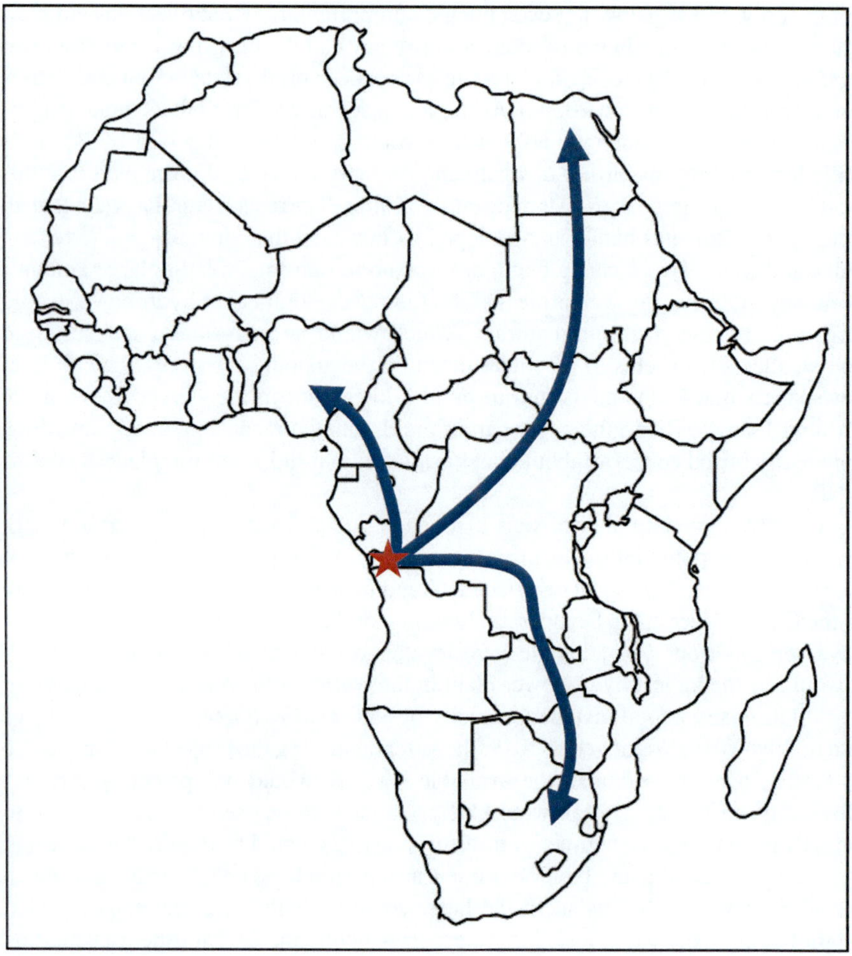

Fig. 27.1 Grand Inga and main transmission lines. HPP site indicated by red star. TL are not exact corridors. *Source* Drawn based on https://afdb-org.jp/wp-content/themes/meteo/pdf/67.%20DRC_Minister%20NEMOYATO_Mr.%20Ibalanky.pdf

they would not interrupt fish migrations, they would not submerge valuable terrestrial habitats, and they would not require any resettlement. In spite of that, and not surprisingly, as mentioned in Sect. 21.2.1, a first such project which was recently identified in Switzerland already met opposition, mainly on the grounds of its impact on the landscape (Häne 2019).

Climate change, as mentioned in Sect. 8.2, may point in yet another direction where dams—although not necessarily hydropower, but mostly reservoirs for storing drinking and irrigation water—may become more important. The general forecast of the climate models points out two main developments, namely, (i) increasing temperatures, and (ii), related to that, increasing frequency of extreme conditions;

while it is never possible to identify one single event as a certain effect of climate change, present development seems to clearly indicate that this second part is just as real as the first one. An example for that could be the situation in Turkey in the late summer of 2021: while massive forest fires ravaged large areas on the Mediterranean coast, the north of the country suffered from floods caused by extraordinarily heavy rainfalls. One prognosis for specific regions of the world is that overall rainfall may not really change by much, but distribution may become much more erratic, short bouts of heavy rainfall being followed by extended periods of drought. This can mean increasingly difficult conditions mainly for agriculture. Making storage available for retaining water may be a solution for this problem (Bossard 2009: 119).

Finally, there is technological development. Replacing old hydromechanical and electric equipment (turbines and generators) with new, more efficient ones, is a potential which certainly has to be—and is being—used whenever an existing hydropower plant needs to be refurbished. Other technologies may be developed and may be applicable in certain situations. Chaudhari et al. (2021) have studied the possibility of using in-steam turbines instead of dams in rivers in the Amazon basin and they come to the conclusion that about 63% of the energy to be generated from conventional hydropower could be obtained by using such turbines, which do not require storage, and this at about 50% of the cost (in terms of USD/kWh), and with considerably less impact on the river. Technological development, however, is not limited to more efficient and/or new types of turbines, but also includes the use of alternative energy sources, mainly solar and wind. As mentioned in Sect. 14.3, these are often presented as solutions for sustainable energy generation, and thus as an alternative to conventional hydropower projects. However, energy storage is required for these alternatives, and pumped storage still seems to be the best option for large-scale developments of that type.

27.1.3 Increasing Resistance

When Crutzen (2002) proposed to call our era "Anthropocene", one point that he mentioned where mankind is changing the world at a global scale, in addition to massive increase of energy consumption ensuing emissions of GHGs and causing climate change, was that "dam building and river diversion have become commonplace". Thieme et al. (2021), analysing the status of rivers worldwide in comparison with identified potential sites for dams and hydropower plants, come to the conclusion that over 260,000 km of rivers, including Amazon, Congo, Irrawaddy and Salween mainstream rivers, would lose free-flowing status if all dams were built. Numerous organisations, on a local, regional, national or international level, are engaged in activities against dam projects; the most well-known and most active among theme is International Rivers (https://www.internationalrivers.org/, formerly International Rivers Network, IRN). A number of books with a critical attitude towards dam projects are listed in Sect. 1.2, and numerous cases of and reasons for such criticism are provided throughout this book. Resistance against dam projects does certainly not come as a surprise. As with any increasingly rare resource, getting the remaining part will be more difficult. The reasons are numerous and generally well known. In addition to the increasing scarcity of suitable sites—and certainly in part triggered by this—there is the concern about impacts on rivers as natural habitat and on other environmental services provided by free-flowing rivers, impacts on biodiversity, increasing pressure on indigenous populations, resettlement, etc. These are all concerns that need to be taken seriously, and this means that planning will have to be done in an increasingly careful manner. While state-of-the-art engineering, including all technical advances, will remain important, the involvement of other disciplines may or will in all likelihood get even more relevant than it is already. This in turn means that the ESIA, including all the other IAs shortly discussed in Sect. 3.13, will continue to play its role, and that this process as well needs to be developed and improved. This is the topic of the following section.

27.2 Improving ESIA

27.2.1 The Weaknesses of ESIA

The weaknesses of ESIA—real as well as assumed ones—have been addressed under different headings, and this will not all be repeated here (see e.g. Sect. 3.11.5). The main issue is again that different stakeholders have different expectations, which leads to the view from one side that ESIAs are just "toothless paper tigers", or are prepared by project proponents in a way as to whitewash (or, in this case, rather greenwash?) their projects instead of achieving what they are supposed to achieve, namely, proving that a project must not go ahead because of its inacceptable environmental and/or social impacts, while the opposite side sees the process and the reports just as a means for obstructing development and progress, and at best as only an additional hurdle to be taken, making the project development process more lengthy and more expensive. As I hope I have been able to show, these views are not always unfounded: ESIA reports can be prepared in a way as to avoid the hot topics, by not accurately identifying major impacts, which then, if control is not carried out as it should, can lead to an official acceptance of a project without insisting on required impact mitigation or, in the worst case, not identifying impacts which should actually have been considered as a no-go for the project. On the other hand, it is also possible that a report is prepared with the obvious aim to kill a project, by overemphasising the importance of certain impacts or by actually constructing impacts, as mentioned e.g. by making statements like "numerous endemic species will disappear", without any substantiation of the point. I also hope that I have been able to demonstrate that neither of these expectations actually does justice to the role of the ESIA process: the study, and finally the report, has to identify and discuss facts, and has to propose the required and appropriate mitigation measures, and the approval process then has, based on these facts, to come to a decision on whether or not the project should go ahead, and if yes under what conditions.

Still, if there is a case of such a misuse of the process, the main reason for this is ultimately incompetence at one level or another, a situation which admittedly can be aggravated by pressure from one side or another. Impact assessment is in any case always also a learning progress, and it certainly needs continuous improving. The intention here is not to demonstrate what needs to be done for bringing the process up to a standard where it will be prefect in any way; however, a few points will be raised which hopefully can make a contribution to this way of constant improvement.

27.2.2 The Formalist and the Pragmatist

Basically, there are two possible approaches for preparing and carrying out an impact assessment study. One is to look at the TOR, the relevant guidelines and other requirements to be followed, and based on this to make a list of activities (which can be translated into sections of the report to be prepared or a number of specific reports required as annexes or complements of the main report), and then working through this list point by point. The other possibility is to take the project as a starting point, figuring out what type of changes it will cause in the environment, and then focus on those that are considered as relevant. The formalist will tend to choose approach one, while the pragmatist will rather prefer approach two. The question is: which of these possibilities is more likely to lead to the required result? The formalist approach has the advantage that it will ensure that none of the points required will be left out: if TOR or guidelines say that this or that is required, it will be done. The risk here is—and this risk is very real—that this will lead to, or at least encourage, the checklist-minded procedure e.g. by a competent authority, where the focus of the evaluation will be more on the fulfilment of the formal requirements, ticking off the list of these requirements, rather than on the content, i.e. the relevance of the topics and the quality of assessment carried out and mitigation proposed. The advantage for the pragmatic approach is that most probably all really relevant points will have been properly identified, and adequate mitigation measures will have been formulated. The risk of this approach is that then, although the impact assessment as such may be correct, some formal requirements have not been met. There is a high probability that in such a case the draft report will come back from the review by the competent authority stating that this or that topic was not addressed, a shortcoming which needs to be corrected.

The consequence resulting from this dilemma is quite obvious: both approaches are important, and both have their role to play. If the TOR, to use that example again, state that a baseline noise study has to be done, then so it is, it has to be done even if it is very clear that this aspect, from and impact and mitigation point of view, is totally irrelevant. Just as important, the relevant impacts have to be identified and assessed, and the required mitigation measures to be proposed.

Still, it is not always that easy and clear-cut, and often again because of the fact that in a given situation not everybody will necessarily come to the same conclusion. One example can be the following. IFC Performance Standard 6 on biodiversity conservation defines the so-called critical habitat (see also Sect. 6.3.1) and the importance of habitats identified as such. Guidance Note 6, which provides indications on how to deal with the requirements of PS6 mentions several times a critical habitat assessment (as e.g. in GN115, p. 36: "This will be based on the client's risks and impacts identification process, including the ESIA, biodiversity assessment and/or critical habitat assessment."), which is not mentioned in PS6, and which is also not clearly defined in GN6. In this case, the pragmatist may come to the conclusion that there is a requirement to first check whether the project in question could affect any habitats which, according to IFC criteria, would qualify as critical habitat, and if yes, that the required baseline description, impact assessment and mitigation has to be provided in the ESIA report. The formalist, however, will see this as a requirement to carry out a formal Critical Habitat Assessment, i.e. a separate study provided as a separate report. In the end, the outcome might be the same, but the approach may have been rather different.

27.2.3 Improving the Process

First of all, and in spite of its potential or actual weaknesses, the ESIA process must absolutely be maintained. It is still the only instrument for addressing environmental and social impacts of dam and hydropower projects, and of any other major project that can have negative impacts. But it is equally necessary to constantly try to improve it, in order to reach a higher standard. It has to be mentioned here immediately that with this request another problem is created, very similar to one mentioned above: it will be difficult if not impossible to reach a commonly accepted agreement about what such a "higher standard" should be, the opinions on this will necessarily go in opposed directions. The only thing that can be done here is to give some indications on what I think would be required for reaching this goal.

Looking at ESIA reports, one can get the impression that they are evolving in one specific direction: they get increasingly more bulky, with a more complex and complicated structure, and with an increasing number of lengthy and very detailed annexes. It would be an interesting task to analyse a number of such reports in detail for answering the question whether they have only increased in volume, or whether quality has increased in the same proportions; quality here would have to stand for clarity, relevance of content, any mainly quality of the proposed mitigating measures. Having said that, I'm fully aware of the fact that this would not be an easy task, and that once again, beauty—and this holds for quality as well—lies in the eyes of the beholder. One question could certainly be looked into: how much of the baseline investigation did actually help in identifying impacts, how much of that led to relevant mitigation measures, and how much was just unnecessary contributions; the example of plankton investigations in rivers, mentioned in Sect. 3.12, comes to mind. In any case, the principle originally formulated by Robert Goodland that "EA reports should be concise and focused on the significant environmental issues" (Sect. 3.4.2) should be followed more closely.

One main reason for this development is certainly to be found in the type of TORs one receives for first preparing a proposal and then the report. There are two types of such TORs, both have their set of problems. The first is the TOR prepared e.g. by WB or any other organisation looking for an ESIA consultant. They typically provide a short overview of the situation, an equally short description of the project including some information on the project site, and then a list of tasks to be completed. The consultant is then expected to prepare a proposal which will cover all relevant aspects (including everything covered by the ominous and frequently used sentence "which contains, but is not limited to …") and propose a budget which will cover the entire study. The main issue here is that with this necessarily limited information it is very difficult to get a clear picture of (relevant) impacts to be expected, and of the effort required for actually getting the data required for adequately addressing the important questions. The second type of TORs, and here I'm tempted to label that as "TORs", are the those one can receive from the "competent authorities", i.e. the Environmental Protection Agencies, i.e. the body which will examine the report, and which ultimately has to issue (or refuse!) the environmental permit (which can be named differently depending on the country), including the decision on which mitigation measures are considered as conditions for project implementation. These official TORs are usually issued after the submission of a first, preliminary, report from the consultant, which usually includes proposed TOR, if deemed necessary including modifications of the original ones. Unfortunately, rather often what one gets in such a situation is not at all a list of to-dos adapted to the project and its specific condition, but a standardised, sometimes very detailed, table of content for the ESIA report to be prepared, which does not at all reflect the situation. One such "TOR" we had received rather recently was exactly such a structure to be followed (and it had to be followed point by point: checklist!), which very obviously had been prepared for one specific project (this was clear because the name of this project was still included in a few places), and which was now being applied for each project, irrespective of the type, and certainly irrespective of the environmental and social situation in the project area. Needless to say, it was a complete catalogue of topics which might arise in any of the imaginable cases.

The way for improving this situation, as I see it, would be to insist on and increase the importance of the screening and scoping phase of an ESIA study; as it is now, this unfortunately is just listed as a requirement in the process, but unfortunately it does not really have much weight. Screening, as used here, does not only mean to check whether for the project in question an ESIA is required or not; this, in most if not all cases, is a point that is easily clarified. As a rule, national ESIA procedures include a list of projects which have to follow the assessment process, by simply including "hydropower projects with more than xy MW installed capacity" in this list. What is unfortunately rarely if ever the case is to carry out an environmental and social screening in the earliest possible stage of project development, i.e. during the identification of potential dam sites. It is usually not very difficult, in such an early stage, to already identify, without any detailed study, whether a potential dam site would be likely—or even absolutely certain—to create major environmental or social problems (and: yes, here again: the definition of "major" ...); such projects should not even make it into the list of potential dam sites. This type of early screening, unfortunately, is not done very often (see Sect. 5.5).

A scoping process is usually carried out in the case of a serious ESIA for a dam project. The main objectives of this phase are quite clear:

- To get a good knowledge on the project, having now access to plans etc. which were not available before.
- Getting a first, but usually quite good and certainly very relevant impression of the conditions in the project area, based on short site visits which have to include the most important specialists of the ESIA team.
- Getting a first impression of (potential) project impacts, which will allow to group them into the following categories:
  - No impact or impact certainly irrelevant; this issue will not be investigated any further.
  - Impact of low intensity or otherwise of little importance; will be mentioned in the report, however, no additional effort like field work will be done in this respect.
  - Impact potentially, probably or certainly of importance, but sufficient information available for adequately assessing it; no detailed investigation required.
  - Impact potentially, probably or certainly of importance, and sufficient information is not available for adequately assessing it; here, field work will be required (and, most probably, its extent can now be estimated).
- One additional point may have to be mentioned here: if the screening and scoping exercise reveals an impact of a nature, magnitude or importance that might be a no-go for the project, this should be the point in time to discuss this issue with the relevant stakeholders. This might even lead to the decision that no further planning efforts are required, or that the possibility for a major modification of the project has to be evaluated first.

On the basis of these results, it would then be possible to formulate really relevant terms of reference. This then would also need to be reflected in the TOR finally issued (or accepted) by the authority, and the later evaluation process would have to be based on these, and not on a standardised checklist. Unfortunately as said, this is very often not the case. And there is another issue that has to be mentioned here: complex impact assessment studies are often accompanied by a steering committee, which supervises the process and "provides guidance". Here, some of the institutions often playing that role, the World Bank prominently among them, have something that I am tempted to call a "bad habit": every now and then, and sometimes very often, the persons carrying out this role as supervisors change. This, in many cases, may be inevitable for a number of reasons. The unfortunate consequence of this is that very often these newcomers, who obviously are not familiar with the development the assignment has taken so far, and who apparently very often also do not have the time to study the documents which have been produced (and accepted) in the process, need to justify their presence by seeing something that so far everybody else has overlooked. To give one possibly somewhat extraordinary example: in a meeting with such a steering committee, where the draft report of a rather comprehensive ESIA was discussed, one of the members, who had been brought in literally in the last moment, asked me "why does your study include an audit of the construction site?"; here, the answer was very simple and straightforward: because this is one of the tasks listed in the TOR. It is certainly true that usually, when an ESIA for a hydropower project is done, there is as yet no construction site, but this was not an ordinary case, and this person had obviously not had the time to familiarise himself with project and TOR. However, the situation is not always that simple, and this can lead e.g. to a decision that a specific point which was clearly identified in the scoping phase as irrelevant, now suddenly needs more attention. It can be, very certainly, that something defined at an early stage has to be modified, for whatever reasons, but in principle conditions of the final TOR should be respected.

27.2.4 Institutional Strengthening, Capacity Building and Training

These three points are certainly of high importance, especially in low- and middle-income countries. The position of the government agencies responsible for environmental protection, including those departments in charge of management of protected areas, often have a weak position in comparison with e.g. a ministry of energy, and probably very often also suffer from a lack of required financial means. Institutional strengthening, but also capacity building in the form of training courses in various disciplines, as well as on-the-job training by involvement in different types of studies is certainly required. ESIAs mandated by various organisations often include training in two different forms, namely, as formal training courses to specific subjects, and as on-the-job training by e.g. involving staff of specific government agencies in field work of different types. This is a step in the right direction, but it is certainly not sufficient.

Institutional strengthening and capacity building: In many places, one of the weakest links in the chain are the environmental agencies, i.e. the "competent authority" which has the task to receive, check and comment upon ESIAs, and finally come to a decision. This weakness can have more than one reason. One is certainly the fact that a Ministry of Environment has a weak position relative to other players involved, especially the usually much more powerful Ministry of Energy. However, another issue is very often the fact that officers in such agencies, in addition to having little or sometimes as good as no means to do their job, also are not really well trained, and, just as importantly, lack the experience which would be required for actually doing a good job. Adequate education is certainly a basic requirement, but institutional strengthening, on-the-job training etc. is also required.

Training: When I started my career, it was often made quite clear to me by older and more experienced colleagues from the engineering departments that "we the engineers" know exactly what needs to be done, and that all the relevant aspects are being taken care of. This situation has changed considerably. Future engineers today also have to or can follow courses on environmental and social effects of projects; I had the opportunity to teach such courses for a number of years, and my impression was that the participants were interested in the topics raised. It is certainly the case that today young engineers are much more open-minded and ready to discuss with a partner from a different faculty in order to try and find better solutions. Of course, better solutions can only be reached when all parties involved know what they are talking about and want to find solutions. This means that not only the training of engineering students needs to be constantly updated and improved, the same is true for the counterparts in these discussion, i.e. the people who are on the environmental and social side of the table, and this may be an even more difficult task, since specialists from very different faculties are likely to be involved in such processes, among them many who during their studies had never even thought about such a possibility. Therefore, in order to raise not only awareness, but to actually improve knowledge, in many places, Switzerland certainly among them, about 40 years ago special curricula were created under the general heading of "environmental engineering" (see Sect. 3.5). This is certainly a good thing, at least as long as there is awareness, and not least among those who chose this subject, about two potential dangers. Students following these courses get a very broad knowledge of the relevant fields in environment and environmental protection. This can give them the impression that they know everything there is to know about the environment, not being aware that maybe they are good generalists with a good overview, but not necessarily an expert in any of the subjects at hand. The important point here is to know when a specialist is required. The other problem is that, while "environmental", they are still basically engineers, and this can (and not too seldom does) mean that they have the impression that for every problem there is an engineering solution. As pointed out in a paper (Zwahlen 1992), many engineers, understandably, have a problem to accept "no, this cannot be done", because it leaves them with the impression of not having done their job.

27.2.5 Introducing Alternative Processes

One basic question is whether the ESIA is an adequate instrument for addressing all the pressing issues and for finding adequate solutions, whether additional, more comprehensive, better procedures should be added to the process, or whether such alternative procedures should actually replace the EISA, especially in view of the latter's numerous shortcomings. As expressed above (Sect. 27.2.3), the ESIA is an important instrument for project evaluation and must be maintained. In the course of the last 30 years or so, steps in the direction of improving it by adding other instruments have been taken, mainly by adding a number of other IAs, as described in Sect. 3.13.

One major effort for improving the entire process has been taken by IHA by launching its Sustainability Assessment Protocol in 2008 (IHA 2010). According to this document, "The Hydropower Sustainability Assessment Protocol is a sustainability assessment framework for hydropower development and operation. It enables the production of a sustainability profile for a project through the assessment of performance within important sustainability topics." It contains tools for assessing a project at four different stages, namely Early Stage, Preparation, Implementation and Operation, and it defines a number of criteria (not limited to environmental and social aspects) to be used in the process. The application of the protocol on a specific project mainly relies on three basic sources of information, namely:

- Any studies, reports etc. prepared in the framework of project development, as well as other written information of relevance, like e.g. development policies for relevant sectors.
- Discussions with stakeholders, in individual meetings as well as in group discussions.
- Short site visits for getting a first direct impression of the situation on site.

A formal assessment is carried out by a group of certified assessors, whereby not necessarily all members of this group will be involved in all the activities related to the assessment.

I had the occasion to participate as an observer in the application of the Protocol on a number of projects, with the aim of giving my opinions, comments and observations on this instrument, and I shortly explain here the most important points thereof; not everybody was entirely happy with all of my comments.

First of all: if applied properly, it is a very good instrument, giving a thorough analysis and evaluation of the project, clearly identifying strong as well as potentially weak points of the project development process, in this way providing a good basis for improvements where required. From the evaluation that was made in Sulawesi, Indonesia, I just want to mention two very specific points to illustrate potential benefits of an application of the protocol. After a discussion of the project which included representatives from all relevant government agencies, namely land planning, energy, irrigation, environment, social security and possibly some more, one engineer from the department of irrigation told me that this meeting had been very important for him since before he had "never been aware of the fact that in the water sector there were other interests than irrigation". While this in its absolute form may be a rather surprising statement, it nevertheless highlights the necessity for following an interdisciplinary approach, of entering into a dialogue with other stakeholders, and of exchanging information. The second point was something that we had learned from documents, but which was then made very clear during the site visit: the project in question was a multi-purpose project, the two main objectives being irrigation and hydropower, with a storage reservoir as the most relevant part of it. Unfortunately, just a few years back a group of people had been relocated due to another project in the same region, and the resettlement site chosen for them, where they were now still in the process of getting established, was located in the future reservoir, just a short stretch upstream of the dam site. This is definitely the result of a lack in coordination among different administrative entities, and the Protocol is certainly an instrument which would lead to preventing such an unfortunate outcome from happening.

My concerns were on another level:

- The developer of a hydropower project has the obligation to follow and to be in compliance with national legislation and procedures, including those relative to environmental and social impacts, as well as with international standards and guidelines. This is mandatory, and the processes to follow can be rather lengthy and far-reaching. The Protocol is applied on a voluntary basis, and it is at least imaginable that a developer would hesitate to follow another quite demanding procedure on top of those that he has to by law.
- An assessment according to the Protocol has to be prepared by certified assessors; at least for the time being, this means that most if not all of them will be expat experts in most cases, which makes application a rather costly exercise. The answer to this concern is obviously that it will be necessary to provide training to local specialists, in order to use local expertise for these assessments.
- Looking at the assessment from an environmental and social point of view, it is quite clear that what the Protocol does in this respect basically is the obligation of the ESIA. This means that, if there are weaknesses in the ESIA process, this will be reflected in the assessment (it being based on results from other project related documents), or that it will have to re-do the parts of the ESIA considered as insufficient. Doing the assessment with local experts will require that they are sufficiently qualified, because replacing or supplementing an ESIA procedure of insufficient quality with an equally low-quality protocol assessment would not really improve the situation. This in turn raises the question whether in such a case it would not be preferable—and more important and possibly more efficient—to focus on capacity building for national ESIA agencies, in order to bring them up to standards.

These points are certainly not arguments against using the Protocol. However, considering them I have my doubts whether the Protocol will actually be applied to the extent that it could—or maybe should—be applied.

27.3 Improving Implementation

Del dicho al hecho, hay mucho trecho

Spanish proverb

From said to done is a long way

or: Easier said than done.

27.3.1 The Result Is Important

It is probably not wrong to assume that everybody concerned about environmental and social impacts would agree with the view that ultimately not the good, well-written, concise yet comprehensive ESIA report is the most important step in project development, but the final outcome, the result achieved. A good planning process is an important component, but even the best of planning is not of much use if implementation is not up to standards. It is the same with legislation; an official of the state electricity company in DRC (Democratic Republic of the Congo) once told me, when on the occasion of a site visit we discussed issues related to the ROW of a transmission line, "agriculture in the wayleave of the line is forbidden by law, but we do not implement it". Such a law is definitely useless, just like decreeing legally protected areas for the conservation of biodiversity without giving them any real protection on the ground. In this latter case, it might even be counter-productive: it could give a potential poacher, in search of e.g. rare species for the illicit pet market, a first indication on where to look. It is certainly the same with the ESIA for HPPs: coming up with perfect mitigation measures is of no use if later on these measures are not being implemented correctly; unfortunately, there are numerous examples for this, as e.g. the "pre-impoundment reservoir area clearing" for Houay Ho HPP (Sect. 17.5).

One important step for assuring proper implementation was the introduction of an ESMMP as part of the ESIA, which does not only describe mitigation measures, but also clearly defines responsibilities as well as a monitoring process. A number of points relevant for a good implementation are addressed here.

27.3.2 Minimising Impacts on Rivers

When talking about hydropower, one of the main concerns from an environmental point of view is the increasing fragmentation of rivers (see Sect. 4.1.2). In this respect, WWF (2021) listed a number of points which should help guiding development in the direction of reducing this type of impact. These points are cited and shortly commented here (*comments in italic font*).

- Avoiding fragmenting rivers by instating formal protections of rivers or by exploring alternative development options, such as non-hydro power renewable energy, like solar and wind. Early, system-wide planning is the best solution to meet development needs and keeps rivers and the benefits that they provide connected and flowing. *What does not work here is the "or": while formal protection, if implemented, will prevent fragmentation of a river, this measure obviously does not contribute in any way to energy supply. Development of alternative renewable energy sources, especially wind and solar, is important, but does not solve the problem of energy storage. Finally, as good and important early and system-wide planning is, this is not automatically a guarantee that rivers will be kept intact; much depends on the priorities of the planners, which are defined to a great extent by socio-economic considerations, where protection of natural systems, rivers among them, may not be a priority.*

- Minimizing the impacts of dams on rivers by siting dams in locations with fewer impacts on people and nature. Technologies such as environmental flows can also help minimize dam impacts by allowing water to move through dams in a manner that mimics natural flows. *This is basically accepted, and is certainly often an aim. However, as mentioned above, most of the best sites have already been used.*
- Restoring rivers through dam removal, an increasingly popular option in the U.S. due to the high costs of dam maintenance and the negative impacts on surrounding ecosystems. *This was discussed in Sect.* 13.4. *Since an increasing number of dams is reaching the end of their life time, decommissioning will be a topic of growing importance; however, dam removal is only one of at least two options. If an HPP is decommissioned, it stands to reason that an alternative will have to be found for filling the gap in energy supply.*
- Offsetting the negative impacts of dams. If one river is severed by a dam, there may be options to protect another river to ensure that similar values are maintained in the region. *This is a highly recommended approach, but unfortunately not one that has often materialised, as mentioned shortly in Sect.* 20.5.

As very often, it is not very difficult to develop good ideas, the implementation is the problem. As Hughes (2009: 59) puts it, "… good ideas are not enough".

27.3.3 You Get What You Pay For

In the international competitive market, one principle has become rather firmly established, and is e.g. usually followed by WB and other funding agencies: choosing the lowest bid. This is certainly a very rational and economically justified behaviour, something that most consumers do every time they go to the supermarket. However, in the dam and dam-related study business, this has a number of risks, as mentioned already e.g. for the difficult case of preparing an offer knowing that if one offers everything that will be required in the end the bidder will be eliminated because the price will be too high (see Sect. 3.6.2). If the lowest bidder is selected, this should clearly be the lowest bidder whose proposal includes everything that will (foreseeably) be required for fulfilling the contract, and which includes everything that is requested by the tender documents. This, however, is not always the case, and knowing this will motivate many proponents to reduce their bid to the bare minimum, excluding a number of items considered as not so important, or at least not including them. Later on, when it turns out that something which in this way had been omitted is still required, the argument of the contractor will be that there is no budget for this item, and that extra payment has to be made available. The big question is here whether in the end one gets the best overall price by having chosen the lowest bidder, or whether a more complete bid, therefore including a higher budget to start with, would ultimately have been given a better result. Difficult to say, but in any case the completeness of a proposal should be checked very carefully, and all incomplete bids should be eliminated before the prices are considered.

27.3.4 Communication

"Do good. Do it well." is the motto of DEVEX, an organisation focusing on development (https://pages.devex.com/about-devex). This is certainly also applicable to the implementation of mitigation measures. One other maxim could also be applied here: "do good and talk about it"; this is the translation of a German phrase ("tue Gutes und sprich darüber "), the origin of which does not seem to be very clear, but which is widely used, and it could also be applied to implementation and monitoring efforts related to HPP development. While it is important to implement mitigation measures and to carry out monitoring in the best possible way, it is also important to communicate the results, as this will help in improving the outcome of future projects. A few issues where information on successful mitigation and monitoring—whereby it has to be considered that information about measures which failed could be just as relevant—would be very valuable are the following:

- **Development of fish fauna and fish stock**: this issue was mentioned several times (Sects. 10.3 and 16.4). As stated there, it is well known that the interruption of a river by a dam and the creation of a reservoir change fish fauna, and with it fisheries, in a fundamental way. Numerous studies document the situation in a river, and numerous others the new situation in the reservoir after impoundment, but, to the best of my knowledge, there is as yet no long-term follow-up of such a development, which covers the situation in the river in its natural state, the changes brought about by construction, and the development in the reservoir until the moment where a new stable situation was reached. A good monitoring program was carried out e.g. during the construction phase of Bujagali HPP, but unfortunately this was abandoned at the time of commissioning, when management responsibility was transferred from the developing to the operating entity (Sect. 10.3). It would be very valuable to have a number of such long-term studies covering the entire period of transition, from a number of cases representing different conditions.
- **Fish migrations**: while much information is available on fish migrations and on the performance of fish passes in temperate and mainly boreal regions, focussing mostly on salmons, this is still not the case for tropical rivers, although—or maybe because—the situation there is much more complex, with a far higher number of migrating species with very different behavioural patterns, sizes and swimming abilities. It might be that a considerable amount of data is already available, in some data storage devices of HPP operators, and for advancing in this difficult situation it would be essential to make such information available more widely. In addition to success stories it would also be of high relevance to be informed

about devices and structures which have failed to produce the expected results, even if one can understand the reluctance to publish such results. Still, it could prevent others from repeating the same mistakes.

- **Reservoir development**: this is an important issue, also related to the question of greenhouse gas (GHG) emissions from reservoirs, and with this once more to the discussion on sustainability of hydropower (Sect. 8.2 and Chap. 17). One aspect of this is the development of the situation in the reservoir concerning oxygen stratification, the other is the question of GHG emissions—and here mainly methane—from the reservoir. It would be of high relevance to have reliable data on these processes and their development starting from impoundment till the moment when a new equilibrium was reached.
- **Residual flow**: the importance of having a residual flow is generally recognised, but remains a controversial point: not enough water to maintain the ecosystem on the one hand, too much water lost for energy generation on the other hand (Chap. 20). There are many rules, but certainly even more different situations, and simply applying a rule may not lead to a satisfactory result. Ever more sophisticated models help to determine the required quantity as well as its seasonal distribution. Still, here as well results from a careful monitoring of a large variety of real settings would be very useful, also for optimising the approach on a case to case basis.
- **Floodplain management**: as described in Sect. 8.2, and in the case study in Sect. 22.5, floodplains are adapted to, and to a great extend depend on, seasonal fluctuations in river discharge, while storage HPP aim at the opposite: reducing the "loss" of spilled water in the high-flow season and increasing productivity in the low-flow season. Careful management of flows also in the interest of maintaining functional floodplains, e.g. by releasing specific flood peaks at well-defined times of the year can go a long way in this direction. How much exactly, when and in what from such releases are required for maximising the benefits needs testing and careful monitoring, a challenging task especially in the case of cascades with HPPs managed by different operators and possibly even in different countries.

A good example of such a long-term monitoring program is provided by SOTEAG (Shetland Oil Terminal Environmental Advisory Group; https://www.soteag.org.uk/), which is monitoring birds and the status of the environment in the surroundings of a terminal and port since more than 40 years, and publishes regular reports.

I would like to encourage my colleagues involved in ESIA preparation, implementation and monitoring to not only do the job, but to talk about it, preferably in the form of publications. This is done widely, but often with a different agenda and from a different point of view, by academics, but practitioners seem to be very reluctant to do this, and if they do, it is mostly done in the form of presentations during congresses, like e.g. the annual ICOLD meetings, the papers appearing in the congress proceedings. This is fine, but one has to face the fact that these presentations as well as the publications mostly reach colleagues, people active in "dam business", but not a wider audience. Even if such publications, very often, are not, strictly speaking, the result of scientific work, but rather reflect applied, practical experience, results from such efforts can still be very valuable especially for evaluating efficiency and effectiveness of mitigation measures, and in this way hopefully contributing to improve them. Quite clearly, willingness of practitioners alone is not sufficient, especially for more complex evaluations of processes of longer duration, like the ones mentioned above; the first requirement is the willingness of project proponents and operators (with the problem of a continuous effort when responsibilities are handed from one to the next) to carry out such studies, even though this necessarily comes at a cost. But especially in the discussions of the sustainability of hydropower such experiences are extremely valuable. It is of little use to have this information in some reports in some secluded archive, or on the hard drives of a few computers—the information must be readily accessible.

References

Bossard, L. (ed.). 2009. *Regional atlas on West Africa. West African studies*, 284 pp. Paris: OECD.

Boyé, H., and M. de Vivo. 2016. *The environmental and social acceptability of dams*. Field actions science reports, special issue 14. http://journals.openedition.org/factsreports/4055.

Chaudhari, S., E. Brown, R. Quispe-Abad, et al. 2021. In-stream turbines for rethinking hydropower development in the Amazon basin. *Nature Sustainability* 4: 680–687. https://doi.org/10.1038/s41893-021-00712-8.

Crutzen, P.J. 2002. Geology of mankind. *Nature* 415: 23. https://doi.org/10.1038/415023a. https://www.nature.com/articles/415023a.

Farinotti, D., V. Round, M. Huss, L. Compagno, and H. Zekollari. 2019. Large hydropower and water-storage potential in future glacier-free basins. *Nature* 575 (7782): 341–344. https://doi.org/10.1038/s41586-019-1740-z.

Gates, B. 2021. *How to avoid a climate disaster*, 257 pp. New York and Toronto: Knopf.

Green, N., B.K. Sovcacool, and K. Hancock. 2015. Grand designs: Assessing the African energy security impolications of the Grand Inga Dam. *African Studies Review* 58 (1): 133–158. https://doi.org/10.1017/asr.2015.

Häne, S. 2019. Grüner Bruderkampf am Gletscher. *Tagesanzeiger*, 09 Aug 2019, 4.

Hughes, J.D. 2009. *An environmental history of the world. Humankind's changing role in the community of life*, 2nd ed., 306 pp. New York: Routledge.

IHA. 2010. *Sustainability assessment protocol*, new ed. 2020, 213 pp. London: International Hydropower Association. https://www.hydrosustainability.org/assessment-protocol.

Perras, A. 2020. China gräbt halb Asien das Wasser ab. *Tagesanzeiger*, 31 Dec 2020. https://www.tagesanzeiger.ch/china-graebt-halb-asien-das-wasser-ab-308409429848.

Scudder, T. 2005. *The future of large dams. Dealing with social, environmental, institutional and political costs*, 389 pp. London: Earthscan.

Thieme, M.L., D. Tickner, G. Grill, et al. 2021. Navigating trade-offs between dams and river conservation. *Global Sustainability* 4 (e17): 1–7. https://doi.org/10.1017/sus.2021.15.

Warner, J., S. Jomantas, E. Jones, M.S. Ansari, and L. de Vries. 2019. The fantasy of the Grand Inga hydroelectric project on the River Congo. *Water* 11: 407. https://doi.org/10.3390/w11030407.

WWF. 2021. Over 160,000 miles of rivers are at risk of losing their free-flowing status due to dams. https://www.developmentaid.org/#!/news-stream/post/105028/rivers-dams?utm_source=Newsletter&utm_medium=Email&utm_campaign=NewsAlert&token=cf608f8b-06e8-11ea-8cc5-52540068df95.

Zwahlen, R. 1992. Failure of irrigation projects and consequences for a different approach: A case study. *Ecological Economics* 5: 163–178.

Annex A
TOC for an EIA

This is an example for a standard Table of Contents provided in a national ESIA guideline. See an alternative structure and discussion of the two approaches in Sect. 3.4.2.

Example from Laos: MONRE 2016: ESIA Guidelines, Annex 13: Table of Contents of ESIA Report.

Comment: this is a copy of the English version of Annex 13 of this guideline; the only changes I made: shortened a few entries in order to fit content on one line and corrected the (obviously wrong) numeration in Sect. 5.5.

Abbreviations and definitions

Executive Summary

Certification of the ESIA

Table of contents

1 **Project overview**

- 1.1 Project presentation and rationale
- 1.2 Related development projects
- 1.3 Presentation on project owner and the consultant preparing the ESIA
- 1.4 Project owner's commitment
- 1.5 Structure of the ESIA

2 **Policy, legal framework and institutional arrangement**

- 2.1 The environmental and social policies of the project owner
- 2.2 Policy and legal framework
- 2.3 International treaties and conventions related to project impacts
- 2.4 GoL's institutional arrangement
- 2.5 International policy, guidelines and standard (if any)
- 2.6 Environmental and social standard of the project

R. Zwahlen, *Assessing the Environmental Impacts of Hydropower Projects*, Environmental Earth Sciences, https://doi.org/10.1007/978-3-030-91185-0

3 The project's description and options

- 3.1 The project's characteristic introduction and description of the options
  - 3.1.1 Background and rationale of the project
  - 3.1.2 The project location
  - 3.1.3 The project's infrastructures
  - 3.1.4 The project's options
- 3.2 Comparison and selection of the options
  - 3.2.1 Methodology
  - 3.2.2 Technical explanation on the option selected
    - 3.2.2.1 Pre-construction phase
    - 3.2.2.2 Construction phase
    - 3.2.2.3 Operation phase
    - 3.2.2.4 Decommissioning phase
- 3.3 Design of details

4 Environmental and social description

- 4.1 Determination of the study's limitation
- 4.2 Physical components
  - 4.2.1 Meteorological conditions (including climate change)
    - 4.2.1.1 Methodology
    - 4.2.1.2 Description
  - 4.2.2 Topographical features
  - 4.2.3 Geographical conditions/tectonic plates
  - 4.2.4 Soil
  - 4.2.5 Natural hazard
  - 4.2.6 Hydrology
  - 4.2.7 Erosion and sedimentation
  - 4.2.8 Groundwater and surface water quality
  - 4.2.9 Mineral resources
  - 4.2.10 Sound and vibration
  - 4.2.11 Air quality
  - 4.2.12 Mapping
- 4.3 Biological components
  - 4.3.1 Terrestrial ecology/wildlife
  - 4.3.2 Forest and vegetation coverage
  - 4.3.3 Aquatic organisms and animals
  - 4.3.4 Wetlands
  - 4.3.5 Protected Area
  - 4.3.6 Biodiversity
  - 4.3.7 Mapping
- 4.4 Socio-economic components
  - 4.4.1 Administration and limitation
  - 4.4.2 Land use
  - 4.4.3 Social profile

4.4.3.1 Communities
4.4.3.2 Population
4.4.3.3 Education
4.4.3.4 Housing conditions
4.4.3.5 Vulnerable group
4.4.3.6 Ethnic groups
4.4.3.7 Gender equality
4.4.3.8 Religion
4.4.3.9 Social and political organization

4.4.4 Health profile
4.4.4.1 Mortality and morbidity rate
4.4.4.2 The incidence of infectious diseases
4.4.4.3 The incidence of chronic diseases
4.4.4.4 Diet and nutrition
4.4.4.5 Mental health and well-being
4.4.4.6 Access to health services
4.4.4.7 Access to drinking water, sanitation and waste management
4.4.4.8 Others

4.4.5 Economic profile
4.4.5.1 Employment
4.4.5.2 Traditional production systems
4.4.5.3 Household income
4.4.5.4 Cost of living
4.4.5.5 Land ownership
4.4.5.6 Local businesses
4.4.5.7 Existing production activities
4.4.5.8 Other

4.4.6 Public facilities
4.4.6.1 Roads
4.4.6.2 Navigation and ports
4.4.6.3 Airport
4.4.6.4 Transmission line
4.4.6.5 Electricity
4.4.6.6 Pipes
4.4.6.7 Hospitals/health centres and clinics
4.4.6.8 Schools and educational institutions
4.4.6.9 Nursery and kindergarten
4.4.6.10 Temple
4.4.6.11 Cemetery

4.4.7 Water source, usage and supply
4.4.8 Energy source, usage and supply
4.4.9 Unexploded ordnance
4.4.10 Mapping
4.4.11 Local community concerns

4.5 The cultural component
4.5.1 Archaeology
4.5.2 Temple, monument
4.5.3 Ethnic group
4.5.4 Mapping
4.6 The spatial component
4.6.1 Ancient remains
4.6.2 Interesting spots
4.6.3 Landscape
4.6.4 Mapping

5 Environmental and social impact assessment and mitigation

5.1 Methodology
5.1.1 Scope of assessment
5.1.2 Geographical scope: study area
5.1.3 Temporary area
5.1.4 Methodology
5.1.4.1 Impact assessment and mitigation
5.1.4.2 Presentation of each sub-plan of ESMMP
5.1.4.3 Presentation of each site specific-plan of ESMMP
5.1.5 Mapping
5.1.6 Requirement of model development
5.1.6.1 Air quality
5.1.6.1.1 Methodology
5.1.6.1.2 Results
5.1.6.1.3 Mapping
5.1.6.2 Surface water quality
5.1.6.2.1 Methodology
5.1.6.2.2 Results
5.1.6.2.3 Mapping
5.1.6.3 Groundwater quality
5.1.6.3.1 Methodology
5.1.6.3.2 Results
5.1.6.3.3 Mapping
5.1.6.4 Noise
5.1.6.4.1 Methodology
5.1.6.4.2 Results
5.1.6.4.3 Mapping
5.1.6.5 Other
5.2 Impact identification
5.3 Impact mitigation measures and residual impact
5.3.1 Biological impact
5.3.1.1 Air quality
(1) Pre-construction phase
(2) Construction phase

(3) Operation phase
(4) Decommissioning phase
5.3.1.2 Surface water quality
5.3.1.3 Groundwater quality
5.3.1.4 Erosion and sedimentation
5.3.1.5 Water source
5.3.1.6 Fish and fish habitat
5.3.1.7 Ecosystem
5.3.1.8 Mammals, amphibians and reptiles
5.3.1.9 Other
5.3.2 Social impact assessment
5.3.2.1 Community and services
(1) Pre-construction phase
(2) Construction phase
(3) Operation phase
(4) Decommissioning phase
5.3.2.2 Economic development
5.3.2.3 Employment
5.3.2.4 Education and training
5.3.2.5 Resettlement, land acquisition and compensation
5.3.2.6 Demographic change
5.3.2.7 Public health and nutrition
5.3.2.8 Occupation, health and safety
5.3.2.9 Gender
5.3.2.10 Ethnic group
5.3.2.11 Vulnerable groups
5.3.2.12 Changes in land use
5.3.2.13 Traditional livelihoods and production systems
5.3.2.14 Access to natural resources
5.3.2.15 Local business
5.3.2.16 Existing production activities
5.3.2.17 Communication and transport
5.3.2.18 Water source and water supply
5.3.2.19 Natural hazard and climate change risk
5.3.2.20 Other
5.3.3 Cultural impact assessment
5.3.3.1 Ancient remains
(1) Pre-construction phase
(2) Construction phase
(3) Operation phase
(4) Decommissioning phase
5.3.3.2 Cultural heritage
5.3.3.3 Traditional and livelihood values
5.3.3.4 Other
5.3.4 Assessment of impact on scenery

5.3.4.1 Ancient remains
(1) Pre-construction phase
(2) Construction phase
(3) Operation phase
(4) Decommissioning phase
5.3.4.2 Interesting spot
5.3.4.3 Specific landscape
5.3.4.4 Other
5.4 Risk assessment
5.4.1 The context of quality risk assessment
5.4.2 Methodology
5.4.3 Outcome of quality risk assessment
(1) Pre-construction phase
(2) Construction phase
(3) Operation phase
(4) Decommissioning phase
5.5 Cumulative impact assessment
5.5.1 Methods and guidance
5.5.2 Identification of ecosystem components
5.5.3 Specifying area and timeline
5.5.4 CIA
5.5.5 Development of management framework

6 **ESMMP**

6.1 Sub-plans of ESMMP
6.2 Site-specific plans

7 **Public consultation and information dissemination**

7.1 Introduction
7.2 Guidelines and methodology
7.3 Stakeholders
7.4 PAP involvement
7.5 Conducting consultation and dissemination activities
7.6 Results of consultation during the preparation and review of the TOR and scoping report
7.7 Results of consultation during the preparation of the ESIA report
7.8 Results of consultation during the review process of the ESIA report
7.9 Public involvement activities during project construction
7.10 Public involvement activities during project construction or transferring the project

Reference

Appendices (separate copies if required)

Appendix 1: mapping

- Project description
- Environmental description
- Mitigation measures

Appendix 2: project feature description

- Diagram showing operation process
- Map of project overview
- Other

Appendix 3: report from expert(s)

- Fishery study
- Erosion and sedimentation study
- Socio-economic study
- Other

Appendix 4: modelling report (if required)

- Air quality
- Surface water quality
- Groundwater quality
- Noise
- Other

Appendix 5: public consultation and dissemination

- Minutes of the meeting with the list of participants

Annex B
List of Mekong Fish Species from ESIAs

This is a detailed list of species recorded in the ESIAs mentioned and discussed in Sect. 10.3.1.

See Table B.1.

About 20 of the species listed in the table are not listed in Rainboth (1993). This is mostly due to the fact that these species have been identified as new species, or even new genera, since the publication of this book. This reflects the fact that even now the fish fauna of the Mekong area is not yet fully known.

R. Zwahlen, *Assessing the Environmental Impacts of Hydropower Projects*, Environmental Earth Sciences, https://doi.org/10.1007/978-3-030-91185-0

Table B.1 Fish species recorded in the ESIAs for five Mekong mainstream dam projects

Family	No.	Scientific name	Synonym[1]	XHPP	XHPP	PBHPP	PLHPP	XHPP	SHPP	SHPP	LPHPP
				2010	2013a	2013b	2017	2018a	2018b	2018c	2019
Ambassidae	1	Parambassis apogonoides			1						1
	2	Parambassis siamensis		1	1					1	
	3	Parambassis wolffii	Parambassis wolffi			1					
Bagridae	4	Bagrichthys macracanthus			1			1			
	5	Hemibagrus filamentus			1	1	1	1	1		1
	6	Hemibagrus nemurus		1	1	1	1	1	1		1
	7	Hemibagrus wyckii	Mystus wycki			1	1				
	8	Hemibagrus wyckioides	Mystus wyckioides	1	1	1	1	1			1
	9	Leiocassiss siamensis		1							
	10	Mystus albolineatus				1					
	11	Pseudomystus siamensis	Leiocassis siamensis	1	1				1		1
Balitoridae	12	Balitora sp.		1							
	13	Homaloptera smithi			1						
	14	Schistura bucculenta		1	1						
	15	Xenenthodon cancila	Xenentodon cancila	1	1						
Channidae	16	Channa gachua	Channa orientalis		1		1	1	1		1

(continued)

Table B.1 (continued)

Family	No.	Scientific name	Synonym[1]	XHPP	XHPP	PBHPP	PLHPP	XHPP	SHPP	SHPP	LPHPP
				2010	2013a	2013b	2017	2018a	2018b	2018c	2019
	17	Channa striata			1		1	1	1		
Cichlidae	18	Oreochromis niloticus[2]			1		1		1	1	1
Clariidae	19	Clarias batrachus			1		1				
	20	Clarias gariepinus			1		1				
	21	Clarias sp.[3]					1				
Clupeidae	22	Clupeichthys aesarnensis				1			1	1	
	23	Tenualosa thibaudeaui			1						
Cobiitidae	24	Acanthopsis sp.[3]	Acantopsis		1	1					
	25	Acanthopsis rungthipae	Acantopsis rungthipae								1
	26	Acanthopsoides gracilentus				1					
	27	Syncrossus helodes	Botia helodes		1						1
	28	Syncrossus hymenophysa			1						
	29	Yasuhikotakia lecontei	Botia lecontei								1
	30	Yasuhikotakia sp.[3]									1

(continued)

Table B.1 (continued)

Family	No.	Scientific name	Synonym[1]	XHPP	XHPP	PBHPP	PLHPP	XHPP	SHPP	SHPP	LPHPP
				2010	2013a	2013b	2017	2018a	2018b	2018c	2019
	31	Yasuhikotakia modesta	Botia modesta		1						
Cyprinidae	32	Amblyphrarhyngodon chulahornae	A. chulabornae			1					
	33	Amblyrhynchichthys micracanthus			1			1			
	34	Amblyrhynchichthys truncatus			1					1	
	35	Bangana behri						1			
	36	Bangana elegans		1							
	37	Barbichthys laevis						1	1		
	38	Babichthys nitidus	Barbichthys laevis								1
	39	Barbonymus altus	Barbodes altus	1	1	1		1	1		1
	40	Barbonymus gonionotus[4]	Barbodes gonionotus	1	1	1	1		1	1	
	41	Barbodes gonionotus[4]	Barbonymus gonionotus								1
	42	Barbodes schwanenfeldi[4]									1
	43	Barbonymus schwanenfeldii[4]	Barbodes schwanenfeldi		1						
	44	Cirrhinus cirrhosus		1	1						
	45	Cirrhinus molitorella			1			1			1

(continued)

Table B.1 (continued)

Family	No.	Scientific name	Synonym[1]	XHPP	XHPP	PBHPP	PLHPP	XHPP	SHPP	SHPP	LPHPP
				2010	2013a	2013b	2017	2018a	2018b	2018c	2019
	46	Cirrhinus jullieni									1
	47	Cosmochilus harmandi			1			1			1
	48	Crossocheilus reticulatus		1	1			1			
	49	Cyclocheilichthys apogon			1					1	1
	50	Cyclocheilichthys armatus		1	1	1	1	1		1	1
	51	Cyclocheilichthys enoplos			1	1	1			1	
	52	Cyclocheilichthys lagleri									1
	53	Cyclocheilichthys furcatus		1	1	1		1	1		1
	54	Cyclocheilichthys repasson		1	1						1
	55	Cyclocyalichthys sp.[5]					1				
	56	Cyprinus carpio[6]			1	1	1	1			1
	57	Dangila spilopleura				1					
	58	Danio albolineatus	Brachydanio albolineatus		1			1			
	59	Danio fangfangae	Devario fangfangae	1							

(continued)

Table B.1 (continued)

Family	No.	Scientific name	Synonym[1]	XHPP	XHPP	PBHPP	PLHPP	XHPP	SHPP	SHPP	LPHPP
				2010	2013a	2013b	2017	2018a	2018b	2018c	2019
	60	Discherodontus ashmeadi		1							
	61	Epalzeorhynchos munense			1			1			1
	62	Esomus metallicus		1	1	1					
	63	Garra cambodgiensis		1	1						
	64	Garra cyclostomata		1							
	65	Garra fascicauda									1
	66	Hampala dispar		1	1						
	67	Hampala macrolepidota		1	1						1
	68	Hemibarbus maculatus[7]		1							
	69	Henicorhynchus cuadimaculatus		1							
	70	Henicorhynchus lobatus			1	1		1			1
	71	Henicorhynchus siamensis		1	1	1	1			1	
	72	Henicorhynchus ornatipinnis			1						1

(continued)

Table B.1 (continued)

Family	No.	Scientific name	Synonym[1]	XHPP	XHPP	P3HPP	PLHPP	XHPP	SHPP	SHPP	LPHPP
				2010	2013a	2013b	2017	2018a	2018b	2018c	2019
	73	Hypsibarbus lagleri			1			1			
	74	Hypsibarbus malcolmi			1						1
	75	Hypsibarbus pierrei			1			1			1
	76	Hypsibarbus vernayi			1	1		1	1	1	1
	77	Hypsibarbus wetmorei			1			1	1		1
	78	Hypsibarbus sp.[3]					1				
	79	Labeo barbatulus			1			1			
	80	Labeo chrysophekadion	Morulius chrysophekadion		1	1					1
	81	Labeo rohita						1			1
	82	Labiobarbus leptocheilus	Dangila cuvieri	1	1			1			1
	83	Labiobarbus lineatus	Dangila labiata							1	1
	84	Labiobarbus siamensis	Dangila spilopleura		1	1		1			1
	85	Labiocheilus davisi	Lobocheilos davisi								1
	86	Labiocheilus melanotaenia	Lobocheilos melanotaenia								1
	87	Luciosoma setigerum			1						

(continued)

Table B.1 (continued)

Family	No.	Scientific name	Synonym[1]	XHPP	XHPP	PBHPP	PLHPP	XHPP	SHPP	SHPP	LPHPP
				2010	2013a	2013b	2017	2018a	2018b	2018c	2019
	88	Macrochirichthys macrochirus			1						1
	89	Mekongina erythrospila		1	1			1			1
	90	Mystacoleucus atridorsalis			1	1					
	91	Mystacoleucus greenwayi			1						
	92	Mystacoleucus lepturus		1					1		1
	93	Mystacoleucus marginatus	M. chilopterus	1	1	1	1	1	1	1	1
	94	Neolissocheilus stracheyi		1							
	95	Onychostoma fusiforme		1							
	96	Onychostoma gerlachi			1						
	97	Onychostoma meridionale									1
	98	Opsarius koratensis		1	1				1		
	99	Opsarius pulchellus		1	1						1

(continued)

Table B.1 (continued)

Family	No.	Scientific name	Synonym[1]	XHPP	XHPP	PBHPP	PLHPP	XHPP	SHPP	SHPP	LPHPP
				2010	2013a	2013b	2017	2018a	2018b	2018c	2019
	100	Osteochilus hasselti	O. vittatus						1	1	
	101	Osteochilus lini							1		
	102	Osteochilus melanopleurus			1						
	103	Oxygaster pointoni								1	
	104	Parachela karmandi	Paralaubuca harmandi			1	1				
	105	Parachela oxygassteroides	Parachela oxygastroides			1	1				
	106	Parachela siamensis		1		1	1				1
	107	Paralaubuca barroni							1		
	108	Paralaubuca harmandi							1	1	
	109	Paralaubuca riveroi			1	1					
	110	Paralaubuca typus			1				1	1	
	111	Poropuntius deauratus				1					
	112	Poropuntius laoensis		1	1	1					
	113	Poropuntius kremfi	P. krempfi								1
	114	Propuntius sp.[3]									1
	115	Probarbus jullieni						1			1
	116	Puntioplites falcifer		1	1			1			1
	117	Puntioplites proctozystron	P. proctozysron	1	1	1		1	1	1	1

(continued)

Table B.1 (continued)

Family	No.	Scientific name	Synonym[1]	XHPP	XHPP	PBHPP	PLHPP	XHPP	SHPP	SHPP	LPHPP
				2010	2013a	2013b	2017	2018a	2018b	2018c	2019
	118	Puntius brevis							1	1	
	119	Puntius orphoides									1
	120	Puntius stoliczkanus	Pethia stoliczkana		1						
	121	Raiamas guttatus	Raiamus guttatus	1	1				1	1	
	122	Rasbora aurotaenia								1	
	123	Rasbora borapetensis				1					
	124	Rasbora dusonensis	R. argyrotaenia	1						1	
	125	Rasbora paviana								1	
	126	Rasbora paviei		1						1	
	127	Scaphiodonichthys acanthopterus		1	1						
	128	Scaphognathops bandanensis			1						1
	129	Scaphognathops stejnegeri			1			1			
	130	Scaphognathops theunensis[8]									1
	131	Sikukia flavicauda	S. flavicaudata								1
	132	Sikukia gudgeri		1	1	1		1	1	1	1
	133	Sikukia stejnegeri				1					

(continued)

Table B.1 (continued)

Family	No.	Scientific name	Synonym[1]	XHPP	XHPP	PBHPP	PLHPP	XHPP	SHPP	SHPP	LPHPP
				2010	2013a	2013b	2017	2018a	2018b	2018c	2019
	134	Systomus orphoides		1							
	135	Thynnichthys thynnoides								1	1
	136	Tor sinensis									1
	137	Tor tambroides		1	1						
Dasyatidae	138	Dasyatis laosensis	Hemitrygon laosensis		1			1			1
Datnioididae	139	Datnioides undecimradiatus			1						
Eleotridae	140	Oxyeleotris marmorata		1	1		1	1			1
Gobiidae	141	Gobiopterus chuno							1	1	
	142	Papuligobius ocellatus	Rhinogobius ocellatus		1				1		
	143	Rhinogobius mekongianus		1	1						
	144	Rhinogobius sp.[3]									1
Gyrinocheilidae	145	Gyrinocheilus aymonieri			1						
	146	Gyrinocheilus pennocki			1						
Hemiramphidae	147	Dermogenys siamensis	D. pusilla		1	1			1		
Mastacembelidae	148	Macrognathus semiocellatus			1						

(continued)

Table B.1 (continued)

Family	No.	Scientific name	Synonym[1]	XHPP	XHPP	PBHPP	PLHPP	XHPP	SHPP	SHPP	LPHPP
				2010	2013a	2013b	2017	2018a	2018b	2018c	2019
	149	Macrognathus siamensis					1		1		
	150	Mastacembelus armatus			1						1
	151	Mastacembelus favus					1		1		
Nandidae	152	Pristolepis fasciata			1			1	1		1
Notopteridae	153	Chitala blanci			1			1			
	154	Chitala ornata			1			1	1		
	155	Notopterus notopterus		1	1		1		1		1
Osphronemidae 9	156	Trichogaster microlepis	Belontiidae						1		1
	157	Trichogaster trichopterus							1		
Pangasiidae	158	Helicophagus wanndersii	H. waandersii		1			1			
	159	Helicophagus leptorhynchus									1

(continued)

Table B.1 (continued)

Family	No.	Scientific name	Synonym[1]	XHPP	XHPP	PBHPP	PLHPP	XHPP	SHPP	SHPP	LPHPP
				2010	2013a	2013b	2017	2018a	2018b	2018c	2019
	160	Pangasianodon hypophthalmus			1		1	1			
	161	Pangasius bocourti			1			1			
	162	Pangasius conchophilus			1			1			
	163	Pangasius krempfi						1			
	164	Pangasius larnaudii	P. larnaudiei		1						
	165	Pangasius macronema			1	1		1			
	166	Pangasius sanitwongsei						1			
	167	Pseudolais micronemus	Pangasius micronemus					1		1	
	168	Pseudolais pleurotaenia	Pangasius pleurotaenia		1			1		1	1
Schilbeidae	169	Clupisoma sinense	Laides sinensis		1						
	170	Laides hexanema			1						
	171	Laides longibarbis			1			1			1
	172	Luciosoma bleekeri[10]			1						

(continued)

Table B.1 (continued)

Family	No.	Scientific name	Synonym[1]	XHPP	XHPP	PBHPP	PLHPP	XHPP	SHPP	SHPP	LPHPP
				2010	2013a	2013b	2017	2018a	2018b	2018c	2019
Siluridae	173	Belodontichthys truncatus			1	1		1			
	174	Belodonichthys dinema									1
	175	Hemisilurus heterorhynchus			1						
	176	Hemisilurus mekongensis			1			1			
	177	Kryptopterus cryptopterus			1					1	1
	178	Kryptopterus cheveyi			1		1				1
	179	Kryptopterus limpok									1
	180	Micronema hexapterus							1		
	181	Ompok bimaculatus				1					
	182	Ompok euginiatus	O. eugeneiatus								1
	183	Phalacronotus apogon			1			1			1
	184	Phalacronotus bleekeri		1	1	1		1			1
	185	Phalacronotus micronemus	Micronema micronema	1	1			1			
	186	Wallago attu			1	1					
Sisoridae	187	Bagarius bagarius			1			1			1

(continued)

Table B.1 (continued)

Family	No.	Scientific name	Synonym[1]	XHPP	XHPP	PBHPP	PLHPP	XHPP	SHPP	SHPP	LPHPP
				2010	2013a	2013b	2017	2018a	2018b	2018c	2019
	188	Bagarius suchus			1			1			
	189	Bagarius yarrelli			1	1		1	1	1	1
	190	Glyptothorax lampris			1						
	191	Glyptothorax laoensis	G. laosensis	1							1
	192	Glytothorax fuscus		1							
Soleidae	193	Brachirus harmandi	Euryglossa harmandi		1				1		
	194	Euryglossa harmandi		1							
Sundasalangidae	195	Sundasalanx praecox			1						
Tetraodontidae	196	Auriglobus nefastus	Chonerhinos nefastus		1				1		
	197	Monotreta fangi				1					
	198	Monotreta turgidus	Pao turgidus	1							
	199	Monotrete cochinchinensis	Pao cochinchinensis		1						
	200	Tetraodon cambodgiensis[4]	Pao cambodgiensis		1				1		
	201	Pao cambodigensis[4]									1

(continued)

Table B.1 (continued)

Family	No.	Scientific name	Synonym[1]	XHPP	XHPP	PBHPP	PLHPP	XHPP	SHPP	SHPP	LPHPP
				2010	2013a	2013b	2017	2018a	2018b	2018c	2019
	202	Tetraodon suvattii	Pao suvattii		1				1		
Total				54	119	43	27[11]	57	41	30	80
No. of species recorded for the first time in this series				54	85	14	5	7	8	3	26

Source XHPP: Xayaburi HPP: 2010: EIA, TEAM, 2010; 2013: Fish monitoring during construction; 2018a: fish monitoring after commissioning; PBHPP: Pak Beng HPP: 2013: EIA, NCG, 2015. PLHPP: Pak Lay HPP: 2017: EIA, NCG, 2018. SHPP: Sanakham HPP, EIA, NCG, 2018: 2018b: dry season; 2018b: wet season. LPHPP: Luang Prabang HPP: 2019: EIA, Pöyry, 2020

Notes 1. Synonyms: indicated are synonyms used (as valid names) either in Rainboth (1996) or in fishbase (https://www.fishbase.se/search.php); this is neither a complete list of synonyms, nor does a listing of a name in this column necessarily mean that this is the "correct" name, i.e. the one presently recognised as valid. It is used here mainly to show that depending on the reference used for identifying fish species, one can end up with different names for the same species

2. Exotic species; often raised in fish ponds, economically important, originally from Africa

3. Specimens identified only to the genus level; can be identical with the (or one of the) species listed from the dame genus

4. Species listed in different samplings under a different synonym (see note No. 1). The cases marked as [3] or [4] are likely to reduce somewhat the total number of recorded species

5. The genus name *Cyclocyalichthys* mentioned for Pak Lay HPP (POWERCHINA, 2017) cannot be identified in any of the other sources used; it is assumed that this is a misspelling, but it was not possible to come to a conclusion on what name was meant

6. Exotic species originally from Europe and Western Asia. Widely bred pond fish

7. Northern Asian (Mongolia, Siberia) species, introduced in many other places. Is known to have caused problems in some rivers where it was introduced

8. This is a species that was identified as an endemic in the Nam Theun river (whence the specific name). Could be a misidentification, or else it would indicate that it is more widely distributed than assumed

9. Wrong family, belongs to the family Belontiidae; the other possibility, that it is a species of the Osphronemidae which was misnamed, seems unlikely

10. *Luciosoma bleekeri* is a species of the Cyprinidae family; unknown whether it was misplaced in the list or misnamed

11. The ESIA for Pak Lay HPP (POWERCHINA, 2017) mentions that a total of 54 fish species were identified; however, the report does not provide a list of species, and only 27 are mentioned in the text

Index

R. Zwahlen, *Assessing the Environmental Impacts of Hydropower Projects*, Environmental Earth Sciences, https://doi.org/10.1007/978-3-030-91185-0

T